# Theory and Practice of Archaeological Residue Analysis

Edited by

Hans Barnard
Jelmer W. Eerkens



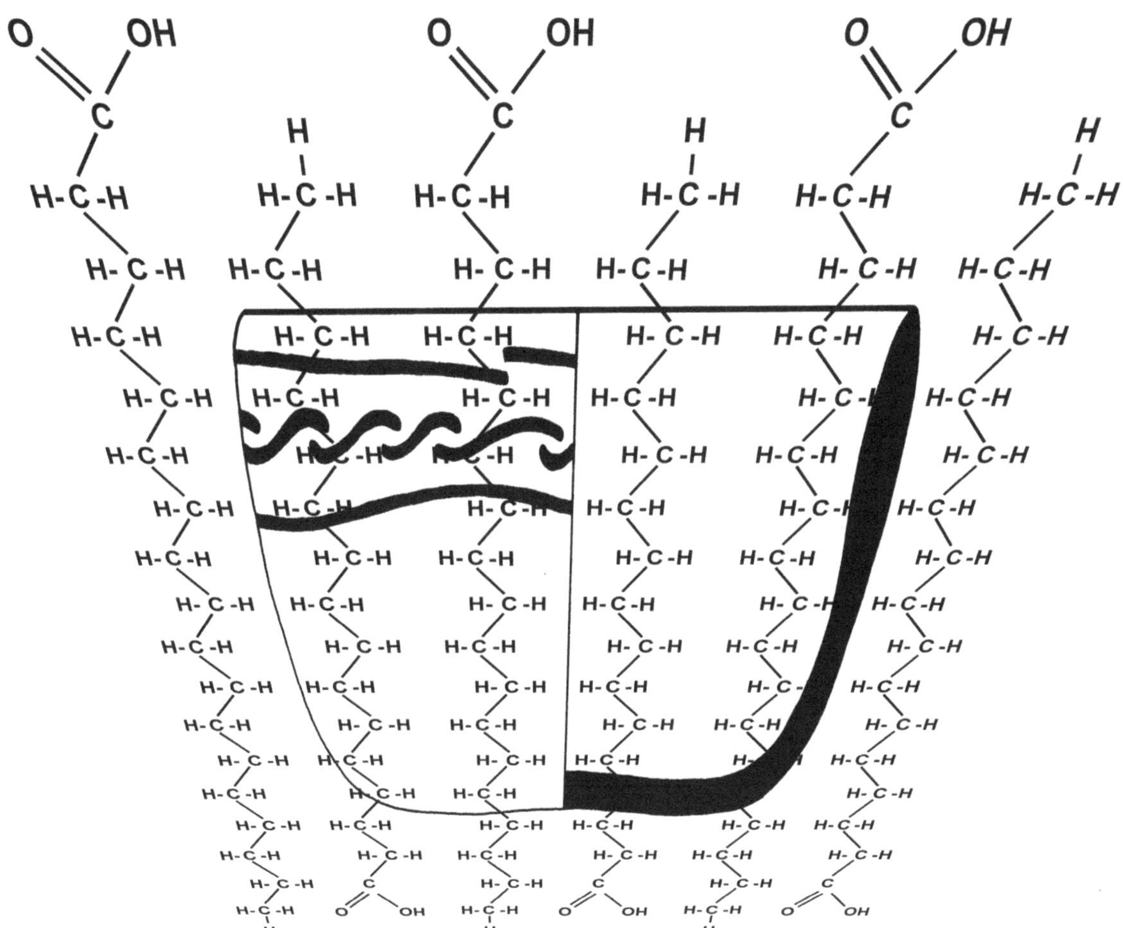

BAR International Series 1650
2007

Published in 2016 by
BAR Publishing, Oxford

BAR International Series 1650

*Theory and Practice of Archaeological Residue Analysis*

ISBN 978 1 4073 0084 9



BAR Publishing is the trading name of British Archaeological Reports (Oxford) Ltd. British Archaeological Reports was first incorporated in 1974 to publish the BAR Series, International and British. In 1992 Hadrian Books Ltd became part of the BAR group. This volume was originally published by Archaeopress in conjunction with British Archaeological Reports (Oxford) Ltd / Hadrian Books Ltd, the Series principal publisher, in 2007. This present volume is published by BAR Publishing, 2016.

Printed in England

BAR
PUBLISHING

BAR titles are available from:

BAR Publishing
122 Banbury Rd, Oxford, OX2 7BP, UK
EMAIL info@barpublishing.com
PHONE +44 (0)1865 310431
FAX +44 (0)1865 316916
www.barpublishing.com

# Table of Contents

## Table of Contents

# CHAPTER ONE
# Introduction
## J.W. Eerkens and H. Barnard

Jelmer Eerkens, Associate Professor; Department of Anthropology, University of California, Davis; One Shields Avenue; Davis, CA 95616; USA; <jweerkens@ucdavis.edu> and Hans Barnard, Research Associate; Cotsen Institute of Archaeology, University of California, Los Angeles; P.O.-Box 951510; Los Angeles, CA 90025; USA; <wendrich@barnard.nl>.

Organic residues include a broad range of materials that can be analyzed at a macro-, micro- or molecular level. They represent the carbon-based remains (in combination with H, N, O, P and S) of fungi, plants, animals and humans. Organic residues have been extracted and studied from a variety of archaeological materials including ceramics, flaked stone, bone, coprolites, cooking stones, grinding stones, pigments and wood from shipwrecks. A wide range of biomolecules has been isolated including lipids, proteins, starches, DNA and plant lignin. Commonly studied residues include blood proteins from the surfaces of stone tools, fatty acids absorbed into the walls of cooking pots and hafting adhesives or sealants made from resin or pitch. These different residue sources have been studied by archaeologists to varying degrees, based on their ability to withstand decomposition in different depositional environments as well as the requirements of the analytical methods. Thus, lipids preserved in cooking pots are relatively well-studied because of their relative propensity to survive archaeologically and their amenable behavior in laboratory conditions. Techniques to isolate, identify and interpret proteins and peptides from similar contexts are still being developed while DNA from food residues has, to the best of our knowledge, not yet been studied.

Organic residue analysis is a relatively new technique to archaeology (Figure 1). The chapters of this volume bring together scholars from across the globe and attest to the diverse range of analytical methods, material types, spatio-temporal cultural units and research questions to which organic residue analysis has been applied (Table 2). They are partly the proceedings of a symposium on this subject, held on 31 March 2005 in Salt Lake City (Utah) during the 70$^{th}$ Annual Meeting of the Society for American Archaeology (Table 1), and partly the result of our invitation to contribute forwarded to many active in this field.

The study of ancient organic residues is inherently an interdisciplinary endeavor and judging by co-authorships in published articles, commonly brings together researchers from a range of scientific fields. We feel such collaborations are important if not critical to the development of the field (Barnard et al. 2007). While archaeologists are in a unique position to interpret and give context to the results of residue studies *vis à vis* their experience with the artifacts, sites and regions that they study, they are often less familiar with biochemical techniques and the complex molecules found in many residues. Technicians and biochemists can provide important information by providing context to analytical results, but usually outside of an anthropological theoretical framework. Together they may fully appreciate the archaeological significance of the findings and identify the potential sources of the various biomolecules, be they contaminants, byproducts of decomposition or intact compounds. This interaction between scholars trained in archaeology with those trained in organic chemistry has resulted in the interdisciplinary field of 'archaeochemistry'. The formalization of this discipline in recent years now allows undergraduate and graduate students to tailor their education such that they will fit comfortably at the intersection of both archaeology and organic chemistry, with complementary training in analytical techniques. Programs specializing in this field have recently been established in both Europe and North America.

### Growth of Residue Studies

While most researchers would probably agree that organic residue analysis is an active field that is increasing in popularity and scope, we were interested in quantifying the growth of the field. To achieve this end, we performed a simple search of *Anthropology Plus*, a large database of the anthropological literature, including all major and many minor archaeology journals as well as edited volumes. We searched for all literature containing the words 'residue' or 'residues' in either the title or keywords. This search retrieved over 300 unique works. After winnowing titles that were obviously not related to organic residue analysis in archaeology (such as 'The modern Chinese family: ideology, revolution and residues'), we had compiled a list of 268 entries. This list is certainly not exhaustive of all archaeological studies of organic residues, but we feel it represents a fairly unbiased cross-section of self-identified works through the last 50 years. Indeed, many of the authors of chapters in this volume are represented in the list.

## Theory and Practice of Archaeological Residue Analysis

| Session | Author(s) | Affiliation(s) | Title |
|---|---|---|---|
| Morning session (Symposium 21): Theory | Hans Barnard | Cotsen Institute of Archaeology, University of California, Los Angeles (USA) | Theory and Practice of Archaeological Residue Analysis and the 'Round Robin' Experiment |
| | Jim Cassidy | Department of Anthropology, University of California, Santa Barbara (USA) | Subsistence Change during the Final Neolithic in the Russian Far East as revealed by Fatty Acid Residue Analysis |
| | Marlize Lombard and Lyn Wadley | Natal Museum and the University of the Witwatersrand (South Africa) | Blind Testing for the Recognition of Residues using Light Microscopy: Results and Lessons learnt |
| | Robert Lusteck | Department of Anthropology, University of Minnesota (USA) | Residues of Maize in North American Pottery: What Phytoliths can add to the Story of Maize |
| | Sean Rafferty | Department of Anthropology, University at Albany (USA) | The Archaeology of Alkaloids |
| | Eleanora Reber | Department of Anthropology, University of North Carolina, Wilmington (USA) | The Well-Tempered Pottery Analysis: Residue and Typological Analysis of Potsherds from the Lower Mississippi Valley |
| Afternoon session (Symposium 46): Practice | Jelmer Eerkens | Department of Anthropology, University of California, Davis (USA) | Gas Chromatography Mass Spectrometry (GC/MS) Analysis of Fatty Acids in Ancient Potsherds |
| | Dana Beehr and Stan Ambrose | Department of Anthropology, University of Illinois, Urbana (USA) | Reconstructing Mississippian Diet in the American Bottom with Stable Isotope Ratios of Pot Sherd Residues |
| | Marcus Forster, Carl Heron, Ben Stern, Oliver Craig and Søren Andersen | University of Bradford (Great Britain), University of Rome (Italy) and the National Museum of Denmark | The Contents of Late Mesolithic/Neolithic Ceramics from Denmark |
| | Michael Gregg | Department of Anthropology, University of Toronto (Canada) | Survival of Organic Residues in Pottery from Southwest Asia during the Early Holocene |
| | Hanneke Hoekman-Sites | Department of Anthropology, Florida State University (USA) | Using Residue Analysis to Confirm Trade Connections at Pella, Jordan |
| | Mary Malainey | Department of Anthropology, Brandon University (Canada) | Fatty Acid Analysis of Archaeological Residues: Procedures and Possibilities |
| | Micala Rider, Paul Fish, William Longacre, Matthew Young and Mark Malcomson | Department of Anthropology, University of Arizona (USA) | Residue Analysis of Fatty Acids preserved in Pottery Sherds: Method of Interpretation to Account for the Possible Pitfalls in Analysis |
| | Ran Boytner | Cotsen Institute of Archaeology, University of California, Los Angeles (USA) | Discussant |

Table 1: List of the presentations of the sessions on the 'Theory and Practice of Archaeological Residue Analysis' (Salt Palace Convention Center, 31 March 2005) during the 70th Annual Meeting of the Society for American Archaeology in Salt Lake City (Utah). For abstracts see www.archbase.org/residue/.

| Chapter | Material | Residue | Method | Focus |
|---|---|---|---|---|
| Chapter 2: Lusteck and Thompson | Ceramics | Phytoliths | Microscopy | The 'story of maize'; phytolith composition of different varieties of maize |
| Chapter 3: Lombard and Wadley | Stone tools | Starches and cellulose | Microscopy | Description of methodology; applications for tool use studies |
| Chapter 4: Shishlina *et al.* | Ceramics and human stomachs | Phosphates, phytoliths and pollen | Photocalorimetry and microscopy | Different burial treatments; seasonality in burials |
| Chapter 5: Barnard *et al.* | Ceramics | Lipids | GC/MS | Description of methodology |
| Chapter 6: Regert | Ceramics | Lipids | IR, DI-MS and chromatography | Classification of pots by use |
| Chapter 7: Malainey | Ceramics | Lipids | GC/MS | Decomposition processes |
| Chapter 8: Eerkens | Ceramics | Lipids | Simulation | Decomposition processes |
| Chapter 9: Oudemans and Boon | Ceramics | Lipids | GC/MS | Preservation, decomposition and contamination issues |
| Chapter 10: Cassidy | Ceramics | Lipids | GC/MS | Seasonality of sites and diet reconstruction |
| Chapter 11: Hoekman-Sites | Ceramics | Lipids | GC/MS and IR | Use of pots; trade; core-periphery models |
| Chapter 12: Reber | Ceramics | Lipids | GC/MS and carbon isotope analysis | Relating form and function of pots |
| Chapter 13: Solazzo and Erhardt | Ceramics | Lipids | GC/MS | Decomposition processes; comparing the interior and exterior of pots |
| Chapter 14: Raffery | Ceramics | Alkaloids | GC/MS | Exploration of possibilities; research agenda |
| Chapter 15: Beehr and Ambrose | Ceramics | Total organic residue | Carbon and nitrogen isotopes analysis | Development of a fast technique; contamination issues |
| Chapter 16: Barnard *et al.* | Ceramics | Total organic residue | A range of methods | Blind test comparison between different laboratories |
| Chapter 17: Barnard *et al.* | Ceramics | Proteins | A range of methods | Description of methodology |

Table 2: Overview of the subject matter of the chapters in this volume. GC/MS = combined gas chromatography mass spectrometry; IR = infrared spectroscopy; DI-MS = direct inlet mass spectrometry.

Figure 1 shows the results of our search, tabulating residue studies by the year in which they appeared in print. As can be seen, after a low level of articles in the 1960s and 1970s, the figure shows an uncanny linear rise in the number of publications beginning in the early 1980s and reaching a peak in 2004, though with a few lighter years from 2000-2003. The slight dip in publications in 2005 can most likely be explained by the fact that the *Anthropology Plus* database is not yet fully updated. Not surprisingly, articles pertaining to organic residues are highly clustered in terms of where they appear. Nearly 90% of the articles in the database are in journals. Two journals in particular, the *Journal of Archaeological Science* (n=84) and *Archaeometry* (n=29), account for over 40% of the literature in our database. Both journals are highly technical and reach a specific audience within the archaeological community. Many of the articles on residue analysis appearing in these journals fall into four basic categories: reporting on the application of new analytical techniques to ancient residues; reporting the extraction of new types of biomolecules in residues; application of a previously used technique to a new artifact type; or attempts to work out analytical kinks in previously used techniques, often through experimental studies. All these trends suggest that residue studies are still fairly specialized and have not often been used to address substantive theoretical or prehistoric issues. Such a pattern is not unexpected for a relatively new technique, but one that we hope and expect will change in the coming years. To examine further the development of organic residue studies, we narrowed our analysis to just these two journals as we were interested whether residue studies were occupying an increasing or a decreasing, or a relatively stable, relative proportion of all publications.

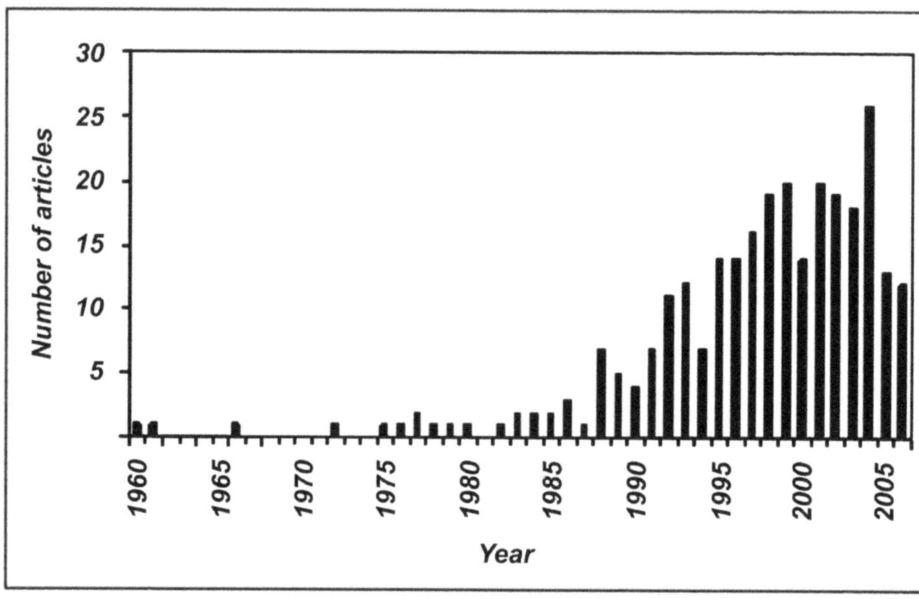


Figure 1: Number of articles on archaeological organic residue analysis in the *Anthropology Plus* database, by year since 1960 (n=268).

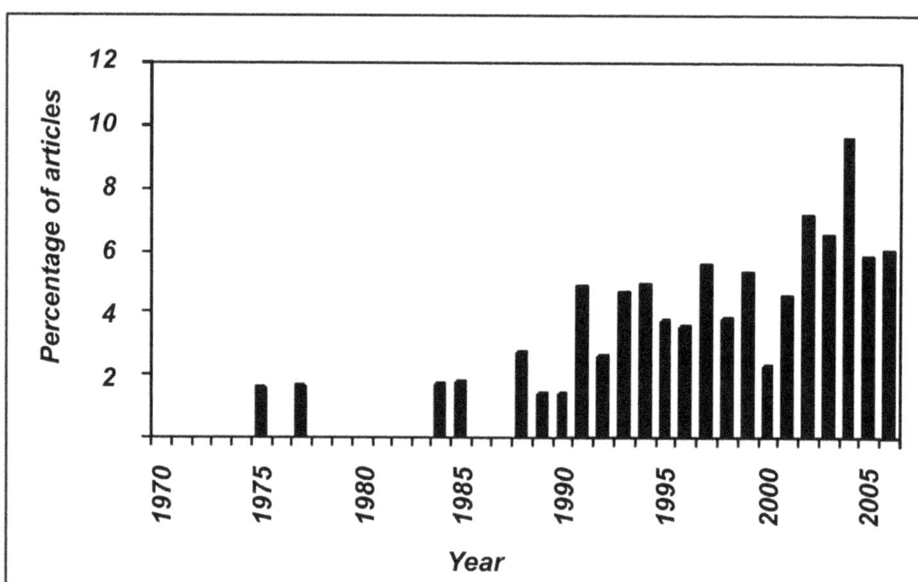


Figure 2: Publications on archaeological organic residue studies as a percentage of all articles in the *Journal of Archaeological Science* and *Archaeometry*, by year since 1970 (n=113).

Figure 2 shows that the study of archaeological organic residues is clearly on the rise within the *Journal of Archaeological Science* and *Archaeometry*, comprising less than 2% of all articles in the 1980s, but steadily increasing to 10% of all articles by 2004. Thus, by all accounts, organic residue studies are increasing in both an absolute and relative sense in the archaeological sciences, indicating the growing number of individuals involved in this type of research. The current volume reflects this growing interest in residue analyses.

Overall, we find these trends encouraging, but we also see room for improvement. It is clear that a growing number of archaeologists and biochemists appreciate the potential of archaeological organic residue analysis, and significant effort is being invested to develop these techniques. Much is being learned methodologically, for example, about how to extract residues, what kinds of residues seem to preserve in what kinds of ancient materials, how to avoid contamination, and the like. At the same time, in the increasing range of techniques being applied to ancient residues, there seems to be little effort to standardize methodologies or cross-check results across different laboratories and techniques (but see Barnard et al. 2007). It is our impression that there are several different 'teams' and 'schools' of analysts (some larger, some smaller in number) that seem to be working more or less independently on their own sets of unique problems, using their own sets of specific methods. In our view too little effort is made to cross-check the results and exchange detailed information about the possibilities and problems of the various techniques.

## Scope of the Current Volume

To bring residue analysis out of predominantly the technical journals, and into a broader range of venues, it will be necessary to address some of these shortcomings. The chapters in this volume attempt to do just so, to move beyond the documentation of a new technique, a new biomolecule, or a new artifact type, and use residue analysis as a means to answer anthropological questions or to address important shortcomings in the identification of residue materials (such as decomposition).

A range of analytical techniques are represented in the volume for the identification of organic residue materials, including microscopy (Lusteck and Thompson, Lombard and Wadley, Shishlina et al.), infrared spectroscopy (Regert, Hoekman-Sites) and light element mass spectrometry (Beehr and Ambrose). Yet, the dominant technique represented in this volume chapters is gas chromatography (GC), often coupled to mass spectrometry (Chapter 5, Malainey, Oudemans and Boon, Cassidy, Reber, Solazzo and Erhardt, Rafferty). Some authors have combined multiple complimentary techniques (Shishlina et al., Regert, Hoekman-Sites, Chapter 16).

There is a clear emphasis on ceramic materials within this volume, a trend that is reflected in organic residue studies in general, though perhaps not quite as strongly. Only one of the chapters (Lombard and Wadley) examines stone tools, while another (Sishlina et al.) investigates soil from burials where the stomach would have been and a final (Eerkens) examines residue preservation in the abstract. The remainder (thirteen) of the chapters focuses primarily on pots and pipes. The emphasis on ceramics, and primarily pots within that category, must in part stem from scholarly tradition. Residues adhering to the surfaces of pots, as well as those absorbed within the walls of pots, have been shown to survive in a number of different burial contexts and in some cases over thousands of years. Because pot sherds are clearly related to specific aspects of human activities like cooking, eating, storage and transport, and have been well-studied in their own right, there are a number of advantages from an anthropological perspective to their additional study through organic residue analysis. Methods for extracting and analyzing such residues are well-established and have been published in numerous studies. In this respect, analyzing residues from pot sherds is a less risky endeavor, favoring parallel studies in different spacio-temporal cultural contexts. Other artifact categories, such as fire-cracked rock, do not have this history, though research is beginning in this area (Quigg et al. 2001; Buonasera 2005).

A major issue in most studies, including the majority of the contributions in this volume, is a means to separate the various compounds that comprise a residue to help identify source materials. In theory, plant or animal sources contain unique compounds that can be used to identify their presence in an ancient residue. Such compounds are generally referred to as 'biomarkers' and their identification in residues will provide an unambiguous identification of a particular source material. Isolating and identifying such biomarkers from a residue is therefore an important step in the identification of original source materials. Lipid biomarkers, however, are uncommon or readily decompose into less diagnostic byproducts. One approach commonly used in the chapters of this volume to overcome this problem is to identify source materials by the relative proportion of different non-diagnostic lipids, in the same way that obsidian sourcing or provenance analysis is done using the concentration of different elements. Potentially better biomarkers are proteins (or peptides) that are more or less directly related to DNA, which is specific for each species of plants and animals (Appendix II). However, techniques to isolate, identify and interpret proteins from archaeological contexts are still being developed (Chapter 17).

Some authors use microscopy to identify and separate the larger compounds of a residue (Lusteck and Thompson, Lombard and Wadley, Shishlina et al.). However, visual techniques can be time consuming, particularly when quantification of thousands of compounds is necessary. The dominance of gas chromatography in the chapters of this volume, indeed in organic residue studies in general, stems in part from the ease with which this instrument can separate compounds into different classes (fatty acids, acylglycerides, terpenoids, etc.). These can often be further divided into finer and finer classes based on their specific structure. For example, fatty acids can be categorized by length of their carbon chain (C20:0 vs. C22:0), the number of double bonds (C20:0 vs. C20:1), the specific position of those double bonds within the carbon chain (C20:1ω9 vs. C20:1ω11), as well as various kinds of branched structures for certain fatty acids. Different derivitization techniques and instrument settings allow fine-tuning of the chromatographic method to separate, isolate and quantify these different classes of compounds. Such fine-tuning requires analyzing each sample multiple times; each round providing important information that helps to refine subsequent extraction and derivitization methods or instrument settings.

A theme that runs through many of the chapters is the effect of decomposition on the identification of source materials (Shishlina et al., Malainey, Eerkens, Oudemans and Boon, Solazzo and Erhardt, Beehr and Ambrose). As the chapters testify, simulating and experimentally replicating decomposition processes is a field that still needs much research. This is an area where both food scientists (e.g., Frankel 1998), and paleontologists (e.g., Collins et al. 2002) have ventured, but their timescales are usually different from those relevant to archaeology. In this respect, archaeologists will have to work out many of these issues alone. Organic residues in an archaeological context will change over time, not only with respect to the proportion of the components, because of selective leaching and oxidation, but also at a molecular level (saturation, deamination, agglutination, etc.). This means that molecules may no longer match their modern equivalent, while many of the modified molecules may not be present in the available databases (Chapters 16 and 17). The only way to resolve this issue is for archaeologists to assemble a specific archaeological database. This could be combined with, or modeled after the spectral database of the Users' Group for Mass Spectrometry and Chromatography (MaSC), a group of scholars and scientists working in the fields of conservation and preservation studies (http://www.mascgroup.org/).

Occasionally, the results of organic residue research on archaeological artifacts are published in biochemical journals, such as the *Bulletin of the Chemical Society of Japan* (Shimoyma et al. 1995), *Fresenius' Journal of Analytical Chemistry* (Casoli et al. 1995), or *Tetrahedron Letters* (Hansel et al. 2004). Indeed, biochemists sometimes use archaeological samples to address biochemical rather than archaeological research questions (e.g., Stankiewicz et al. 1995; van Bergen et al. 1999). These journals are understandably reluctant to publish the details of the biochemistry background and methods, which they assume to be common knowledge among their readers, but they also frequently fail to publish the archaeological discussion, presumably because this is outside the scope of the journal. Valuable information thus fails to make its full impact in the archaeological *and* biochemical communities at large. Archaeologists, on the other hand, sometimes fail to appreciate all the intricacies and limitations of chemical analysis. This was among the factors that fueled the heated debates on the interpretation of blood residues on stone tools (Fiedel 1996; Newman et al. 1997), and the indications for cannibalism among Native Americans (Diamond 2000; Dongoske et al. 2000). Again the Internet, with its ever increasing number of on-line articles and almost unlimited search facilities, may be instrumental in the closing of this rift between the harder and the softer sciences.

A final issue in need of attention is the relationship between the encountered organic residues and the function of vessels within a society (Regert, Reber, Beehr and Ambrose). It is always assumed that the residue found in a ceramic vessel is in some way related to the original use of the vessel. It remains unclear, however, if the residue represents the first food to come into contact with the ceramic matrix, after which the available binding sites are saturated, or the last, if older residues are continually replaced by new ones, or a combination of all food ever to have been inside the vessel, if the molecules that make up the residue compete for the available binding places (Barnard et al. 2007). Are the organic residues the remains of the first or the last use of a vessel, or rather an amalgam of residues from many use-phases? How do seasonal, social or ritual factors influence the formation and preservation of organic residues in ceramic vessels? Most ancient and modern societies have very strict views, albeit mostly subconscious, on what to eat, and on how, when and where to eat it. To know such rules would be very helpful beforehand, but they may prove very difficult to infer from the study of ceramic vessels alone. On the other hand, the custom to reserve specific vessels for certain foodstuffs or to serve meals in individual portions, each in their own container, is certainly not universal. And then there are also vessels that are not used for food at all, and food that is not prepared in vessels. All these matters must be considered when associating residues with vessel form and function. They can only be addressed by carefully combining the results of biochemical analyses with archaeological, historical and ethnographical data.

The chapters of this volume represent a cross-section of current research on ancient organic residues, a rapidly expanding field in the archaeological sciences. A range of techniques are applied to a range of biomaterials, though a clear focus on chromatographic techniques and lipids is evident. A less diverse range of material types are present, with an overwhelming focus on ceramics. Although there is much work to be done, residue studies have come a long ways and have shed new light on a range of anthropological issues and questions. With this volume we hope to have added a few building blocks to the bridge that is being constructed between archaeology and biochemistry, and to have contributed some topics to the discussion between archaeologists and biochemists, indispensible for this endeavor to be successful.

## References

**Barnard,** H., S.H. Ambrose, D.E. Beehr, M.D. Forster, R.E. Lanehart, M.E. Malainey, R.E. Parr, M. Rider, C. Solazzo and R.M Yohe II (2007). Mixed Results of Seven Methods for Organic Residue Analysis Applied to One Vessel with the Residue of a Known Foodstuff. *Journal of Archaeological Science* **34**: 28-37.

**Buonasera**, T. (2005). Fatty Acid Analysis of Prehistoric Burned Rocks. A Case Study from Central California. *Journal of Archaeological Science* **32**: 957-965.

**Casoli**, A., P. Mirti and G. Palla (1995). Characterization of Medieval Proteinaceous Painting Media using Gas Chromatography and Gas Chromatography-Mass Spectrometry. *Fresenius' Journal of Analytical Chemistry* **352**: 372-379.

**Collins**, M.J., C.M. Nielsen-Marsh, J. Hiller, C.I. Smith, J.P. Roberts, R.V. Prigodich, T.J. Wess, J. Csapò, A.R. Millard and G. Turner-Walker (2002). The survival of Organic Matter in Bone. A Review. *Archaeometry* **44,3**: 383-394.

**Diamond**, J.M. (2000). Talk of Cannibalism. *Nature* **407**: 25-25.

**Dongoske**, K.E., D.L. Martin and T.J. Ferguson (2000). Critique of the Claim of Cannibalism at Cowboy Wash. *American Antiquity* **65,1**: 179-190.

**Fiedel**, S.J. (1996). Blood from Stones? Some Methodological and Interpretative Problems with Blood Residue Analysis. *Journal of Archaeological Science* **23**: 139-147.

**Frankel**, E.N. (1998). *Lipid Oxidation.* Ayr: Oily Press.

**Hansel**, F.A., M.S. Copley, L.A.S. Madureira and R.P. Evershed (2004). Thermally Produced ω-(o-alkylphenyl)alkanoic Acids Provide Evidence for the Processing of Marine Products in Archaeological Pottery Vessels. *Tetrahedron Letters* **45**: 2999-3002.

**Newman**, M.E, R.M. Yohe, B. Kooyand and H. Ceri (1997). 'Blood' from Stones? Probably. A Response to Fiedel. *Journal of Archaeological Science* **24**: 1023-1027.

**Quigg**, M.J., M.E. Malainey, R. Przybylski and G. Monks (2001). No Bones about it. Using Lipid Analysis of Burned Rock and Groundstone Residues to examine Late Archaic Subsistence Practices in South Texas. *Plains Anthropologist* **46**: 283–303.

**Shimoyma**, A., N. Kisu, K. Harada, S. Wakita, A. Tsuneki and T. Iwasaki (1995). Fatty Acid Analysis of Archaeological Pottery Vessels Excavated in Tell Mastuma, Syria. *Bulletin of the Chemical Society of Japan* **68**: 1565-1568.

**Stankiewicz**, B.A., J.C. Hutchins, R. Thomson, D.E.G. Briggs and R.P. Evershed (1997). Assesment of Bog-body Tissue Preservation by Pyrolysis-Gas Chromatography/Mass Spectrometry. *Rapid Communications in Mass Spectrometry* **11**: 1884-1890.

**van Bergen**, P.F., I. Poole, T.M.A. Ogilvie, C. Caple and R.P. Evershed (1999). Evidence for Demethylation of Syringyl Moieties in Achaeological Wood using Pyrolysis-Gas Chromatography/Mass Spectrometry. *Rapid Communications in Mass Spectrometry* **14**: 71-79

# CHAPTER TWO
# Residues of Maize in North American Pottery: What Phytoliths can add to the Story of Maize

**R.K. Lusteck and R.G. Thompson**

Robert Lusteck, Department of Anthropology, University of Minnesota; 395 Hubert H. Humphrey Hall; Minneapolis, MN 55455 and Robert Thompson; Interdisciplinary Archaeological Studies, University of Minnesota; 395 Hubert H. Humphrey Hall; Minneapolis, MN 55455 (USA). We would like to thank Michael Michlovic and Fred Schneider for the opportunity to work on the Shea Site food residues. Michael Michlovic also provided the opportunity to work on the residues from 21CY39 and 21WL1. The Anthropology Department of the University of Minnesota provided funding for the AMS date on Shea Site food residues. Guy Gibbon has made many thoughtful comments on various versions of this chapter. We would like to thank Susan Mulholland, who, along with George Rapp, developed the taxonomy of grass phytoliths used in this paper.

Most of the chapters in this volume discuss the recovery of molecular residues from archaeological artifacts such as pottery. Chemical residue analysis has become an increasingly important and promising tool to archaeologists. We would like for archaeologists to not overlook an equally important source of information, the systematic recovery of physical residues from pottery. The encrustations on pottery, whether visible to the naked eye or not, are a veritable treasure trove of data and should be treated as an archaeological feature. Plants cooked in pottery often leave behind microscopic remains such as starch grains, pollen grains and phytoliths. Our work focuses on the recovery and identification of opal phytoliths from cooking residues that can be used to identify plant taxa. Opal phytoliths are formed in a number of plants. Most groundwater systems contain low concentrations of silicic acids that are absorbed through the root system of plants. As the plant loses water by transpiration, silica is exuded into or between the cells where it polymerzies into a silica imprint of the shape of these cells (Piperno 1988; 2006).[1] After the plant has died and decayed such imprints, known as opal phytoliths or silica phytoliths, can survive for thousands or even millions of years.

One problem with traditional microbotanical research is that it has relied almost exclusively on sediment samples (Pearsall 2000). While sediment samples have their place and can be very useful, there is the constant issue of tying the phytoliths to a secure cultural context. As a context for archaeological discovery, residues are superior to other contexts, such as soils. Food residues incorporate phytoliths into their matrix, be it the encrustations found on the inside and outside of pottery, dental calculus that accumulates on teeth (of both human and non-human species), the working surfaces of stone tools, or the undigested waste in a coprolite (Thompson and Mulholland 1994; 2006). Residues are stable and do not facilitate movement or intrusions. As a context of discovery, one can reasonably expect to find that the plants present in the residues, for example inside a pot, were placed there purposely (Mulholland 1987; 1993). Residues are also amenable to accelerated mass spectrometry (AMS) radiocarbon dating, allowing us to obtain good dates on our cultural contexts. Once the soil contact layer is removed and the organic portion of the food residue matrix is dissolved away, the phytoliths remain as a secure cultural record of and important part of the paleodiet (Hart et al. 2003; Thompson et al. 2005; Thompson 2006).

Phytoliths have been studied for some time but have really come to hold their own in archaeology in the past few decades (Piperno 1988). Just like pollen types, there are different phytolith forms that have different taxonomic utility, ranging generally from family to genus. For example, many members of the Cucurbitaceae family produce phytoliths that can be identified at the level of genera (Piperno et al. 2002). One plant family, the Poaceae or grass family, has been shown to produce an abundance of phytoliths in the leaf, stem and inflorescence (Mulholland 1989; Piperno 1988; Piperno 2006). The Poaceae family is of particular interest to many archaeologists because it includes all of the staple annual grasses on which many agricultural systems were, and still are, based, such as wheat (*Triticum* spp.), rice (*Oryza* spp.) and maize (*Zea mays*). These grasses produce a very common phytolith in their inflorescence (the flowering or reproductive parts of the plant). These are commonly referred to as rondels because of their general round to oval appearance. In fact, Mulholland and Rapp (1992) characterize these shapes as indicative of grasses.

---

[1] Silicic acid is the name of a group of chemical compounds with the general formula $(SiO_x(OH)_{4-2x})_n$ Very small amounts of simple silicic acids of geological origin, such as metasilicic acid ($H_2SiO_3$) and disilicic acid ($H_2Si_2O_5$), will dissolve in groundwater; ocean water contains small amounts of orthosilicic acid ($H_4SiO_4$). Common opal is amorphous $SiO_2{\cdot}nH_2O$ (hydrated silicon dioxide) with a water content of 5-20% (Iler 1979).

More recent research into these phytolith forms has shown that the particular characteristics of rondels are under genetic control (Piperno et al. 2001; Staller and Thompson 2002; Thompson 2006). Although this has not yet been fully explored in most grasses, it is somewhat understood in maize and its progenitor, teosinte (*Zea mays* ssp. *parviglumis*). A gene known as 'teosinte glume architecture' (tga1) was found to control the deposition of silica in the inflorescence of Zea species (Dorweieler et al. 1993; Dorweiler and Doebley 1997). As the name suggests, tga1 directly controls the overall form of the inflorescence glumes, the small leaf structures that encapsulate the fertilized seeds in grasses. These are the pieces that are commonly stuck between teeth when eating corn on the cob. In the transition from teosinte to maize, the glumes have become greatly reduced in size so that they no longer cover the seed, but rather cradle it where it attaches to the cob.

The reorganization caused by the maize allele of this gene is one of the key changes in the transition from teosinte to maize (Doebley 2004). It was further recognized that tga1 is a regulatory gene, meaning it interacts with other genes and the environment to create subtle variations in expression, unique to each lineage. As tga1 interacts with other genes during cob growth, the amount of silica deposition as well as the morphology of the forms are impacted. While rondel forms are found in many grasses, examination of a 'population' of rondels from any maize lineage reveals a 'fingerprint' in the silica deposition. howver, it has proven exceedingly difficult to determine how best to identify such fingerprints.

We are using a new term in phytolith analysis, the 'phenotypic profile' (Lusteck 2006). The phenotypic profile is defined as those phytoliths produced by a plant which, when statistically analyzed, are diagnostic at some level. This profile varies among plant types, but is consistent at some level, be it the species, variety etc. These profiles can be used to deal with the problem of phytolith redundancy. Although many plants produce the same types of phytoliths (redundancy), the suite of phytoliths that constitute each profile are unique. For the purposes of this research, although all grasses produce rondel phytoliths, the phenotypic profile for maize is significantly different from those of other grasses. Furthermore, as individual lineages pick up idiosyncrasies in their genetic code, the profile of a lineage of maize grown in one area will come to look different than the profiles of other maize populations.

Using the phenotypic profile, we can assign groups to genetically related lineages of maize. There are numerous problems with using actual genetics for such a study, most importantly the difficulty of obtaining viable DNA samples from charred maize remains (Janicke-Deprés et al. 2003; Janicke-Deprés and Smith 2006). The phenotypic profile produced by phytoliths in maize cobs is directly controlled by genetics and can serve as an adequate proxy. The profile cannot detect all genetic change, clearly, only those that in turn affect the deposition of silica. However, this seems to be a sufficiently sensitive characteristic to pick up major changes. And the level of the lineage is still much more informative than a typology based on row-number.

Like many cereals, maize has lost the ability to easily disarticulate the seeds and will not continue to grow in an abandoned area as many other domesticates will. Without the direct intervention of humans to harvest and disperse the seeds, maize will disappear (Hart 1999). Therefore, if maize remains are found, one can reasonably assume that humans were directly involved. Unlike many other domesticates, maize is very distinct from its progenitor. At a microscopic scale the differences may be subtle, size ranges and characteristics of microremains from maize and teosinte may overlap (Russ and Rovner 1989), but the macrosopic phenotype is always very distinct.

Perhaps most importantly, however, is the social significance that is commonly associated with maize. Many archaeologists have failed to critically examine maize use and the context of maize adoption into a culture. This has biased the archaeological history of maize in many regions; archaeologists have looked for proof of preconceived notions of maize use, instead of letting the data dictate their interpretations (Thompson et al. 2005). For example, if maize was primarily adopted as a staple in New York, it should appear ubiquitously. This does not occur before around 900 CE, yet maize phytoliths dating to 2270 BP have been recovered, suggesting a much different history of use (Hart et al. 2003). Maize may have been introduced as a source of alcohol or, as has recently been suggested in Mexico, a source of sugar from stalk quids (Smalley and Blake 2003). It is necessary to keep an open mind regarding the possible uses of maize. Maize often has special significance within cultures well before becoming a staple. The social context of maize is different than other crops in both modern and prehistoric cultures. In Peru, maize chicha (beer) was used to maintain reciprocal ties and political alliances during the Inca expansion (Hastorf and Johannessen 1994). In the Caribbean, Newsom and Deagan (1994) propose that maize may have been limited to elite or ritual contexts. In fact, in many cultures, maize is intimately tied with ceremonial and religious contexts (Cowan 1985; Swanton 1946).

## Materials and Methods

In this report we present a reanalysis of vessels from the Shea Site (32CS101) on the Maple River, a single vessel from the Dahnke Site (32CS29) at the junction of the Sheyenne and Red Rivers, another single vessel from the Wilford Site (21WL1) along the Red River, and two vessels from 21CY39 near the Buffalo River (Figure 1). Together, the analyses at these sites demonstrate the potential of phytolith assemblages to identify maize lineages in secure cultural context. Additionally, AMS dating of the same residues from the Shea Site provides precise data on the time of maize utilization along with carbon isotope data reflecting its proportion in the residues. Although applied to only a few sites on the Northeastern Plains, the data recovered from food residues indicate that maize has a longer and more complex history in the Northeastern Plains than previously envisioned.

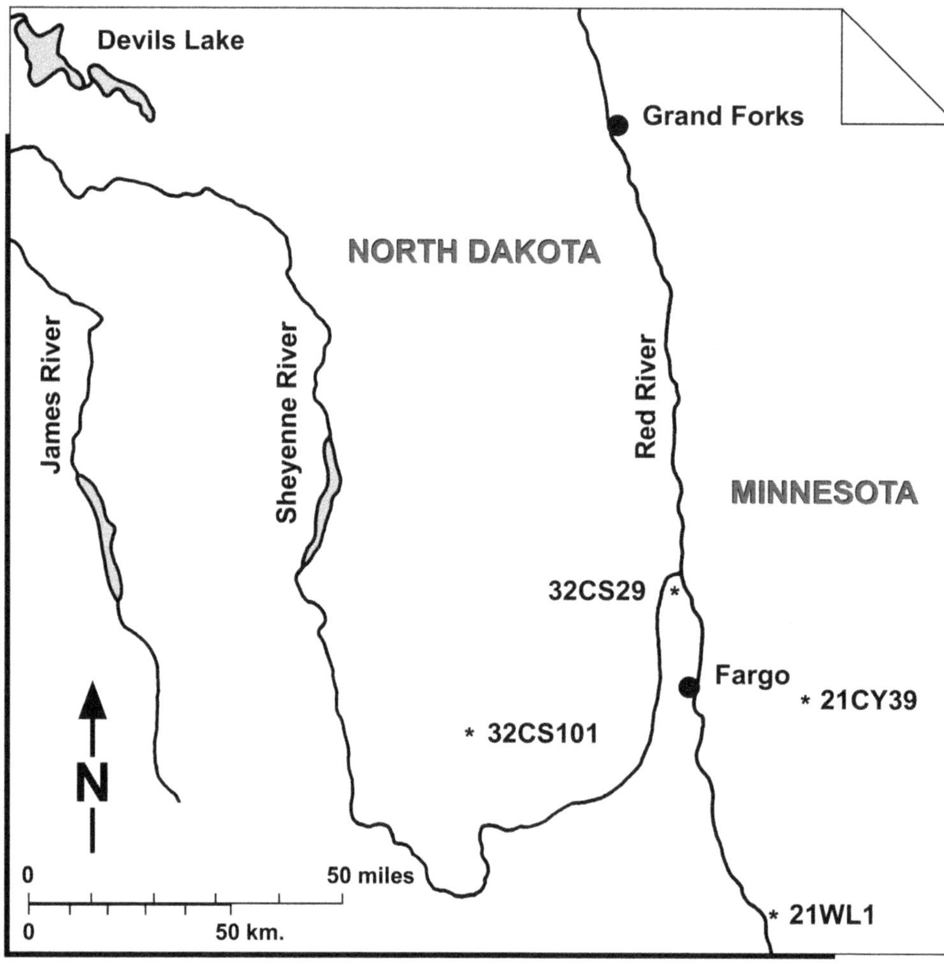


Figure 1: Map of the Northeastern Plains, in North Dakota and Minnesota, showing the location of the sites mentioned in the text (32CS101 = Shea Site; 32CS29 = Dahnke Site; 21WL1 = Wilford Site).

The new paradigm that we are exploring uses a phenotypic profile of phytoliths taken from the edible portions of plants to establish an analog to which archaeologically derived assemblages can be compared. The comparisons are made using square chord distance statistics. Developed in palynology (the study of pollen and spores), the square chord distance creates a dissimilarity matrix by comparing every sample/analog against every other sample/analog while accounting for small and large sample sizes and noise.

In an ideal situation, a crust of residue is visible on the inside of a pot sherd. To lessen the possibility of outside contamination, all sherds were gently washed with a solution of Alconox. The outside of the residue was scraped with a razor blade under running water to remove phytoliths from soil contact. The remaining residue was removed with dental tools. In the less ideal situation, we employed a sonic toothbrush to remove less obvious residues. Either way, the residue sample was placed in nitric acid at 90°C for at least 24 hours to remove the organic matrix. Samples were rinsed and centrifuged three times with distilled water and twice with isopropyl alcohol. A few drops of the final supernatant were pipetted onto slides and secured with Permount.

One hundred rondel phytoliths per sample were identified and individually photographed (at 400x using a DMLB light microscope with digital camera), measured and placed into a taxonomy originally developed by Mulholland and Rapp (1992), modified by Thompson (2000), and Thompson and Lusteck (2000). The presence of processes and the characters of both dorsal and ventral faces were used to classify the shapes of phytoliths (see Thompson 2006 for a detailed discussion). Measurements include the maximum width, maximum length and length/width (aspect) ratio. These measurements were taken using ScionImage digital imaging software. Each of these profiles was compared against a database of over 100 analog samples, from both modern and archaeological contexts using the square chord distance (Overpeck et al. 1985, Prentice 1980). The output of this procedure is a dissimilarity matrix. The SYSTAT and PAST statistical packages provided useful ways of presenting and evaluating these results using Ward's cluster analysis.

| Sample | Provenience | Provided by | Remarks |
|---|---|---|---|
| Shea1 | S23 W23 5-30 cm South Wall Balk | Michlovic and Schneider | Residue sample |
| Shea2 | N5+6 W12 Bastion Area | Michlovic and Schneider | Residue sample |
| Dahnke-Reinke Sandy Lake Sherd | Dahnke Site | Thompson | Residue sample |
| Wilford Site Sherd | Wilford Site | Michlovic | Residue sample |
| 21CY39A | S45 W88 | Michlovic | Residue sample |
| 21CY39B | Feature 2 | Michlovic | Residue sample |
| Mandan sweet corn samples A, C | Grown in Grand Forks, ND | Schneider | Experimental gardens |
| Mandan sweet corn sample B | Grown in Ames, IA | Schneider | University of Iowa greenhouse |
| Arikara flint corn | Grown in Grand Forks, ND | Schneider | Experimental gardens |
| Devil's Lake Sioux flint samples A, B | Grown in Grand Forks, ND | Schneider | Experimental gardens |
| Mandan red flour corn samples A, B | Grown in Grand Forks, ND | Schneider | Experimental gardens |
| Mandan blue flour corn | Grown in Grand Forks, ND | Schneider | Experimental gardens |
| Mandan yellow flour corn | Grown in Grand Forks, ND | Schneider | Experimental gardens |
| Dakota flint corn | Grown in Grand Forks, ND | Schneider | Experimental gardens |
| Mandan clay red corn samples A, B | Grown in Madison, WI | Iltis | University of Wisconsin herbarium |
| Mandan black flour corn samples A, B | Grown in Madison, WI | Iltis | University of Wisconsin herbarium |
| Northern flint corn | Grown in Boston, MA | Reber | Experimental gardens |
| Iroquois flour corn | Grown in Albany, NY | Reber | Experimental gardens |
| Shoepeg dent | Grown in St. Paul, MN | Weiblen | University of Minnesota herbarium |

Table 1: Overview of our maize and residue samples.

Schneider (2002) recently reviewed the slowly accumulating evidence for early horticulture on the Northeastern Plains, which collectively contradicts the long held assumption that the sparse recovery of plant remains is representative of a lack of horticultural activities. The limited recovery of data relating to horticulture, and plant use in general, is rather seen as a result of the history of investigations and inadequate research design. Nevertheless, as Schneider points out:

> Over the past 30 years, however, evidence has gradually accumulated for the presence of cultigens and the inferred presence of local horticulture at prehistoric sites, primarily non-earthlodge village sites, outside of the Missouri River valley in the adjacent northeastern Plains subarea. (Schneider 2002, 33).

Schneider elaborates on five issues which cloud the recognition of horticulture at sites in this region:

1. A biased view of the nature of the sites, based on their small size and readily apparent accumulations of faunal remains;
2. the belief that the climate of the region would not support crop production;
3. failure to use appropriate field techniques for the recovery of archaeobotanical remains;
4. lack of familiarity with early historic accounts of native horticulture;
5. lack of experiments to demonstrate the potential of native horticultural techniques to the region (Schneider 2002).

Bias in the interpretation of archaeological sites from the subarea was discussed in detail in Schneider's paper. The immediately apparent form of most of the area's sites, faunal remains and stone tools, creates an impression of hunting camps. As an historical outgrowth of the acceptance of this initial interpretation of the sites, the paradigm and research designs needed to investigate the potential importance of horticulture, and plant use in general, have not been developed and implemented as a regular practice.

Even with intensive flotation and pollen sampling built into investigations from the planning stages, as was the case at the Shea Site, the recovery of plant remains can be sparse. The soils at the Shea Site consist of well-drained sands and silts that do not provide a good environment for plant preservation. Few plant macrofossils were recovered from the site, providing a very patchy record of plant utilization. Kernels of maize and fragments of maize cob cupules were among the plant remains recovered. Bison scapula hoes, recovered from several contexts at the site, suggest that maize horticulture may have been an important activity. At the time of publication of the final report, the investigators concluded, on the basis of available data, that maize was an unknown portion of the diet of these late prehistoric villagers (Michlovic and Schneider 1993).

Because of the lack of data on the process of adopting and increasing dependence on maize horticulture, the literature seems to present fully agricultural villages appearing as if from nowhere. The underlying hypothesis is that crops and their producers migrated to the region as a fully developed complex. Only a systematic emphasis on the recovery of data pertaining to plant use can test this assumption. The recovery of phytolith assemblages from food residues should be a fundamental part of site analysis, especially since it can frequently supply information on plant use at sites in which preservation conditions preclude the use of other, more familiar techniques.

The provenience of the food residue and modern maize samples are listed in Table 1. One residue sample from Shea1 provided sufficient organic residue for AMS radiocarbon dating (provided by Beta Analytic). The procedure also involved the determination of $^{13}C/^{12}C$ ratios, commonly used as indicators for the presence of $C_4$ plants, such as maize.

## Results

A comparative collection of maize cob chaff phytoliths was first compared to itself, allowing the identification of typical distances between samples of the same and different lineages of maize (Table 2). Using the entire database, including the measurements of length and width, samples of maize of the same varieties had distance indices from 0.32 to 1.02. As Ball and Brotherson (1992) demonstrated that the size of phytoliths is influenced by environmental factors to a greater degree than other aspects of morphology, a second comparison of maize varieties was conducted using only morphological data with length and width measurements eliminated (Table 3, Figure 2).

In this table, the results are easier to interpret. Maize samples of the same varieties had a distance of about 0.40, and closely related varieties in the 0.50 range. The Devil's Lake Sioux flint samples (DLSFA and DLSFB) showed the most variation, with a 0.70 dissimilarity index.[2] The two varieties from the northeast showed greater distance from the Northeastern Plains samples. All of these maize varieties fall under the Northern flint category described by Doebley et al. (1986). These varieties are genetically very similar. They have lost about 50% of the genetic variation found in the maize varieties of the southwest from which they are thought to

[2] 'Flint' refers to maize with kernels characterized by a hard layer enclosing the soft endosperm; Northern flint is distinct subgroup of maize (*Zea mays* var. *indurata*).

derive, but the phytolith assemblages produced by each type can be distinguished.

Each of the food residue samples contained abundant biogenic silica and each sample yielded the 100 rondel phytoliths required for statistical comparison. The diatoms recovered from each residue sample are common to shallow fresh water, and are often found in association with shallow water plants. The distance matrix generated for the residue samples is given in Table 4. At least two distinct varieties within the Northern flint complex have been detected in the residues recovered at the Shea Site. This is significant for the interpretation of subsistence patterns at the Shea Site and the Northeastern Plains in general. The cultivation of more than one variety of maize indicates that maize cultivation was developed to the extent that different maizes, possibly grown for different purposes or at different times of year, were present at the site. The residue sample from N5+6 W12, Bastion Area, Trench A clustered with a group of races sampled for this analysis. It showed considerable variation and may even represent a mixture of races. The residue sample from Sheal clustered closely with one variety of maize, Mandan sweet corn. This variety of maize consistently produces a unique assemblage of rondel phytoliths in its cupule and glume, both in the field and greenhouse grown samples.

| | MYF | MBLFA | MBLFB | MCRA | MCRB | MSCA | MSCB | MSCC | MBF1 | MBF2 | MBF3 | MRFA | MRFB | ARFL | DLSFA | DLSFB | DAKF | NFLINT | IRFLOUR | DENT |
|---|---|---|---|---|---|---|---|---|---|---|---|---|---|---|---|---|---|---|---|---|
| MYF | - | 67 | 78 | 133 | 120 | 83 | 130 | 139 | 71 | 92 | 111 | 115 | 170 | 99 | 75 | 118 | 143 | 106 | 168 | 181 |
| MBLFA | | - | 32 | 119 | 93 | 101 | 146 | 172 | 61 | 55 | 79 | 112 | 158 | 90 | 80 | 100 | 138 | 118 | 150 | 181 |
| MBLFB | | | - | 107 | 93 | 116 | 161 | 191 | 77 | 61 | 74 | 127 | 183 | 97 | 85 | 113 | 141 | 118 | 150 | 178 |
| MCRA | | | | - | 69 | 200 | 268 | 295 | 172 | 135 | 160 | 251 | 279 | 162 | 134 | 137 | 207 | 203 | 244 | 260 |
| MCRB | | | | | - | 186 | 273 | 299 | 163 | 125 | 154 | 238 | 293 | 174 | 130 | 173 | 226 | 215 | 248 | 270 |
| MSCA | | | | | | - | 68 | 74 | 73 | 128 | 160 | 100 | 113 | 132 | 100 | 144 | 156 | 103 | 167 | 139 |
| MSCB | | | | | | | - | 61 | 113 | 192 | 214 | 106 | 91 | 177 | 162 | 184 | 156 | 147 | 181 | 148 |
| MSCC | | | | | | | | - | 126 | 205 | 233 | 121 | 96 | 188 | 172 | 205 | 174 | 164 | 194 | 170 |
| MBF1 | | | | | | | | | - | 83 | 87 | 89 | 117 | 97 | 85 | 103 | 116 | 85 | 129 | 148 |
| MBF2 | | | | | | | | | | - | 68 | 120 | 175 | 78 | 86 | 104 | 153 | 132 | 147 | 179 |
| MBF3 | | | | | | | | | | | - | 116 | 190 | 76 | 89 | 89 | 138 | 99 | 96 | 187 |
| MRFA | | | | | | | | | | | | - | 102 | 96 | 89 | 142 | 152 | 114 | 104 | 132 |
| MRFB | | | | | | | | | | | | | - | 166 | 147 | 177 | 131 | 154 | 124 | 156 |
| ARFL | | | | | | | | | | | | | | - | 73 | 113 | 137 | 111 | 112 | 168 |
| DLSFA | | | | | | | | | | | | | | | - | 92 | 148 | 107 | 116 | 169 |
| DLSFB | | | | | | | | | | | | | | | | - | 126 | 114 | 129 | 136 |
| DAKF | | | | | | | | | | | | | | | | | - | 132 | 138 | 172 |
| NFLINT | | | | | | | | | | | | | | | | | | - | 108 | 159 |
| IRFLOUR | | | | | | | | | | | | | | | | | | | - | 165 |
| DENT | | | | | | | | | | | | | | | | | | | | - |

Table 2: Dissimilarity Index of modern maize lineages. MYF = Mandan yellow flour; MBLF = Mandan black flour; MCR = Mandan clay red; MSC = Mandan sweet corn; MBF = Mandan blue flour; MRF = Mandan red flour; ARFL = Arikara flint; DLSF = Devil's Lake Sioux flint; DAKF = Dakota flint; NFLINT = Northern flint; IRFLOUR = Iriquois flour; DENT = Shoepeg dent (Table 1).

The AMS date (Beta-77427) provided a measured radiocarbon date of 390 +/- 60 BP. The conventional radiocarbon date was 450 +/- 60 BP. The $^{13}C/^{12}C$ ratio was -21.6‰. This is a good indicator of the use of maize (a $C_4$ grass) in a pottery vessel (Hastorf and DeNiro 1985). Two independent experiments have shown that carbon isotope values provide a relative measure of the amount of maize present in a food residue sample (Hart et al. 2003). Based on the results of these experiments the carbon isotope ratio present in the food residue would indicate that the original food represented was more than 50% maize by weight.

At the Dahnke and Wilford Sites single vessels were examined. At the Dahnke Site, a straight-necked Sandy Lake rimsherd, common across the northern wooded area of Minnesota and the Northeastern Plains, yielded residue for the study. At the Wilford Site another common late prehistoric vessel was recovered. Phytolith assemblages from both sites represent Northeastern Plains maize varieties. At 21CY39, along the Buffalo River, maize phytolith assemblages were recovered from two pottery vessels. Just as at the Dahnke and Wilford Sites, Northeastern Plains varieties of maize were represented. Radiocarbon dates from secure cultural contexts at this site place the represented occupation at about 1000 BP.

The detection of more than one variety of maize at the Shea Site has important implications for the interpretation of the social importance of maize on the Northeastern Plains. With at least two varieties of maize present, it can be said that maize use was likely more than an incidental addition to the diet. In addition, Will and Hyde (1917) have provided a detailed description of the use of sweet corn, noting that it was usually used alone. This may explain why the phytolith assemblage recovered from food residues is so closely related to the assemblages recovered from sweet corn cobs.

We know that Native Americans used multiple kinds of corn. There is historic evidence from the Iroquois of the Northeast, to the Mandan of the Plains, to the Pueblos of the Southwest of keeping types of maize separated. The phenotypic profile is currently the best way to get at these emic types. While we cannot discover traits like color, we can discover traces of the actual lineages that were selected for by prehistoric people by using the phenotypic profile as a proxy. Unlike previous studies that mask maize variety in typologies based upon row number or kernel characteristics, the phenotypic profile as a type of phytolith assemblage approach can separate out individual lineages of maize and identify variation in context of use.

| | MYF | MBLFA | MBLFB | MCRA | MCRB | MSCA | MSCB | MSCC | MBF1 | MBF2 | MBF3 | MRF | MRFB | ARFL | DLSF | DLSFB | DAKF | NFLINT | IRFLOUR | DENT |
|---|---|---|---|---|---|---|---|---|---|---|---|---|---|---|---|---|---|---|---|---|
| MYF | - | 45 | 50 | 89 | 65 | 40 | 52 | 45 | 39 | 64 | 73 | 51 | 43 | 74 | 54 | 91 | 74 | 69 | 80 | 107 |
| MBLFA | | - | 20 | 82 | 53 | 60 | 74 | 78 | 31 | 31 | 41 | 44 | 53 | 62 | 57 | 76 | 67 | 83 | 67 | 112 |
| MBLFB | | | - | 77 | 58 | 66 | 78 | 80 | 36 | 32 | 35 | 45 | 56 | 58 | 52 | 79 | 59 | 78 | 53 | 99 |
| MCRA | | | | - | 50 | 94 | 113 | 120 | 83 | 80 | 85 | 106 | 85 | 91 | 71 | 66 | 85 | 111 | 91 | 115 |
| MCRB | | | | | - | 75 | 104 | 100 | 69 | 61 | 72 | 82 | 70 | 96 | 54 | 84 | 83 | 117 | 74 | 110 |
| MSCA | | | | | | - | 40 | 31 | 37 | 79 | 91 | 64 | 38 | 83 | 59 | 109 | 84 | 73 | 96 | 119 |
| MSCB | | | | | | | - | 38 | 60 | 99 | 117 | 62 | 53 | 98 | 87 | 127 | 100 | 97 | 112 | 118 |
| MSCC | | | | | | | | - | 58 | 103 | 120 | 79 | 51 | 100 | 91 | 137 | 105 | 99 | 130 | 130 |
| MBF1 | | | | | | | | | - | 48 | 46 | 44 | 40 | 61 | 55 | 78 | 64 | 63 | 71 | 103 |
| MBF2 | | | | | | | | | | - | 33 | 51 | 53 | 53 | 62 | 76 | 68 | 92 | 67 | 110 |
| MBF3 | | | | | | | | | | | - | 58 | 73 | 50 | 56 | 63 | 65 | 69 | 43 | 120 |
| MRF | | | | | | | | | | | | - | 40 | 55 | 43 | 102 | 82 | 78 | 62 | 114 |
| MRFB | | | | | | | | | | | | | - | 59 | 42 | 99 | 73 | 83 | 79 | 108 |
| ARFL | | | | | | | | | | | | | | - | 52 | 95 | 72 | 77 | 57 | 109 |
| DLSF | | | | | | | | | | | | | | | - | 74 | 79 | 72 | 48 | 111 |
| DLSFB | | | | | | | | | | | | | | | | - | 76 | 95 | 88 | 94 |
| DAKF | | | | | | | | | | | | | | | | | - | 81 | 87 | 100 |
| NFLINT | | | | | | | | | | | | | | | | | | - | 69 | 127 |
| IRFLOUR | | | | | | | | | | | | | | | | | | | - | 126 |
| DENT | | | | | | | | | | | | | | | | | | | | - |

Table 3: Dissimilarity Index of maize phytolith assemblages based on morphological data only. MYF = Mandan yellow flour; MBLF = Mandan black flour; MCR = Mandan clay red; MSC = Mandan sweet corn; MBF = Mandan blue flour; MRF = Mandan red flour; ARFL = Arikara flint; DLSF = Devil's Lake Sioux flint; DAKF = Dakota flint; NFLINT = Northern flint; IRFLOUR = Iriquois flour; DENT = Shoepeg dent (Table 1).

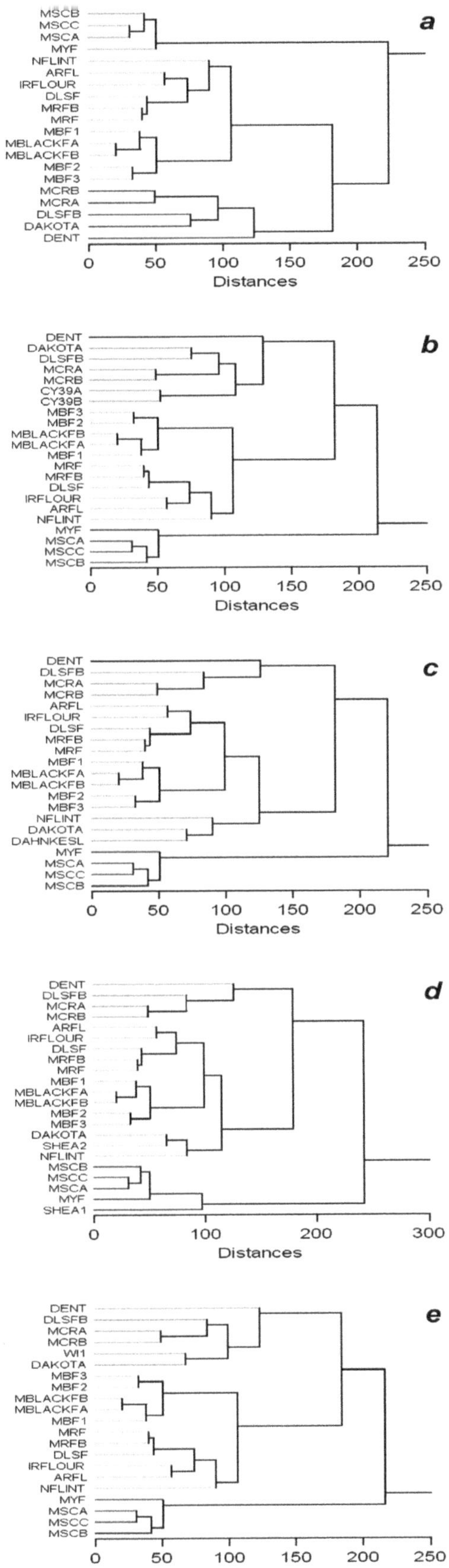


Figure 2: Cluster diagrams illustrating square chord distances comparing a) modern maize varieties; b) modern maize varieties and residues from 21CY39; c) modern maize varieties and residues from the Dahnke Site (32CS29); d) modern maize varieties and residues from the Shea Site (32CS101); and e) modern maize varieties and residues from 21WL1.

The presence of more than one lineage is not the only indicator of the importance of maize at the Shea Site. The carbon isotope data is of great significance. At the time that Beta Analytic produced the AMS date and associated isotope ratio study, only Hastorf and DeNiro (1985) had published on the nature of isotope data recovered from a food residue. According to this publication, a carbon isotope ratio of -21.6‰ would indicate a marginal presence of maize. However, independent experiments, conducted by Hart et al. (2003) show that an isotope ratio of -21.6‰ is an indicator that maize was probably more than 50% by weight of the food entering the pot. The results presented here from a single pot are not sufficient to redefine the role of maize in Northeastern Plains villages, but they are sufficient to demonstrate the need for more detailed study of the importance of maize to these people.

The Dahnke and Wilford Sites, along with 21CY39, demonstrate two important notions. First, remains of maize are present in the Northeastern Plains in sufficient amounts to be located at many sites. Second, remains of maize are present in sites that Schneider refers to as artifact scatters, considered the remains of hunting camps. In fact, at the Wilford Site no other evidence of plant use was recovered. The radiocarbon date of 21CY39, about 1000 CE, shows that maize was present quite early at the site, and was consumed by the producers of Blackduck pottery, who are not usually considered maize producers.

## Conclusions

Collectively, these studies show that maize was potentially more important to the peoples of the Northeastern Plains Village Complex and other late perhistoric people than was previously thought. More than one lineage of maize was present and maize may have represented a greater portion of the diet than previous studies indicate.

| | MYF | MBLFA | MCRB | MSCA | MSCB | MSCC | MBF1 | MRFB | DLSF |
|---|---|---|---|---|---|---|---|---|---|
| **CY39A** | 58 | 60 | 55 | 57 | 72 | 71 | 56 | 47 | 69 |
| **CY39B** | 68 | 51 | 72 | 75 | 79 | 97 | 54 | 61 | 66 |
| **Shea 1** | 55 | 78 | 92 | 50 | 51 | 48 | 63 | 61 | 104 |
| **Shea 2** | 66 | 65 | 73 | 84 | 101 | 108 | 62 | 61 | 53 |
| **Dahnke** | 62 | 66 | 92 | 53 | 66 | 74 | 59 | 58 | 80 |
| **Wilford** | 69 | 61 | 29 | 83 | 108 | 99 | 77 | 70 | 75 |

Table 4: Dissimilarity Index of maize phytolith assemblage recovered from food residues compared to maize varieties based on morphological characteristics.

The limited recovery of data has profound implications for the study of plant use on the Northeastern Plains and in the Americas as a whole. Phytolith analysis has the potential to answer long-standing questions about the introduction and diversification of maize in North America. Genetic and archaeological studies demonstrate that maize was domesticated in or near the Balsas River Valley of Southwestern Mexico, and was present from Manitoba to Argentina by the time of arrival of Columbus (Bennetzen et al. 2001; Smith 1998). Much of the story of the movement and social significance of maize between its initial adoption and its eventual distribution to the limits of its growth regime remain unknown. Other plants, such as wild rice (*Zizania* spp.), can also be traced using phytoliths (Thompson et al. 1995), as can squash and beans. Paterson et al. (1995) noted the convergent patterns of evolution for all domesticated grasses, so this technique should in fact be useful worldwide. Food residues, rather than being simply amorphous carbon, contain morphologically intact and taxonomically useful phytolith assemblages. Archaeologists studying lifestyle, social structure, household and engendered activities, as well as shifts in subsistence activities, should incorporate the investigation of food residues in their analyses on a regular basis.

## References

**Ball**, T.B. and J.D. Brotherson (1992). Effect of Varying Environmental Conditions on Phytolith Morphometries in Two Species of Grass (*Bouteloua curtipendula* and *Panicum virgatum*). *Scanning Electron Microscopy* **6**: 1163-1181.

**Bennetzen**, J., E. Buckler, V. Chandler, J. Doebley, J. Dorwheiler, B. Gaut, M. Freeling, S. Hake, E. Kellogg, R. S. Peothig, V. Walbot and S. Wessler (2001). Genetic Evidence and the Origin of Maize. *Latin American Antiquity* **12**: 84-86.

**Doebley**, J.F. (2004). The Genetics of Maize Evolution. *Annual Review of Genetics* **38**: 37-59.

**Doebley**, J.F., M.M. Goodman and C.W. Stuber (1986). Exceptional Genetic Divergence of Northern Flint Corn. *American Journal of Botany* **73**: 64-69.

**Dorweiler**, J., A. Stec, J. Kermicle and J.F. Doebley (1993). Teosinte Glume Architecture-1-A Genetic Locus Controlling a Key Step in Maize Evolution. *Science* **262**: 233-235.

**Dorweiler**, J. and J.F. Doebley (1997). Developmental Analysis of Teosinte Glume Architecture 1. A key locus in the Evolution of Maize (Poaceae). *American Journal of Botany* **84**: 1313-1322.

**Hart**, J.P. (1999). Maize Agriculture Evolution in the Eastern Woodlands of North America. A Darwinian Perspective. *Journal of Archaeological Method and Theory* **6,2**: 137-180.

**Hart**, J.P., R. Thompson and H.J. Brumbach (2003). Phytolith Evidence for Early Maize in the Northern Finger Lakes Region of New York. *American Antiquity* **68**: 619-640.

**Hastorf**, C.A. and M.J. DeNiro (1985). Reconstruction of Prehistoric Plant Production and Cooking Practices by a New Isotope Method. *Nature* **315**: 489-491.

**Hastorf**, C.A. and S. Johannessen (1994). Becoming Corn-Eaters in Prehistoric America. In S. Johannessen and C.A. Hastorf (eds.). *Corn and Culture in the Prehistoric New World*. Boulder: Westview Press, pp. 427-444.

**Iler**, R.K. (1979). *The Chemistry of Silica. Solubility, Polymerization, Colloid and Surface Properties, and Biochemistry*. New York: Wiley.

**Jaenicke-Després**, V.R., E.S. Buckler, B.D. Smith, T.M Gilbert, A. Cooper, J. Doebley and S. Pääbo (2003). Early Allelic Selection in Maize as revealed by Ancient DNA. *Science* **302**: 1206-1208.

**Jaenicke-Després**, V.R. and B.D. Smith (2006). Ancient DNA and the Integration of Archaeological and Genetic Approaches to the Study of Maize Domestication. In J.E. Staller, R.H. Tykot and B.F. Benz (eds.). *Stories of Maize I. Multidisciplinary Approaches to the Prehistory, Biogeography, Domestication, and Evolution of Maize (*Zea mays *L.)*. New York: Academic Press, pp. 83-95.

**Lusteck**, R. (2006). The Migrations of Maize into the Southeastern United States. In J.E. Staller, R.H. Tykot and B.F. Benz (eds.). *Stories of Maize I. Multidisciplinary Approaches to the Prehistory, Biogeography, Domestication, and Evolution of Maize (*Zea mays *L.)*. New York: Academic Press, pp. 521-528.
**Michlovic**, M. and F. Schneider (1993). Shea Site. A Prehistoric Fortified Village on the Northeastern Plains. Plains. *Anthropologist* **38**: 117-138.
**Michlovic**, M. and F.E. Swenson (1998). Northeastern Plains Village Pottery. North Dakota. *History* **65**: 11-25.
**Mulholland**, S.C. (1987). *Phytolith Studies at Big Hidatsa, North Dakota*. University of Minnesota, Minneapolis Ph.D. thesis.
**Mulholland**, S.C. (1993). A test of phytolith analysis at Big Hidatsa, North Dakota. In D. Pearsall and D. Piperno (eds.). *Current Research in Phytolith Analysis. Applications in Archaeology and Paleoecology*. Philadelphia: MASCA, pp. 131-145.
**Mullholland**, S.C. and G. Rapp, Jr. (1992). A Morphological Classification of Grass Silica Bodies. In G. Rapp, Jr. and S.C. Mulholland (eds.). *Phytolith Systematics. Emerging Issues*. New York: Plenum Press, pp. 65-89.
**Overpeck**, J.T., T. Webb III and I.C. Prentice (1985). Quantitative Interpretation of Fossil Pollen Spectra. Dissimilarity Coefficients and the Method of Modern Analogs. *Quaternary Research* **23**: 87-108.
**Paterson**, A., Y. Lin, Z. Li, K. Schertz, J. Doebley, S. Pinson, S. Liu, J. Stansel and J. Irvine (1995). Convergent Domestication of Cereal Crops by Independent Mutations at Corresponding Genetic Loci. *Science* **269**: 1714-1719.
**Pearsall**, D.M. (2000). *Paleoethnobotany. A Handbook of Procedures*. San Diego: Academic Press.
**Piperno**, D.R. (1988). *Phytolith Analysis. An Archaeological and Geological Perspective*. San Diego: Academic Press.
**Piperno**, D.R., I. Holst, A.J. Ranere, P. Hansell and K.E. Stothart (2001). The Occurrence of Genetically Controlled Phytoliths from Maize Cobs and Starch Grains from Maize Kernels on Archaeological Stone Tools and Human Teeth, and in Archaeological Sediments from Southern Central America and Northern South America. *The Phytolitharien* **13**: 1-7.
**Piperno**, D.R. (2006). *Phytoliths. A Comprehensive Guide for Archaeologists and Paleoecologists*. New York: Alta Mira Press.
**Piperno**, D.R., I. Holst, L. Wessel-Beaver and T.C. Andres (2002). Evidence for the Control of Phytolith Formation in Cucurbita Fruits by the Hard Rind (Hr) Genetic Locus. Archaeological and Ecological Implications. *Proceedings of the National Academy of Science U.S.A.* **99**: 10923-10928.
**Prentice**, I.C. (1980). Multidimensional Scaling as a Research Tool in Quaternary Palynology. A Review of Theory and Methods. *Review of Palaeobotany and Palynology* **31**: 71-104.
**Russ**, J. and I. Rovner (1989). Stereological Identification of Opal Phytolith Populations from Wild and Cultivated Zea. *American Antiquity* **54**: 784-792.
**Schneider**, F. (2002). Prehistoric Horticulture in the Northeastern Plains. *Plains Anthropologist* **47**: 33-50.
**Smalley**, J. and M. Blake (2003). Sweet Beginnings. *Current Anthropology* **44,5**: 675-703.
**Smith**, B.D. (1998). The Emergence of Agriculture. Second Edition, Scientific American Library, New York. 2001 Documenting Plant Domestication. The Consilience of Biological and Archaeological Approaches. *Proceedings of the National Academy of Sciences U.S.A.* **98**: 1324-1326.
**Staller**, J. and R. Thompson (2002). A Multidisciplinary Approach to Understanding the Initial Introduction of Maize into Coastal Ecuador. *Journal of Archaeological Science* **29** :33-50.
**Swanton**, J.R. (1946). *Indians of the Southeastern United States. Bureau of American Ethnology Bulletin 137*. Washington D.C.: Smithsonian Institution.
**Thompson**, R.G. (1993). *Phytolith Analysis of Food Residues from Prehistoric Pottery from the Mantaro Valley, Peru*. Paper presented at the Northeast Andeanist Conference. Pittsburgh, 20 November 1993.
**Thompson**, R.G. (2000). *Phytolith Analysis of Food Residues from Yutopian, Argentina*. Unpublished report prepared for Dr. Joan Gero, American University.
**Thompson**, R.G. (2006). Tracing the Diffusion of Maize into South America through Phytolith Analysis of Food Residues. In M.A. Zeder, D.G. Bradley, E. Emshwiller, and B.D. Smith (eds.). *Documenting Domestication. New Genetic and Archaeological Paradigms*. Berkeley: University of California Press, pp. 82-96.
**Thompson**, R.G., J.P. Hart, H.J. Brumbach and R. Lusteck (2005). Phytolith Evidence for Twentieth-Century B.P. Maize in Northern Iroquoia. *Northeast Anthropology* **68**: 25-40.
**Thompson**, R.G., R. Kluth and D. Kluth (1995). Brainerd Ware Pottery Function Explored Through Opal Phytolith Analysis of Food Residues. *Journal of Ethnobiology* **15**: 305.
**Thompson**, R.G. and R. Lusteck (2000). *Phytolith Analysis of Food Residues in Selected Colorado Ceramics*. Unpublished report prepared for Dr. Richard Krause, University of Alabama.
**Thompson**, R.G. and S. Mulholland (1994). The Identification of Corn in Food Residues on Utilized Ceramics at the Shea Site (32CS101). *The Phytolitharian* **8**: 7-11.
**Will** G.F. and G. E. Hyde (1917). *Corn Among the Indians of the Upper Missouri*. Lincoln: University of Nebraska Press.

# CHAPTER THREE
# Micro-Residues on Stone Tools: The Bigger Picture from a South African Middle Stone Age Perspective
## M. Lombard and L. Wadley

Marlize Lombard; Natal Museum; Private Bag 9070; Pietermaritzburg; 3200 South Africa; <mlombard@nmsa.org.za> and Lyn Wadley; Archaeology Department; School of Geography, Archaeology and Environmental Studies; University of the Witwatersrand; Private Bag 3; WITS 2050 Johannesburg, South Africa. We are indebted to all students and colleagues who participated in the Rose Cottage Cave and Sibudu Cave excavations or contributed to Ancient Cognition and Culture in Africa (ACACIA) research projects. Our appreciation goes to the Department of Archaeology at the University of the Witwatersrand for the use of their microscopes and digital microphotography equipment. The research of Marlize Lombard is funded by the Palaeontological Scientific Trust (PAST) and supported by the Natal Museum. Lyn Wadley received funds from the National Research Foundation of South Africa and the support of the University of the Witwatersrand. The opinions expressed here, or any oversights, are those of the authors and are not to be attributed to the funding agencies or supporting organizations.

Our stone tool micro-residue analysis was developed within the bigger framework of Middle Stone Age research in South Africa. Progress in our methodology was partly influenced by addressing the problems encountered during a series of four blind tests, two of which were entirely field-based. This resulted in a more secure strategy for distinguishing plant and animal residues, and we have made advances in the identification of incidental as opposed to use-related residues. A multi-stranded approach improved our chances of correctly identifying and interpreting residues on archaeological stone tools. Focused micro-residue analyses and the interpretation of results can now be used to gain detailed knowledge of Middle Stone Age human behavior regarding hunting and butchery activities, as well as variations in hafting technologies and the functional application of ochre. Micro-residue analyses applied to tools from the post-Howiesons Poort, Howiesons Poort and Still Bay technocomplexes contribute towards global research that investigates human behavioral evolution. This chapter aims to contextualize the development of our micro-residue research.

The scope of this chapter cannot facilitate an in-depth discussion. By highlighting some aspects that influenced the direction of our research we hope to create an impression of our current framework. The first South African Ph.D. thesis on stone tool residue studies was completed in 2000 (Williamson 2000a) under the direction of Lyn Wadley and the late Tom Loy. This study introduced the basic principles as well as the potential of residue analysis for the analysis of the Stone Age in the sub-continent where an approximately 2.5 million year old tool making tradition is recognized. Sub-Saharan Africa is a region with biological and cultural continuity, so the African record is indispensable for human behavioural evolution (Kuhn and Hovers 2006). It is against this background that Lyn Wadley began general Middle Stone Age research, while Marlize Lombard started exploring questions about Middle Stone Age hunting and hafting behavior. These explorations, and associated research questions and projects, resulted in our approach to micro-residue analysis gaining its distinctive direction and momentum. Experimentation, modern replication and blind testing became integral parts of our research design (Lombard et al. 2004; Lombard and Wadley 2007; Rots and Williamson 2004; Wadley 2005a; b; 2006; Wadley and Lombard, in press; Wadley et al. 2004a), without losing sight of the main goal: to improve our understanding of past human behavior.

The micro-residue work is deeply embedded in archaeological questions that have arisen during the course of excavations first at Rose Cottage Cave, Free State, and at Sibudu Cave, KwaZulu-Natal (Gibson et al. 2004; Lombard 2004; 2005a; Wadley et al. 2004b; Williamson 1996; 1997; 2000b; 2004; 2005). For over a decade, stone tools have been collected from these sites with the specific intention of subjecting them to residue analysis. Rose Cottage Cave appears to have been occupied, perhaps intermittently, over a period of about 90 ka (Pienaar 2006; Valladas et al. 2005) and Sibudu Cave has many finely separated post-Howiesons Poort Middle Stone Age layers that have been dated to between about 37 ka and 60 ka old by optically stimulated luminescence (OSL) (Jacobs 2004; Wadley 2005c; Wadley and Jacobs 2004).[1] Below the post-Howiesons Poort layers, dating to about 60 ka ago, there are two meters of Sibudu Cave deposits with Howiesons Poort and Still Bay technocomplexes, for which OSL dates are not yet available. Both sites have yielded hundreds of thousands stone tools from long sequences containing multiple industries (Clark 1997a; b; 1999; Harper 1997; Villa et al. 2005; Wadley 1992; 1996;

[1] One ka (kilo-annum) is 1000 years, 1000 ka (1000,000 years) equals one Ma (Mega-annum), which replaces the previously used mya (million years ago).

1997; 2000a; b; 2001a; b; 2005c). These stone tools occur in varying contexts over time. At both sites, but particularly at Sibudu Cave, we have extraordinary evidence for environmental change and for change in the use of features such as hearths. It is important to take a holistic view and position micro-residue analysis within the broader framework of data such as fauna, plant and geological sediments (Allott 2004; Cain 2004; Plug 2004, in press; Schiegl et al. 2004; Wadley 2004). The micro-residue research on stone tools is closely linked to our multi-disciplinary approach of the other archaeological material and is intended to complement it. Within this context we are consistently attempting to improve our methodology and are exploring the potential uses of stone tools and investigating whether these were composite, hafted tools.

## How Blind Tests improved our Method

Blind testing was the impetus for the Society for American Archaeology (SAA) Symposium (Salt Lake City, 31 March 2005) of which this publication is an outcome (Barnard et al. 2007). Even though our tests, based on the morphological identification of residues on stone tools using light microscopy, are essentially different from the other tests presented at this Symposium, they adequately illustrate how blind testing can be used to strengthen analytical methodology. In 2004 we published the protocols and results of our first two blind tests (Wadley et al. 2004a). The original aim of the tests was to assess the ability, of Marlize Lombard, to identify a variety of plant and animal residues using light microscopy. Problems arising during the testing process made it clear that greater value might be gained from the lessons learnt about the methodology and the direction for future micro-residue research. Addressing the problems identified during our first tests stimulated new research, and subsequently two more tests were conducted, this time totally field-based (Lombard and Wadley 2007). The series of four blind tests have facilitated the development of a multi-stranded approach that provides a cautious, but secure strategy for the interpretation of archaeological residues.

Our first test highlighted the problems of a morphological distinction of certain plant and animal residues. Most mistakes in the identification during the test were made on faunal material and this suggested that some archaeological faunal residues might in the past have been erroneously interpreted as plant material (Lombard and Wadley 2007). This issue is of particular interest in our broader archaeological investigations because Sibudu Cave has large faunal assemblages but, prior to 2003, we found relatively few animal residues on stone tools and this seemed counter-intuitive. Errors in identification could have contributed to the perception that plant materials were more often processed than animal materials, or that animal residues did not preserve as well. The research of Marlize Lombard continued with the recognition of animal residue types and comparison of these with a variety of plant residues. An important outcome was the confirmation that birefringence (the double refraction of incident light) is not an exclusive characteristic of cellulose plant residues; certain faunal tissues are also highly birefringent when viewed with cross-polarized incident light.[2] Another outcome was the recognition that there are morphological similarities between certain plant and animal residues, such as ordered cell structures, cell shapes, the characteristics of fibers and the color and translucency of residues (see the color plate for some examples of ancient and modern micro-residues). Degradation, such as expected on archaeological tools, can make the distinction between plant and animal residues even more difficult.

The tests have also confirmed that not all residues on stone tools are associated with the use of the tool. Contaminants collect easily as a result of mistakes or unintentional coincidences (Wadley et al. 2004a) and during deliberate field-based replications and exposure to dust (Lombard and Wadley 2007). In our opinion the residue analysts need to give as much attention to the recognition of potential contaminants as to other residues. In addition to inclusions from ancient dust, archaeologically recovered tools may have accumulated other kinds of contaminating residues post-depositionally, prior to excavation. Microscope slides with dust samples, or soil samples from archaeological contexts, may help to establish records of the microscopic morphological appearance of such contaminants and they can be compared with putative use-related residues on stone tools (Lombard, 2006b; Lombard and Wadley 2007).

Our third test highlighted the problems associated with the types of rock on which residues may occur. It illustrates that whitish, light reflective and refractive rocks (such as quartz and quartzite) could reduce the

[2] We do not use a standard polarizing microscope, but a metallurgical microscope with incident light, which does not pass through the sample but is reflected, and rotating analyzer and polarizer filters to cross-polarize the light. Cross-polarization of light is obtained when the analyzer filter, between the sample and the eyepiece, is at a right angle with the polarizer filter, between the light source and the sample. This situation is sometimes referred to as 'crossed Nicols' or XPL. Many materials affect the plane of polarized light and will therefore light up when viewed in such light. Their specific behavior may assist the identification of such materials, including some organic residues. This technique is based on the field of optical mineralogy (Nesse 2003).

accuracy of residue identifications. Residues with dark colors like ochre, animal tissue, resin and bark are clearly visible on these raw materials, but colored inclusions in the rocks may be misleading. Another observation made during the analysis of quartz tools is that some residues tend not to adhere to the smooth glass-like surfaces and accumulate only in cracks and crevices. This makes detailed plotting and quantification of the residues on these rock types challenging. The difficulty of detecting whitish, translucent and birefringent residues such as fat, bone, silica skeletons or starch grains on quartz and quartzite should be acknowledged; such residues can easily be overlooked if the analyst is not aware that extra care should be taken. More analytical work is needed on these rock types. Until this has been done we should approach archaeological tools made on these rocks with caution.

Addressing the mistakes of the first three tests resulted in a perfect score for our fourth test. However, it does not follow that archaeological residues can necessarily be interpreted with the same accuracy. A margin of error must be expected, but we consider this margin reduced through our explorations. As a result of our experience with the series of four blind tests we suggest that using multi-stranded evidence is a satisfying solution for distinguishing between some plant and animal residues, and for isolating residues that are potential contaminants. We believe that minuscule fragments of plant tissue, fibers, starch grains, feathers or hair can only be securely interpreted as use-residues when there are supportive strands of evidence. This applies to archaeological tools as well as to experimental ones. Supportive evidence could come from repetitive clustering of a variety of plant residue types, such as tissue, fibers and starch grains in attendance with other plant residues such as wood, bark cells, resin or plant exudates. The same principle holds for faunal residues, which can be most convincingly identified when hair, animal tissue, fat, bone, blood and collagen (or several of these residue types) are found as associated residues. Single elements of any residue type provide less confident identifications and interpretations (Lombard and Wadley 2007).

A single residue, or residue type, seldom allows identification of the use of a stone tool. Our cumulative replication, experimentation, testing and archaeological work on more than a thousand tools (Delagnes et al. in press; Gibson et al. 2004; Lombard 2003; 2004; 2005; 2006a; b; c; Lombard et al. 2004; Rots and Williamson 2004; Williamson 1996; 1997; 2000; 2004; 2005), shows that when use-related residues preserve, they often preserve collectively. Not once has only one residue type, or one single residue, been observed on used replicated tools or on ancient tools from Sibudu Cave, which were unquestionably used because of tell-tale traces of wear and macro-fractures. This being said, we acknowledge that the organic preservation, and therefore also the preservation of the organic residues, at Sibudu Cave is probably exceptional (Allott 2004; Plug 2004; Wadley 2003; Wadley and Jacobs 2004). Comparable residues might not be present on tools from other sites or other contexts (Lombard 2004; 2005; 2006a; b).

In addition to improving our method, the tools made and used for our blind tests and other replication work have become valuable reference material within our existing comparative collection. The importance of comprehensive comparative material cannot be overstated. Microscope slides and residues on replicated stone tools are ideal for three-dimensional comparison and residues stored in this way are better for reference purposes than photographs. By monitoring the condition of these replicated specimens at regular intervals the potential degradation or preservation of residues can be studied as well (Lombard and Wadley 2007).

The blind tests and the problems addressed during the testing process have greatly influenced the development of our current approach to micro-residue analysis and the interpretation of results (Lombard and Wadley 2007; Wadley et al. 2004). This approach is well illustrated in the work on wear by Rots et al. (2006). We now have a far more secure strategy for distinguishing plant and animal residues and we have made advances in the identification of incidental as opposed to use-related residues. We believe, furthermore, that the blind test outcomes and field-based replication work have improved our chances of correctly identifying and interpreting micro-residues on archaeological stone tools.

For the comprehensive interpretation of archaeological micro-residue data a spatial analysis method was developed. Tools are divided into portions according to their morphology (Lombard 2004; 2005; 2006b). Each portion includes a dorsal and ventral side; all residues are plotted and counted in relation to these portions. This method highlights the possible existence of distribution patterns. It needs to be stressed that distribution patterns can only be recognized when micro-residue types are repeatedly recorded. Although this method cannot be considered an accurate quantification of the residues, it does provide a realistic reflection of the actual distribution and concentrations of preserved residues on the tools. The recovered data can be compared to existing notions of function or hafting arrangements gained from technological observations, previous analyses of the traces of use or ethnographic and archaeological examples. Functional or hafting interpretations at assemblage-level can be attempted when data from about 20 tools of similar morphology and context are analyzed and micro-residue distribution patterns occur repeatedly. These data can also be statistically tested for coincidental distribution patterns (Lombard 2004; 2005; 2006b).

## Zooming Out

The Middle Stone Age in southern Africa is currently a focal point of global research interest investigating the origins of modern human behavioral (Conard 2005; d'Errico 2003; d'Errico et al. 2003; Henshilwood and Marean 2003; Henshilwood and Sealy 1997; Hovers and Belfer-Cohen 2006; Klein 2000; McBrearty and Brooks 2000; Wadley 2001). Refined dating methods, fine-grained typological and technological analyses as well as carefully controlled excavations are providing information with increased resolution for this period (Henshilwood 2005; Henshilwood, Sealy et al. 2001; b; Henshilwood and Sealy 1997; Jacobs et al. 2003a; b; Tribolo et al. 2005; Valladas et al. 2005; Villa et al. 2005; Wadley 2005c; Wadley and Jacobs 2004; in press; Wurz et al. 2003; Wurz 2005; Wurz et al. 2005). Sites are gradually yielding cultural objects that were previously only associated with Later Stone Age or Upper Paleolithic behavior. Such objects include the intentionally marked ostrich eggshell from Diepkloof, Western Cape (Parkington et al. 2005) and the shell beads, engraved bone and engraved ochre from Blombos Cave, southern Cape (d'Errico et al. 2001; Henshilwood et al. 2002; 2004). Also central to current Middle Stone Age behavioral research are questions on the origins of projectile technology (Brooks et al. 2006; Shea 2006), the origins and development of hafting technologies (Ambrose 2001; d'Errico 2003; Mazza et al. in press; Mellars 1996; Rots and Van Peer 2006) and the symbolic and functional application of ochre (Ambrose 1998; Barham 1998; 2002; Barton 2005; d'Errico 2003; Godfrey-Smith and Ilani 2004; Hovers et al. 2003; Knight et al 1995; Van Peer et al. 2004; Wadley 2005a; b; 2006; Wadley et al. 2004b; Watts 1998; 2002; Wreshner 1980; 1982). It is within these last three focus areas that our micro-residue research is able to contribute detailed empirical data.

A multi-analytical functional study suggests that points from the post-Howiesons Poort layers at Sibudu Cave, dated to between about 50 ka and 60 ka ago, were predominantly used as hafted spear tips. The quantification, plotting and chi-square statistical tests of the distribution patterns of 440 residue occurrences on the 24 whole points show that the distribution of the residue types cannot be considered coincidental (Lombard 2004; 2005). The traces indicate the use of wooden shafts thus revealing a wood working industry for which there exists little additional evidence in the southern African Middle Stone Age. The stone points were probably glued to the shafts with an adhesive and then lashed with plant twine for added strength during impact. The evidence indicates that the spears were most likely used as thrusting or throwing spears. This impression is supported by the technological data of Villa et al. (2005) whose measurements of points from a layer within the post-Howiesons Poort sequence fall within the range of thrusting or throwing spears.

Further indicators for effective hunting during this phase at Sibudu Cave come from the associated faunal assemblage analyzed by Plug (2004). The age profiles of the Sibudu Cave samples show that the number of bones from juvenile or very old animals is not exceptionally high. Most animals are adults, with some sub-adults. Bones damaged by carnivores are scarce in relation to the size of the sample. These factors indicate that the people using the cave during the post-Howiesons Poort hunted rather than scavenged and targeted large animals in their prime on a regular basis (Plug 2004). This illustrates how data from three different analytical procedures (faunal analysis, technological analysis and functional analysis of stone tools) on material from Sibudu Cave is used to crosscheck and strengthen the interpretation that stone tipped spears were used as effective hunting weapons between 50 ka and 60 ka ago.

Based on their small size and apparent standardization, it has always been speculated that backed tools from the Howiesons Poort were hafted (Deacon 1989; 1993; Deacon 1995; Wurz 1999). Initial micro-residue evidence for the hafting of such tools from the Howiesons Poort at Rose Cottage Cave, dating from about 60 ka to 68 ka ago, was provided by Gibson et al. (2004). The distribution and positions of plant tissue, plant fibers and white starchy residue on these backed tools were interpreted as possible indications for hafting. A total of 48 backed tools of various types were analyzed. Backed blades appear to have been hafted laterally, segments (also called crescents or lunates) might have been placed transversely into their hafts, while obliquely backed blades were possibly hafted with their short axis in the haft (Gibson et al. 2004). This study, as well as other residue analyses on extensive post-Howiesons Poort tool samples from Sibudu Cave indicated that the distribution of ochre might be associated with other hafting traces (Wadley et al. 2004b; Lombard 2004). Subsequent experimentation and replication work conducted by Lyn Wadley established that the inclusion of ochre in the recipe could result in more effective adhesives (Wadley 2005a; b; 2006).

Spatial analysis and quantification were used to establish the relationship between ochre and resin on a sample of 53 Howiesons Poort segments from Sibudu Cave. During the analysis, ochre was documented 502 time and resin 585 times (Lombard 2006b). Most of the ochre (80%) and resin occurrences, 80% and 87% respectively, were located on the backed portions usually associated with hafting. The results are interpreted as compelling direct evidence that the tools were hafted and that ground ochre was used in the adhesive recipe. It supports the hafting evidence for backed tools from the Howiesons Poort at Rose Cottage Cave and previous observations about the association of ochre with Middle Stone Age hafting technology (Wadley et al. 2004b). Our experimental and micro-residue research program on ochre and its functional application in the production of Middle Stone Age adhesives is intended as an expansion of our current understanding of the versatility and value of ochre in prehistory. It is not intended as an alternative hypothesis for its possible symbolic role.

A similar trend for the quantity and distribution of ochre and resin residues on the Sibudu segments was observed on a sample of 30 non-quartz Howiesons Poort segments from Umhlatuzana Rock Shelter, about 100 km southwest of Sibudu Cave in KwaZulu-Natal. However, when a sample of 25 quartz segments from Umhlatuzana Rock Shelter was analyzed, using the same methodology, resin occurred 269 times and ochre only 43 times. Although both residues were concentrated on the backed edges, 68% of the quartz segments have resin but no ochre on them. The same is true for only 23% of the non-quartz sample from the shelter and 10% of the Sibudu Cave sample. The quartz and crystal quartz segments are not only generally smaller than those made on hornfels and dolerite, but they are also less elongated. Based on these morphological attributes we have suggested that they could have been hafted differently from the larger, longer segments produced on other raw materials (Delagnes et al. in press). Quartz is very hard, Mohs' scale 7, with smooth and glass-like surfaces (Bishop et al. 2001).

During our replication and blind testing we found that residues do not readily adhere to the hard, smooth surfaces of quartz to the same degree that they adhere to more porous, courser grained materials (Lombard and Wadley 2007). It is therefore feasible to consider that a different, possibly more 'sticky', adhesive recipe may have been used for hafting quartz tools. This hypothesis needs to be tested with focused replication work, but the data could indicate that Howiesons Poort people applied different adhesive recipes for different hafting requirements. The micro-residue analysis conducted on a small sample of crystal quartz backed tools from Sibudu Cave showed repeated combinations and concentrations of animal residues (animal tissue, bone, collagen, fat and blood) on the portions that were not hafted (Delagnes et al. in press). These tools were used to process animal products. It is not yet clear whether the composite tools, of which the quartz pieces were components, were used for hunting or butchery purposes. However, a macro-fracture study conducted on a range of backed Howiesons Poort tools from Klasies River Cave 2 showed that some of these tools could have been used as tips for hunting weapons (Lombard 2005b; Wurz and Lombard, in press).

Recently a study was also conducted on a sample of Still Bay pointed tools from Sibudu Cave that are possibly about 75 ka old. The sample size is still small, so that assemblage-level studies, similar to those conducted on the Howiesons Poort segments, are not yet possible. However, a detailed tool-by-tool analysis made it possible to test existing hypotheses and generate new working hypotheses for the functions and hafting technologies of these tools (Lombard 2006a). Wadley (2006) suggested that double-pointed, bifacial points with asymmetrical bases from Sibudu Cave were not intended to be reversible in their hafts, but that the bases were pointed to facilitate a type of hafting that was favored at the time. The micro-residue distribution patterns and other traces of use on the two available points of this type support this suggestion. Furthermore, both tools show signs of having been used as knives for butchering activities. It is possible that the asymmetrically pointed bases were an adaptation to facilitate the effective hafting of the tools as knives (Lombard 2006a). This also supports Minichillo's (2005) suggestion that some Still Bay points from the Cape were used as knives. There is one triangular bifacial point, similar in morphology to post-Howiesons Poort points. Macro-fractures as well as the distribution of animal residues on some of the distal broken point fragments show that they could have been used for hunting. The current working hypothesis is that the asymmetrical points with pointed bases were hafted as knives while symmetrical, triangular points were possibly hafted as hunting weapons. Continued work at Sibudu Cave and other sites with Still Bay assemblages will test this hypothesis.

## Conclusion

The aim of this chapter is to illustrate how stone tool micro-residue analysis in South Africa has evolved in close correlation with local archaeological excavations, experimental research projects and global research trends. The methodology and the reference collections, which were partly developed as a result of our blind test experiences, enable us to generate quantitative data with a reduced margin of error on the assemblage-level. Such data can be used to compare the results of assemblages from different contexts at the same site, or assemblages from different sites. One example is the comparison of the ochre and resin distribution patterns on non-quartz and quartz Howiesons Poort segment assemblages from KwaZulu-Natal.

Both the Howiesons Poort and Still Bay technocomplexes are central to the debate about the emergence of modern cognitive behavior and capacities (Ambrose, in press; Deacon 1989; 2001; Deacon and Wurz 1996; Henshilwood et al. 2001a, Henshilwood et al. 2001b; Lombard 2005c; Minichillo, in press; Wurz 1999; Wurz and Lombard, in press;). Our current work is starting to provide detailed, empirical evidence for hunting and butchery activities as well as insight into the complexities of hafting technologies practiced at KwaZulu-Natal sites during these periods. The interpretation of the results of these focused research projects informs on cognitive and technological skills and planning abilities. It shows that more than 60 ka ago, people understood the properties of various raw materials and tool shapes, and probably adapted their hunting, butchery and adhesive technologies accordingly. Continued work in KwaZulu-Natal at Sibudu and Umhlatuzana, and at other sites, which we hope will have equally well-preserved residues and other traces of use, can provide more detailed data about human behavior, contributing to the bigger picture of the process of modernization.

## References


**Allott**, L. (2004). Changing Environments in Oxygen Isotope Stage 3. Reconstructions using Archaeological Charcoal from Sibudu Cave. *South African Journal of Science* **100**: 179-184.

**Ambrose**, S.H. (1998). Chronology of the Later Stone Age and Food Production in East Africa. *Journal of Archaeological Science* **25**: 179-184.

**Ambrose**, S.H. (2001). Palaeolithic Technology and Human Evolution. *Science* **291**: 1748-1753.

**Ambrose**, S.H. (in press). Howiesons Poort Lithic Raw Material Procurement Patterns and the Evolution of Modern Human Behaviour. *Journal of Human Evolution.*

**Barham**, L. (1998). Possible Early Pigment Use in South-central Africa. *Current Anthropology* **39**: 703-710.

**Barham**, L. (2002). Systematic Pigment Use in the Middle Pleistocene of South-central Africa. *Current Anthropology* **43**: 181-190.

**Barnard**, H., S.H. Ambrose, D.E. Beehr, M.D. Forster, R.E. Lanehart, R.E. Parr, M.E. Malainey, M. Rider, C. Solazzo and R.M Yohe II (2007). Mixed Results of Seven Methods for Organic Residue Analysis Applied to One Vessel with the Residue of a Known Foodstuff. *Journal of Archaeological Science* **34**: 28-37.

**Barton**, L. (2005). Origins of Culture. Functional and Symbolic Uses of Ochre. *Current Anthropology* **46**: 499.

**Bishop**, A.C., A.R. Woolley and W.R. Hamilton (2001). *Minerals, Rocks and Fossils*. London: George Philip's.

**Brooks**, A.S., L. Nevell, J.E. Yellen and G. Hartman (2006). Projectile Technologies of the African MSA. Implications for Modern Human Origins. In E. Hovers and S.L. Kuhn (eds.). *Transitions before the Transition. Evolution and Stability in the Middle Palaeolithic and Middle Stone Age*. New York: Springer, pp. 233-255.

**Cain**, C.R. (2004). Notched, Flaked and Ground Bone Artefacts from Middle Stone Age and Iron Age Layers at Sibudu Cave. *South African Journal of Science* **100**: 195-197.

**Clark**, A.M.B. (1997a). The Final Middle Stone Age at Rose Cottage Cave. A Distinct Industry in the Basutolian Ecozone. *South African Journal of Science* **93**: 449-458.

**Clark**, A.M.B. (1997b). The MSA/LSA Transition in Southern Africa. New Technological Evidence from Rose Cottage Cave. *South African Archaeological Bulletin* **52**: 113-121.

**Clark**, A.M.B. (1999). Late Pleistocene Technology at Rose Cottage Cave. A Search for Modern Behaviour in an MSA Context. *African Archaeological Review* **16**: 93-119.

**Conard**, N.J. (2005). An Overview of the Patterns of Behavioural Change in Africa and Eurasia during the Middle and Late Pleistocene. In F. d'Errico and L. Backwell (eds.). *From Tools to Symbols. From Early Hominids to Modern Humans*. Johannesburg: Witwatersrand University Press, pp. 295-332.

**Deacon**, H.J. (1989). Late Pleistocene Palaeoecology and Archaeology in the Southern Cape, South Africa. In P. Mellars and C.B. Stringer (eds.). *The Human Revolution. Behavioural and Biological Perspectives of the Origins of Modern Humans*. Edinburgh: Edinburgh University Press, pp. 547-564.

**Deacon**, H.J. (1993). Southern Africa and Modern Human Origins. In M.J. Aitken, C.B. Stringer and P.A. Mellars (eds.). *The Origins of Modern Humans and Impact of Chronometric Dating*. Princeton: Princeton University Press, pp. 104-117.

**Deacon**, H.J. (1992). Southern Africa and Modern Human Origins. *Philosophical Transactions of the Royal Society, London* **B 337**: 177-183.

**Deacon**, H.J. (1995). An Unsolved Mystery at the Howieson's Poort Name Site. *South African Archaeological Bulletin* **50**: 110-120.

**Deacon**, H.J. (2001). Modern Human Emergence. An African Archaeological Perspective. In P.V. Tobias, M.A. Raath, J. Moggi-Cecci and G.A. Doyle (eds.). *Humanity from African Naissance to Coming Millennia*. Johannesburg: University of the Witwatersrand Press, pp. 213-222.

**Deacon**, H.J. and S. Wurz (1996). Klasies River Main Site, Cave 2. A Howiesons Poort Occurrence. In G. Pwiti and R. Soper (eds.). *Aspects of African Archaeology*. Harare: University of Zimbabwe Publications, pp. 213-218.

**Delagnes**, A., L. Wadley, P. Villa and M. Lombard (2006). Crystal Quartz Backed Tools from the Howiesons Poort at Sibudu Cave. *Southern African Humanities* **18,1**: 43-56.

**d'Errico**, F. (2003). The Invisible Frontier. A Multiple Species Model for the Origin of Behavioural Modernity. *Evolutionary Anthropology* **12**: 188-202.

**d'Errico**, F., C. Henshilwood, G. Lawson, M. Vanhaeren, A.-M. Tillier, M. Soressi, F. Bresson, B. Maureille, A. Nowell, J. Lakarra, L. Backwell and M. Julien (2003). Archaeological Evidence for the Emergence of Language, Symbolism, and Music. An

Alternative Multidisciplinary Perspective. *Journal of World Prehistory* **17**: 1-70.

**d'Errico**, F., C.S. Henshilwood and P. Nilssen (2001). An Engraved Bone Fragment from ca. 75 Kya Middle Stone Age Levels at Blombos Cave, South Africa. Implications for the Origin of Symbolism. *Antiquity* **75**: 309-318.

**Gibson**, N.E., L. Wadley and B.S. Williamson (2004). Residue Analysis of Backed Tools from the 60 000 to 68 000 Year-old Howiesons Poort Layers of Rose Cottage Cave, South Africa. *Southern African Humanities* **16**: 1-11.

**Godfrey-Smith**, D.I. and S. Ilani (2004). Past Thermal History of Hematite Fragments from Qafzeh Cave deduced from Thermal Activation Characteristics of the 110°C TL Peak of Enclosed Quartz Grains. *Revue d'Archéomètrie* **28**: 185-190.

**Harper**, P.T.N. (1997). The Middle Stone Age Sequences at Rose Cottage Cave. A Search for Continuity and Discontinuity. *South African Journal of Science* **93**: 470-475.

**Henshilwood**, C.S. (2005). Stratigraphic Integrity of the Middle Stone Age Levels at Blombos Cave. In F. d'Errico and L. Backwell (eds.). *From Tools to Symbols. From Early Hominids to Modern Humans*. Johannesburg: Witwatersrand University Press, pp. 441-458.

**Henshilwood**, C.S., F.E. d'Errico, C.W. Marean, R.G. Milo and R. Yates (2001b). An Early Bone Tool Industry from the Middle Stone Age at Blombos Cave, South Africa. Implications for the Origins of Modern Human Behaviour, Symbolism and Language. *Journal of Human Evolution* **41**: 631-678.

**Henshilwood**, C.S., F. d'Errico, R. Yates, Z. Jacobs, C. Tribolo, G.A.T. Duller, N. Mercier, J.C. Sealy, H. Valladas, I. Watts and A.G. Wintle (2002). Emergence of Modern Human Behaviour. Middle Stone Age Engravings from South Africa. *Science* **295**: 1278-1280.

**Henshilwood**, C.S. and C.W. Marean (2003). The Origin of Modern Human Behaviour. Critique of the Models and their Test Implications. *Current Anthropology* **44**: 627-651.

**Henshilwood**, C.S., F. d'Errico, M. Vanhaeren, K. van Niekerk and Z. Jacobs (2004). Middle Stone Age Shell Beads from South Africa. *Science* **384**: 1278-1280.

**Henshilwood**, C.S. and J.C. Sealy (1997). Bone Artefacts from the Middle Stone Age at Blombos Cave, Southern Cape, South Africa. *Current Anthropology* **38**: 890-895.

**Henshilwood**, C.S., J.C. Sealy, R.J. Yates, K. Curz-Uribe, P. Goldberg, F.E. Grine, R.G. Klein, C. Poggenpoel, K.L. van Niekerk and I Watts (2001a). Blombos Cave, Southern Cape. Preliminary Report on the 1992-1999 Excavations of the Middle Stone Age Levels. *Journal of Archaeological Science* **28**: 421-448.

**Hovers**, E. and A. Belfer-Cohen (2006). 'Now you see it, now you don't'. Modern Human Behaviour in the Middle Palaeolithic. In E. Hovers and S.L. Kuhn (eds.). *Transition before the Transition. Evaluation and Stability in the Middle Palaeolithic and Middle Stone Age*. Springer: New York, pp. 295-303.

**Hovers**, E., S. Ilani, O. Bar-Yosef and B. Vandermeersch (2003). An Early Use of Colour Symbolism. Ochre Use by Modern Humans in Qafzeh Cave. *Current Anthropology* **44**: 491-522.

**Jacobs**, Z., G.A.T. Duller and A.G. Wintle (2003a). Optical Dating of Dune Sands from Blombos Cave, South Africa II. Single Grain Data. *Journal of Human Evolution* **44**: 613-625.

**Jacobs**, Z., A.G. Wintle and G.A.T. Duller (2003b). Optical Dating of Dune Sands from Blombos Cave, South Africa I. Multiple Grain Data. *Journal of Human Evolution* **44**: 599-612.

**Jacobs**, Z. (2004). Development of Luminescence Techniques for Dating Middle Stone Age Sites in South Africa. University of Wales Ph.D. thesis.

**Klein**, R.G. (2000). Archaeology and the Evolution of Human Behaviour. *Evolution of Anthropology* **9**: 17-36.

**Knight**, C., C. Powers and I. Watts (1995). The Human Symbolic Revolution. A Darwinian Account. *Cambridge Archaeological Journal* **5**: 75-114.

**Kuhn**, S.L. and E. Hovers (2006). General Introduction. In E. Hovers and S.L. Kuhn (eds.). *Transitions before the Transition. Evolution and Stability in the Middle Palaeolithic and Middle Stone Age*. New York: Springer, pp. 1-11.

**Lombard**, M. (2003). Lithic Rings from Honingklip, Goergap, Olieboomspoort and Windsorton. A Functional Interpretation based on Microscopy. *South African Archaeological Bulletin* **58**: 6-13.

**Lombard**, M. (2004). Distribution Patterns of Organic Residues on Middle Stone Age Points from Sibudu Cave, KwaZulu-Natal, South Africa. *South African Archaeological Bulletin* **59**: 37-44.

**Lombard**, M. (2005a). Evidence for Hunting and Hafting during the Middle Stone Age at Sibudu Cave, KwaZulu-Natal, South Africa. A Multianalytical Approach. *Journal of Human Evolution* **48**: 279-300.

**Lombard**, M. (2005b). A Method for Identifying Stone Age Hunting Tools. *South African Archaeological Bulletin* **60**: 115-120.

**Lombard**, M. (2005c). The Howiesons Poort of South Africa. What we know, What we think we know, What we need to know. *Southern African Humanities* **17**: 33-55.

**Lombard**, M. (2006a). First Impressions on the Functions and Hafting Technology of Still Bay Pointed Artefacts from Sibudu Cave. *Southern African Humanities* **18,1**: 27-41.

**Lombard**, M. (2006b). Direct Evidence for the Use of Ochre in the Hafting Technology of Middle Stone Age Tools from Sibudu Cave, KwaZulu-Natal. *Southern African Humanities* **18,1**: 57-67.

**Lombard**, M., L. Parsons and M.M. van der Ryst (2004). Middle Stone Age Lithic Point Experimentation for Macro-fracture and Residue Analyses. The First Set of Experiments and Preliminary Results with Reference to Sibudu Cave Points. *South African Journal of Science* **100**: 159-166.

**Lombard**, M., and L. Wadley (2007). The Morphological Identification of Micro-residues on Stone Tools using Light Microscopy. Progress and Difficulties based on Blind Tests. *Journal of Archaeological Science* **34**: 155-165.

**Mazza**, P.P.A., F. Martini, B. Sala, M. Magi, M.P. Colombini, G. Giachi, F. Landucci, C. Lemorini, F. Modugno and E. Ribechini (in press). A New Palaeolithic Discovery. Tar-hafted Stone Tools in a European Mid-Pleistocene Bone-bearing Bed. *Journal of Archaeological Science.*

**McBrearty**, S. and A.S. Brooks (2000). The Revolution that wasn't. A New Interpretation of the Origin of Modern Human Behaviour. *Journal of Human Evolution* **39**: 453-563.

**Mellars**, P. (1996). *The Neanderthal Legacy*. Princeton: Princeton University Press.

**Minichillo**, T.J. (2005). Middle Stone Age Lithic Study, South Africa. An Examination of Modern Human Origins. University of Washington Ph.D. thesis.

**Minichillo**, T.J. (in press). Raw Material Use and Behavioural Modernity. Howiesons Poort Lithic Foraging Strategies. *Journal of Human Evolution.*

**Nesse**, W.D. (2003). *Introduction to Optical Mineralogy*. New York: Oxford University Press.

**Parkington**, J., C. Poggenpoel, J.-P. Rigaud and P.-J. Texier (2005). From Tool to Symbol. The Behavioural Context of Intentionally Marked Ostrich Eggshell from Diepkloof, Western Cape. In F. d'Errico and L. Backwell (eds.). *From Tools to Symbols. From Early Hominids to Modern Humans*. Johannesburg: Witwatersrand University Press, pp. 475-492.

**Pienaar**, M. (2006). Dating the Stone Age at Rose Cottage Cave, South Africa. University of Pretoria M.A. thesis.

**Plug**, I. (2004). Resource Exploitation: Animal Use during the Middle Stone Age at Sibudu Cave, KwaZulu-Natal. *South African Journal of Science* **100**: 151-158.

**Plug**, I. (2006). Aquatic Animals and their Associates from the Middle Stone Age Levels at Sibudu Cave. *Southern African Humanities* **18,1**: 289-299.

**Rots**, V. and P. van Peer (2006). Early Evidence of Complexity in Lithic Economy. Core-axe Production, Hafting and Use at Late Middle Pleistocene Site 8-B-11, Sai Island (Sudan). *Journal of Archaeological Science* **33**: 360-371.

**Rots**, V., L. Pirnay, P. Pirson and O. Baudoux (2006). Blind Tests shed Light on Possibilities and Limitations for Identifying Stone Tool Prehension and Hafting. *Journal of Archaeological Science* **33**: 935-952.

**Rots**, V. and B.S. Williamson (2004). Microwear and Residue Analysis in Perspective. The Contribution of Ethnographical Evidence. *Journal of Archaeological Science* **31**: 1287-1299.

**Schiegl**, S., P. Stockhammer. C. Scott and L. Wadley (2004). A Mineralogical and Phytolith Study of the Middle Stone Age Hearths in Sibudu Cave. *South African Journal of Science* **100**: 185-194.

**Shea**, J.J. (2006). The Origins of Lithic Projectile Point Technology. Evidence from Africa, the Levant, and Europe. *Journal of Archaeological Science* **33**: 823-846.

**Tribolo**, C., N. Mercier and H. Valladas (2005). Chronology of the Howiesons Poort and Still Bay Techno-complexes. Assessment and New Data from Luminescence. In F. d'Errico and L. Backwell, L. (eds.). *From Tools to Symbols. From Early Hominids to Modern Humans*. Johannesburg: University of the Witwatersrand Press, pp. 493-511.

**Valladas**, H., L. Wadley, M. Mercier, C. Tribolo, J.L. Reyss and J.L. Joron (2005). Thermoluminescence Dating on Burnt Lithics from Middle Stone Age Layers at Rose Cottage Cave. *South African Journal of Science* **101**: 169-174.

**Van Peer**, P., V. Rots and J.-M. Vroomans (2004). A Story of Colourful Diggers and Grinders. The Sangoan and Lupemban at Site 8-B-11, Sai Island, Northern Sudan. *Before Farming* **2004/3**: 1-28.

**Villa**, P., A. Delagnes and L. Wadley (2005). A Late Middle Stone Age Artefact Assemblage from Sibudu (KwaZulu-Natal). Comparisons with the Middle Palaeolithic. *Journal of Archaeological Science* **32**: 399-422.

**Wadley**, L. (1992). Rose Cottage Cave. The Later Stone Age Levels with European and Iron Age Artefacts. *South African Archaeological Bulletin* **47**: 8-12.

**Wadley**, L. (1996). The Robberg Industry of Rose Cottage Cave, Eastern Free State: The Technology, Spatial Patterns and Environment. *South African Archaeological Bulletin* **51**: 46-74.

**Wadley**, L. (1997). Where have all the Dead Men gone? Stone Age Burial Practices in South Africa. In L. Wadley (ed.). *Our Gendered Past. Archaeological Studies of Gender in Southern Africa*. Johannesburg: University of the Witwatersrand Press, pp. 107-133.

**Wadley**, L. (2000a). The Early Holocene Layers of Rose Cottage Cave, Eastern Free State. Technology, Spatial Patterns and Environment. *South African Archaeological Bulletin* **55**: 18-31.

**Wadley**, L. (2000b). The Wilton and Pre-ceramic Post-classic Wilton Industries at Rose Cottage Cave and their Context in the South African Sequence. *South African Archaeological Bulletin* **55**: 90-106.

**Wadley**, L. (2001a). Preliminary Report on Excavations at Sibudu Cave, KwaZulu-Natal. *Southern African Humanities* **13**: 1-7.

**Wadley**, L. (2001b). What is Cultural Modernity? A General View and a South African Perspective from Rose Cottage Cave. *Cambridge Archaeological Journal* **11**: 201-221.

**Wadley**, L. (2004). Vegetation Changes between 61 500 and 26 000 Years ago. The Evidence from Seeds in

Sibudu Cave, KwaZulu-Natal. *South African Journal of Science* **100**: 167-173.

**Wadley**, L. (2005a). Putting Ochre to the Test. Replication Studies of Adhesives that may have been used for Hafting Tools in the Middle Stone Age. *Journal of Human Evolution* **49**: 587-601.

**Wadley**, L. (2005b). Ochre Crayons or Waste Products? Replications compared with MSA 'Crayons' from Sibudu Cave, South Africa. *Before Farming* **2005/3**: 1-12.

**Wadley**, L. (2005c). A Typological Study of the Final Middle Stone Age Tools from Sibudu Cave, KwaZulu-Natal. *South African Archaeological Bulletin* **60**: 51-63.

**Wadley**, L. (2006). Revisiting Cultural Modernity and the Role of Ochre in the Middle Stone Age. In H. Soodyall (ed.). *The Prehistory of Africa. Tracing the Lineage of Modern Man*. Johannesburg: Jonathan Ball Publishers, pp. 49-63.

**Wadley**, L. and Z. Jacobs (2004). Sibudu Cave, KwaZulu-Natal. Background to the Excavations of Middle Stone Age and Iron Age Occupations. *South African Journal of Science* **100**: 145-151.

**Wadley**, L. and Z. Jacobs (2006). Sibudu Cave. Background to the Excavations, Stratigraphy and Dating. *Southern African Humanities* **18,1**: 1-26.

**Wadley**, L. and M. Lombard (in press). Small Things in Perspective: The Contribution of our Blind Tests to Micro-residue Studies on Archaeological Stone Tools. *Journal of Archaeological Science*.

**Wadley**, L., M. Lombard and B.S. Williamson (2004a). The First Residue Analysis Blind Tests. Results and Lessons Learnt. *Journal of Archaeological Science* **31**: 1491-1450.

**Wadley**, L., B.S. Williamson and M. Lombard (2004b). Ochre in Hafting in Middle Stone Age Southern Africa. A Practical Role. *Antiquity* **78**: 661-675.

**Watts**, I. (1998). The Origin of Symbolic Culture. The Middle Stone Age of Southern Africa and Khoisan Ethnography. University College London Ph.D. thesis.

**Watts**, I. (2002). Ochre in the Middle Stone Age of Southern Africa. Ritualised Display or Hide Preservative? *South African Archaeological Bulletin* **31**: 5-11.

**Williamson**, B.S. (1996). Preliminary Stone Tool Residue Analysis from Rose Cottage Cave. *Southern African Field Archaeology* **5**: 36-44.

**Williamson**, B.S. (1997). Down the Microscope and Beyond. Microscopy and Molecular Studies of Stone Tool Residues and Bone Implements from Rose Cottage Cave. *South African Journal of Science* **93**: 458-464.

**Williamson**, B.S. (2000a). Prehistoric Stone Tool Residue Analysis from Rose Cottage Cave and other Southern African sites. University of the Witwatersrand Ph.D. thesis.

**Williamson**, B.S. (2000b). Direct Testing of Rock Painting Pigments for Traces of Haemoglobin at Rose Cottage Cave, South Africa. *Journal of Archaeological Science* **27**: 755-762.

**Williamson**, B.S. (2004). Middle Stone Age Tool Function from Residue Analysis at Sibudu Cave. *South African Journal of Science* **100**: 174-178.

**Williamson**, B.S. (2005). Subsistence Strategies in the Middle Stone Age at Sibudu Cave. The Microscopic Evidence from Stone Tool Residues. In F. d'Errico and L. Backwell (eds.). *From Tools to Symbols. From Early Hominids to Modern Humans*. Johannesburg: University of the Witwatersrand Press, pp. 513-524.

**Wreschner**, E.E. (1982). Red Ochre, the Transition between Lower and Middle Palaeolithic and the Origin of Modern Man. In A. Ronen (ed.). *The Transition from Lower to Middle Palaeolithic and the Origin of Modern Man. British Archaeological Reports International Series 151*: Oxford: BAR Publishing, pp. 35-39.

**Wurz**, S. (1999). The Howiesons Poort Backed Artefacts from Klasies River. An Argument for Symbolic Behaviour. *South African Archaeological Bulletin* **54**: 38-50.

**Wurz**, S. (2005). Exploring and Quantifying Technological Differences between the MSA I, MSA II and Howieson's Poort at Klasies River. In F. d'Errico and L. Backwell (eds.). *From tools to Symbols. From Early Hominids to Modern Humans*. Johannesburg: University of the Witwatersrand Press, pp. 418-440.

**Wurz**, S., N.J. Le Roux, S. Gardner and H.J. Deacon (2003). Discriminating between the End Products of the Earlier Middle Stone Age Sub-stages at Klasies River using Biplot Methodology. *Journal of Archaeological Science* **30**: 1107-1126.

**Wurz**, S. and M. Lombard (in press). 70 000-year-old Geometric Backed Tools from the Howiesons Poort at Klasies River, South Africa. Were they used for hunting? In K. Seetah and B. Gravina (eds.). *Bones for Tools, Tools for Bones. The Interrelationship of Lithic and Bone Raw Materials*. Cambridge: McDonald Institute of Archaeological Research Monographs.

**Wurz**, S., P. van Peer, N. Le Roux, S. Gardener and H.J. Deacon (2005). Continental Patterns in Stone Tools. A Technological and Biplot-based Comparison of Early Late Pleistocene Assemblages from Northern and Southern Africa. *African Archaeological Review* **22**: 1-25.

8
10
12
14
7
9
11
13

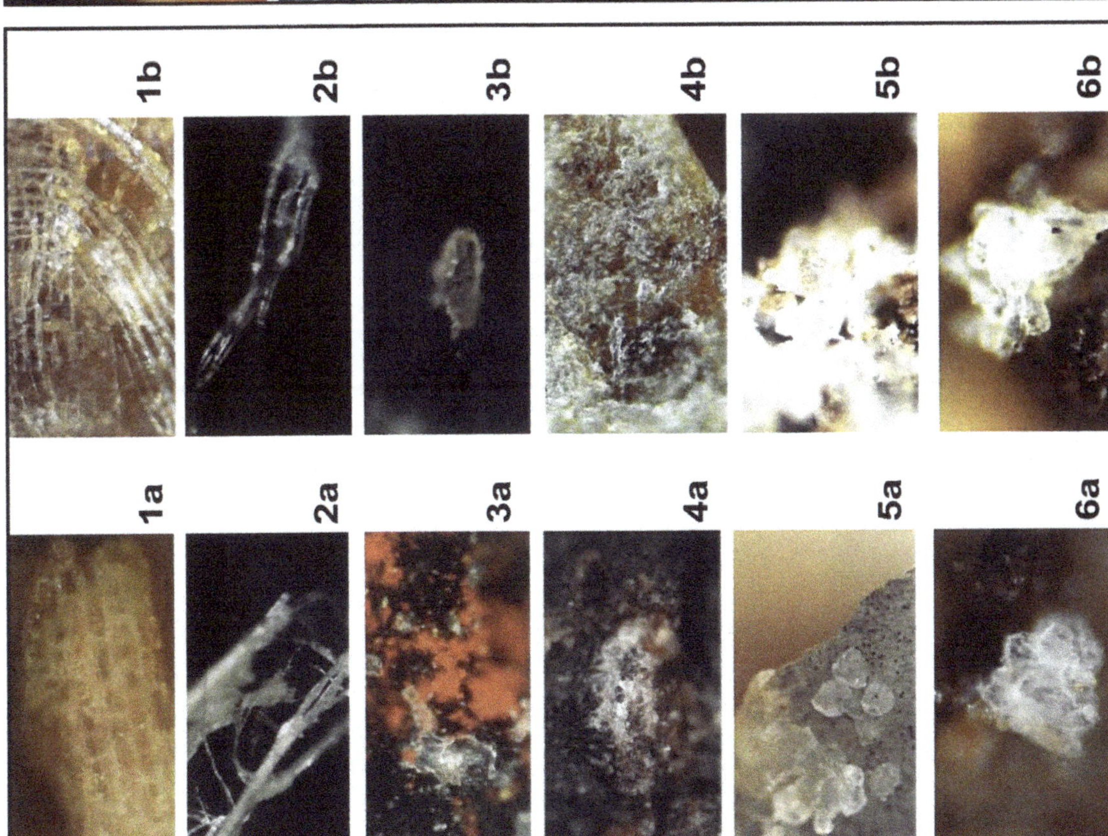
1b
2b
3b
4b
5b
6b
1a
2a
3a
4a
5a
6a

## Caption to the Color Plate

Figures 1 to 3 illustrate the similarity between some vegetal and faunal residues. The images in column 'a' are of vegetal residues, and the residues opposite them in column 'b' are of faunal residues with similar characteristics. All micrographs were taken using cross-polarised light.
**1a**: Inner epidermal cells of wet wood, *Combretum zeyheri*, on the edge of a replicated tool used to scrape the bark off a green stem (200x);
**1b**: striated muscle tissue deposited on the tool as a result of cutting beef (200x);
**2a**: plant fibres on the edge of a replicated tool used to scrape a fibrous leaf, *Sanseviera pearsonii* (200x);
**2b**: collagen fibre on the edge of a replicated tool used to cut fatty cartilage of an *Aepyceros melampus* carcass (200x);
**3a**: unidentifiable plant tissue on the edge of a replicated tool used to strip bark off a *Grewia flava* stem;
**3b**: bone fragment on a replicated tool used to cut bone and cartilage of an *Aepyceros melampus* carcass.

Figures 4 to 6 illustrate how residues from modern replication projects can assist in the interpretations of faunal residues on archaeological tools. The images in column 'a' are modern replicated residues, while the images in column 'b' represent faunal residues found on Middle Stone Age points from Sibudu Cave (between 50,000 and 60,000 years old).
**4a**: Faunal tissue on a spearhead used to stab a *Connochaetes taurinus* carcass (100x);
**4b**: faunal tissue on the edge of a Middle Stone Age point (200x);
**5a**: fat cells from modern beef (100x);
**5b**: dehydrated fat cells on a Middle Stone Age point (100x);
**6a**: rumpled bone or sheet collagen on a replicated tool used to cut bone and cartilage of an *Aepyceros melampus* carcass (500x);
**6b**: rumpled bone or sheet collagen on a Middle Stone Age point (500x);

Figures 7 and 8 show resin surfaces not used for hafting, while images 9 and 10 show resin with woody residue of the hafts adhering to the surfaces.
**7**: *Acacia karroo* resin with woody residue in the resin, indicated by the arrow (100x);
**8**: *Acacia karroo* resin with plant fibre in the resin, indicated by the arrow (100x);
**9**: *Acacia karroo* resin with woody residue on the surface (100x); insert (200x): woody residue;
**10**: *Acacia karroo* resin with woody residue and fibres on the surface (200x).

Figures 11 to 14 illustrate the combinations of residues occurring on tools that were used to replicate tasks.
**11**: Residues as a result of cutting a *Combretum zeyheri* branch (200x); a: resinous deposit, b: plant fibre, c: woody residue; insert (500x): d: starch grain in the woody residue, e: vessel in woody residue;
**12**: residues as a result of scraping the fibrous leaf of *Sanseviera pearsonii* (200x); a: plant fibre, b: leaf cell structure, c: plant tissue;
**13**: residues as a result of scraping the outside of an *Aepyceros melampus* hide (200x); a: hair, b: collagen, c: animal tissue;
**14**: residues as a result of scraping the inside of an *Aepyceros melampus* hide (200x); a: collagen fibre, b: animal tissue, c: collagen.

# CHAPTER FOUR
# Methods of Interpreting Bronze Age Vessel Residues: Discussion, Correlation and the Verification of Data

**N.I. Shishlina, A.V. Borisov, A.A. Bobrov and M.M. Pakhomov**

Natalia Shishlina; State Historical Museum; Krasnaya Ploshchad 1/2; Moscow 103012; Russia; <shishlin@rol.ru>, Alexander Borisov; Institute of Physical, Chemical and Biological Problems of Soil Science; Russian Academy of Sciences; Institutskaya 2; Pushchino, Moscow 142290; Russia; <a.v.borisov@rambler.ru>, Anatoly Bobrov; Faculty of Soil Science, Moscow State University; Leninskiye Gory; Moscow 119899; Russia; <anatoly-bobrov@yandex.ru> and Michael Pakhomov; Vyatka State University; Krasnoarmeyskaya 26; Kirov, Nizniy Novgorod 610002; Russia; <mp1234@yandex.ru>. This work was done with the support of National Geographic, grant number 7432-03, and the Russian Fund of Fundamental Research (RFFI), grant number 05-06-80112.

Food is an important element of any material culture and is closely linked with economy. Changes in climate or economy, or the spread of new technologies will have its impact on existing dietary systems. The use of different products and their specific combinations depend on regional peculiarities, trade exchange, seasonality of the economical cycle, welfare of the population and way of life as well as psychological and religious arrangements (Arutyunov et al. 1995). The identification of these dietary systems may provide additional information on the nature of the adaptation of a nomadic steppe population, such as the Majkop, Catacomb, Yamnaya and Srubnaya Culture period population of the Caspian steppe during the Bronze Age (3800-1550 BCE), to ecological niches as well as the nature of human dependence on the environment.

Many techniques can be used to reconstruct dietary systems. In this chapter special attention is given to the identification of vessel residues, which can provide direct evidence of the type of food used (Schoeninger and Moore 1992; Demkin 1997; Demkin and Demkina 2000; Demkin et al. 2000). A similar approach, pollen and phytoliths analyses, was used to identify residues from in pottery from ancient Peru (Jones 1993). Our work aimed to identify vessel residues using different methods. The accuracy of each technique was assessed by correlating the results of two or three different methods.

One of our methods assumed that after death the remains of the 'last meal' remains in the stomach of the deceased and is later deposited in the abdominal cavity of the skeleton. The composition of the last meal can verify, or falsify, the results of vessel residue analysis. Similar research has been done previously at other archeological sites (Shishlina 2001a; Berg 2002).

## Cultural Context

The area of our research is the northwestern Caspian steppe (Figure 1). This region has many Bronze Age 'kurgans' (burial grounds) dating to the Majkop (3800-3200 BCE), Yamnaya (3000-2500 BCE), Early Catacomb (2600-2300 BCE), Catacomb (2500-2100 BCE) and Srubnaya (1800-1550 BCE) periods. Paleoenvironmental research of the area identified climatic changes and links between these changes and the pastoral economy (Shishlina 2001b). During the Majkop Culture Period the area experienced a relatively mild and humid climate. The local economy was based on raising animals in spring and summer. Yamnaya Culture Period groups were the first to occupy the entire territory along the low lying river valleys and lake shores, as well as on the nearby watershed plateaus. Their seasonal economy was based on raising domesticated animals and short distance pasture routes. Around 2400-2000 BCE (calibrated) the climate became more arid which led to the exploitation of all ecological niches by the Catacomb Culture Period population. Their seasonal routes extended several hundred km. The Srubnaya Culture Period people were pastoralists as well and their economy closely resembled that of the Catacomb Culture Period.

All the above cultures are characterized by the vessels placed into their graves during funerary rituals. Many types of such vessels have been recognized: pots with a high neck, jugs with one or two handles, bowls as well as relatively simple vessels (Figure 2-2, 3, 5). Ethnographic comparisons with modern Eurasian steppe nomads, the Kazakhs, Kalmyks and Mongols (Zhitetsky 1893; Zhukovskaya 1988; Tomilov and Kadyrova 1997) show that usually a vessel with ritual food is placed into a grave. This can contain porridge, milk products, soft drinks or a clear meat soup. The main purpose of our research was to identify the residues in Bronze Age grave vessels using several approaches, to propose an algorithm of processing samples, to reconstruct ancient recipes and to compare the data results with data on the gender and age of the deceased.

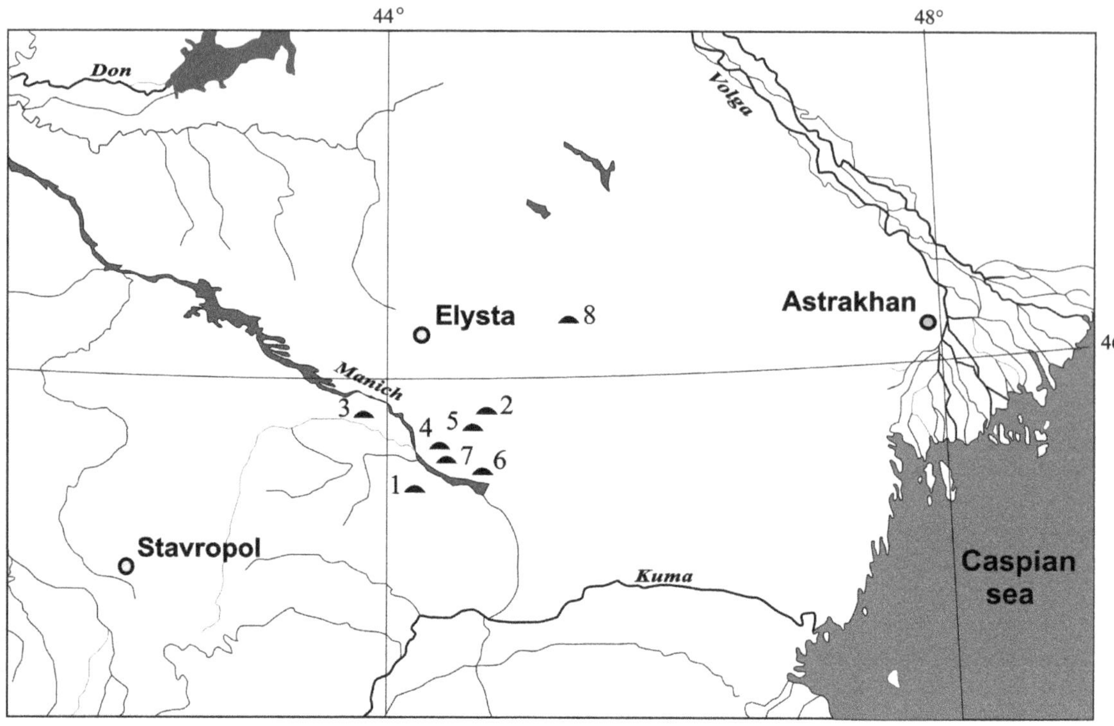


Figure 1: Map showing the location of the sites mentioned in the text. 1: Sharackhalsun; 2: Mandjikiny-1; 3: Baga-Burul; 4: Ostrovnoy; 5: Mu-Sharet 1 and 4; 6: Zunda-Tolga-1; 7: Zunda-Tolga-3; 8: Chilgir-1.

## Methodology

All vessels from ancient graves were full of soil when found. The lay-out of the actual graves, pits and catacombs, and their superstructure, made of wood and plant materials erected on the upper layers of the graves, or wooden doors (Figure 2-1, 4) indicates that after the funeral ceremony the grave remained free of soils for some time. Gradually the plant and wood ceiling rotted and soil poured into the grave, eventually filling the whole space including the vessels. Steppe animals, such as mice and ground squirrels also contributed to the filling of the pots. By that time the food would have been almost completely decomposed. But we suppose that different components of foodstuffs were left inside the pot, at the bottom and on the walls near the bottom. Sometimes a visible layer of dark-brown residue was preserved. In some way the soil from the upper parts of the grave appeared to have 'cured' the layer of food remains on the vessel bottom.During the excavation of the grave and the recovery of its contents, all vessels containing soil were carefully wrapped, to prevent pollen and other alien objects from contaminating the sample, and brought to a field laboratory. Contamination was a concern as our excavations were usually carried out in the spring or the summer, the period of blossoming flowers and trees. Cracked and broken vessels were therefore excluded from the study. Soil samples were taken from the lower internal parts of the vessel using a wooden pallet knife and a rigid flat brush. A control sample was taken from the upper part of the vessel. The weight of each sample was 20-30 g. The rest of the soil in the vessels was subjected to archaeobotanical flotation.

The soil sample from the bottom of the vessels was analyzed to identify the composition of the food placed inside this pot during the funeral ceremony. It is possible that used vessels were entered into the graves and, if so, it could be that remains of the food consumed before the funeral ritual were also preserved inside such vessels. It is unclear if the pots were carefully cleaned and, if so, how often.

As part of the excavation of the skeletons sometimes a sample was taken from the stomach area, between the lumbar backbone and the lower ribs, with the help of a spoon. All skeletons were either supine or put to rest on their side. Analysis of the contents of such samples may identify the composition of the last meal of the deceased. Control soil samples were taken from under the skull of the skeleton and from the bottom of the grave. All soil samples were stored in paper bags.

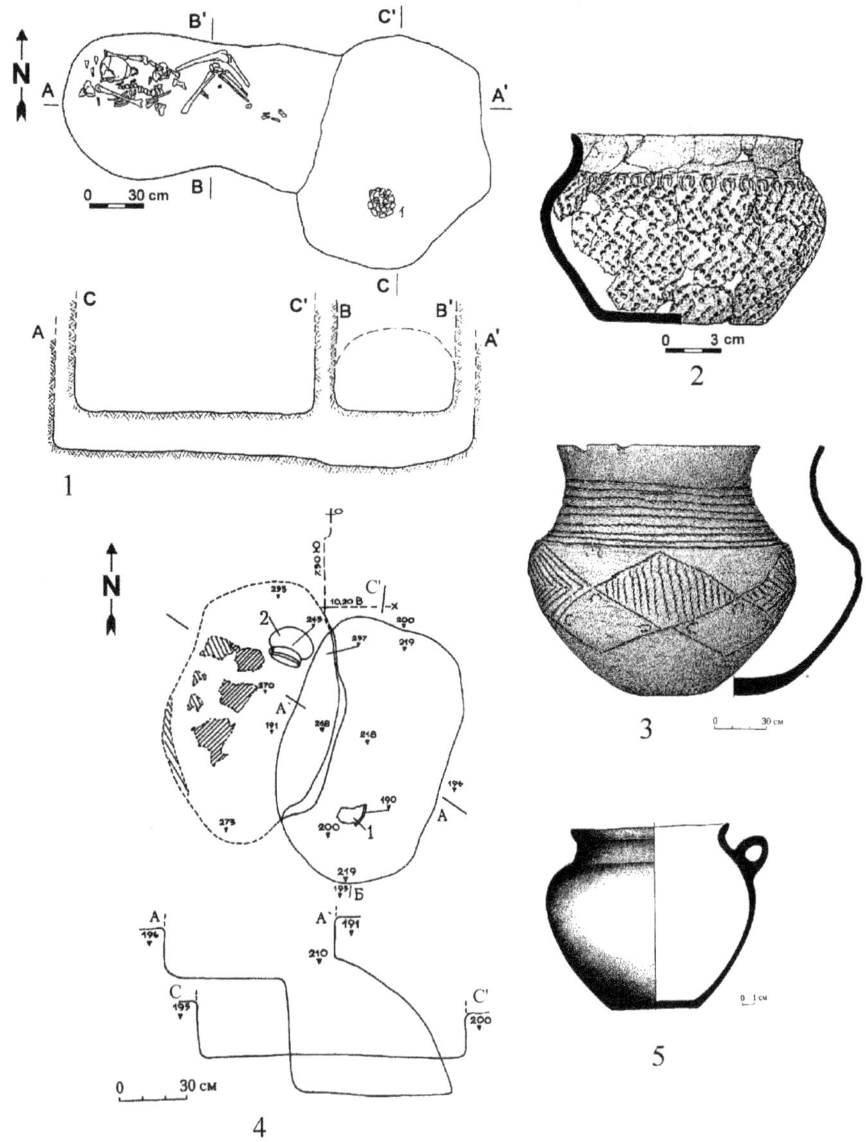


Figure 2: Graves and vessels from the Caspian steppe. 1: Kurgan 5, Grave 6 at Baga-Burul; 2: vessel dating to the Early Catacomb Culture Period from Baga-Burul; 3: vessel from Kurgan 6, Grave 6 at Ostrovnoy; 4: Kurgan 3, Grave 8 at Ostrovnoy; 5: vessel dating to the Catacomb Culture Period from Ostrovnoy.

## Determination of Phosphate Residues

The amount of phosphate in ancient pottery depends more on the organic residues inside the vessels than geological factors. Based on ethnographical sources, we assumed that funeral vessels contained cereals, meat or meat broth, dairy products or simply water. These products are characterized by different phosphorus residues, hemp and poppy seeds having the highest values. The phosphate residue from wheat, barley and oats is twice as high as that from beef and pork. Milk has the lowest phosphorus residue (Table 1). The phosphate residue in the soil in the upper parts of the vessels should not be influenced by the presence of organic substances on the bottom of the vessels. A different concentration of phosphates towards the bottom of the vessel may indicate that these lower levels were influenced by phosphates from foodstuffs. Criteria to identify different foodstuffs were developed from the ratio of phosphate residues in soil from the lower and from the upper parts of the vessel. Phosphates were extracted from the soil samples in 1% $(NH_4)_2CO_3$ adjusted to pH = 9.0. This will dissolve calcium monophosphates and diphosphates, as well as small amounts of organic phosphorus compounds and calcium triphosphates.

| $\Delta P_2O_5$ (mg/100 g soil) | Probable source |
|---|---|
| < 2 (mostly 0-0.5) | Water (in pots and pitchers) |
| 2-8 (mostly 4-7) | Dairy foods (in jugs) or meat broth (in pots) |
| 8-15 (mostly 10-12) | Cereals (in pots) |
| > 20 (mostly around 20) | Narcotic substance (in jugs) |

Table 1: Food identification criteria.

From each air-dried soil sample 5 g was passed through a 1 mm mesh sieve placed on a glass retort with 100 ml 1% $(NH_4)_2CO_3$ after which the retort was shaken for one hour. The extract was filtered and 20 ml was transferred into a 100 ml measuring retort. If the filtrate was not colored, ammonium acetate was neutralized by $H_2SO_4$ in the presence of β-dinitrophenol until the filtrate turned yellow. Then 10 ml $H_2SO_4$ and 10 ml 2% ammonium molibdate were added. The retort was filled with deionized water and shaken vigorously after which 3-4 drops SnCl were added. After 10-15 min minutes the phosphate residue was determined by photocalorimetry. Colored filtrates were first discolored. For this 20 ml filtrate was placed in a 100 ml measuring retort and 5-6 ml 27% $H_2SO_4$ was added. The solution was brought to the boil and 0.1% $KMnO_4$ was added slowly until the filtrate turned pink. Then 10 ml 2% ammonium molibdate and 3-4 drops SnCl were added. This was vigorously shaken and topped off with deionized water. The phosphate residue was determined by photocalorimetry.

## Archaeobotanical Analyses

Soil samples were suspended in water (10 ml for each gram of soil) and sieved, through a 0.5 mm mesh, to separate the organic matter from coarse mineral grains. For phytolith analysis the suspension was boiled for one hour in 20% $H_2O_2$ after which 50 ml 10% $HNO_3$ was added and the sample washed with distilled water and sieved again. Between 100-300 phytoliths were counted for each sample under 200x and 400x magnification. Scanning electron microscopy (SEM) served clarify the morphology of the phytoliths.

Sampel preparation for pollen analysis was based on the separator approach with an acetolysis protocol as proposed by Grichuk and Zaklinskaya (1948). The function of acetolysis is to dissolve cellulose, hemicellulose and chitin. A secondary effect is the darkening of the pollen grains that allows an easier examination of the apertures and surface patterns. Because of their nature, the soil samples from the graves have their peculiarities. Rather than the natural vegetable background of the environment their pollen composition is determined by the intentions of the persons who participated in the funeral ceremony and placed vessels with food in the grave. It is therefore not necessary to calculate the statistic distribution of the pollen grains in the soil from the vessel, as is usually done for stratified soil samples. Likewise this is not necessary for samples from the stomach area of the skeleton. The pollen composition of our samples can be considered decontaminated pollen spectra indicating the suite of plants used for cooking.

There were few pollen grains in the samples taken from the vessels. It proved important to know not only the phenology of the steppe plants, to identify as many of the pollen grains as possible, but to also be able to identify other materials such as siliceous remains of non-thrashed grains of gramineous plants, chitinous remains of the microplates of fish scales (Figure 6), spores of mildew mushrooms that parasitize on cereals and ticks that parasitize on the skin and hair of animals. It also proved important to compare the pollen composition of the soil samples taken from bottom of the vessels with the control samples taken from the same grave (from under the skull, from the bottom of the grave and from the upper part of the same vessel).[1]

Modern edible plants from the area under investigation were collected (Figure 3). Wild cereals were collected in the seed stage in order to investigate the morphology of siliceous remains of non-thrashed grains of gramineous plants. Morphology of these siliceous remains is very diverse and could only be identified by comparing the results of the analysis of fossil siliceous remains with modern wild cereals.

After preparing the soil samples for pollen and phytolith analysis, the rest of the soil from the vessel was floated and passed through a sieve with a 0.2 cm mesh. All organic remains, including bones, in this fraction were identified.

---

[1] Our previous research on over 300 similar samples has shown that soil samples from the upper part of the vessel cannot satisfactory be used as control. None of the samples from this context had any pollen, or other organic materials such as phytoliths or spores. Samples from the bottom of the grave and from under the skull appeared to be more comparable control samples.

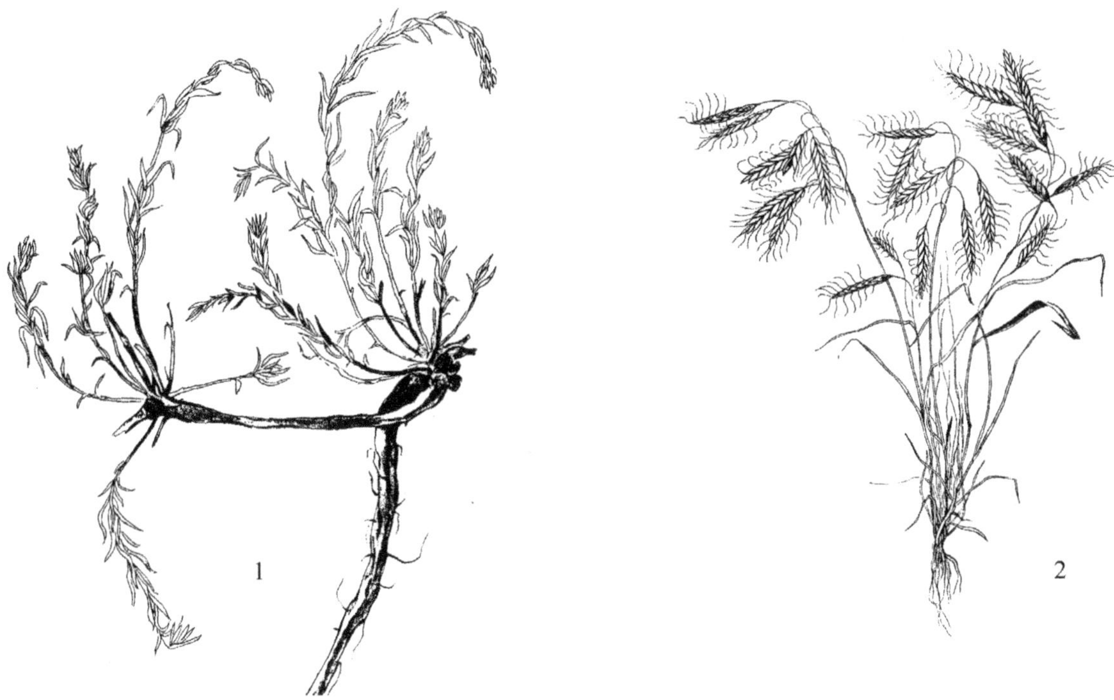


Figure 3: Two of the steppe plants used for food. 1: wormwood (*Artemisia*); 2: barley (*Hordeum*).

| Context (kurgun, grave) | Pollen analysis | Phytolith analysis | Phosphate residue |
|---|---|---|---|
| | **Early Maikop Culture Period, 4000-3700 BCE (calibrated)** | | |
| | *Sharachalsun burial ground* | | |
| **k.2, g.17** amphora, near the arm of a child | Very few pollen grains of *Asteraceae, Cichoriaceae, Varia, Rumex, Poaceae, Chenopodi aceae, Pinus, Picea, Abies*; many remains of grains of gramineous plants, most likely wild barley (*Hordeum*). | --- | 0.2 - water |
| **k.5, g.7** jug, near the north wall of the grave behind an adult skeleton | Very few pollen grains of *Varia, Liliaceae, Chenopodiaceae, Cichoriaceae, Ephedra, Poaceae, Polypodiaceae,* more often pollen of *Pinus, Poceae*, remains of grains of gramineous plants were found. | --- | 0.5 - water |
| | **Yamnaya Culture Period, 3000-2500 BCE (calibrated)** | | |
| | *Mandjikiny-1 burial ground* | | |
| **Under k.9** | Moderate concentrations of *Ephedra, Chenopodiaceae, Asteraceae, Fabaceae, Pinus, Alnus,* a lot of cists and rhisoids of mould, many black non-plant organic remains (meat?). | --- | 0.1 - water |
| | **Early Catacomb Culture Period, 2600-2300 BCE (calibrated)** | | |
| | *Mandjikiny-1 burial mound* | | |
| **k.12, g.2** jar, near the breastbone of a male skeleton (45-55 years) next to a female skeleton (40-55 years) | Abundance of *Chenopodiaceae, Artemisia, Asteraceae, Fabaceae,* often *Liliacae, Iridaceae, Tulipa,* abundance of siliceous remains of non-thrashed grains of gramineous plants. | --- | 13.6 - cereals (porridge) |

| | | | |
|---|---|---|---|
| *Baga-Burul burial ground* | | | |
| **k.5, g.6** pot, near the legs of a male skeleton (45-55 years) | Very few pollen grains of *Poaceae, Asteraceae, Artemisia, Chenopodiacea,* remains of plant lice (*Rhopalosiphoninus*) that parasites on plants, including gramineous plants, are quite common. | Few phytoliths. | 0.7 - water |
| **Catacomb Culture Period, 2500-2000 BCE (calibrated)** | | | |
| *Ostrovnoy burial ground* | | | |
| **k.3, g.8**, pot, on the floor of a cenotaph | Concentration of goosefoot (*Chernopodium),* wormwood, plantain plants (*Plantaginaceae*), mixed grass, very rarely pollen of pine and spruce. One spore of wood fern (*Dryopteris*) and abundant siliceous remains of non-thrashed grains of gramineous plants. | --- | 0.80 - water |
| **k.3, g.10** pot, near the head of a child | --- | Phytoliths of different shape, non-identified plant remains with large pores. | 0.80 - water |
| **k.3, g.23** amphora, near the bones of legs (from a child?) | --- | Rare phytoliths. | 0.0 - water |
| **k.3, g.24** pot, near the bones of a one year old child | --- | Rare phytoliths. | 0.66 - water |
| **k.3, g.32** jar, near the skull of a female skeleton (40-45 years) | Pollen of goosefoot (*Chernopodium*), *Ephedra*, chicory, wormwood, asters, catchfly (*Silene*), legumes. In rare cases, spores of fern and pollen of pine. Abundant siliceous remains of scales of wild gramineous plants. | Rare phytoliths. | 0.40 - water |
| **k.6, g.6** pot, near the skull of a female skeleton (30-35 years) | Very few pollen grains of *Liliaceae, Silenaceae*; rare *Chenopodiaceae* as well as *Betula, Alnus*. Few siliceous remains of scales of grains of gramineous plants (*Hordeum*?). | Background residues of phytoliths. | 0.1 - water |
| **k.6, g.10** pot, in front of a male skeleton (35-45 years) | Very few pollen grains of *Chenopodiaceae, Artemisia,* rare *Pinus, Abies,* a lot of siliceous remains of scales of grains of gramineous plants. | Background residues of phytoliths. | 1.0 - water |
| *Mu-Sharet-1 burial ground* | | | |
| **k.5, g.2** pot, near knees of a female skeleton (25-40 years) | --- | Background residues of phytoliths. | 0.3 - water |
| **k.6, g.1** jar, near the legs of a female skeleton (13-40 years) | --- | Background residues of phytoliths. Oval, flattened artifacts of unclear genesis (non-phytoliths) were found. | 3.5 - clear soup |
| **k.6, g.1** pot, near the legs of a female skeleton (13-40 years) | --- | Background residues of phytoliths. | 0.0 - water |
| *Mu-Sharet-4 burial ground* | | | |
| **k.12, g.4** pot, near the skull of a male skeleton (50-60 years) | Very few pollen grains of *Chenopodiaceae, Artemisia, Asteraceae, Varia*; rare pollen of *Alnus, Pinus, Betula* (trees). There are chitinous remains of microscopic *Arthropoda* and microscopic cilia of grains of gramineous plants. | --- | 13.4 - porridge |

| | | | |
|---|---|---|---|
| | *Zunda-Tolga-1 burial ground* | | |
| **k.10, g.2** pot, near a female skeleton (50-60 years) | Abundance of pollen grains, siliceous remains of scales of grains of gramineous plants | --- | 1.1 - water |
| | *Zunda-Tolga-2 burial mound* | | |
| **k.1, g. 4** pot, near the legs of an adult skeleton | Abundance of pollen grains, predominance of bits of stamen with undecomposed and underdeveloped pollen. In very rare cases, pollen of *Artemisia, Chenopodiaceae*; in one case, *Polypodiaceae.* | Phytoliths of gramineous plant, wormwood, hemp. | 26.2 - narcotics |
| **k.1, g. 5** pot, near adult male and female skeletons | Abundant siliceous remains of scales of grains of gramineous plants (*Hordeum, Helictotrichon*), few pollen grains of *Poaceae, Chenopodiaceae*, chitinous remains of insects | Phytoliths of gramineous plant. | --- |
| **k.1, g.5** pot, near the wall of a grave with a male and a female skeleton | Few pollen grains of *Varia, Chenopodiaceae, Artemisia, Ephedra* (in rare cases), *Asteraceae. Poaceae* is found quite frequently. A lot of chitinous remains of microplates of fish scales, morphologically different. | --- | 71.3 - narcotics |
| **k.2, g.2** near a male skeleton (20-25 years) | --- | Phytoliths of gramineous plants. | 16.1 - cereal (porridge) |
| | *Zunda-Tolga-3 burial mound* | | |
| **k.1, g.1** pot, near the legs of a child (about 10 years) | Very few pollen grains of *Chenopodiaceae, Ephedra*; in one case, *Carpinus, Corylus.* In contrast to other subsoils, with dark to black organics, the organics are of a golden color, maybe of animal origin. Primitive parasites, such as echinococcus, traces of grains of gramineous plants, siliceous remains of scales of grains of gramineous plants. | Phytoliths of gramineous plants, fossil conductive tissue, concentration of fossil ball-like phytoliths | 5.0 - clear soup |
| **k.1, g.3** teenager (17-20 years) | Pollen of *Varia, Chenopodiaceae, Artemisia, Poaceae, Polygonaceae,* chitinous remains of ticks. | --- | 3.4 - clear soup |
| **k.1, g.5** teenager (16 years) | Pollen of *Chenopodiaceae, Asteraceae, Fabaseae, Varia,* siliceous remains of scales of grains of gramineous plants. | Phytoliths of *Cannabis,* gramineous plants. | 39.5 - narcotics |
| **k.1, g.6** near a female skeleton | Abundance of pollen grains of *Varia, Chenopodiaceae, Artemisia,* very seldom *Abies, Pinus,* no grains. | --- | 1.0 - water |
| | *Mandjikiny-1 burial ground* | | |
| **k.14, g.5** | Few pollen grains of *Chenopodiaceae, Artemisia, Asteraceae,* spores of *Asplenium, Polypodiaceae, Botrychium lunaria* | --- | 0.8 - water |
| **k.14, g.7** adult ritual brazier | A lot of siliceous remains of scales from grains of gramineous plants as well as *Varia, Chenopodiaceae, Artemisia, Rosaceae, Brassicaceae, Labiatae* (mint, sage). | Phytoliths of *Cannabis,* gramineous plants. | --- |
| | *Baga-Burul burial mound* | | |
| **k.1, g.2** pot, near the skull of an adult skeleton | Few pollen grains of *Poaceae, Asteraceae, Chenopodiaceae, Silenaceae, Cichoriaceae, Geraneaceae.* In very rare cases, siliceous remains of grains of gramineous plants, abundance of fragments of animal tissue. | --- | 12.3 - porridge |
| **k.5, under g.11** pot | Few pollen grains of *Ephedra, Poaceae, Asreraceae, Artemisia.* | Background residues of phytoliths. | --- |
| **k.5, g.5** pot, near the legs of a male skeleton (45-55 years) | Very few pollen grains, but large spots with underdeveloped pollen of *Varia* (from buds), in very rare cases, pollen of *Malva.* | Few phytoliths. | 0 - water |

| | | | |
|---|---|---|---|
| **k.5, g.5** turnip-shaped large pot, near the skull of a male skeleton (45-55 years) | Spots with concentration of pollen grains of *Asteraceae, Silenaceae, Artemisia.* | Large numbers of brown stick-shaped phytoliths that cannot be used to identify the plant. | 12.1 - porridge |
| **k.5, g.5** pot, near the legs of a male skeleton (45-55 years) | Very few pollen grains of *Artemisia, Varia,* in rare cases *Alnus, Pinus.* | --- | 0 - water |
| **k.5, g.7** bowl, on the floor of a cenotaph | Very few pollen grains of statice (*Plumbaginaceae*), *Silenaceae, Artemisia, Asteracea, Poaceae, Cichoriaceae, Ehpedra,* very seldom *Chenopodiaceae, Polipodiacea, Alnus, Pinus,* few remains of grains of gramineous plants. | Few phytoliths. | --- |
| **k.5, g.11** pot, near the head of a female skeleton (25-35 years) | Practically no pollen grains (in rare cases, *Chenopodiaceae, Poaceae).* A quite common find are chitinous limbs, maybe of plant lice, common siliceous cilia of grains of gramineous plants. The vessel appears to have contained cooked foodstuffs. | Phytoliths from the upper part of the soil. | 5.1 - clear meat soup |
| *Chilgir burial ground* | | | |
| **k.5, g.1** | Very few pollen grains of *Chenopodiaceae,* very seldom *Apiaceae, Ephedra, Asteraceae* | Few phytoliths. | --- |
| k.3, g.2 near the skeleton on a child (about 9 years) | Abundance of pollen grains of *Poceae, Chenopodiaceae,* very seldom *Artemisia, Liliaceae, Alnus, Pinus.* | Few phytoliths. | --- |
| **Srubnaya Culture Period, 1800-1550 BCE (calibrated)** | | | |
| *Chilgir burial ground* | | | |
| **k.2, g.1** | Abundance of pollen grains of *Asteraceae and Linosyris.* | Few phytoliths. | --- |
| **k.2, g.4** | Practically no pollen grains, often siliceous cilia of grains of gramineous plants, insect remains. | Few phytoliths. | --- |

Table 2 (on this and the previous pages): Comparison of the data on residue in Bronze Age vessels from the northeastern Caspian steppe. The measurements to calculate the phosphorus residues were taken on samples from the bottom and from the upper part of the vessel (k: kurgan, g: grave).

## Results

For this project 190 soil samples from vessels and 50 from the stomach area of skeletons were processed. Here we discuss the results of 38 vessels from Bronze Age graves in which the residues were identified by at least two methods (Table 2). Samples from 22 vessels were processed in two ways; samples from 16 vessels in three. In addition to these analyses, all soil samples were floated. Vessels that contained the bones of domesticated animals appeared to be placed in the category 'Dairy foods or meat broth' after phosphate residue analysis (Table 1). Lime seed (*Tilia*) was found in one vessel, a turnip-shaped pot, from Baga-Burul (kurgan 1, grave 5). This is probably instrusive. The residue data was later combined with data on the type of the vessel, its location in the grave and the gender and age of the deceased. Control soil samples from the bottom of the grave and from under the skull were processed and compared with the data in Tables 2 and 3.

## Interpretation of the Results

As a whole, there appeared to be a good correlation between the data obtained by different methods. Vessels with few phytoliths and little pollen were considered to have contained water, a finding corroborated by phosphate residue analysis (for example, in kurgan 5, grave 6 at Baga-Burul and in kurgan 3, graves 10 and 23 at Ostrovnoy). A vessel from kurgan 12, grave 2 at Mandjikiny preserved a relatively high phosphate residue, interpreted as cooked cereal (porridge). Numerous pollen grains mixed with the remains of cereals were identified in the residue in this vessel.

Abundant pollen grains of different plants were identified in a vessel from kurgan 1, grave 4 at Zunda-Tolga, including pollen and phytoliths from the narcotic plants wormwood (*Artemisia lerchiana*, Figure 4a, b and c) and hemp (*Cannabis sativa*, Figure 4d, e and f). According to our phosphate residue analyses this vessel probably contained an herbal infusion, probably used for its narcotic properties. Phytoliths of *Cannabis* and gramineous plants (oats, Figure 5) as well as the remains of grains of gramineous plants were found in a vessel from kurgan 1, grave 5 at Zunda-Tolga. This vessel had a very high concentration of phosphates as well and the residue was interpreted as a narcotic infusion. Phytolith of cereals were also identified in a pot from kurgan 2, grave 2 at Zunda-Tolga. This vessel had a high level of $\Delta P_2O_5$ in its lower part (16.1 mg/100 g) and its residue was interpreted as porridge. Non-organic materials, probably lipids of non-identified origin were identified in a pot from kurgan 6, grave 1 at Mu-Sharet. The value of $\Delta P_2O_5$ suggests that it once contained a clear soup.

| **Context** | **Pollen and phytolith identification** | **Vessel residue** |
|---|---|---|
| | **Early Catacomb Culture Period** | |
| **k.5, g.6** male (45-55 years) | Many cists and spores of mushrooms, pollen of *Asteraceae, Poaceae, Chenopodiaceae,* pollen of *Sparganium* (water plant), seldom pollen of *Almus, Ulnus.* | There was water in the vessel, pollen of *Poaceae, Asteraceae, Artemisia, Chenopodiace* as well as remains of plant lice (*Rhopalosiphoninus*). |
| | **Catacomb Culture Period** | |
| **k.1, g.2** adult skeleton | Large spots of concentration of undeveloped pollen of *Varia*, seldom remains of grains of gramineous plants. | Vessel was filled with cooked meat, or cooked grain and meat. |
| **k.5, g.5** male (45-55 years) | Many pollen grains of soorrel (*Rumex*), *Rosaseae,* seldom *Asteraceae, Artemisia, Rutaceae (Tetradiclis), Scrophulariaceae,* remains of grains of gramineous plants. | Vessel 1: very few pollen grains of *Artemisia, Varia,* in rare cases *Alnus, Pinus;* vessel 2: water and abundance of flower buds, or infusion of buds. |
| **k.5, g.11** pregnant female (25-35 years) | Very few pollen grains, a lot of remains of grains of gramineous plants. | Vessel contained cooked food made of grain, whole or ground flour, spots of decomposed starch grains, clear soup. |

Table 3: Organic residues in the stomach area of some of the skeletons from Baga-Burul and in vessels from the same graves.

One vessel, found in kurgan 1, grave 1 at Zunda-Tolga, is remarkable. There was an organic residue of animal origin in this pot. Primitive parasites, such as echinococcus, had been preserved. These parasitize in the intestines of cattle, dogs, fox, etc. There are parts and whole examples of these microscopic blood sucking organisms. Maybe, the vessel was filled with cattle meat or liver that was not properly cooked. The sample also had traces of grains of gramineous plants (siliceous remains of the scales of grains of gramineous plants) as well as phytoliths of cereals. The value of $\Delta P_2O_5$ indicates that the vessel was used for a clear soup. A vessel from another grave in the same kurgan had chitinous microscopic remains of a tick that parasitized on the liver and in the fur of animals. There were no phytoliths but the value of $\Delta P_2O_5$ again indicates a clear soup.

But there were also findings with a poor correlation. Some vessels, from Sharachalsun and Ostrovnoy, preserved siliceous remains of scales of the grains of gramineous plants and a lot of pollen grains of various steppe plants, a residue identified as porridge. Phosphate analyses, however, indicated only water. A vessel from kurgan 1, grave 6 at Zunda-Tolga had much pollen from *Artemisia* and other steppe plants and was interpreted as an herbal infusion. Again the value of $\Delta P_2O_5$, indicated only water. We believe that the reason of such divergence may be the following. In some cases the phytoliths and pollen can be introduced into the vessels by the activity of shrews. We must suppose that during the funeral ritual vessels were covered by lids. However, no lids of metal, clay or stone dating from the Bronze Age were found in this region. Such lids may therefore have been made of wood or the pots may have been covered by mats or cork. These may have introduced phytoliths and pollen grains, which are obvioulsy not related to the food, in the vessel. The use of pine tar or pitch to seal ceramic vessels is well-documented and these may also have introduced pollen grains in the vessel (Jones 1993).

The highest value of $\Delta P_2O_5$ was in a pot from kurgan 1, grave 5 at Zunda-Tolga. This is suggestive for a narcotic infusion. No phytoliths and very few pollen grains were found in the vessel, but instead many chitinous remains of microplates of the scales of fresh water fish (Figure 6). It is clear that the vessel contained a fish soup or stew, prepared of at least two or three species, which may explain the high value of $\Delta P_2O_5$. This is a clear example where pollen analysis helped us to correct an error in the interpretation of the data.

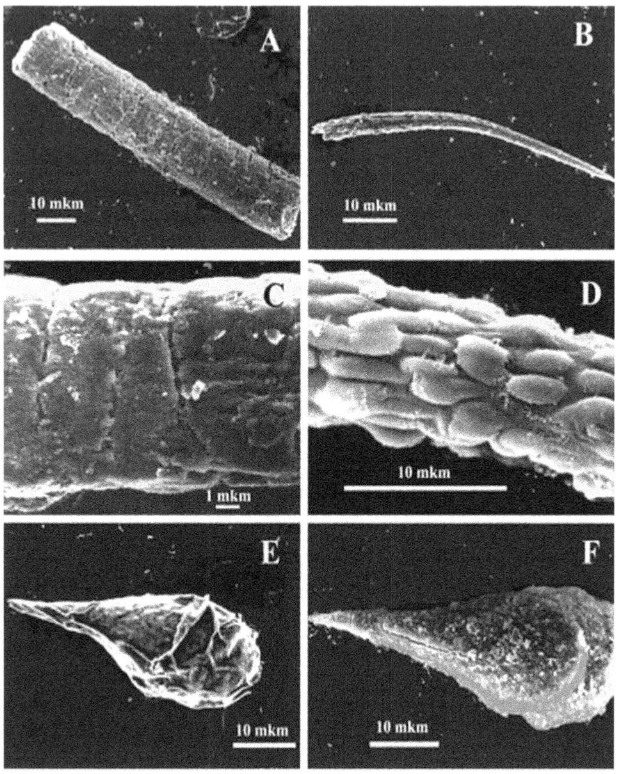


Figure 4: Phytoliths from vessel residues. A, B and C: white wormwood (*Artemisia lerchiana*); D, E and F: cannabis (*Cannabis sativa*).

We also checked the residue of vessels by identification of the 'last meal' in samples from the stomach area of some of the skeletons excavated at Baga-Burul (Table 3). Correlation of the findings with the data from residue analysis obtained interesting results. The two deceased in kurgan 1, grave 1 and kurgan 5, grave 5, both relatively old, appeared to have consumed some vegetable drink before death. These could very well have been medicinal drinks. Many remains of wild cereals were found in the stomach area of a woman of 25-35 years old who apparently died during child birth. In the pot placed inside her grave many poorly preserved wild steppe cereals could be identified.

The control samples from the bottom of graves and from under the skull differed dramatically from the samples described above. Soil samples from the bottom of the graves are characterized by phytoliths of reed, sedge, feather-grass and cane, steppe plants often used to make vegetable fibers and plant mats. Soil samples from under the skull are sometimes characterized by the presence of several steppe flowers and plants, including *Chenopodiaceae*, *Tilia*, *Betula*, *Pinus*, *Liliaceae*, *Linaceae*, *Rumex*, *Asteraceae* and *Artemisia*. Other samples, however, did not contain any pollen grains. We believe that the compositions of these control samples are indicative for the season in which the graves were constructed (Shishlina and Pakhomov 2002).

## Discussion and Recommendations

Correlation of vessel residue data obtained by different methods allows us to propose the following algorithm of investigation. The first task is to identify $\Delta P_2O_5$ to get information about the most likely vessel residue. If $\Delta P_2O_5$ is low (2-3 mg/100 g soil) there was most likely water in the pot. Taking into consideration that the source of phosphate could only have been some vegetable or animal food, no additional pollen or phytolith analyses are needed.

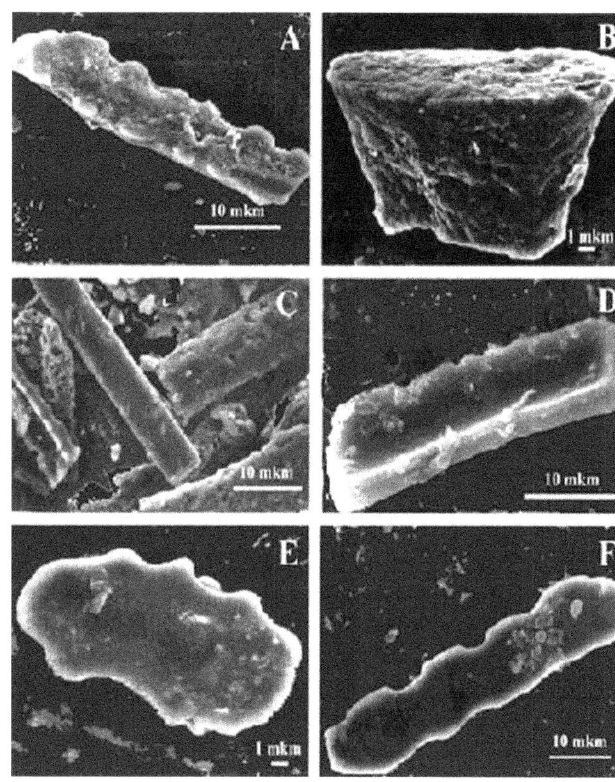


Figure 5: Phytoliths of steppe gramineous plants from Bronze Age Caspian vessels.

If $\Delta P_2O_5$ is higher than 3 mg/100 g soil, it is best to perform both pollen and phytolith analysis. When $\Delta P_2O_5$ is 3-8 mg/100 g soil and phytolith and pollen analyses show low background concentrations of pollen grains and phytoliths, we may identify the residue as an animal or milk product. More precise identification of animal and milk products is not possible with the methods described here. As is obvious from other chapters in this volume a biochemical approach, for example the identification of lipids, may help identify animal and milk food more precisely.

If $\Delta P_2O_5$ exceeds 10 mg/100 g soil and there are high concentrations of pollen grains and phytoliths of wild cereals, we suggest that the residue in the vessel is porridge. A very high level of $\Delta P_2O_5$ against a background of few pollen grains but many micro-remains of fish, on the other hand, can be reconstructed as fish soup. A very high level of $\Delta P_2O_5$ against a high concentration of pollen grains and phytoliths may be interpreted as an herbal infusion, probably used for its medicinal or narcotic properties. Again, these need to be verified biochemically to obtain a more reliable interpretation.

We strongly recommend the floatation of the rest of the soil from the vessels in order to identify other remains, such as seeds, fruits as well as animal and fish bones. Where relevant, the analysis of soil samples from the stomach area of skeletons will provide valuable additional information.

## Conclusion

Identification of vessel residues combined with the stomach contents of the dead helped us to reconstruct ritual and daily food of the Bronze Age nomads who roamed the northwestern Caspian steppe around 4000-2000 BCE. Vegetable food predominated all vessels and wild steppe cereals (*Hordeum*, *Helictotrichon*) appeared to have played an important role. These cereals were gathered and cooked. No domesticated cereals were identified in vessel residues. Clearly, the local pastoral groups of the Majkop, Catacomb and Srubnaya Culture Periods did not practice farming and did not trade domesticated cereals. Other pastoral groups of the Catacomb Culture Period did use domesticated cereals, such as wheat (Korpusova and Lyashko 1981), but probably only in regions with better farming conditions, such as the Asov Sea steppe area, the valleys of the Lower Don and the Middle Volga areas or the Black Sea steppe area. Ethnographic parallels show that during the summer many nomadic groups of the Eurasian steppe still collect steppe plants, including wild cereals.

It is interesting that a similar pattern was observed, by excavation of the Sharakchalsun burial grounds, for the steppe population during the Majkop Culture Period. During this period (around 3500 BCE) groups of farmers maintained numerous piedmont and river valley pastures in the North Caucasus (Korenevsky 2004). During the summer few migrated north, to the steppe, using wild steppe cereals rather than domesticated cereals for food. The steppe population who exploited all ecological niches of the northwestern Caspian steppe during the Catacomb Culture Period did not consist of farmers, but rather of pastoralists using many steppe plants for food and drink. It was probably also the Catacomb Culture Period population who first employed psychotropic plants like wormwood (*Artemisia*) and hemp (*Cannabis*).

We started the analysis of the stomach area to identify the plants used to embalm the dead (Kamenetsky 1995). According to Herodotus, the Scythians who occupied the Eurasian steppe during the first millennium BCE used grass and seeds, placed inside the stomach, for the embalming of their kings (Herodotus, VI,71). Those plants were identified as cyperus, seeds of celery and aniseed (Kamenetsky 1995). However, we did not find these in the stomach area of the Bronze Age skeletons. Instead, we think that the content of the stomach area represents the last meal of the deceased and should be compared to residue in the vessels from the same area.

Herbal infusions prepared of some of the numerous steppe grasses and flowers appeared to have been medicinal or narcotic. Our findings may add to the discussion on the recipe of the legendary 'Soma drink', the Indo-European 'Drink of the Gods overcoming the death' (Gamkrelidze 1984, 822-823; Sherratt 1995). In Ancient India such an infusion was prepared with the help of soma, a plant that is not yet identified. Soma was soaked in water, after which it was ground with the help of squeezing stones, filtered through a screen made of sheep skin and mixed with water, milk and barley. The result was stored in wooden receptacles. Drinking 'soma', which must have been hallucinogenic, involved a special ritual (Toporov 1982, 462). A similar drink is known from Iran where it is named 'chaoma'. This is made of mixed grasses and often contains hemp, henbane, rhubarb or ephedra. These grasses are ground and mixed with milk. The ceremonial vessels from the temple at Gonur-depe in Turkmenistan, dating to the Margiana period (second millennium BCE) probably contained a similar drink as pollen and seeds of hemp, poppy and ephedra were identified in these (Meyer-Melikyan 1998; Meyer-Melikyan and Avetov 1998). We identified hemp, ephedra, goose-foot and wormwood in both vessel residues and stomach area samples from the Catacomb Culture Period. This suggests that the recipe of such a hallucinogenic drink dates back to the third millennium BCE. It proves that hemp was used for funeral rituals and possibly other *rites des passage* as early as the Catacomb Culture Period. Hemp and fragrant steppe plants, such as mint and sage, were also burned in ritual braziers and the Scythians are known to have smoked hemp (*Cannabis*).

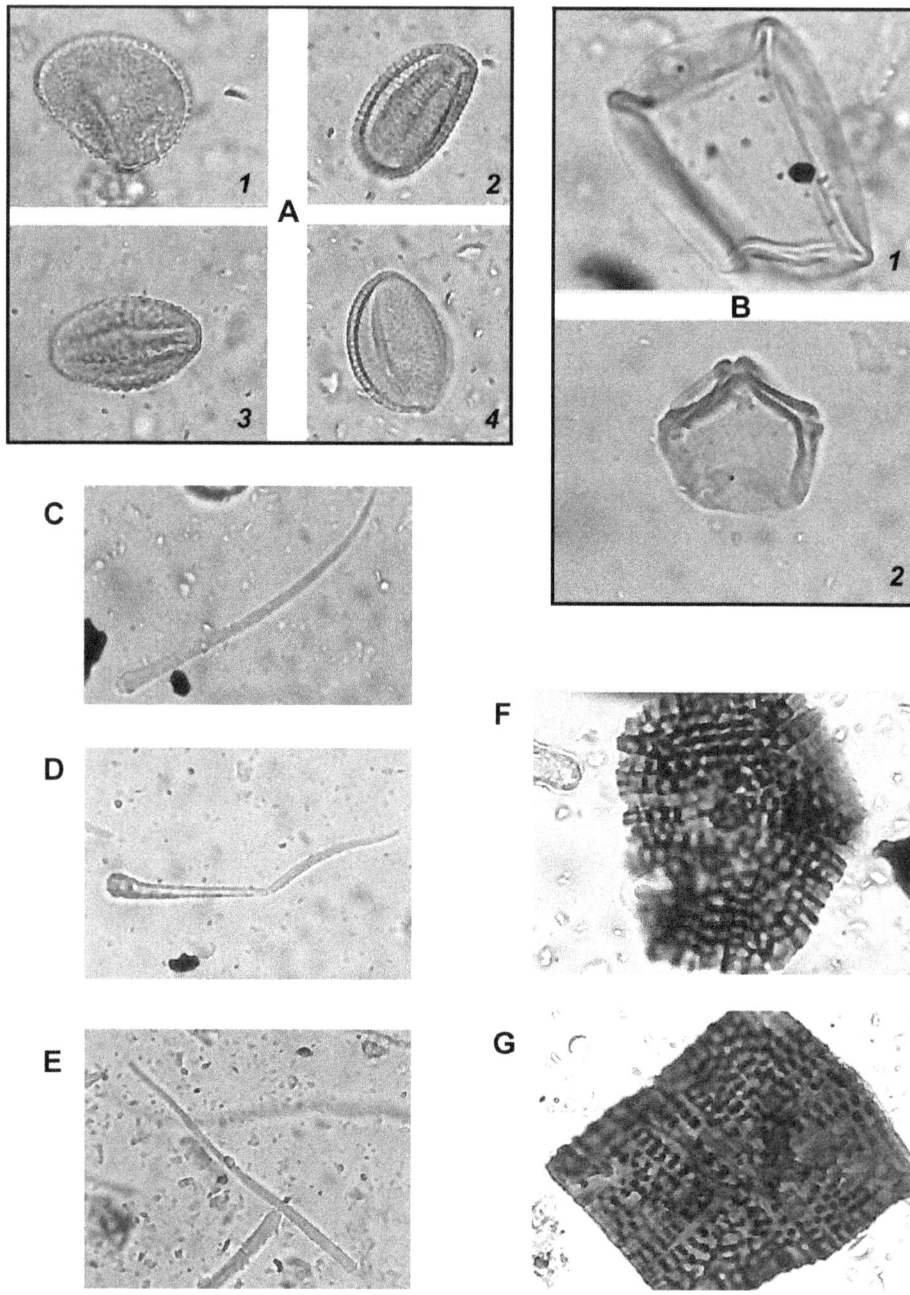


Figure 6: A:1: pollen of *Liliaceae*; A2 and 4: pollen of *Brassicaceae*; A3: pollen of *Salix*; B1: pollen of *Poaceae*; B2: pollen of *Alnus*; C, D and E: siliceous remains of the scales of gramineous plants; F and G: chitinous remains of the microplates of fish scales.

It is important to note that although the residues in some vessels were interpreted as a clear meat soup or a clear soup of wild cereals and steppe grasses, fish soup was only identified in tombs of the Catacomb Culture Period. There appeared to be no clear correlation between the contents of the vessels and the gender or age of the deceased. Graves constructed in the winter usually did not contain any vessels, or vessels that appeared to have contained only water. More work needs to be done to better understand these seasonal differences as well as the gender and age distribution of staple and ritual foods of the Bronze Age nomads on the Caspian steppe.

## References


**Arytynov**, S.A., G.A. Sergeeva and V.P. Kobychev (1995). *Material Culture. Food and Dwellings. Peoples of the Caucasus. Volume 4.* (Materialnaya kultura. Pisha I zhilishe. Narody Kavkaza). Moscow: Nauka (in Russian).

**Berg**, G.E. (2002). Last Meals. Recovering Abdominal Content From Skeletonized Remains. *Journal of Archaeological Science* **29**: 1349-1365.

**Demkin**, V.A. (1997). *Soil Science and Archaeology* (Paleopochvovedeniye I archaeologiya). Pushino: Nauka (in Russian).

**Demkin**, V.A. and T.S. Demkina (2000). Possibility to Determine Funeral Vessels Residue Dating Back to the Bronze and Early Iron Ages (O vozmozhnosty opredeleniya pogrebalnoy pishi v keramicheskich sosudach iz kurganov bronzovogo I zheleznogo vekov). *Ethnographic Review* **4**: 73-81 (in Russian).

**Demkin**, V.A., T.S. Demkina and A.V. Borisov (2000). Steppe Kurgans open New Secrets (Stepnye kurgany otkrivayut noviye tainy). *Priroda* **3**: 31-36 (in Russian).

**Gamkrelidze**, T.V. and V.V. Ivanov (1984). *Indo-European Language and Indo-Europeans* (Indoevropeisky yazyk i Indoevropeytsy). Tbilisi: Mnitsierba (in Russian).

**Grichuk**, V.P. and E.V. Zaklinskaya (1948). *Analyses of Buried Pollen and Spores and it Use in Paleogeography* (Analiz iskopaeymoy piltsy i spor i ego primeneniye v paleogeographii). Moscow: Geografiz (in Russian).

**Jones**, J.O. (1993). Analyses of Pollen and Phytoliths Residue from a Colonial Period Ceramic Vessel. Current Research in Phytolith Analysis. Application in Archaeology and Paleoecology. In D.M. Pearsall and D.R.Piperno (eds.). *MASCA Research Papers in Science and Archaeology. Volume 10.* Philadelphia: University of Pennsylvania, pp. 31-35.

**Kamenetsky**, I.S. (1995). Scythian Elbambing of Dead Tsars (O balzamirovanii umershich tsarey u skifov). In R.M. Munchayev (ed.). *Historical-archaeological bulletin. Volume 1.* Moscow-Armavir: Nauka, pp. 70-76 (in Russian).

**Korenevsky**, S.N. (2004). *The Ancient Farmers and Cattle Breeders of the Northern Caucasus* (Drevneyshiye zemledeltsi i skotovody Predkavkaziya). Moscow: Nauka (in Russian).

**Korpusova**, V.N. and S.N. Lyashko (1970). The Catacomb Culture Grave with Wheat from the Crimea (Katakombnoye pogrebeniye s pshenitsey v Krimu), *Soviet Archaeology* **3**: 38-45 (in Russian).

**Schonoeninger**, M.J. and K. Moore (1992). Bone Stable Isotopes Studies in Archaeology. *Journal of World Prehistory* **6,2**: 247-295.

**Meyer-Melikyan**, N.R. (1998). Analyses of Floral Remains from Togolok 1. In V.I. Sarianidi (ed.). *Margiana and Protozoroastrism.* Athens: Kapon Editions, pp. 198-201.

**Meyer-Melikyan**, N.R. and N.A. Avetov (1998). Analysis of Floral Remains in the Ceramic Vessel from the Gonur Temenos. In V.I. Sarianidi (ed.). *Margiana and Protozoroastrism.* Athens: Kapon Editions, pp. 202-204.

**Sherrat**, A. (1997). *Cups that Cherred. The Introduction of Alcohol to Prehistoric Europe. Economy and Society in Prehistoric Europe.* Prinstone: Princeton University Press, pp. 376-402.

**Shishlina**, N.I. (2001a). Diet Systems of Bronze Age Nomads based on Ethnographic and Archeological Data (Sistema pitaniya kochevnikov epochi bronzi po dannym etnographii i archeologii). In V.L. Yegorov (ed.). *Papers of the State Historical Museum. Volume 126.* Moscow: OOO Poltex, pp. 273-290 (in Russian).

**Shishlina**, N.I. (2001b). The Seasonal Cycle of Grassland use in the Caspian Sea Steppe during the Bronze Age. A New Approach to an Old Problem. *European Journal of Archaeology* **4,3**: 346-366.

**Shishlina**, N.I. and M.M. Pakhomov (2002). Pollen Investigation of Soil Samples from the Ostrovnoy Burial Ground in Kalmykia. In N.I. Shishlina and E. Tsytskin (eds.). *The Ostrovnoy Burial Mound. Results of the Complex Investigation of the Archaeological Sites of the North-West Caspian Sea Steppe.* Moscow: Elista Tisso-Poligraf, pp. 186-195 (in Russian).

**Toporov**, V.N. (1982). Soma. In S.A.Tokarev (ed.). *Myth of the Peoples. Volume 2.* Moscow: Nauka, pp. 462-463.

**Tomilov**, N.A. and L.M. Kadyrova (1997). Folk medical treatment of the Siberia Tatars (Narodnyy sposoby lecheniya u sibirskich tatar). *Ethnographic Review* **5**: 122-131 (in Russian).

**Zhitetsky**, I.A. (1893). *Papers on the Astrakchan Kalmyks Mode of Life. Ethnographical Notes of 1884-1886* (Ocherky byta astrachanskich kalmykov. Etnograficheskiye nablyudeniya, 1884-1886). Moscow: Tipografia M.G.Volchaninova (in Russian).

**Zhukovskaya**, N.L. (1988). *Categories and Symbols of the Mongols Tradidional Culture* (Kategorii I simvoly traditsionnoy kultury mongolov). Moscow: Nauka (in Russian).

# CHAPTER FIVE
# An Introduction to Archaeological Lipid Analysis by Combined Gas Chromatography Mass Spectrometry (GC/MS)
## H. Barnard, A.N. Dooley and K.F. Faull

Hans Barnard, Research Associate; Cotsen Institute of Archaeology; University of California, Los Angeles; P.O.-Box 951510; Los Angeles, CA 90025; USA; <wendrich@barnard.nl>; Alek Dooley; AppliedBiosystems GmbH; Frankfurter Straße 129[B]; 64293 Darmstadt; Germany and Kym Faull PhD, Director; Pasarow Mass Spectrometry Laboratory; The NPI-Semel Institute of Neuroscience and Human Behavior and The Department of Psychiatry and Biobehavioral Sciences; University of California, Los Angeles; 1217 Young Hall; Los Angeles, CA 90095; USA. We would like to thank Jelmer Eerkens and Elizabeth Mullane, for their comments on earlier versions of this chapter, and Willeke Wendrich, for her on-going encouragement and financial support.

As evident from the contributions to this volume, there are several approaches available to examine archaeological organic residues. Many studies concentrate on the interrogation of lipids isolated from archaeological materials while others examine alkaloids or proteins. In time, analyses may include all these classes of compounds, and perhaps others, but currently investigators concentrate on one class of compounds because of limited resources. The reasons behind the focus on lipids are that lipids are relatively stable compounds that are well-studied and easy to analyze (Murphy 1993). Furthermore, the use of methods similar to those employed by other archaeologists enables comparison between different sites and periods. Organic residues have been isolated from ancient objects, using a variety of protocols, including the ceramic matrix of unglazed pottery vessels with a wide range of ages and provenances (Charters et al. 1995; Condamin 1976; Eerkens 2002; Evershed et al. 1991; Gerhardt et al. 1990; Hill et al. 1985; Malainey et al. 1999; Mills and White 1989; Regert et al. 1998; Oudemans and Boon 1991; Patrick et al. 1985; Shimoyama et al. 1995; Skibo and Deal 1995; Stern et al. 2000).

One obstacle encountered by those eager to enter this field is that most publications understandably concentrate on the interpretation of the results and less on the details of the methodology. Another difficulty is the frequent use of jargon, often in the form of abbreviations and acronyms. In this chapter, we explain the basics behind the techniques used to analyze lipids, in archaeological samples, by GC/MS (see also the glossary at the end of this chapter).

## Lipids and Fatty Acids

Lipids are a diverse group of organic molecules that includes, among others, fatty acids, fats (including triacylglycerols), waxes, steroids (including cholesterol) and terpenoids. They are largely hydrophobic ('water-hating') molecules, with small polar parts that are hydrophilic ('water-loving'). Lipids will therefore not readily dissolve in water, but rather form micelles or membranes with the hydrophilic parts on the outside and the hydrophobic parts clustering on the inside in an attempt to exclude water from the hydrophobic core (Figure 1, Berg et al. 2002; Voet and Voet 2004).

Lipids are ubiquitous in nature and are present in nearly all foodstuffs. As they appear to get trapped in the ceramic matrix of unglazed pottery, and remain there intact for centuries, they are potential targets for archaeological residue analysis. Downsides to lipid residue analysis are that individual lipids are not specific to one foodstuff and can potentially be introduced to vessels during all stages of use. In addition, there is not one method of analysis suitable for all types of lipids.

The most commonly analyzed class of lipids are fatty acids. These are strings of $CH_2$-groups, making up the hydrophobic part of the molecule, with an acidic (hydrophilic) COOH-group attached to one end (to the $\alpha$-carbon).[1] A mix of several systems of nomenclature for these is in use, adding to the confusion of those new to the field. The systematic names of a series of saturated fatty acids, based on the number of C-atoms in Greek, and their synonyms are given in Table 1.

---

1 The C-atoms of fatty acids are numbered from the COOH-group, number 2 being the $\alpha$-carbon. The last C-atom, of the final $CH_3$-group, is referred to as the $\omega$-carbon (Figure 1).

| Systematic name | Synonyms | | Formula | Mass |
|---|---|---|---|---|
| Dodecanoic acid | C12:0 | Lauric acid, Vulvic acid | $HOOC-(CH_2)_{10}-CH_3$ | 200 |
| Tridecanoic acid | C13:0 | --- | $HOOC-(CH_2)_{11}-CH_3$ | 214 |
| Tetradecanoic acid | C14:0 | Myristic acid | $HOOC-(CH_2)_{12}-CH_3$ | 228 |
| Pentadecanoic acid | C15:0 | --- | $HOOC-(CH_2)_{13}-CH_3$ | 242 |
| Hexadecanoic acid | C16:0 | Cetylic acid, Palmitic acid | $HOOC-(CH_2)_{14}-CH_3$ | 256 |
| Heptadecanoic acid | C17:0 | Margaric acid, Margarinic acid | $HOOC-(CH_2)_{15}-CH_3$ | 270 |
| Octadecanoic acid | C18:0 | Stearic acid, Steric acid | $HOOC-(CH_2)_{16}-CH_3$ | 284 |
| Nonadecanoic acid | C19:0 | --- | $HOOC-(CH_2)_{17}-CH_3$ | 298 |
| Eicosanoic acid | C20:0 | Arachic acid, Arachidic acid | $HOOC-(CH_2)_{18}-CH_3$ | 312 |
| Heneicosanoic acid | C21:0 | --- | $HOOC-(CH_2)_{19}-CH_3$ | 326 |
| Docosanoic acid | C22:0 | Beheric acid | $HOOC-(CH_2)_{20}-CH_3$ | 340 |
| Tricosanoic acid | C23:0 | --- | $HOOC-(CH_2)_{21}-CH_3$ | 354 |
| Tetracosanoic acid | C24:0 | Lignoceric acid | $HOOC-(CH_2)_{22}-CH_3$ | 368 |
| Pentacosanoic acid | C25:0 | --- | $HOOC-(CH_2)_{23}-CH_3$ | 382 |
| Hexacosanoic acid | C26:0 | Cerinic acid, Cerotic acid | $HOOC-(CH_2)_{24}-CH_3$ | 396 |

Table 1: Details of a series of saturated fatty acids. Mass is the integer molecular mass in Daltons (Table 5).

| Systematic name | Synonyms | Formula | Mass |
|---|---|---|---|
| 9-tetradecenoic acid | C14:1, Myristoleic acid, Myristelaidic acid (trans) | $HOOC-(CH_2)_7-CH=CH-(CH_2)_3-CH_3$ | 226 |
| 9-hexadecenoic acid | C16:1, Palmitoleic acid, Palmitelaidic acid (trans) | $HOOC-(CH_2)_7-CH=CH-(CH_2)_5-CH_3$ | 254 |
| 9-octadecenoic acid | C18:1, Oleic acid, Elaidic acid (trans) | $HOOC-(CH_2)_7-CH=CH-(CH_2)_7-CH_3$ | 282 |
| 11-eicosenoic acid | C20:1, Gondoic acid, trans-Gondoic acid | $HOOC-(CH_2)_9-CH=CH-(CH_2)_7-CH_3$ | 310 |
| 13-docosenoic acid | C22:1, Erucic acid, Brassidic acid (trans) | $HOOC-(CH_2)_{11}-CH=CH-(CH_2)_7-CH_3$ | 338 |
| 15-tetracosenoic acid | C24:1, Selacholeic acid, trans-Selacholeic acid | $HOOC-(CH_2)_{13}-CH=CH-(CH_2)_7-CH_3$ | 366 |

Table 2: Details of a selection of mono-unsaturated fatty acids. Mass is the integer molecular mass in Daltons (Table 5). The single syllable that distinguishes the systematic name of an unsaturated fatty acid from that of its saturated counterpart is underlined.

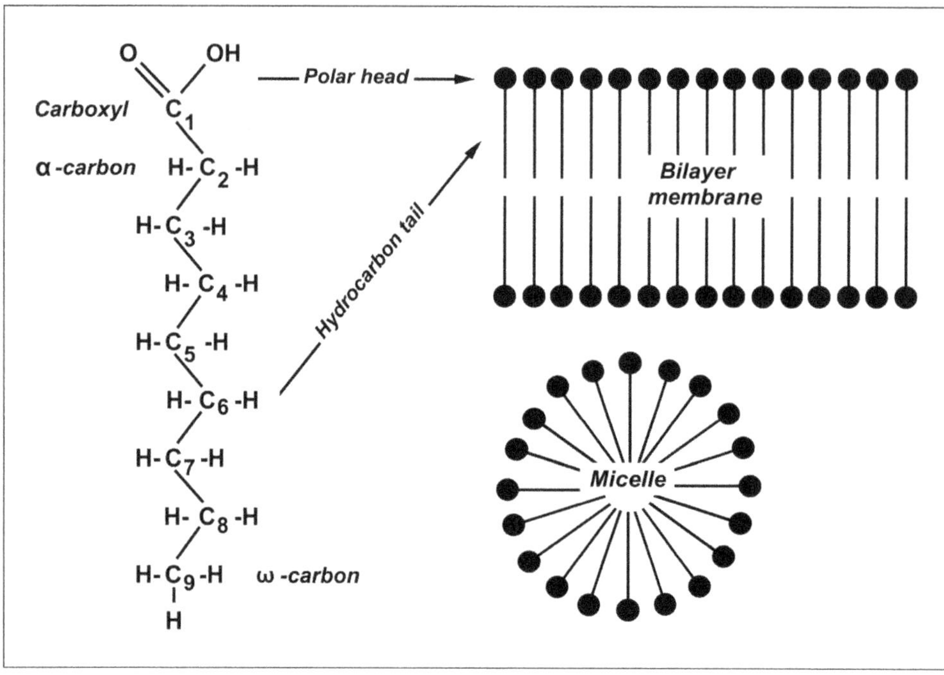


Figure 1: Schematic representation of a fatty acid (on the left) and the way in which such molecules aggregate to form a membrane or a micelle when in an aqueous environment.

Often used is the shorthand notation giving the number of C-atoms in the molecule, followed by the number of double bonds after a colon. Note that C14:0 in this context has nothing to do with the isotope $^{14}C$ used for radiocarbon dating.[2]

Naturally occurring fatty acids usually have an even number of C-atoms, hexadecanoic (C16:0) and octadecanoic (C18:0) acid being two of the most common. This is also reflected in the lack of trivial names for most odd-chain fatty acids. In unsaturated fatty acids two adjacent H-atoms are replaced by another bond between the two C-atoms, forming a double bond.[3] The most common naturally occurring mono-unsaturated fatty acids, with one double bond, are given in Table 2. These are fatty acids with an even number of C-atoms and nine C-atoms on one side of the double bond.[4] Note that the systematic names of mono-unsaturated fatty acids are only one vowel (underlined in Table 2) different from that of their saturated counterparts.

[2] In contrast to the most common $^{12}C$-atoms, which have a nucleus of 6 protons and 6 neutrons, the rare radioactive $^{14}C$-atoms (about $10^{-10}$% of all C-atoms in the atmosphere) have 6 protons and 8 neutrons. Their slow decay into $^{14}N$ (half-life 5730 years) is used for radiocarbon dating (Stott et al. 2003). The ratio of the naturally occurring stable isotope $^{13}C$, with 6 protons and 7 neutrons, to $^{12}C$ is also used for archaeological residue analysis (Ambrose 1993; Mottram et al. 1999; Morton and Schwarcz 2004, cf. Appendix 1).

[3] The common '*cis*' (or Z) configuration has the two parts of the molecule on either side of the double bond pointing in the same direction, giving the molecule a pronounced bend, whereas in the unusual '*trans*' (or E) configuration these parts point in opposite directions, shaping the molecule more like the saturated form. In the shorthand notation the double bond can be indicated by a Δ, for instance C22:1 *cis*-$\Delta^{13}$ being erucic acid.

[4] During their synthesis in higher organisms, fatty acids are elongated by two C-atoms at a time. Mammals do not have enzymes to introduce double bonds beyond the ninth C-atom and have to synthesize such fatty acids from the linoleate (9,12-octadecadienoic acid, C18:2) or linolenate (9,12,15-octadecatrienoic acid, C18:3) in their diet. These essential fatty acids are also referred to as ω-6 and ω-3 fatty acids, respectively, after the distance between the last double bond and the ω-carbon.

| Systematic name | Synonym | Formula | Mass |
|---|---|---|---|
| Ethanedioic acid | Oxalic acid, Oxeric acid | HOOC-COOH | 90 |
| Propanedioic acid | Maloric acid | HOOC-$CH_2$-COOH | 104 |
| Butanedioic acid | Amber acid, Succiric acid | HOOC-$(CH_2)_2$-COOH | 118 |
| Pentanedioic acid | Glutaric acid | HOOC-$(CH_2)_3$-COOH | 132 |
| Hexanedioic acid | Adiperic acid, Adipic acid | HOOC-$(CH_2)_4$-COOH | 146 |
| Heptanedioic acid | Pileric acid, Pimeic acid | HOOC-$(CH_2)_5$-COOH | 160 |
| Octanedioic acid | Cork acid, Suberic acid | HOOC-$(CH_2)_6$-COOH | 174 |
| Nonanedioic acid | Anchoic acid, Azelaic acid, Azelairic acid, Lepargylic acid | HOOC-$(CH_2)_7$-COOH | 188 |
| Decanedioic acid | Ipomic acid, Sebacic acid, Seracic acid | HOOC-$(CH_2)_8$-COOH | 202 |
| Undecanedioic acid | --- | HOOC-$(CH_2)_9$-COOH | 216 |
| Dodecanedioic acid | --- | HOOC-$(CH_2)_{10}$-COOH | 230 |
| Tridecanedioic acid | Brassylic acid | HOOC-$(CH_2)_{11}$-COOH | 244 |
| Tetradecanedioic acid | --- | HOOC-$(CH_2)_{12}$-COOH | 258 |

Table 3: Details of a series of dicarboxylic fatty acids. Mass is the integer molecular mass in Daltons (Table 5). Fatty acids with 6-10 C-atoms are collectively referred to as medium-chain fatty acids.

Unsaturated fatty acids, and especially poly-unstaturated fatty acids (with more than a single double bond), are more abundant in food of vegetable origin (Table 6). Mono-unsaturated fatty acids can be oxidized into dicarboxylic fatty acids, which have a COOH-group on both terminal C-atoms. Like unsaturated fatty acids, dicarboxylic fatty acids are more abundant in oils and fats of vegetable origin.[5] Details of a series of dicarboxylic fatty acids are given in Table 3. Small quantities of branched fatty acids, with one or more $CH_3$-groups attached to the central chain (for instance 3,7,11,15-tetramethyl-hexadecanoic or phytanic acid),[6] are present in many plants and animals (especially ruminants and fish), and in much larger concentrations in many bacteria.[7] One, two or three fatty acids can be attached by an ester-bond ($R_1$-C-O-CO-$R_2$),[8] to glycerol ($CH_2OH$-HCOH-$CH_2OH$) forming a monoacylglycerol (MAG), a diacylglycerol (DAG) or a triacylglycerol (TAG) respectively, in general referred to as fats.[9] These ester-bonds can be broken, and the fatty acids released, by strong alkaline (caustic) metal hydroxides like potassium hydroxide (KOH) or sodium hydroxide

5 As the auto-oxidation of unsaturated fatty acids is a slow process, the age of oil paintings can be approximated from the ratio between unsaturated and dicarboxylic fatty acids (Surowiec et al. 2004).

6 In humans, the inability to metabolize phytanic acid will lead to an accumulation of toxic levels in the brain and other tissue, a disorder known as Refsum's disease. The primary treatment is to avoid food that contains phytanic acid, such as dairy products, lamb and fatty fish.

7 Most odd-chain and branched fatty acids originate from micro-organisms and enter the food chain by symbiosis, such as the bacterial activity in the digestive tract of ruminant animals.

8 Where not relevant for the understanding of the argument, $R_1$ and $R_2$ are used to symbolize any configuration of the rest of the molecule.

9 Triacylglycerols are also referred to as triglycerides or triacylglycerides. The fatty acids making up a DAG or TAG are not necessarily the same and, with few exceptions, most naturally occurring fats are a mix of many different MAGs, DAGs and TAGs.

(NaOH), a reaction referred to as saponification (Figure 2).[10]

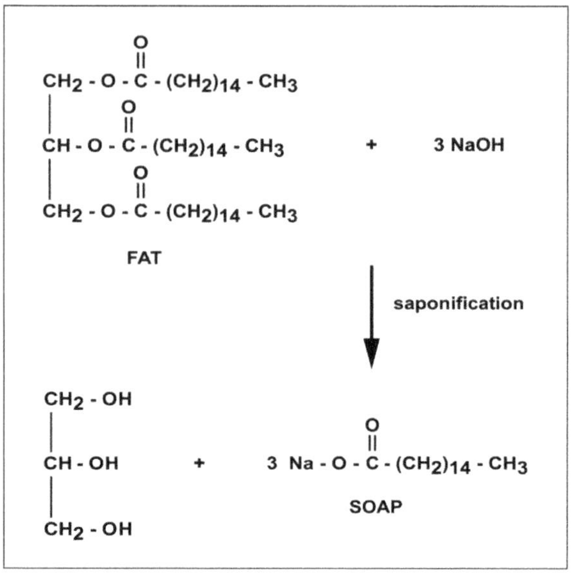


Figure 2: Saponification is the boiling of a lipid in a strong base, such as the reaction of triacylglycerol (fat) with potassium hydroxide (KOH) or sodium hydroxide (NaOH) resulting in the release of glycerol and fatty acid salts (soaps).

## Extraction of an Archaeological Sample

The first step to study the ancient lipids in archaeological pottery is to separate them from the ceramic matrix by extracting them in a suitable solvent, for instance in a 2:1 (by volume) mix of chloroform ($CH$-$Cl_3$) and methanol ($CH_3$-$OH$). Lipids will readily dissolve in this but, as these solvents can only take up fixed amounts of specific lipids (Table 4), the ratio of lipids in solution may differ from that in the residue.

Although not strictly necessary (Gerhardt et al. 1990), grinding a small fragment of the vessel into a fine powder, with an aluminum-oxide mortar and pestle or similar device, and then adding the solvent, will maximize contact between the residue and the solvents. Contamination can be prevented by wearing gloves at all times and by removing the surfaces of the sherd that may have been in contact with all kinds of pollutants (Evershed et al. 1990; Heron et al. 1991). Depending on the thickness of the vessel, a 2 x 2 cm fragment will usually produce sufficient material for 4-6 distinct extractions. As absorption of lipids by the ceramic matrix will not be homogenous throughout the wall of the vessel, taking samples from different portions of the vessel (rim, shoulder, body or base) may yield additional information (Charters et al. 1995; Stern et al. 2000).

The pottery powder can be stored in sterile glass vials, preferably with Teflon lined caps. To protect the sample from further oxidation, and other unwanted reactions, it is advisable to store it at -20°C, or under a non-reactive gas (like nitrogen), and preferably both, whenever possible at any stage during the analysis.

The basic way to extract the organic residue from the pottery powder is as follows. A fixed amount of pottery powder, usually 100-500 mg, is transferred into a clean glass test tube with a measured amount of solvent, usually 1-5 ml. This suspension is then firmly mixed and sonicated. With a laboratory sonicator ultrasound is applied which stimulates the residues into solution. The suspension is then separated, with a laboratory centrifuge or a filter, after which the liquid is transferred into a second test tube. This procedure can be repeated several times, on the same pottery powder and combining the resulting extract, to dissolve as much of the residue as possible. Next, the solvent is removed by evaporation, either in a vacuum-centrifuge or under a gentle steam of nitrogen to preserve more of the volatile compounds. Numerous variations to this basic method are described to increase its yield, either in terms of number or variety of molecules extracted (Malainey et al. 1999; Stern et al. 2000).

[10] In biological systems saponification is performed by enzymes known as lipases (present in high concentration in pancreatic juice and the venom of certain snake species).

| | **Dodecanoic** acid = C12:0 | **Tetradecanoic** acid = C14:0 | **Hexadecanoic** acid = C16:0 | **Octadecanoic** acid = C18:0 |
|---|---|---|---|---|
| **Methanol** ($CH_3$-OH) | 1200 | 173 | 37 | 1 |
| **Chloroform** (CH-$Cl_3$) | 830 | 325 | 151 | 60 |
| **Cyclohexane** ($C_6H_{12}$) | 680 | 215 | 65 | 24 |
| **Acetone** ($CH_3$-CO-$CH_3$) | 605 | 159 | 54 | 15 |
| **Acetonitrile** ($CH_3$-CN) | 76 | 18 | 4 | 0.9 |
| **Water** ($H_2O$) | 0.055 | 0.020 | 0.007 | 0.003 |

Table 4: Solubility of four common saturated fatty acids in six often-used solvents (in grams fatty acid per liter solvent at 20°C).

## Additional Treatment of the Sample

The dry residue is taken up in a small amount, for instance 100 µl, of solvent. This can be the same as was used for the extraction or one deemed more suitable for the next steps in the analysis, such as ethyl-acetate ($CH_3$-COO-$C_2H_5$), acetyl-acetate or iso-octane.[11] Adding potassium or sodium hydroxide (KOH or NaOH respectively) to the suspension will saponify the fats thus increasing the concentration of free fatty acids (Figure 2). Alternatively, saponification can be induced at the previous stage, with the pottery powder still present. Addition of a known amount of a known compound, an internal standard, will enable the approximation of concentration of the compounds found in the residue when the height of its peak on the chromatogram is compared with those of the molecules in the sample.[12] Such an internal standard does not facilitate qualitative analysis. Note that any additional handling of the sample will increase the chances of contamination.

In order for more lipids to pass through a gas chromatograph it is advisable to make them less polar and more thermally stable by replacing the active hydrogen of the COOH-group by a non-polar group, a process referred to as derivatization. Methods often used include esterification (methylization) and silylation. In esterification the active hydrogen is replaced with a methyl-group ($CH_3$) turning the fatty acids into fatty acid methyl esters (FAMEs). Replacing the active hydrogen with a trimethylsilyl (TMS) group (Si-($CH_3)_3$) is referred to as silylation. A large number of derivatizing agents are commercially available, including methanolic HCl and diazomethane for esterification,[13] and N,O-bis (trimethylsilyl) trifluoroacetamide (BSTFA), with or without 1% trimethylchlorsilane (TMCS),[14] for silylation of the sample. For these reactions to take place it is necessary to heat the sample, often to 60°C for as long as one hour, which will at the same time stimulate unwanted reactions, including oxidation. To reduce manipulation of the sample, the residue can be extracted into an agent that both saponifies and derivatizes the sample while it evaporates in the gas chromatograph (Stern et al. 2000), such as tetramethyl-ammonium hydroxide (TMAH) or trimethyl-trifluorotolyl-ammonium hydroxide (TMTFTH).

---

11 Acetyl-acetate is 3-oxobutanoic acid ethyl ester, $CH_3$-CO-$CH_2$-COO-$C_2H_5$. Iso-octane is 2,2,4-trimethyl pentane, $(CH_3)_3$-$CH_2$-CH-$(CH_3)_2$.

12 Both non-polar compounds, such as tetratriacontane ($CH_3$-$(CH_2)_{32}$-$CH_3$), and polar compounds, such as hexacosanoic acid (C26:0) or benzoic acid (phenylcarboxylic acid), are used as internal standards.

13 Because diazomethane ($CH_2N_2$) is explosive, toxic and carcinogenic it is not stored but prepared in small quantities immediately prior to use.

14 TMCS is added to BSTFA to aid the derivatization of obstructed functional groups.

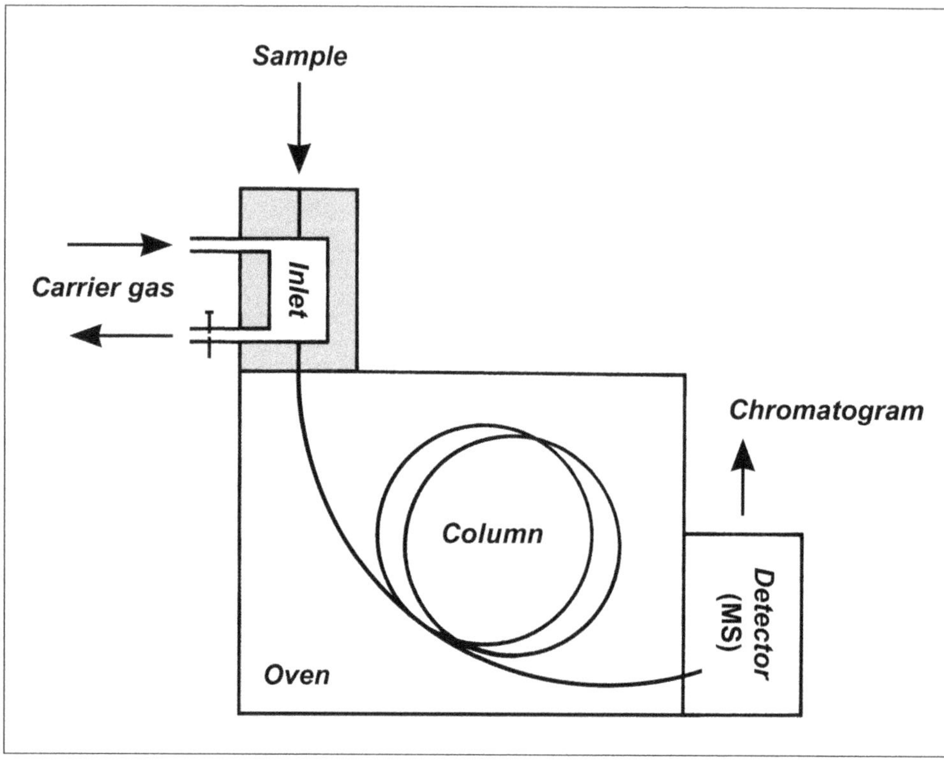


Figure 3: Schematic drawing of the GC-part of a GC/MS. The sample is injected into the top-left and carried by a gas through the column towards the detector that may be a mass spectrometer (MS).

## Gas Chromatography (GC)

Chromatography was first developed as a technique to study complex organic mixtures by separating them in a column of calcium carbonate. The differences in the speed of migration of the various components through the column separated them into bands with different colors ('chroma' means color in Greek).[15] A number of approaches have since been developed based on this principle. All include a 'mobile phase', which carries the components from the sample to where they are detected, and a 'stationary phase', where different components are retained at different rates. In a gas chromatograph the migration takes place in stream of inert gas (He or $N_2$) while the identification of the emerging components can be done by mass spectrometry. The GC-part of the instrument consists of a long (15-60 m), narrow (0.1-0.5 mm) glass column through which a steady flow of carrier gas is maintained,[16] the mobile phase. This column is coated on the inside with a thin layer of liquid to which some molecules in the sample will be attracted, the stationary phase. The beginning of the column is attached to an evaporation chamber, the sample inlet. The column and sample inlet are in ovens of which the temperatures can be carefully controlled. The end of the column is connected to a detector where the abundance of molecules emerging from the column is recorded,[17] or to the inlet of the MS-part of the instrument (Figure 3).

Typically 1 μl of the sample is injected into the sample inlet that is heated to a fixed high temperature (200-300°C). Solvents, sample and additives (such as excess derivatization agent and internal standard) will

[15] The term 'chromatography' was coined by the Russian biologist Mikhail Semyonovich Tsvet (also transliterated as Tswett, Zwet or Cvet), at the very beginning of the 20th century, in the reports on his work on plant pigments.

[16] Gas chromatographs can be operated with either a constant flow of carrier gas, which means the pressure will drop as the temperature of the GC-column rises, or with constant pressure, which means the flow will rise as the temperature rises.

[17] A variety of detectors are used, including conductivity detectors (TCD), electron capture detectors (ECD), flame ionization detectors (FID) and photo-ionization detectors (PID), each with their own specific advantages and disadvantages.

quickly vaporize to be carried onto the column by the carrier gas. A variable outlet allows regulation of the amount of sample actually passing onto the column. To obtain maximum resolution, samples with a high concentration of constituents are split into a part going onto the column and a part exiting the instrument without being analyzed. Samples with a low concentration of constituents, which will include many archaeological samples, can be fed onto the column splitless.

If the temperature of the column is 10-20°C below the boiling point of the solvents they condense in the column forming a solvent slug. When the temperature in the oven is slowly raised, and the solvent starts to evaporate, the molecules in the sample will concentrate in this solvent slug. Focusing the sample using this cold-on-column injection technique will result in sharper peaks in the chromatogram. Once the sample is on the column the mobile phase (the carrier gas) and the stationary phase (the coating inside the column) will compete for the molecules in the sample (Figure 7). For each molecule the outcome is dependent on the temperature inside the column. As the temperature in the column is slowly raised (Figure 4), the various components of the sample will one by one leave the stationary phase and travel with the carrier gas to the end of the column.

After reaching the end of the column the molecules emerge and are detected, in a GC/MS by the MS-part of the instrument. The results of this are plotted against the retention time, which is the time that the stationary phase retained the molecules on the column. The intensity of the most abundant molecule is assigned a value of 100%, the abundance of other molecules is plotted relative to this. On the resulting chromatogram the total ion current (TIC), which is essentially the total number of molecules detected, is also indicated (Figure 5).

To achieve optimum resolution it may be necessary to analyze a sample several times at different split-rates, especially with inherently unpredictable archaeological samples. A rich sample at a low split rate will produce saturated peaks, which could obscure smaller peaks, while a sample that produces no response after a split run, may do so when run splitless. Large molecules (>500-1000 Da), or molecules that are too polar or insufficiently thermally stable, will not pass through a gas chromatograph. They will have to be investigated with other separation techniques, such as high pressure liquid chromatography (HPLC)[18] or electrophoresis,[19] which will not be discussed here. Once separated in this way, further study of the molecules in the sample can be attained by mass spectrometry.

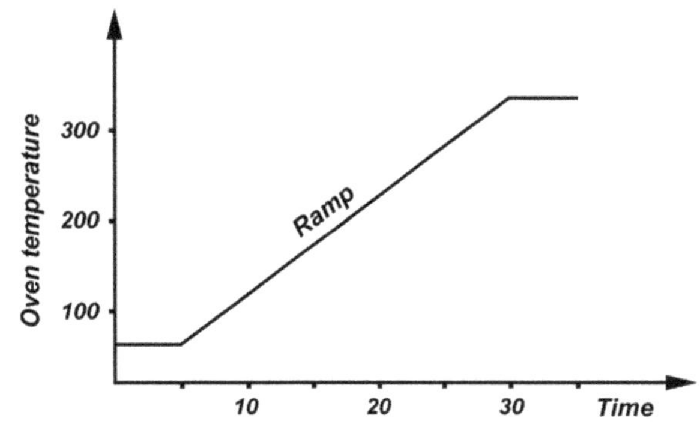


Figure 4: Typical profile of the temperature in the oven of a GC/MS instrument during the analysis (temperature in °C, time in minutes). On the ramp the temperature is raised 12°C/min for 25 min, from 50°C to 350°C.

[18] In high pressure (or high performance) liquid chromatography a sample is carried through a column (the stationary phase) by a solvent (the mobile phase) at high pressure. For organic molecules a non-polar stationary phase (usually silica based, sometimes enriched with carbohydrates or lipids) is often used with a polar mobile phase (acetonitrile, methanol or water). This is referred to as reversed phase HPLC (RP-HPLC) as the first liquid chromatography was performed with a non-polar solvent on a polar column (normal phase HPLC or NP-HPLC) by Mikhail Tsvet at the beginning of the 20$^{th}$ century. UHPLC is ultra high pressure liquid chromatography.

[19] Electrophoresis techniques are based on the differences in migration speed of molecules in an electric field (the mobile phase). The stationary phase can be an agarose or acrylamide gel (gel electrophoresis), or a buffer solution in a capillary (capillary electrophoresis). These techniques are often used for the study of large organic molecules, such as proteins or DNA. Identification of these can be accomplished by UV absorbance, fluorescence or luminenscence, or by autoradiography, antibody-antigen reactions or mass spectrometry.

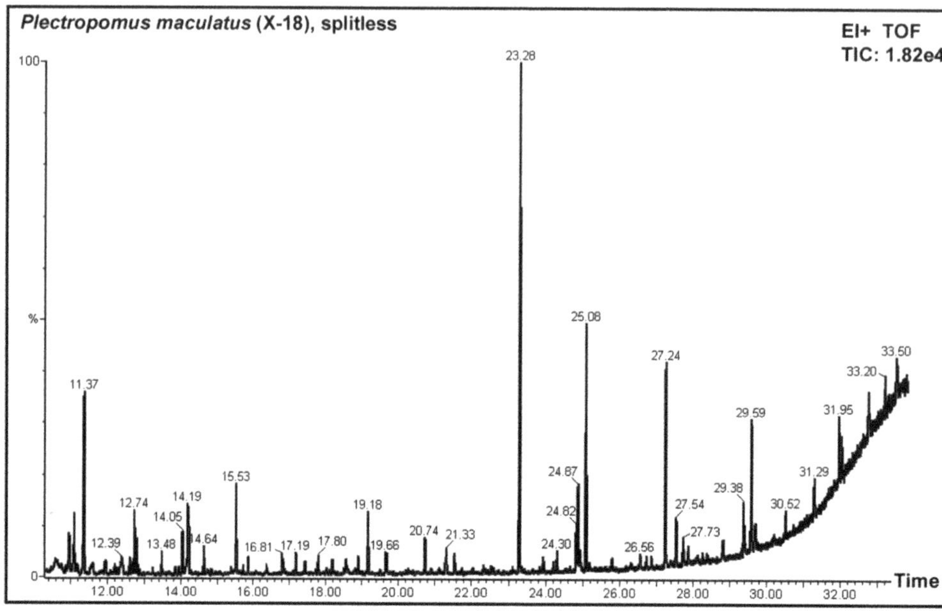


Figure 5: Part (10-33 min) of the chromatogram of the residue left in the matrix of a new, unglazed and unseasoned ceramic vessel after cooking a Red Sea *Plectropomus maculatus* (spotted coral grouper or red coral trout). Each peak, marked with its retention time (abscissa), represents a component of the residue (Figure 13). The tallest peak, at 23.28 min in this example, has a relative abundance of 100% (ordinate). $EI^+$: electron impact positive ionization; TOF: time-of-flight mass analyzer; TIC: total ion current (in arbitrary units).

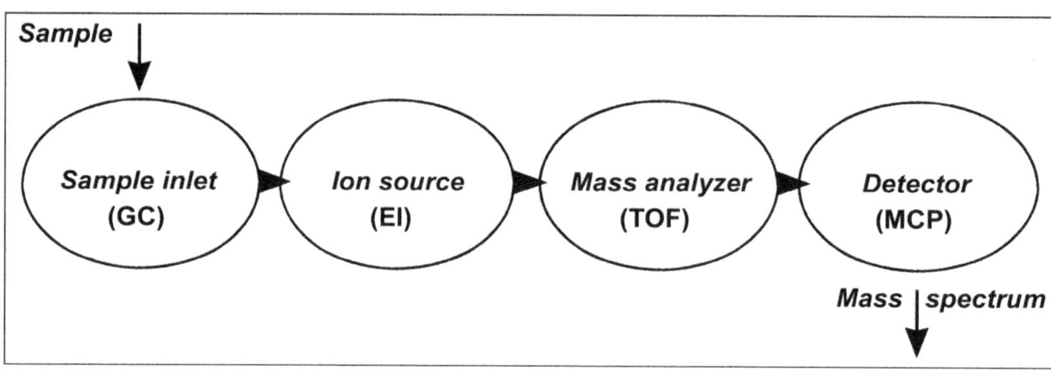


Figure 6: Schematic representation of a mass spectrometer. One possible set-up is a gas chromatograph (GC), an electron impact ion source (EI), a time-of-flight mass analyzer (TOF) and a microchannel plate detector (MCP) respectively.

## Mass Spectrometry (MS)

Mass spectrometry refers to a variety of methods to accurately measure the mass, or rather the mass to charge ratio (m/z), of ions (charged molecules).[20] Knowing the mass of a molecule can enable, and will certainly aid, its identification. All mass spectrometers consist of a sample inlet, an ion source (where the molecules in the sample are ionized), a mass analyzer (where ions are separated according to their m/z) and a detector (Figure 6). Mass analyzers separate ions with a different m/z by applying electro-magnetic forces, requiring the molecules in the sample to be ionized (charged) so that they will respond to such forces. Molecules that do not accept ionization or decompose completely when ionized will escape analysis.

20 Mass spectrometric techniques were first described by J.J. Thomson (1899), A.J. Dempster (1918) and F.W. Aston (1919).

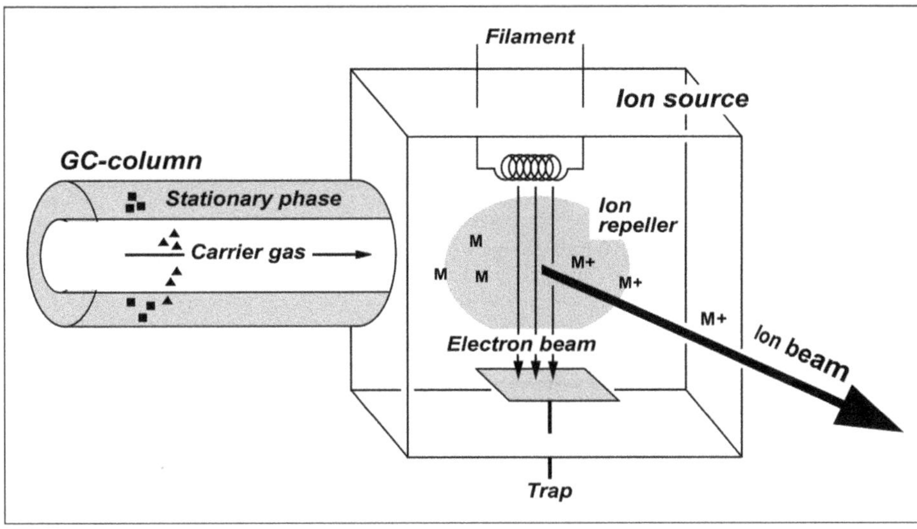


Figure 7: Schematic representation of the interface between a GC-column and an EI ion source. The molecules represented by squares are still in the stationary phase, while those represented by triangles are just moving into the mobile phase. Molecules represented by M have already been carried into the ion source where they are bombarded by high energy electrons (usually 70 eV). Molecules that lose an electron ($M^+$) are pushed into the analyzer of the mass spectrometer by a positively charged ion repeller.

A number of methods to ionize molecules have been developed,[21] each with their own specific advantages and disadvantages. Ionization of a molecule can be achieved by inserting or extracting an electron, a proton ($H^+$) or a small ion.[22] This will obviously change the mass of the original molecule, although the mass of an electron is insufficiently large to be significant.[23] The ionization process might cause molecules to fracture. Analysis of the resulting fragments provides additional information that is helpful in identifying the original molecule. This is especially the case when molecules are ionized in a beam of high energy electrons (electron impact ionization, EI), resulting in positive ions when an electron is knocked out of the molecule (Figure 7). The fragmentation patterns that are the result of this are reproducible, much like a fingerprint, which allows comparing the mass spectra generated by an unknown compound with known spectra in a digital library, such as that compiled by NIST/EPA/NIH.[24] The combination of a mass spectrometer, that can produce potentially identifiable mass spectra, with a gas chromatograph, which can separate complex compounds with high resolution, is a powerful tool to study unknown substances such as archaeological organic residues.[25]

21 These include atmospheric pressure chemical ionization (APCI), electron impact ionization (EI), electrospray ionization (ESI), fast atom bombardment (FAB) and matrix assisted laser desorption ionization (MALDI).

22 Ions utilized include $NH_4^+$, $Cl^-$ or $Na^+$ (with integer masses of 18, 35 and 23 Da respectively).

23 Inserting or extracting a proton ($H^+$) will change the mass of a molecule by 1.0078 Da. (Table 5).

24 NIST/EPA/NIH are the United State's National Institute of Standards and Technology, the Environmental Protection Agency and the National Institutes of Health. Their database holds more than 150,000 mass spectra.

25 The resolution of capillary electrophoresis is comparable with that of gas chromatography while the resolution of HPLC is considerably lower. The large effluents of both capillary electrophoresis and HPLC make them easier to combine with ion sources at atmospheric pressure, such as APCI or ESI, losing the analytical advantage of an EI ion source. Further examination can then be done by tandem mass spectrometry (MS/MS).

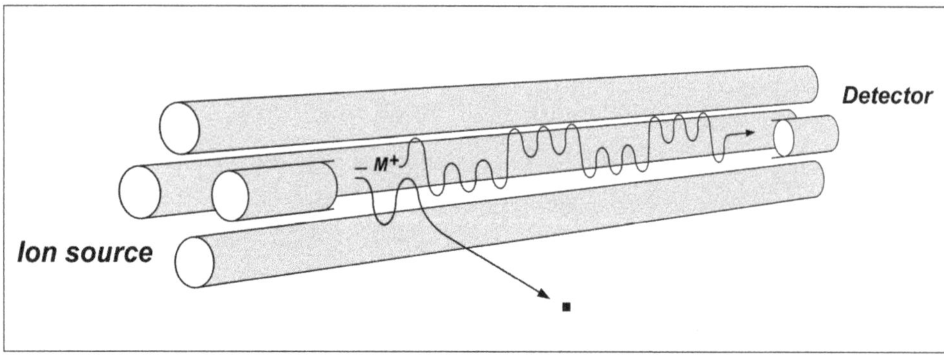


Figure 8: Schematic representation of a quadrupole mass analyzer. Only ions with a specific m/z will follow a stable trajectory ($M^+$), all others (such as ■) exit the space within the quadrupole and will not reach the detector.

After ionization ions of different mass, or charge state, can be separated because of the difference in their behavior in the electro-magnetic fields inside the mass analyzer. As with the ion source, and also the detector, there are different kinds of mass analyzers.[26] Some scan a range of m/z, allowing only ions with one specific m/z to reach the detector, others separate the ions in such a way that they arrive at the detector at different times.[27] All mass analyzers need to be at a high vacuum to avoid collision of the sample ions with other molecules.[28] With the steady influx from the sample inlet this vacuum needs to be constantly maintained.[29] Depending on the type of ion source this is placed outside or within the vacuum surrounding the mass analyzer and the detector.

Analyzers most often used in mass spectrometers include quadrupole and time-of-flight (TOF) mass analyzers. A quadrupole consists of four parallel metal rods, two of which carry a high voltage DC potential while the other two carry a high frequency AC potential (Figure 8).[30] Ions entering the space between these four rods follow a complex path because of the constantly changing electro-magnetic fields. Most will at some point leave the space between the rods or crash into one of the rods. Only ions with a specific m/z will follow a stable trajectory and reach the end of the quadrupole. Changing the voltages on the rods will cause ions with another m/z to reach the detector. Scanning a series of sequential values of m/z must be sufficiently slow to allow ions to travel the length of the rods.

A time-of-flight mass analyzer is basically an empty metal tube with an ion accelerator at its beginning and a ion detector at its end.[31] Upon entering the analyzer ions are accelerated, by an electro-magnetic pulse, after which they are allowed to drift towards the detector. The speed of each ion, and consequently the time it needs to complete this journey,[32] depends on its mass and charge state (m/z). Different ions will therefore reach the detector, where their arrival is recorded, at different times. The m/z of each ion can now be extrapolated from its time of flight.

All detectors are designed to amplify the impact of single ions and turn this into a measurable electric current. This current is then measured and the relative intensity (frequency) of each m/z, during a specific time interval, is plotted on a mass spectrum (Figure 9).[33]

---

26 A variety of mass analyzers are used, including Fourier transform ion cyclotron resonance analyzers (FT-ICR or FTMS), ion traps, magnetic sectors, quadrupoles and time-of-flight analyzers (TOF), each with their own specific advantages and disadvantages.

27 Quadrupoles and magnetic sectors are scanning mass analyzers, while time-of-flight analyzers and ion traps can potentially measure all ions.

28 Many mass analyzers operate at pressures of less than $10^{-3}$ Pascal, which equals about $10^{-8}$ atm, $10^{-5}$ mmHg (torr) or $10^{-7}$ psi.

29 The large effluents of HPLC and capillary electrophoresis, for instance, are difficult to combine with an EI ion source. They are better investigated with atmospheric pressure ion sources (APCI or ESI) followed by tandem mass spectrometry (MS/MS).

30 This frequency is about $10^6$ Hz which is usually referred to as radio frequency (RF).

31 A TOF mass analyzer is typically about 2 m long, sometimes doubled by an electrostatic reflector, which encloses a vacuum of around $10^{-3}$ Pa ($10^{-7}$ psi).

32 The time-of-flight is usually in the 10-100 µsec range.

33 A variety of detectors are used, including array detectors, charge detectors, electron multipliers, Faraday cups, microchannel plate detectors (MCP) and photomultipliers, each with their own specific advantages and disadvantages.

| | Isotopes, mass (abundance) | Mono-isotopic mass (Da) | Average mass (Da) | Integer mass (Da) |
|---|---|---|---|---|
| H | $^{1}H$ = 1.0078 (99.98%)<br>$^{2}H$ = 2.0141 (0.02%) | 1.0078 | 1.0079 | 1 |
| C | $^{12}C$ = 12.0000 (98.90%)<br>$^{13}C$ = 13.0034 (1.10%) | 12.0000 | 12.0110 | 12 |
| N | $^{14}N$ = 14.0031 (99.63%)<br>$^{15}N$ = 15.0001 (0.37%) | 14.0031 | 14.0067 | 14 |
| O | $^{16}O$ = 15.9949 (99.76%)<br>$^{17}O$ = 16.9991 (0.04%)<br>$^{18}O$ = 17.9992 (0.20%) | 15.9949 | 15.9994 | 16 |
| P | $^{31}P$ = 30.9738 (100%) | 30.9738 | 30.9738 | 31 |
| S | $^{32}S$ = 31.9721 (95.02%)<br>$^{33}S$ = 32.9715 (0.75%)<br>$^{34}S$ = 33.9679 (4.21%)<br>$^{36}S$ = 35.9671 (0.02%) | 31.9721 | 32.0655 | 32 |

Table 5: Mono-isotopic, average and integer mass, in Daltons (Da), of the elements most common in organic molecules: hydrogen (H), carbon (C), nitrogen (N), oxygen (O) phosphorus (P) and sulphur (S) (cf. Appendix 1).

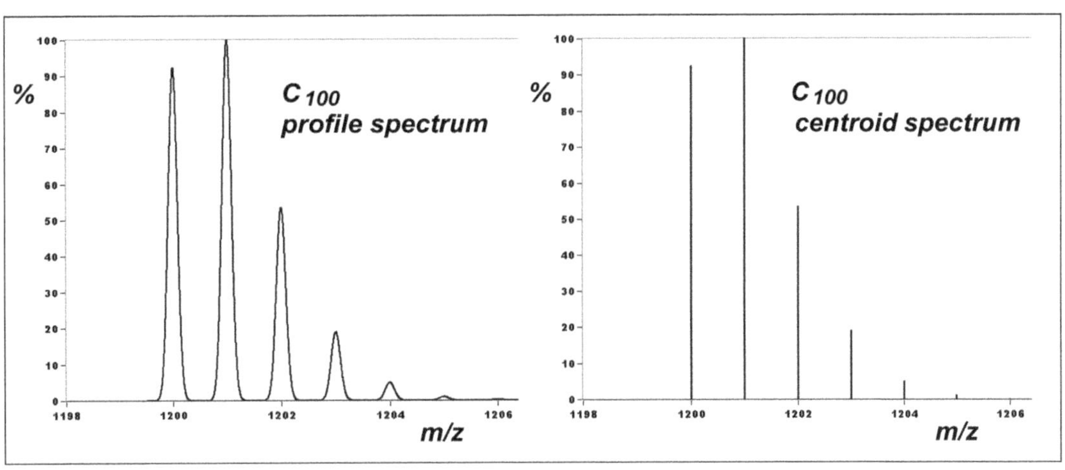


Figure 9: Profile (or continuum, left) and centroid (right) mass spectrum showing the isotopic signal distribution of a hypothetical ion, with a single positive charge, containing only 100 C-atoms ($^{12}C_{99}{}^{13}C_1$ having a relative intensity of 100%).

It has to be stressed that whether or not an ion is recorded by the detector depends on its m/z, not its mass. Ions that carry a double charge will appear to have half their actual mass. Comparatively small organic molecules such as fatty acids, however, usually carry only one charge, especially after the relatively inefficient electron impact ionization ($EI^+$) that typically gives singly charged ions and fragments.

The mass of a molecule is the sum of the mass of all atoms in that molecule. Most elements occur as natural stable isotopes, with different masses, of which the distribution is known (Table 5, Appendix 1). The mono-isotopic mass is the mass of the lightest isotope ($^{12}C$ being 12.00 Daltons by definition), the average mass is the weighted average mass of all isotopes. An organic molecule can therefore not be assigned a single fixed mass, but a predictable series of discretely different masses.

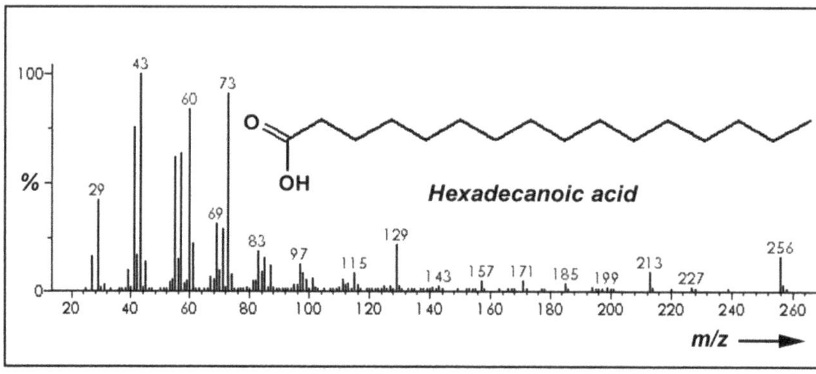


Figure 10: The EI mass spectrum and structure of hexadecanoic acid (palmitic acid, C16:0,) as shown in the NIST/EPD/NIH 02 database. The tallest peak, the base peak, occurs at m/z 43 and is assigned a relative intensity of 100%. The parent molecule has an integer mass of 256 Da (Table 1). The specific fragmentation pattern allows the identification of the molecule.

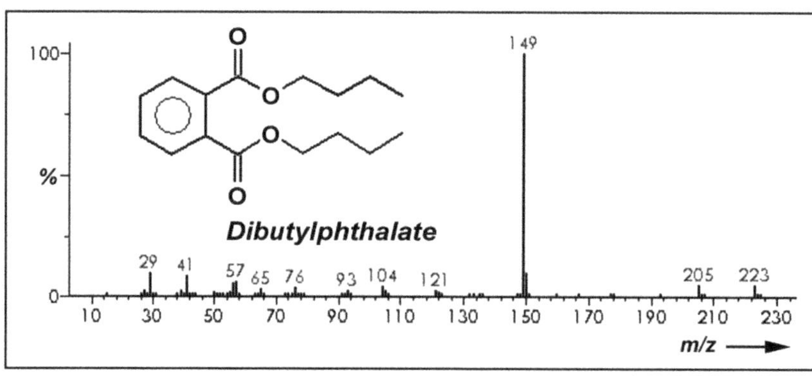


Figure 11: The EI mass spectrum and structure of dibutylphthalate (Celluflex, Genaplast, Polycizer) as shown in the NIST/EPD/NIH 02 database. A base peak at m/z 149 is often reported as indicative for phthalates (Kumar 1999), seen as an indication of contamination of the sample.

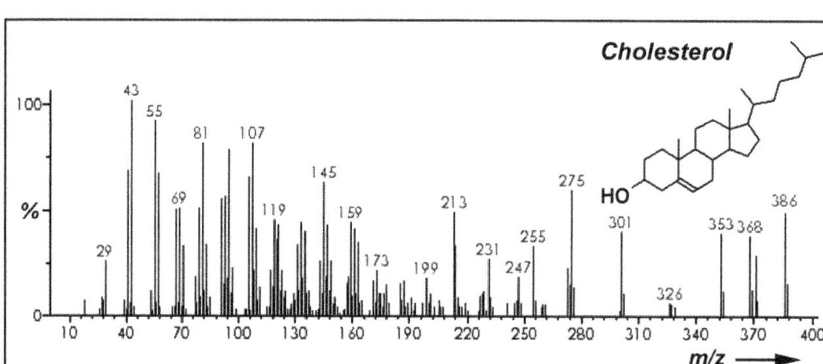


Figure 12: The EI mass spectrum and structure of cholesterol, synthesized by animals but not usually by plants or micro-organisms, as shown in the NIST/EPD/NIH 02 database. The base peak occurs at m/z 43, the parent molecule has an integer mass of 386 Da. The specific fragmentation pattern allows the identification of the molecule.

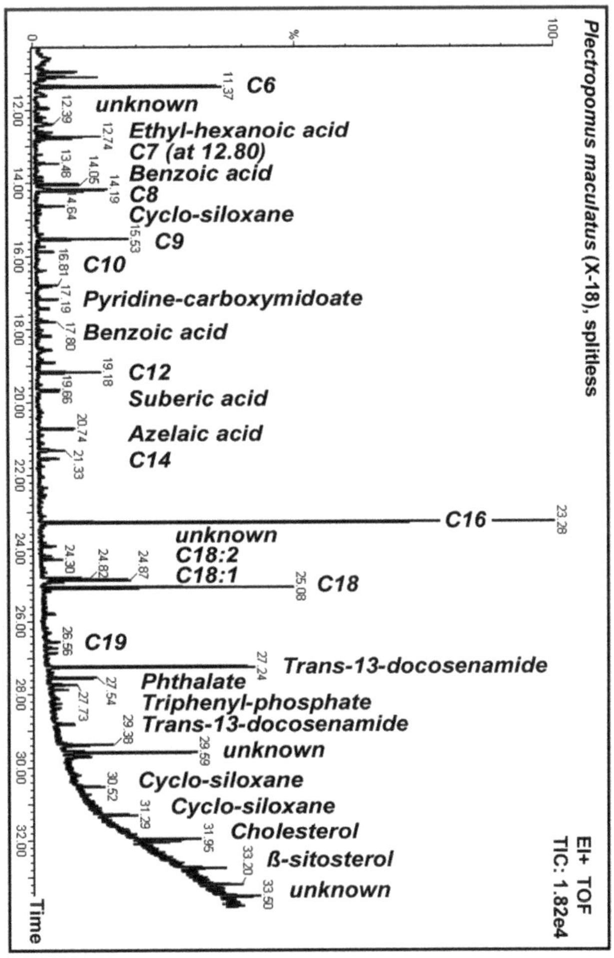


Figure 13: Identification of the peaks in the chromatogram shown in Figure 5 (to make this figure easier to read it is turned 90° clockwise). The saturated and unsaturated fatty acids most likely originate from the fish prepared in the vessel, phthalates (at 27.54 min) may have been introduced by laboratory plastics while cyclo-siloxanes (at 30.52 and 31.29 min) are probably released into the sample by the glass in the gas chromatograph. Two differently derivitized forms of trans-13-docosenamide, with different retention times (at 27.24 and 29.38 min), appear present in the sample. The sterol β-sitosterol (at 33.20 min) was most likely introduced by micro-organisms living off the residue before it was analyzed.

In high-resolution mass spectrometers this will complicate the mass spectrum,[34] but at the same time provide useful information on the composition and charge state of the molecule.[35] Instruments can be set to either display the full spectrum, or a centroid spectrum in which the profile is simplified by a single line with the same height of the original spectrum and passing through its mid-width at half-height (Figure 9).

[34] For instance, a hypothetical molecule of only 100 C-atoms will create at least four peaks on a mass spectrum as only 34% (92% relative abundance) will be mono-isotopic (all $^{12}C$), the majority being $^{12}C_{99}{}^{13}C_1$ (100% relative abundance), 20% will be $^{12}C_{98}{}^{13}C_2$ (54% relative abundance) and 7% will be $^{12}C_{97}{}^{13}C_3$ (19% relative abundance). Analysis of the spectrum will provide additional information on the composition of large molecules (Figure 9).

[35] The distances between the consecutive isotopes allows calculation of the charge state of the molecule. For doubly charged molecules these will be half the distance as for the same molecule singly charged (minus the mass of the ion providing the charge), and for the same molecule triply charged only a third (minus the mass of the ion providing the charge), and so on.

## Interpretation of GC/MS-results

The output of a GC/MS is a combination of a chromatogram (Figures 5 and 13) and a large amount of mass spectra (Figures 10, 11 and 12). Each peak in the chromatogram represents at least one molecule in the sample and a mass spectrum can be created from each of these peaks. Although it is a rare occurrence, chromatographic peaks can represent more than one molecule if the resolution of the GC-column is insufficient to separate them.

With specialized software mass spectra of unknown compounds can be electronically compared with the spectra of known compounds in a digital library.[36] Matches are normally listed in order of their correlation. A visual inspection of the known and unknown mass spectra usually leads to the identification, with a reasonable amount of certainty, of many components in most samples. Note that the molecules extracted from a residue were first derivatized and then ionized, which has changed their molecular structure and mass.[37] Compounds can therefore only be identified when derivatized with the same agents that were used to compile the reference library.

The molecules detected in archaeological organic residues are only those that remain trapped in the ceramic matrix, dissolve in the extraction solvents, survive the sample preparation, pass through the gas chromatograph and ionize in the mass spectrometer. As shown in Figures 5 and 13, many molecules satisfy these criteria. Some, however, must be regarded the result of contamination of the sample. This is obvious in the case of man-made organic molecules like phthalates (Figure 11), added to plastics to keep them flexible, but must also be considered for naturally occurring compounds like anthraquinones, used as dye in textiles and paper, or 13-docosenamide (erucamide), coated on plastics objects to prevent them sticking together. If an internal standard was added it will be possible to approximate the absolute concentration of the different molecules in the sample by comparing the intensity of the peak of the internal standard with those of the now identified components of the sample. Other compounds are interpreted to be biomarkers, molecules more or less specific for certain classes of foodstuffs. Such molecules can be alkaloids (such as caffeine), steroids (such as cholesterol, Figure 12), terpenoids (a large group of mostly polycyclic compounds synthesized by plants, Hanson 2001; Connolly and Hill 2005), but also fatty acids like docosenoic acid (erucic acid, C22:1) and phytanic acid (Hansel et al. 2004). The natural origins of many compounds have been published in the biochemical literature (O'Neil et al. 2001).

A different approach to interpret archaeological organic residues is comparing the abundance of the recovered fatty acids with those of known residues (Table 6). Results can be sometimes obtained with a simple ratio, such as C16:0/C18:0 (palmitic/stearic acid ratio, P/S ratio), or after plotting two ratios in a two-dimensional graph, for example C16:1/C18:1 versus (C15:0+C17:0)/C18:0 on a double logarithmic scale (Eerkens 2005).

It remains unclear whether organic residues represent the first food to come into contact with the ceramics, after which the available binding sites are saturated, or the last, if older residues are continually replaced by new ones. It is also possible that residues represent a combination of all food ever to have been inside the vessel, if the molecules that make up the residue compete for the available binding places. Possible sources of organic residues in archaeological pottery include not only food, but also refuse surrounding a discarded vessel, human remains decaying close to a pot included as a grave gift or micro-organisms breaking down organic residues within the wall of a pot (Figure 13).

Ceramics may also have been employed for industrial purposes, such as the preparation of organic dyes or glues, or as coffins, censers, smoking pipes or to store a multitude of things. Finally, it must be pointed out that many foodstuffs never come into contact with ceramic vessels but are eaten raw, roasted over a fire or prepared and consumed in other ways that do not call for ceramics. The organic residues found in ceramic vessels can therefore not be employed to reconstruct ancient diets, especially not without complementary archaeological or historical data. Further discussion of the archaeologial interpretation of biochemical research is beyond the scope of this introduction, but is the subject of several other chapters in this volume.

---

[36] Programs are available from most manufacturers of GC/MS instruments and NIST/EPD/HIH.

[37] The integer mass of a methyl-group (-$CH_3$), for instance, is 15 Da and that of a TMS-group (trimethylsilyl) is 73 Da.

| | | Meat | Nuts | Berries | Roots | Fish | Greens |
|---|---|---|---|---|---|---|---|
| C16:0 / C18:0 | Fresh | 0-4 | 0-9 | 2-6 | 3-12 | 4-6 | 5-12 |
| | Degraded | 0-7 | 0-18 | 4-12 | 6-24 | 8-12 | 10-24 |
| C16:1 / C18:1 | Fresh | 0.02-0.2 | 0-0.3 | 0-0.08 | 0.05-0.7 | 0.2-0.5 | 0-0.7 |
| | Degraded | 0.08-0.8 | 0-1.2 | 0-0.32 | 0.3-2.8 | 0.8-2.0 | 0-2.8 |

Table 6: P/S ratios (C16:0/C18:0) of different known fresh and degraded organic residues in pottery compared with the C16:1/C18:1 ratios (Eerkens 2005, Malainey et al. 1999). Note that the separation of the different classes of foodstuffs is incomplete.

## Glossary

The explanations in this glossary repeat the descriptions used in this chapter, as they apply to archaeological lipid analysis by GC/MS (see also the glossary at the end of Appendix II). For more comprehensive definitions we refer to biochemical textbooks, such as Berg et al. 2002, Murphy 1993 or Voet and Voet 2004.

**$[M-H]^-$**: negatively charged molecule because of a missing proton (see ion)
**$^{12}C$**: stable C-isotope with a nucleus of 6 protons and 6 neutrons (about 98.9% of all C-atoms)
**$^{13}C$**: stable C-isotope with a nucleus of 6 protons and 7 neutrons (about 1.1% of all C-atoms)
**$^{14}C$**: rare (about $10^{-10}$%) unstable (radio-active) C-isotope with a nucleus of 6 protons and 8 neutrons, its slow decay into $^{14}N$ (half-life 5730 years) is used for radiocarbon dating
**AC**: alternating current (see RF)
**active hydrogen**: in lipids the replaceable hydrogen in the COOH-group (see derivatization)
**analyzer**: see mass analyzer
**APCI**: atmospheric pressure chemical ionization
**array detector**: type of detector used in a mass spectrometer
**average mass**: see mass
**base peak**: the tallest peak of a mass spectrum, assigned a relative intensity of 100%
**biomarker**: a molecule more or less specific for a class of foodstuffs
**branched fatty acid**: fatty acid with one or more $CH_3$-groups attached to the central chain, often synthesized by micro-organisms
**BSTFA**: N,O-bis(trimethylsilyl)trifluoroacetamide, a derivatization agent (see silylation)
**capillary electrophoresis**: electrophoresis using a buffer solution in a capillary as the stationary phase
**carboxyl**: COOH-group
**carrier gas**: see mobile phase
**centroid spectrum**: mass spectrum in which the profile is replaced by a single line, with the same height of the original (profile or continuum) spectrum and passing through its mid-width at half-height
**charge detector**: type of detector used in a mass spectrometer
**chromatogram**: plot of the relative abundance of the components in a sample against their retention time inside a chromatograph
**chromatography**: an analytical technique based on the differences in the speed of migration of the components of complex mixtures through a medium
**cold-on-column injection**: focusing a sample in a GC-column by first condensing it in a solvent slug and then concentrating it by slowly raising the temperature and evaporating the solvent slug
**Da**: Dalton, the unit of atomic mass, $^{12}C$ ($19.9 \times 10^{-27}$ kg) being 12.00 Da by definition
**DAG**: diacylglycerol (see TAG)
**DC**: direct current (see AC)
**derivatization**: replacing the active groups of a molecule with a non-polar group to reduce polarity and increase thermal stability, by esterification (methylization) or silylation
**diazomethane**: $CH_2N_2$, an explosive, toxic, carcinogenic, but efficient and easy to use derivatization agent (see methylization)
**dicarboxylic fatty acid**: fatty acid with COOH-groups on both ends, often the oxidation product of a mono-unsaturated fatty acid
**ECD**: electron capture detector, used in a gas chromatograph
**EI**: electron impact ionization
**$EI^+$**: electron impact positive ionization, which triggers reproducible fragmentation
**electron multiplier**: type of detector often used in a mass spectrometer
**electrophoresis**: separation technique based on the differences in migration speed of molecules in an electric field (the mobile phase), the stationary phase can be a gel (gel electrophoresis) or a buffer solution in a capillary (capillary electrophoresis)
**enzyme**: biocatalyst, a protein that specifically accelerates a chemical reaction, in a biological system, without being used up
**EPA**: Environmental Protection Agency (USA)
**ESI**: electrospray ionization
**esterification**: see methylization

**FAB**: fast atom bombardment ionization
**FAMEs**: fatty acid methyl esters, fatty acids derivatized by esterification (methylization)
**Faraday cup**: type of detector used in a mass spectrometer
**FID**: flame ionization detector, used in a gas chromatograph
**FT-ICR**: see FTMS
**FTMS**: Fourier transform mass spectrometry, or Fourier transform ion cyclotron resonance (FT-IRC) mass spectrometry
**GC/MS**: gas chromatography combined with mass spectrometry
**GC-column**: glass tube inside the oven of a GC/MS, coated with a polar layer (stationary phase) and filled with a steady flow of carrier gas (mobile phase)
**gel electrophoresis**: electrophoresis using an agarose or acrylamide gel as the stationary phase
**HPLC**: high pressure (performance) liquid chromatography in which a sample is forced through a column (stationary phase) by a liquid (mobile phase) at high pressure (see RP-HPLC)
**hydrophilic**: 'water-loving', eager to dissolve in water
**hydrophobic**: 'water-hating', reluctant to dissolve in water
**integer mass**: the mass without decimals of an element
**ion repeller**: charged plate in an ion source that pushes oppositely charged ions into the mass analyzer of a mass spectrometer
**ion source**: part of a mass spectrometer where the molecules in the sample are converted to gas phase ions
**ion**: charged molecule, an extra electron or a missing proton ($H^+$) will result in a negative ion ($M^-$ or $[M-H]^-$ respectively), while a missing electron or an extra proton will result in a positive ion ($M^+$ or $MH^+$ respectively)
**IS**: internal standard, a known amount of a known compound added to a mix of unknown compounds with the purpose of approximating the concentration of these compounds once they are analyzed
**linoleate**: 9,12-octadecadienoic acid (C18:2), an essential fatty acid (see ω-6 fatty acid)
**linolenate**: 9,12,15-octadecatrienoic acid (C18:3), an essential fatty acid (see ω-3 fatty acid)
**lipase**: enzyme that performs saponification (in biological systems)
**lipids**: a diverse group of organic molecules including fatty acids, fats, waxes, steroids and terpenoids
**long-chain fatty acids**: fatty acids with more than ten C-atoms
**m/z**: mass to charge ratio, pronounced 'm over z' (see mass spectrum)
**$M^-$**: negatively charged molecule because of an extra electron (see ion)
**$M^+$**: positively charged molecule because of a missing electron (see ion)
**MAG**: monoacylglycerol (see TAG)
**magnetic sector**: a type of mass analyzer
**MALDI**: matrix assisted laser desorption ionization
**mass analyzer**: part of a mass spectrometer where ions are separated according to their m/z
**mass spectrometry**: an analytical technique based on the accurate measurement of the m/z of ions using differences in their behavior in an electro-magnetic field
**mass spectrum**: plot of the relative intensity (frequency) of ions with different m/z, as measured by a mass spectrometer (see centroid spectrum)
**mass**: the mass of an element can be the mass of its lightest isotope, the mono-isotopic mass, or the weighted average mass of all its isotopes, the average mass (see Da)
**MCP**: microchannel plate detector, used in a TOF mass spectrometer
**Me**: methyl ($-CH_3$)
**medium-chain fatty acids**: fatty acids with six to ten C-atoms
**MeOH**: methanol ($CH_3$-OH)
**methylization**: replacing active groups of a molecule with a methyl-group ($CH_3$) in order to reduce polarity and increase thermal stability (see diazomethane)
**$MH^+$**: positively charged molecule because of an extra proton (see ion)
**micelle**: a sphere of lipids with the hydrophilic parts on the outside and the hydrophobic tails towards the center
**mobile phase**: in gas chromatography, the flow of gas inside the GC-column carrying the molecules evaporating from the stationary phase
**mono-isotopic mass**: see mass
**mono-unsaturated fatty acid**: fatty acid with one double bond, more common in food of vegetable origin, which can be oxidized into dicarboxylic fatty acids
**MS/MS**: tandem mass spectrometry where two mass analyzers are combined in one instrument allowing either to function as a filter
**NIH**: National Institutes of Health (USA)
**NIST**: National Institute of Standards and Technology (USA)
**NP-HPLC**: normal phase HPLC, with a polar stationary phase and non-polar mobile phase, this is the traditional but now less frequently used set-up (see RP-HPLC)
**oleic acid**: *cis*-9-octadecenoic acid (C18:1 *cis*-$\Delta^9$)
**P/S ratio**: the ratio of the abuncance of palmitic acid (C16:0) versus stearic acid (C18:0)
**palmitic acid**: hexadecanoic acid (C16:0)
**palmitoleic acid**: *cis*-9-hexadecenoic acid (C16:1 *cis*-$\Delta^9$)
**parent molecule**: the original molecule in a sample before it is fractured inside a mass spectrometer (by ionization or otherwise)
**photomultiplier**: type of detector used in a mass spectrometer
**phthalates**: a large group of man-made molecules, commonly added to plastics
**phytanic acid**: 3,7,11,15-tetramethyl-hexadecanoic acid which is present in the meat and milk of ruminant animals and the fat of fish
**PID**: photo-ionization detector, used in a gas chromatograph

**poly-unsaturated fatty acid**: fatty acid with two or more double bonds, more common in food of vegetable origin
**QQQ**: triple quadrupole mass spectrometer (see MS/MS)
**quadrupole**: a type of mass analyzer
**RAM**: relative atomic mass or average mass (see mass)
**ramp**: gradually increase of the temperature inside the GC-column to make different components in the sample move from the stationary into the mobile phase
**reflectron**: electrostatic mirror in an TOF mass analyzer that increases the flight path of the ions and improves their separation
**relative abundance**: the abundance of a component in a mixture relative to the most abundant component (see chromatogram)
**relative intensity**: the abundance of a specific ion, with a certain m/z, relative to the most abundant ion in the mass spectrum (see base peak)
**retention time**: time between the introduction of a sample into a separation device (such as a GC-column) and the arrival of the component at the detector (see chromatogram)
**RF**: radio frequency, about $10^6$ Hz, the AC potential carried by two of the four rods of a quadrupole mass analyzer
**RP-HPLC**: reversed phase HPLC, with a non-polar stationary phase and a polar mobile phase, this is the most common set-up today (see NP-HPLC)
**saponification**: the reaction between a metal hydroxide, like KOH or NaOH, with a fat resulting in glycerol and fatty acid salts (soaps)
**saturated peak**: truncated peak in a chromatogram caused by the saturation of the detector
**short-chain fatty acids**: fatty acids with less than six C-atoms
**silylation**: replacing the active groups of a molecule with a TMS-group in order to reduce polarity and increase thermal stability (see BSTFA)
**solvent slug**: solvents with sample condensed in a GC-column (see cold-on-column injection)
**sonication**: the application of ultrasound to stimulate residues into solution
**source**: see ion source
**split injection**: feeding only part of a sample onto a GC-column, rich samples will be split into a part going into the column and a part exiting the instrument without being analyzed (see splitless injection)
**splitless injection**: feeding a complete sample into the GC-column (see split injection)
**stationary phase**: in gas chromatography, the active layer inside the GC-column that will slow down the molecules carried by the mobile phase (see retention time)
**stearic acid**: octadecanoic acid (C18:0)
**TAG**: triacylglycerol, TAGs are also referred to as triglycerides or triacylglycerides
**tandem mass spectrometry**: see MS/MS
**TCD**: conductivity detector, used in a gas chromatograph
**terpenoids**: also referred to as isoprenoids, a large group of mostly polycyclic compounds synthesized by plants for their defense, color or fragrance
**thermal stability**: the ability of a molecule to survive heating and evaporation without decomposition (see derivatization)
**TIC**: total ion current
**TMAH**: tetramethyl-ammonium hydroxide, a reagent that can derivatize a compound in the sample inlet of a gas chromatograph (see TMTFTH)
**TMCS**: trimethylchlorsilane, added to BSTFA to aid the silylation of otherwise obstructed functional groups
**TMS**: trimethylsilyl, Si-$(CH_3)_3$ with an integer mass of 73 Da (see silylation)
**TMTFTH**: trimethyl-trifluorotolyl-ammonium hydroxide, a reagent that can saponify and derivatize a compound in the sample inlet of a gas chromatograph (see TMAH)
**TOF**: time-of-flight, a type of mass analyzer
**UHPLC**: ultra high pressure liquid chromatography (see HPLC)
**unsaturated fatty acid**: fatty acid with one or more double bonds
**z**: charge state
**α-carbon**: the C-atom to which the COOH-group of a fatty acid is attached, also referred to as number 2, the C-atom in the COOH-group being number 1
**ω-3 fatty acid**: fatty acid with three C-atoms between the last double bond and the ω-carbon, such as 9,12,15-octadecatrienoic acid (linolenate C18:3), which mammals have to obtain from their food
**ω-6 fatty acid**: fatty acid with six C-atoms between the last double bond and the ω-carbon, such as 9,12-octadecadienoic acid (linoleate, C18:2), which mammals have to obtain from their food
**ω-carbon**: the C-atom in the terminal $CH_3$-group of a fatty acid

## References


**Ambrose**, S.H. (1993). Isotopic Analysis of Paleodiets. Methodological and Interpretive Considerations. In M.K. Sandford (ed.). *Investigations of Ancient Human Tissue. Chemical Analyses in Anthropology*. Langhorne: Gordon and Breach Science Publishers, pp. 60-130.

**Berg**, J.M., L. Stryer and J.L. Tymoczko (2002). *Biochemistry*. New York: Freeman and Company.

**Charters**, S., R.P. Evershed, P.W. Blinkhorn and V. Denham (1995). Evidence for the Mixing of Fats and Waxes in Archaeological Ceramics. *Archaeometry* **37,1**: 113-127.

**Condamin**, J., F. Formenti, M.O. Metais, M. Michel and P. Blond (1976). The Application of Gas Chromatography to the Tracing of Oil in Ancient Amphorae. *Archaeometry* **18,2**: 195-201.

**Connolly**, J.D. and R.A. Hill (2005). Triterpenoids. *Natural Product Reports* (Royal Society of Chemistry) **22**: 230-248.

**Eerkens**, J.W. (2002). The Preservation and Identification of Piñon Resins by GC-MS in Pottery from the Western Great Basin. *Archaeometry* **44,1**: 95-105.

**Eerkens**, J.W. (2005). GC-MS Analysis and Fatty Acid Ratios of Archaeological Potsherds from the Western Great Basin of North America. *Archaeometry* **47,1**: 83-102.

**Evershed**, R.P., C. Heron and L.J. Goad (1990). Analysis of Organic Residues of Archaeological Origin by High Temperature Gas Chromatography and Gas Chromatography-Mass Spectroscopy. *Analyst* **115**: 1339-1342.

**Evershed**, R.P., C. Heron and L.J. Goad (1991). Epicuticular Wax Components Preserved in Potsherds as Chemical Indicators of Leafy Vegetables in Ancient Diets. *Antiquity* **65**: 540-544.

**Gerhardt**, K.O., S. Searles and W.R. Biers (1990). Corinthian Figure Vases. Non-destructive Extraction and Gas Chromatography-Mass Spectrometry. *MASCA Research Papers in Science and Archaeology* **7**: 41-50.

**Hansel**, F.A., M.S. Copley, L.A.S. Madureira and R.P. Evershed (2004). Thermally Produced ω-(*o*-alkylphenyl)alkanoic Acids Provide Evidence for the Processing of Marine Produces in Archaeological Pottery Vessels. *Tetrahedron Letters* **45**: 2999-3002.

**Hanson**, J.R. (2001). Diterpenoids. *Natural Product Reports* (Royal Society of Chemistry) **18**: 88-94.

**Hill**, H.E., J. Evans and M. Card (1985). Organic Residues on 3000-Year-Old Potsherds from Natunuku, Fiji. *New Zealand Journal of Archaeology* **7**: 125-128.

**Heron**, C., R.P. Evershed and L.J. Goad (1991). Effects of Migration of Soil Lipids on Organic Residues Associated with Buried Potsherds. *Journal of Archaeological Science* **18**: 641-659.

**Kumar**, R. (1999). A Mass Spectral Guide for Quick Identification of Phthalate Esters. *American Laboratory* **35,** November: 32-35

**Malainey**, M.E., R. Przybylski and B.L. Sherriff (1999). Identifying the Former Contents of Late Precontact Period Pottery Vessels from Western Canada using Gas Chromatography. *Journal of Archaeological Science* **26**: 425-438.

**Mills**, J.S. and R. White (1989). The Identification of the Resins from the Late Bronze Age Shipwreck at Ulu Burun (Kaş). *Archaeometry* **31,1**: 37-44.

**Mottram**, H.R., S.N. Dudd, G.J. Lawrence, A.W. Stott and R.P. Evershed (1999). New Chromatographic, Mass Spectrometric and Stable Isotope Approaches to the Classification of Degraded Animal Fats Preserved in Archaeological Pottery. *Journal of Chromatography A,* **833**: 209-221.

**Morton**, J. and H.P. Schwarcz (2004). Paleodietary Implications from Isotopic Analysis of Food Residues on Prehistoric Ontario Ceramics. *Journal of Archaeological Science*, **31**: 503-517.

**Murphy**, R.C. (1993). *Mass Spectrometry of Lipids. Handbook of Lipid Research Volume 7*. New York: Plenum Press.

**O'Neil**, M.J., A. Smith, P.E. Heckelman, J.R. Obenchain Jr., J.A.R. Gallipeau, M.A. D'Arecca and S. Budavari (2001). The Merck Index. An Encyclopedia of Chemicals, Drugs, and Biologicals. Whitehouse Station: Merck & Co. Inc.

**Oudemans**, T.M.F. and J.J. Boon (1991). Molecular Archaeology. Analysis of Charred (Food) Remains from Prehistoric Pottery by Pyrolysis-Gas Chromatography/Mass Spectrometry. *Journal of Analytical and Applied Pyrolysis* **20**: 197-227.

**Patrick**, M., A.J. de Koning and A.B. Smith (1985). Gas Liquid Chromatographic Analysis of Fatty Acids in Food Residues from Ceramics found in the Southwestern Cape, South Africa. *Archaeometry* **27,2**: 231-236.

**Regert**, M., H.A. Bland, S.N. Dudd, P.F. van Bergen and R.P. Evershed (1998). Free and Bound Fatty Acid Oxidation Products in Archaeological Ceramic Vessels. *Proceedings of the Royal Society, London B*, **265**: 2027-2032.

**Shimoyama**, A., N. Kisu, K. Harada, S. Wakita, A. Tsuneki and T. Iwasaki (1995). Fatty Acid Analysis of Archaeological Pottery Vessels Excavated in Tell Mastuma, Syria. *Bulletin of the Chemical Society of Japan* **68**: 1565-1568.

**Skibo**, J.M. and M. Deal (1995). Pottery Function and Organic Residue, an Appraisal. In C. Yeung and W.B. Li (ed.), *Conference on Archaeology in South-East Asia*. Hong Kong, University Museum and Art Gallery: 321-330.

**Stern**, B., C. Heron, M. Serpico and J. Bourriau (2000). A Comparison of Methods for Establishing Fatty Acid Concentration Gradients Across Potsherds. A Case Study Using Late Bronze Age Canaanite Amphorae. *Archaeometry* **42,2**: 399-414.

**Stott**, A.W., R. Berstan, R.P. Evershed, C. Bronk-Ramsey, R.E.M. Hedges and M.J. Humm (2003). Direct Dating of Archaeological Pottery by Compound-specific $^{14}C$ Analysis of Preserved Lipids. *Analytical Chemistry* **75**: 5037-5045.

**Surowiec** I, I. Kaml and E. Kenndler (2004). Analysis of Drying Oils Used as Binding Media for Objects of Art by Capillary Electrophoresis with Indirect UV and Conductivity Detection. *Journal of Chromatography A* **1024 (1-2)**: 245-254.

**Voet**, D. and J.G. Voet (2004). *Biochemistry, Volume 1*. New York: Wiley and Sons.

# CHAPTER SIX

# Elucidating Pottery Function using a Multi-step Analytical Methodology combining Infrared Spectroscopy, Chromatographic Procedures and Mass Spectrometry

## M. Regert

Martine Regert; Unité Mixte de Recherche 171; Centre National de la Recherche Scientifique (CNRS), Ministère de la Culture et de la Communication; Centre de Recherche et de Restauration des Musées de France (C2RMF); 14, Quai François Mitterrand; 75001 Paris Cedex; France; <martine.regert@culture.gouv.fr>. I would like to thank the Ministère de la Recherche (ACI N° 4117 Jeunes chercheurs 2003), the Ministère de la Culture et de la Communication (PNRC 2003) and the GdR ChimArt (CNRS - Ministère de la Culture et de la Communication) for their financial support. I also want to express my gratitude to the archaeologists that allowed me to study the ceramic vessels discussed in this chapter, namely Anne-Marie and Pierre Pétrequin (Chalain), Yves Lanchon (Bercy), Dominique Bosquet (Podrî l'Cortri), Didier Binder (Pertus II), Stéphane Vacher (Grand Aunay) and Thierry Leroy (La Fangade). Sincere thanks are due to Pierre-Alain Gillioz for fruitful discussions and advise during the writing of this chapter.

To determine the function of a wide range of prehistoric ceramic vessels from several recently excavated archaeological sites, an interdisciplinary methodology was developed to link analytical chemistry, geochemistry and archaeology. This method comprised the description of charred surface residues and their distribution in ceramic containers, combined with a multi-step analytical strategy based upon a panel of complementary analytical techniques, namely infrared spectroscopy, direct inlet mass spectrometry and chromatographic procedures. The attribution of the molecular constituents that are detected and identified by these techniques into four categories of markers (biomarkers, anthropogenic transformation markers, natural degradation markers and contaminants) appeared a powerful tool for interpreting the molecular composition in terms of procurement, production and use of natural substances exploited by prehistoric people. Subcutaneous animal fats, dairy products, beeswax, plant oils, pine resins and birch bark tar were demonstrated to be processed in the containers studied. The identification of these substances led to the distinction of ceramic vessels linked with culinary activities from those related to technical activities, such as pottery repairing, coating or waterproofing as well as adhesive production, storage or use.

This chapter is intended to highlight both the analytical strategies developed and the results obtained on pottery function dealing with the characterization of organic remains preserved in archaeological ceramic vessels. I will first provide a general overview of the amorphous organic residues which may be preserved in ceramic containers. After this, the methodology as well as the processes that allow interpretation of chemical data in archaeology are described and explained. Finally, some results obtained on various ceramic vessels from Neolithic, Bronze and Iron Age sites in Belgium and France are presented and discussed.

Understanding pottery function is probably one of the most difficult tasks in archaeology. Until recently, most of the publications dealing with archaeological vessels discuss this matter only briefly, or not at all. Assumptions on pottery function are often quite general and usually limited to broad categories like food processing or the storage or transportation of various items (Rice 1987). The emphasis on culinary uses does not take into account the great diversity and variability of archaeological ceramic containers. This is probably due to the fact that most of the hypotheses on vessel use are primarily based on morphological and physical characteristics of the vessels. Theoretical relationships may obviously be supposed between morphology (shape, size, wall thickness) or physical properties (porosity, durability, resistance to thermal stress) and a particular use. For example, vessels with narrow openings may well have been containers for liquids; vessels with a round base and thin porous walls are more resistant to thermal stress and may have been used for cooking (Rice 1987; Orton et al. 1993). However, pottery function depends on a large number of properties, and their combination, and ethnographic research shows that no one-to-one correlation exists between use and the intrinsic properties of a vessel (Rice 1987, 224; Sigaut 1991). Furthermore, one must note that in most cases confusion exists between 'function' and 'functioning', as discussed by Sigaut (1991). Indeed, 'functioning refers to the way an object works or is worked, function refers to the whole set of ends for which it is put to use' (Sigaut 1991, 21). Defining vessel use by actions like storing, processing or transferring clearly implies functioning. Assessing pottery function must then not only deal with 'how was the vessel used?', but also with 'what did the vessel contain and with what purpose?'.

The only way to answer the second question is to identify the former contents of the vessels wherever these are preserved. Visible residues adhering to ceramic vessels and their location on the vessels were noticed and described as early as the end of the $19^{th}$ century

(Heintzel 1880; 1881); the possibility to chemically identify them had to wait for almost a century and the availability of analytical tools such as chromatographic techniques and mass spectrometry (Condamin et al. 1976; Condamin and Formenti 1978; Evershed et al. 1990; 1991; 1992a; b; Heron and Evershed 1993; Charters et al. 1995; Regert et al. 1999). Coupling chemical identification of the contents of the vessels with archaeological criteria, including both intrinsic (shape, materials, physical properties, use traces) and extrinsic (context of discovery, experimental data, ethnographic data) attributes of the vessels, provides many prospects for better understanding pottery function and functioning (Barnard et al. 2007). With this in mind several series of ceramic vessels from different European regions, archaeological contexts and periods (all from recent and well documented excavations) were investigated.

| Description of the visible organic residue | Putative function of the organic remains |
|---|---|
| Regular thin residues covering all the surfaces of the vessel | Coating agents |
| Black rough residues with a low adherence to the pottery, mostly located on the inner part of the vessel or residues that have spilled on the outside of the vessel (located on the rim and the upper outer parts of the vessel) | Culinary commodities |
| Brown to black residues, hardly adhering to the pottery, located either on the inside or the outside surfaces of the vessel | Adhesive materials |
| Brown to black residues, often quite thick, strongly adhering to the pottery, related to damaged parts of the vessel | Repairing materials |

Table 1: Description of the kinds of visible amorphous organic residues encountered on archaeological ceramic vessels and their putative former function.

Organic matter is known to be particularly well preserved in specific environmental contexts such as anaerobic (lacustrin, fluvial or marine), very dry or frozen sites. At archaeological sites, amorphous organic remains either linked with culinary commodities (dairy products, vegetable oils, fermented beverages) or technical activities (adhesive production, dyeing) have been preserved from the Neolithic and probably earlier. Pottery favors the preservation of either carbonized surface residues or organic matter trapped in the porous ceramic matrix (Rice 1987, 211). Although such remains are usually related to pottery function, some of them may also be witnesses of the final stages of pottery production. Depending on their nature and the time when they were introduced in the pottery life, organic residues represent different aspects of the 'vessel life' (Table 1). Three cases will be briefly discussed: organic remains related to the last stage of pottery making; organic remains that provide information on pottery function and organic remains that may be related to pottery repair.

## Organic Remains and Pottery Production

Several organic materials from animal or plant origin may be added as temper to the clay used for pottery production (Rice 1987, 407). Such materials, including plant fibers, shell, animal hair and dung, are usually not considered 'organic residues' as they are part of the matrix of the vessel. Moreover, because they have been added to clay before the firing process, their molecular composition is usually not preserved and is not identified by organic residue analyses.

Organic substances may also have been used for coating the surfaces of a ceramic vessel to decrease its porosity, increase the heating effectiveness and strength or to achieve a particular aesthetic effect (Arnold 1985, 140; Schiffer 1990; Diallo et al. 1995). Ethnographic research has shown that various substances, such as tree resins, saps, bark decoctions and other organic mixtures, have been used for this purpose (Arnold 1985, 140; Diallo et al. 1995). Although this practice is frequently attested ethnographically, few archaeological clues ascertain the use of organic substances as coating agents. A clear example of an archaeological ceramic vessel undoubtedly coated with an organic agent was excavated at Grand Aunay (Regert et al. 2003a). Coating residues have been identified on this sherd, dating to the Iron Age, covering both its inside and outside surfaces. The residue was quite homogeneous with a slightly brilliant brown color; its thickness varied between 20-200 μm. Coating agents may be detected by observing both the surface aspect of the coating and a section of the sherd under a microscope.

Some authors have also suggested that organic materials could have been used as adhesives at the final stage of pottery making, during the decoration of the vessels. Particularly, Vogt (1949, 50-51) talks about ceramic vessels decorated with white birch bark that has been attached with an organic adhesive.

## Organic Remains and Pottery Function

Generally speaking, most of the amorphous residues adhering to ceramic containers may be related to their use. Except for traces of soot, that are the result of heating over an open fire, these residues may be classified into two main categories. The first category comprises black, brittle and rough residues, usually quite thin but up to 1 mm in thickness, or more, which have low adherence to the ceramics. In most cases, such residues are located on the inner surface of the pottery. Some authors have noticed that they often have a similar distribution: thick residues at the bottom-wall transition and near the top of the vessel (Duplaix-Rata 1997, 734). This particular distribution is interpreted as resulting from convection movements of liquid or viscous commodities that were prepared in the vessels. When the vessel overflowed, thin organic residues will also be preserved on the outside of the pottery, again near the rim. The analysis of such organic remains shows evidence of culinary activities.

The second category includes residues with a color varying from brown to black. These remains are generally strongly attached to the surface, usually on the inside of the container. Their thickness greatly varies and can be more than 1 mm. Such residues are more compact and less brittle than the cooking residues. Their analysis shows that, in most cases, they may be related to adhesive production or storage.

When pots are sufficiently porous, they may have trapped organic matter within their walls. Such matter is not visible, but may be recovered by grinding a sherd and then extracting, analyzing and identifying the organic residue (Evershed et al. 1992a; b). In most cases, the analysis of such absorbed organic remains shows evidence of culinary activities. However, one must note that only certain activities are likely to give rise to the absorption of organic matter in the porous clay matrix of a ceramic vessel. Although the absorption processes have not yet been fully studied, it may be assumed that the more viscous the substances in the vessel, the better the absorption. Heating is also supposed to increase the absorption. The duration of use of a vessel will also influence the amount of organic matter trapped into the ceramic matrix. Consequently, some uses may not leave any trace in the vessel. The case of holed vessels, usually considered as cheese strainers, is a clear example. These particular vessels were most likely used to separate cheese from whey; an activity that does not imply any heating nor the prolonged stay of any liquid or viscous organic matter in the vessels. The probability of organic matter getting trapped in the porous matrix of a cheese strainer must therefore be considered very low (Bourgeois and Gouin 1995). Furthermore, recent studies of Neolithic holed vessels show that some of them may have trapped beeswax (Regert et al. 2001a). All this serves to show that caution has to be observed when interpreting archaeological organic residues; the absence of such remains does not prove that a vessel was not used.

## Organic Remains and Pottery Repair

The last class of organic remains that may be observed at the surface of ceramic vessels are those that are clearly related to damage to the vessel. Such residues are brown to black, often quite thick and strongly attached to the vessel. Usually located along ancient cracks, they were used to join the two parts of the break (Bosquet et al. 2001). In some cases, holes were drilled on either side of the crack. A vegetable or animal string was then passed through the holes, to reinforce the vessel, after which an organic substance, such as birch bark tar or resins, was used to consolidate and waterproof the fracture (Harmeyer et al. 1995).

## Analytical Strategy

Amorphous organic remains always consist of complex molecular mixtures. The method of choice for their analysis therefore usually includes a chromatographic technique to separate the different molecules. But because both sample treatment and chromatographic analytical conditions strongly depend on the chemical properties of the molecular constituents preserved, it is often convenient to first perform a preliminary analysis. Infrared spectroscopy (IR) and direct inlet mass spectrometry (MS) can be used for this purpose. The spectral fingerprints obtained provide a first idea of the substances preserved (Regert and Rolando 2002; Regert et al. 2003b; Scalarone et al. 2003; Colombini et al. 2005). With these data it is then possible to choose the best method of chromatographic analysis. This approach, however, is not always helpful. In particular, culinary samples do not usually provide spectra that can be easily interpreted as the organic matter is not sufficiently concentrated and often mixed with sedimentary matrix. The IR spectra obtained appear more characteristic for the inorganic matter present in the sediment than for the organic matter that has to be identified.

As mentioned, the organic residues preserved in ancient ceramic vessels may be related to their nature and their function, even if this relation is quite general. Depending on the aspect observed, either with the naked eye or under a binocular or a microscope, one of two analytical protocols can be followed: one for adhesive substances, the other adapted to the lipid residues of culinary commodities (Figure 1). In most cases, adhesive residues are quite compact and they contain a high amount of organic matter that is not mixed with clay of ceramic vessels nor sediment. Adhesive investigation comprises a multi-step analytical strategy: IR spectroscopy on a micro-sample usually allows detecting the dominating substance. Pistacia and pine resins, birch bark tar, beeswax but also copal and frankincense can thus be detected (Figure 2).

When substances characterized by bands that do not overlap are present, it may be possible to detect two different materials in a sample. The double peaks at 729-719 $cm^{-1}$ (Figure 2), indicative of long chain hydrocarbons, are easily detected, even when beeswax is mixed with a tar or a resin. As many resins contain terpenoids, with alcohol or carboxylic acid functions, IR fingerprints of different materials, especially after heating or ageing, may be convergent. IR spectroscopy may be followed by direct inlet mass spectrometry using electron impact (EI+) ionization. Birch bark tar, pistacia and pine resins, frankincense, copal as well as beeswax can be detected by this method, by comparing the obtained archaeological spectra with the spectra of known reference materials (Regert and Rolando 2002; Scalarone et al. 2003; Colombini et al. 2005, figure 2).

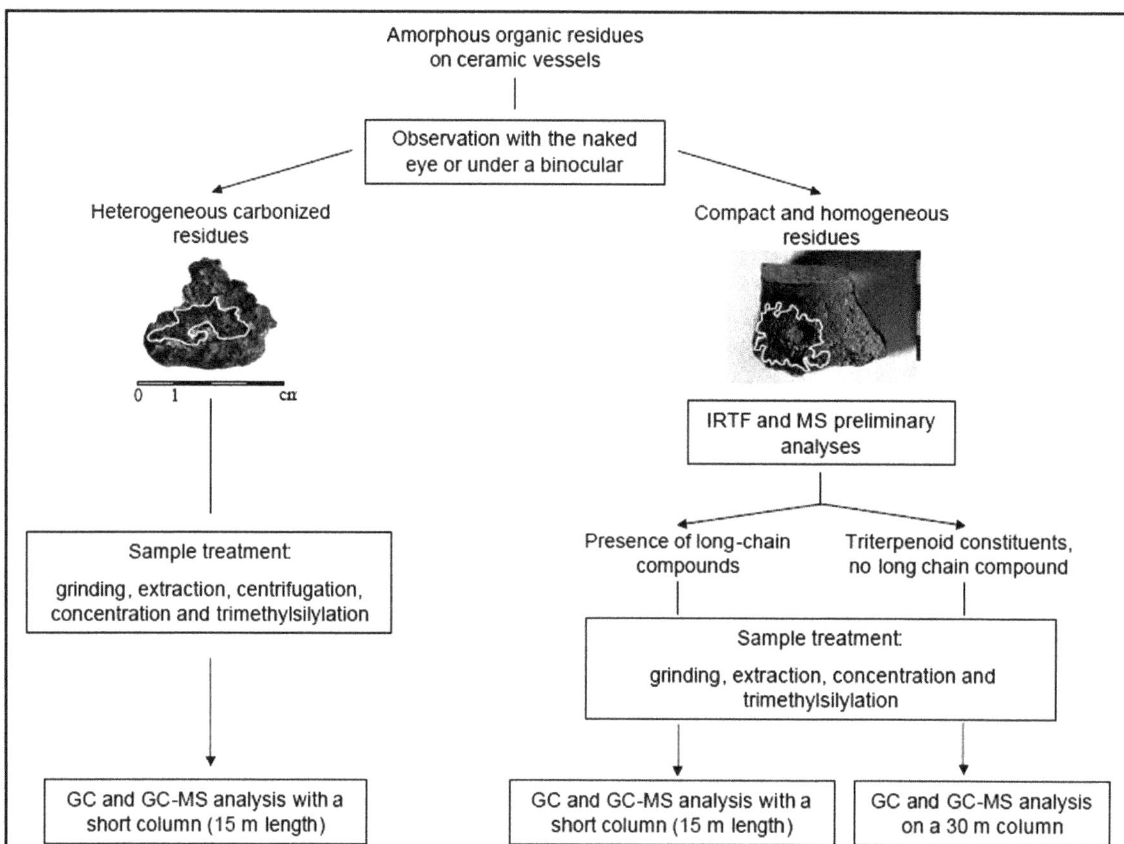


Figure 1: Analytical protocols proposed for archaeological organic residue analysis.

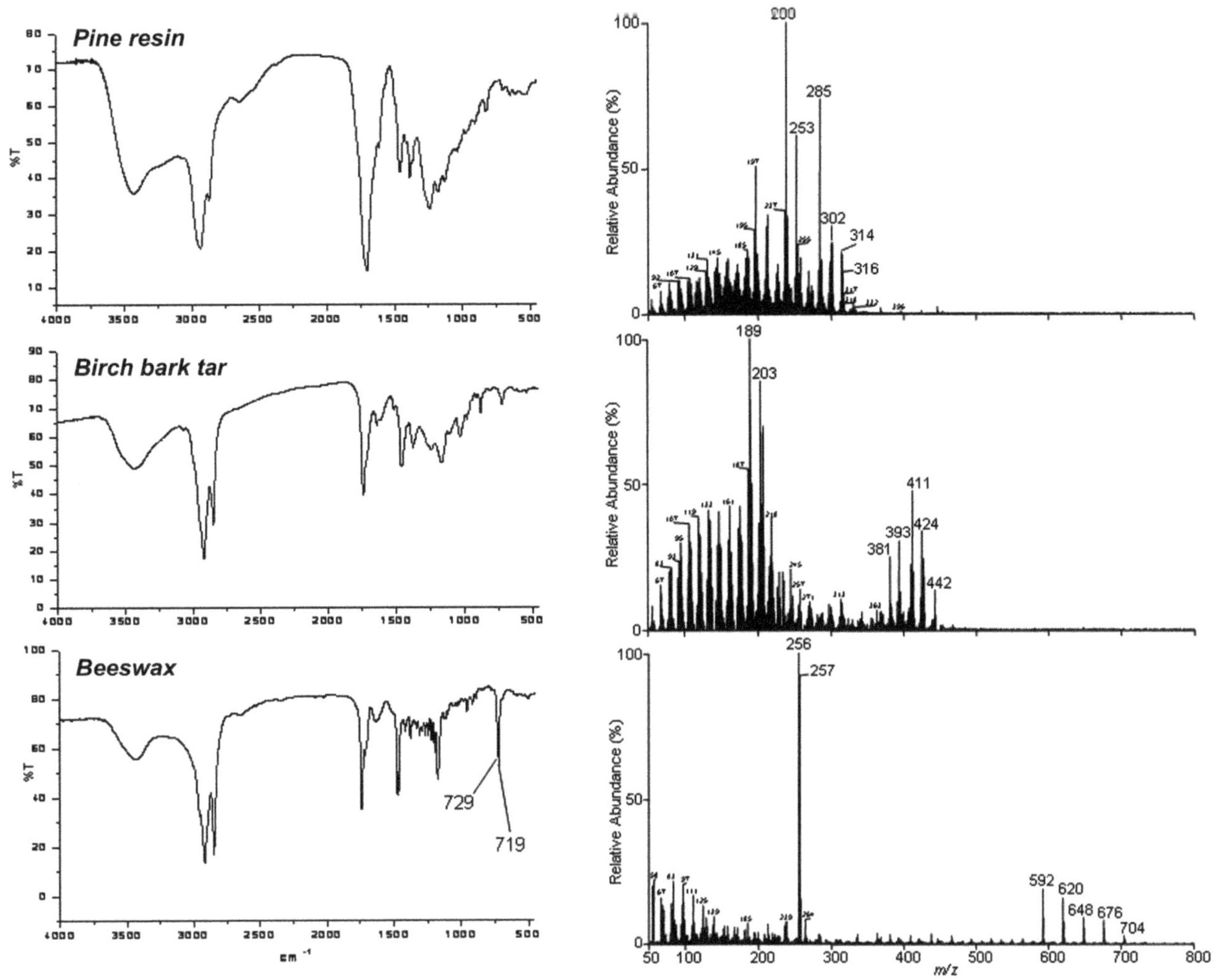


Figure 2: Infrared spectra (left) and direct inlet electron impact ionization mass spectra (right) of pine resin (above), birch bark tar (middle) and beeswax (below).

Depending on the obtained results, sample treatment and chromatographic conditions can now be chosen on the basis of the following three criteria. If the spectral fingerprint shows the presence of a resin or a tar, which are mostly solvent soluble, dichloromethane extraction followed by trimethylsilylation and concentration can be performed. The sample can then be analyzed by a standard method on an apolar column of 30 m, whatever kind of injector is used (Regert et al. 2003a). If beeswax or other products containing triacylglycerols, such as animal fats, are detected the same sample treatment can be used (extraction, derivatization and concentration), but the analysis should be performed on a shorter column (15 m) with a thin film and using a programmed flow. The use of an on-column injector is also recommended. All these parameters are important to allow the elution of the less volatile components (Regert et al. 2005). If resins known to be largely polymerized are detected, such as copal, the sample can be analyzed directly by pyrolysis gas chromatography mass spectrometry (Py/GC/MS) after in situ derivatization.[1]

Residues that look like cooking remains can be directly prepared for GC and GC/MS analysis with a protocol previously developed by other authors. In brief, charred surface residues or a piece of sherd are ground, extracted in a mixture of chloroform and methanol, centrifuged, concentrated and derivatized by trimethylsilylation.

[1] Py/GC/MS is performed on a GC/MS instrument with an attached quartz chamber in which a sample can be quickly heated to a high temperature in a oxygen-free environment. This results in the cleavage of large molecules in the sample into a series of small molecules. These are then introduced in the GC/MS part of the instrument and analyzed (cf. Chapter 5).

Analysis is then performed on a short column to be able to elute low volatile compounds, such as triacylglycerols. Wine residues, when preserved in archaeological context, are often polymerized and Py/GC or Py/GC/MS is often the only way to identify them (Garnier et al. 2003), although liquid chromatography combined with tandem mass spectrometry (LC/MS/MS) has recently also proven successful (Guash-Jané et al. 2004).

Coating agents may be treated as adhesive materials, as described above. However, next to the chemical analysis, part of the sample should be studied under a microscope to assess the thickness of the coating and its relationship with the clay of the vessel to obtain information on the method of coating, like its use in solid or liquid form.

When more information is needed on the molecular structure of some biomarkers, to obtain precise data on their origin, complementary analyses using soft ionization techniques and mass spectrometry can be deployed (Garnier et al. 2002). The study of stable isotopic ratios may also be helpful (Evershed et al. 1997a; Dudd and Evershed 1998; Evershed et al. 2002; Copley et al. 2005 a; b; c; Barnard et al. 2007)

Determination of the different molecular constituents separated and detected by GC and GC/MS is based on their retention time and the comparison of the unknown mass spectra with those in electronic libraries. When the structure of most of the molecular constituents extracted has been determined, it is then necessary to establish relationships between molecular data and to assess pottery function. In other words, it is necessary to combine the chemical data with information on the archaeology and the environment of the site. Most studies have so far concentrated on lipid profiles. This may be explained by their good preservation and relative biological specificity (Eglinton and Logan 1991). It is now possible to identify waxes, animal fats and plant oils, waxes, resins and tars in ceramic vessels. Beer and wine were also investigated (Condamin and Formenti 1978; Badler et al. 1990; Maksoud et al. 1994; McGovern et al. 1996; McGovern 1998; Garnier et al. 2003; Guash-Jané et al. 2004), but polysaccharides (Oudemans and Boon 1991) or proteins (Evershed and Tuross 1996) are still difficult to fully characterize in archaeological contexts (cf. Chapter 17).

## Interpreting Analytical Data

GC and GC/MS analyses provide a set of molecular constituents that have to be related to natural substances, but also to their strategies of procurement and production. The concept of biomarkers, which comes from geochemistry, is very useful, although it has to be adapted to an archaeological use. A biomarker may be defined in geochemistry as a specific molecular skeleton that has some biological specificity in its origin (Philp and Oung 1988). Because geochemistry usually deals with molecules that have been preserved in different sedimentary contexts for very long periods of time, the initial molecular precursors have been submitted to various processes of diagenesis. The biomolecular constituents that are identified are the result of these transformations. Most of the time they are defunctionalized, with only the hydrocarbon skeleton is preserved. When studying molecules preserved in archaeological contexts, with time scales of centuries or at most millennia, it is often possible to identify molecules that have not been transformed and that still have their alcoholic or acidic functional groups. These molecules can be identified as archaeological biomarkers.

Some constituents are the result of human actions, like the mixing or heating of natural raw materials, or of natural modification processes. These constituents should be considered anthropogenic transformation markers and natural degradation markers, respectively. Migration of soil constituents may also alter the molecular profile of the substances in the vessel; these should be considered contamination markers (Regert 1996). These four categories of molecular markers provide different levels of information on the amorphous organic materials studied (Figure 3). A set of biomarkers may be directly linked to its biological origin whereas anthropogenic transformation markers are indicators of the various transformation processes applied to the materials by ancient people. Natural degradation and contamination markers modify the primary profile of the natural substances and thus give evidence for the diagenetic processes and the degree of alteration of the substances studied. For the attribution of each molecule to a specific class of molecular markers knowledge of the chemical composition of natural substances, as well as the geochemistry of the sedimentary matrix in which the samples were preserved, is necessary. Interpretations are also based on the results obtained by the analysis of contemporary reference materials, whenever artificially transformed and altered in the laboratory by, for instance, prolonged storage at relatively high temperatures (Dudd et al. 1998; Regert et al. 1998a; b; 2001a; Chapter 7).

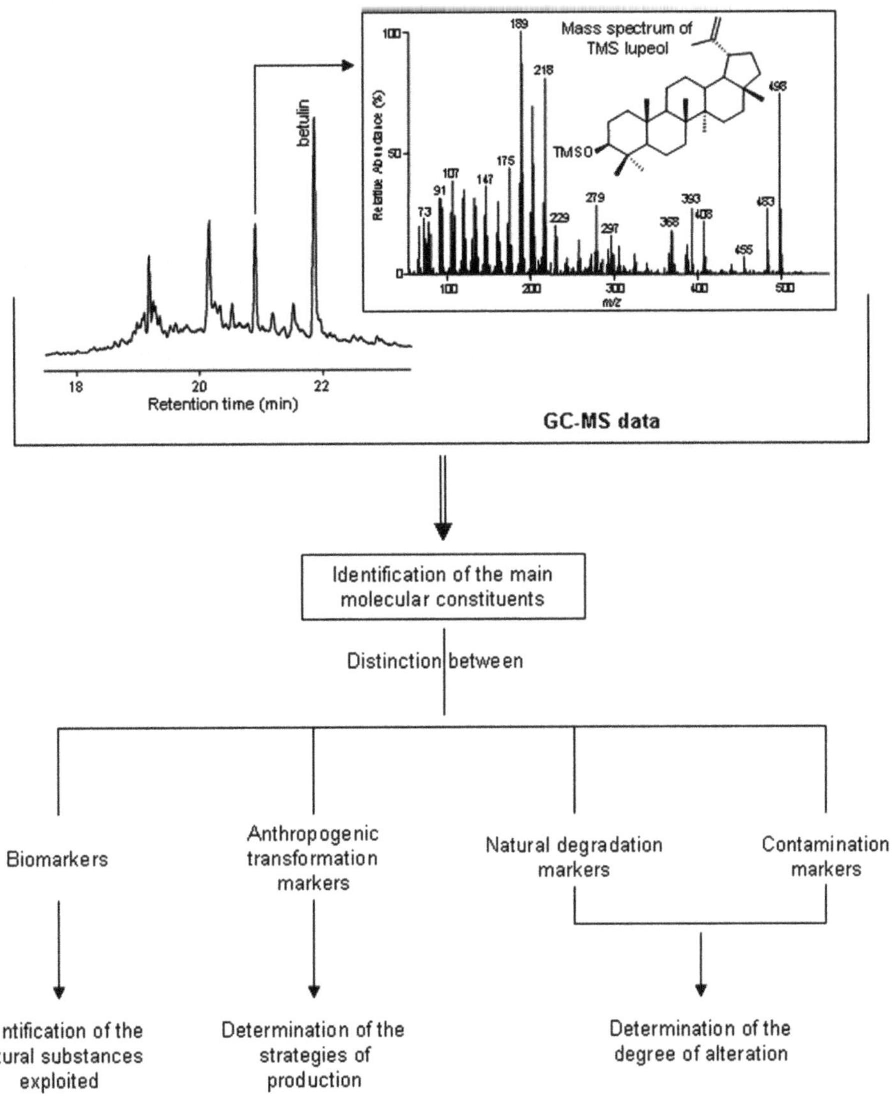


Figure 3: Scheme of the interpretation processes, based on the distinction between four categories of molecular markers, used for the identification of the natural substances processed in archaeological ceramic vessels.

Using the different analytical strategies presented above, including the classification into four categories of the biomarkers, it was possible to identify several natural substances in a series of prehistoric ceramic vessels. The data presented here correspond to the results obtained on about 100 ceramic vessels or sherds from six archaeological sites. Table 2 provides some details on the archaeological sites investigated and the samples that have been studied. When possible, vessels with known dimensions and morphology were selected for organic residue analysis. The ceramic containers from the Neolithic lacustrin sites of Chalain had already been reconstructed and studied. In the other cases, attention focused on sherds with charred residues on their surfaces. It is not yet possible to establish a relationship between content and ceramic shape as ceramic analysis is still on-going. Ninety percent of the samples contained enough lipids (more than 20 µg/g) to provide spectral and chromatographic patterns that could be related to their origin and their degree of alteration or transformation by human actions. The remaining 10% was characterized by a low content of organic matter that did not allow determination of their origin.

| Site | Location | Date | Description | Organic contents |
|---|---|---|---|---|
| Podrî l'Cortri | Belgium | Early Neolithic linear pottery | 10 samples from 8 ceramic vessels near cracks of damaged ceramic vessels | Birch bark tar (4) |
| Bercy | Paris, France | Middle Neolithic | 22 sherds with their associated charred surface residues; 9 samples from carbonized surface residues from 9 different ceramic vessels | Beeswax (5); subcutaneous animal fats (12); dairy products (5); vegetable oil (2); unidentified vegetable substance (2); unknown matter (5) |
| Chalain 3 and 4 | Jura, France | Final Neolithic (around 3000 BCE) | 84 samples from 28 ceramic vessels; charred surface residues and the associated sherds | Animal fats (14); dairy products (7); mixture of animal fats and beeswax (3); unknown matter (4) |
| Pertus II | Méailles, Alpes de Haute Provence, France | Middle Neolithic (first half of the 4th millennium BCE) | 15 sherds with organic residues | Birch bark tar (15) |
| La Fangade | Pond of Thau, Hérault, France | Final Bronze Age (12th-11th century BCE) | One sherd with yellow translucent residue | Pine resin (1) |
| Grand Aunay | Sarthe, France | Second Iron Age (2nd century BCE) | 16 sherds with residues | Birch bark tar (13); mixture of birch bark tar and beeswax (1); beeswax (1); animal fats (1) |

Table 2: Overview of the archaeological samples investigated (Pétrequin 1997; Regert et al. 2001b; Blanchet 1992; Dietsch 1997; Bosquet et al. 2001; Regert et al. 2003a; Regert and Rolando 2002). The numbers in parenthesis in the last column indicate the number of samples.

## Description of the Samples

The organic residues that were analyzed were classified into three groups. Black irregular charred surface residues inside ceramic vessels were supposed to correspond to culinary or related commodities. Such residues were mainly found on the Neolithic sites of Bercy and Chalain, but also on some sherds from the Iron Age site of Grand Aunay. These were directly analyzed, after extraction and trimethylsilylation, by GC and GC/MS. Brown to black compact residues on the inside or outside surface of ceramic vessels were assumed to be related to adhesive production or storage. These were analyzed by IR and MS followed by GC and GC/MS. The Bronze Age site of La Fangade and the Iron Age site of Grand Aunay provided most of these remnants.

A third group of residues, with the same aspect as those described above but located near ancient cracks of ceramic vessels, from the Neolithic sites of Podrî l'Cortri in Belgium and Pertus II in France, were interpreted as evidence for the use of adhesive for the repairing of ceramic vessels in ancient times. Another sherd, from Grand Aunay, preserved a thin surface coating. By combining the results of these observations with the biochemical data, owing to the methodology described above, it was possible to distinguish ceramic vessels linked with culinary commodities from those related to adhesive or resin production and storage. Most of the ancient organic remains could be attributed to the use of the vessels; some samples could be shown to be related to pottery coating and repairing.

Most of the vessels or sherds on which black and heterogeneous amorphous carbonized organic residues were observed, were characterized by the presence of several categories of biomolecular markers, namely triacylglycerols and cholesterol but also diacylglycerols, monoacylglycerols and fatty acids (Figure 4). These constituents were identified by their retention time and their mass spectrum. Some of them are biomarkers that can be ascribed to animals (triacylglycerols and cholesterol) whereas the others may be considered as degradation markers of these molecules (Evershed et al. 1992a; Heron and Evershed 1993; Charters et al. 1995; Evershed et al. 1997a; Dudd et al. 1998; Regert et al. 1998b; Dudd et al. 1999; Regert et al. 1999; Regert et al. 2001b; Evershed et al. 2002; Regert et al. 2003b).

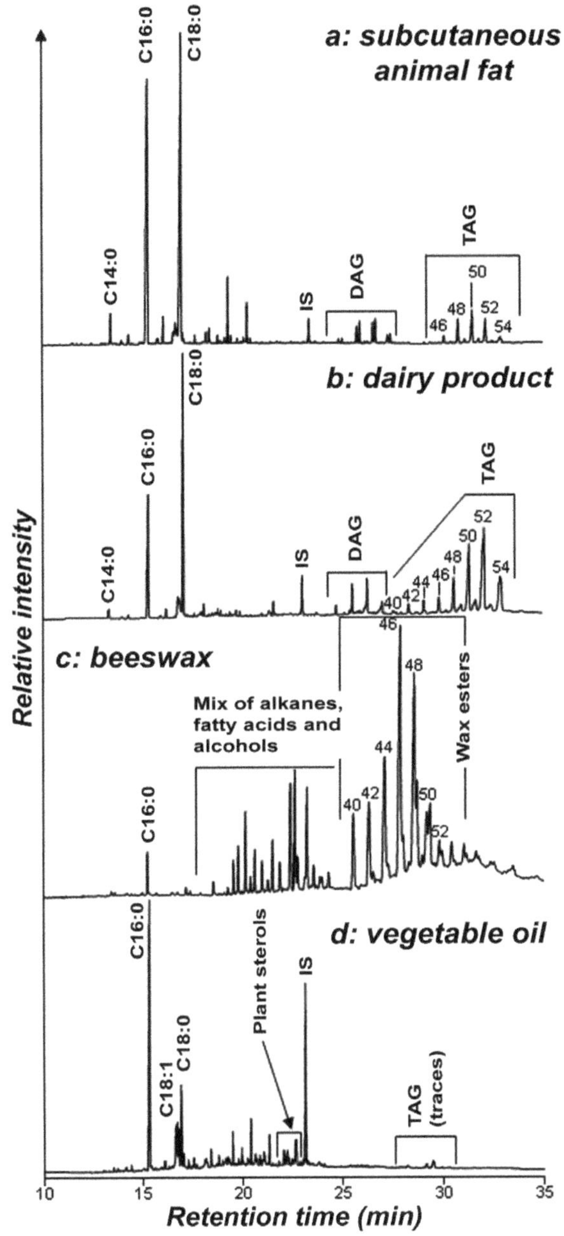


Figure 4: Chromatograms of samples from the sites of Chalain (a and b) and Bercy (c and d) characteristic of animal and plant commodities. IS: internal standard; DAG: diacylglycerols; TAG: triacyglycerols.

Where triacylglycerols were detected, two main distributions were observed: a narrow distribution (Figure 4a), which was attributed to subcutaneous animal fats, and a larger one (Figure 4b), interpreted as evidence for dairy products (Dudd and Evershed 1998; Copley et al. 2005a; b). Such chromatographic profiles were encountered in ceramic vessels from the two Neolithic sites investigated (Bercy and Chalain), but also in one sherd from the Iron Age site of Grand Aunay (Figure 5). The ceramic vessels from these three sites also provided chromatographic fingerprints for beeswax, either pure or mixed with animal fats. The identification of this substance is based on the presence of homologous series of odd-numbered *n*-alkanes, even-numbered fatty acids and palmitate esters with even carbon chains with 40-50 carbon atoms (Heron et al. 1994; Evershed et al. 1997b; Regert et al. 1999; Regert et al. 2001a; b; Evershed et al. 2003; Regert et al. 2005). In four ceramic sherds from Bercy, two different chromatographic profiles may be ascribed to vegetable products that still have to be fully identified. Two samples had a P/S ratio of about 5, consistent with a vegetable oil, together with plant sterols (Figure 4d). Two other samples were characterized by the presence of triterpenoid biomarkers, namely lupeol, α- and β-amyrin and their long chain ester derivatives, clearly indicative of plant materials. Based on these results, the black carbonized organic residues investigated may be classified in six categories (Table 2, Figure 5): subcutaneous animal fats; dairy products; beeswax; beeswax mixed with animal fats; vegetable oils and an unknown triterpenoid plant substance.

If identifying the organic residues in ancient ceramic vessels is intended to be part of a process to understand pottery function, it is necessary to establish a relationship between a residue and a function. Such relationships may be established using the properties and the possible uses known for the materials identified. Combining the presence of animal subcutaneous fats or dairy products, with that of charred surface residues and their distribution in the vessels, it may be assumed that most of these vessels were dedicated to culinary activities, such as the preparation of stews or gruels containing either fatty meat or milk. The characteristics of Chalain pottery, large jars with traces of soot on their outer surfaces, are consistent with this hypothesis (Regert et al. 1999).

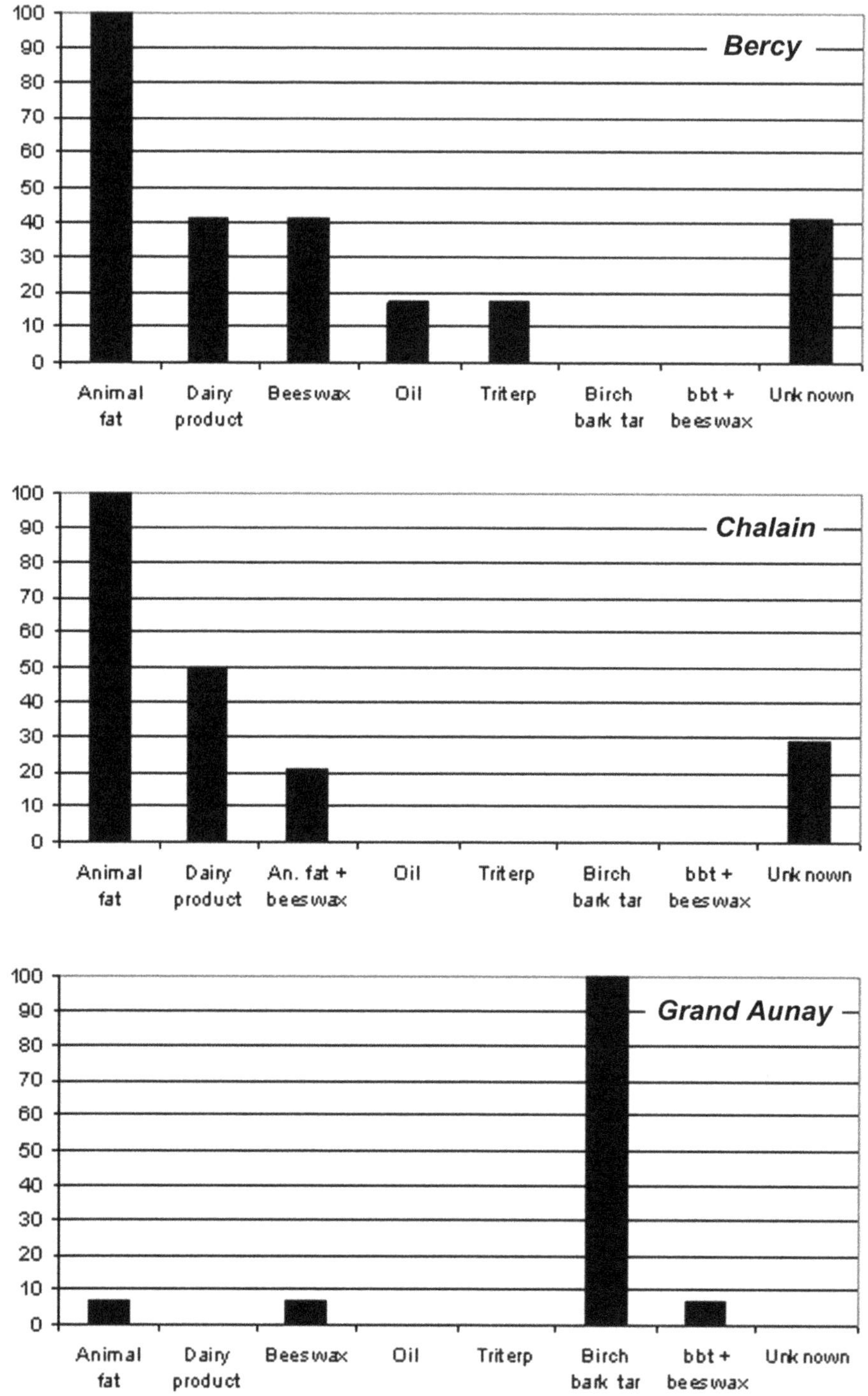


Figure 5: Histograms showing the distribution of the inferred origins of the organic residues at the three different sites discussed in this chapter relative to the most abundant material (100% on the y-axis). An. fat = animal fat; triterp = triterpenoids; bbt = birch bark tar.

The presence of beeswax is more difficult to interpret. When mixed with animal fats, it may be evidence for the use of honey in a stew, and thus for honey recipes that date back to the final Neolithic. In this period, beeswax and honey may not have been separated and traces of beeswax may be proof for the use of honey. On the other hand, beeswax may also have been used for waterproofing ceramic vessels before use. As both the charred surface residues and the organic matter trapped into the porous matrix of the pottery provided the same results, I would say that, certainly in the case of the vessels from Chalain, mixing bee products and animal fats was most likely related to culinary practices. When beeswax is identified as a pure product in a vessel, pottery function is more difficult to asses due to the various uses of this substance. Once again, beeswax may have been used as a waterproofing agent. Alternatively, the container may have been employed to store beeswax until it was used in medicines, cosmetics, adhesives, sealing agents, etc. (Regert et al. 2001a, 550).

Animal products have been identified in ancient ceramic vessels by many authors (Evershed et al. 1992a; Heron and Evershed 1993; Heron et al. 1994; Charters et al. 1995; Evershed et al. 1997a; b; Dudd and Evershed 1998; Dudd et al. 1998; Regert et al. 1998; Dudd et al. 1999; Regert et al. 1999; Craig et al. 2000; 2002; 2003; Regert et al. 2001a; b; Evershed et al. 2002; Evershed et al. 2003; Regert et al. 2003b; Copley et al. 2005a; b; c). The identification of plant material is less common. Evershed et al. (1991) recognized epicuticular leaf wax of *Brassica oleacera* in archaeological ceramic containers. The presence of vegetable oil in two sherds from Bercy has to be compared with the results of the archaeobotanical research at the site (Dietsch 1997). Among the seeds discovered at Bercy, several are from oil-producing plants such as hazelnuts, grapes, dogwood (*Cornus sanguinea*), water lilies (*Nuphar lutea*) and poppy seeds. Because some of these plants, namely dogwood, provide non-edible oils it is difficult to know whether the vessels were dedicated to culinary activities or to other purposes. Further studies are necessary to identify the oils more precisely in order to better determine the function of the ceramic vessels in which their residues were preserved.

## Pottery with Tars or Resins

When the organic residues were either compact and brown to black or yellowish and translucent, they were interpreted as related to adhesive use. Consequently, these were treated according to the analytical strategy described above (IR and MS followed by GC and GC/MS or Py/GC/MS, depending on the expected materials). It was shown that most of these remains contained triterpenoid biomarkers, mainly betulin and lupeol, and anthropogenic transformation markers, such as lupa-2,20(29)-dien-28-ol, of birch bark tar (Binder et al. 1990; Hayek et al. 1990; 1991; Charters et al. 1993; Evans and Heron 1993; Regert and Rolando 1996; Regert et al. 1998a; Aveling and Heron 1998; Grünberg et al. 1999; Regert et al. 2000; Grünberg 2002; Urem-Kotsou et al. 2002; Regert 2004; Regert et al. 2006). This substance was produced as early as the Middle Paleolithic and was used for different purposes during the Neolithic and later. The vessels from Grand Aunay containing this material may have been associated to adhesive production, storage or use. One of the samples studied contained beeswax together with birch bark tar (Figure 6). Beeswax was probably used to improve the plasticity of the adhesive. A sample from the Bronze Age site of La Fangade showed the presence of a pine resin by a set of biomolecular markers derived from abietic acid (sylvic acid) and other diterpenoid constituents (Regert 2004). Although ceramic vessels are often considered as culinary utensils, these results shed new light on pottery function and clearly show that ceramic vessels have always been employed for a wider range of activities.

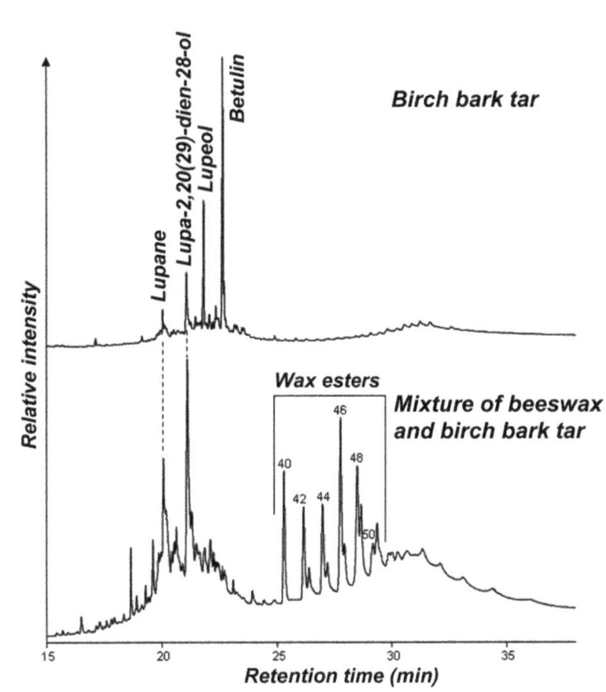


Figure 6: Chromatograms of birch bark tar and of a mixture of birch bark tar and beeswax, obtained from a sample from the Iron Age site of Grand Aunay.

Among the pottery containers and the ceramic sherds discovered at the Neolithic sites of Pertus II and Podrî l'Cortri (Table 2), a series of several sherds were noticed that preserved brown organic residues, systematically located near ancient cracks (Figure 7). Small perforations, near the part where the pottery was damaged, were made in some of these sherds in prehistoric times (Harmeyer et al. 1995, 177). They are generally considered to be holes for fitting the two parts of a vessel together with some flexible bond. All these observations led to the conclusion that the visible organic residues are probably related to one or more steps in the process of pottery repair.

Figure 7: Sherds from the Neolithic sites of Podrî l'Cortri (above) and Pertus II (below) with residues of adhesives located near ancient cracks or repairing perforations.

The analyses by IR, MS and GC followed by GC/MS on samples of the organic residue from these sherds showed that they were characterized by the presence of triterpenoids from the lupane family, indicating the presence of birch bark tar. One sherd from Grand Aunay was covered with a brown coating. The study of a section under the microscope showed that this coating was 50-200 µm thick and that it had penetrated into the ceramic vessel, indicating that it was applied as a warm liquid, or at least a viscous substance. Analysis demonstrated that this material was also made of birch bark tar.

Although most of the amorphous organic remains preserved in ceramic vessels are related to the use of the vessels, these results clearly show that they may also be related to the last stage of pottery production, the coating with birch bark, or to the repairing steps of the vessel during their life time. These results also provide new data on the possible uses of birch bark tar that was, until now, mostly known as a hafting adhesive.

## Conclusion

Amorphous organic remains have been noticed by archaeologists quite early. First mentions of such residues date from the end of the 19th and the beginning of the 20th century (Heintzel 1880; 1881; Cotte and Cotte 1917). However, because no charcoal, seeds or pollen were preserved with these remnants, they could not be studied before the introduction of analytical chemistry in the field of archaeology. Due to the unknown and complex composition of charred organic residues preserved in the surface of ceramic vessels, it was necessary to develop new and efficient methods combining chemical, geological and archaeological techniques. The method described in this chapter relies on the observation of carbonized residues in pottery containers, together with a multi-step analytical strategy based upon the use of a panel of complementary analytical techniques. The analytical data obtained provide compositional information, at a molecular level, on the archaeological samples. The attribution of the molecular constituents to four classes of markers (biomarkers, anthropogenic transformation markers, natural degradation markers and contaminants) is a powerful tool for interpreting the molecular composition in terms of strategies of procurement, production and use of natural substances by prehistoric people.

The study of samples from over 100 sherds or vessels, provided new information pointing not only at culinary purposes, but also at 'technological' activities such as the production and storage of adhesives. A range of several natural substances, either from animal or plant origin, were chemically identified in charred surface residues associated with ceramic vessels or sherds, and also in organic matter trapped in the porous matrix of pottery vessels. Subcutaneous animal fat, dairy products, beeswax, plant oil, birch bark tar and pine resin were demonstrated to be processed. Depending on their aspect, chemical composition and position on the ceramics as well as the shape of the container in which they were identified, these residues could be correlated to different pottery functions. Particularly, a clear distinction could be established between culinary vessels, containing animal fats or dairy products, and ceramics dedicated to adhesive production, storage or use, preserving birch bark tar or pine resin. These last commodities may be evidence for repairing or coating ceramic vessels, especially when they are located near ancient cracks or regularly spread over the surface of the vessel.

In some cases, however, pottery function is more difficult to assess. The mixtures of beeswax and animal fats may be either attributed to culinary recipes, such as honey stew, or to the use of beeswax as a waterproofing agent. Vegetable oils may indicate the storage of oil for culinary purpose as early as the Middle Neolithic, but also the possible exploitation of non-edible oils (dogwood) for purposes such as lighting, waterproofing, tanning etc.

Identifying the former contents of pottery containers is of great importance for establishing the story of the exploitation of natural products through time, especially dairy products and vegetable oils, but also for studying improvements realized in the technological field. The transition from the use of birch bark tar to that of pine resins, in the Bronze Age, and the mixing of birch bark tar with beeswax, in the Iron Age, emphasizes the diversification of the materials used as adhesives when metallurgy was developed. Such evolution may be partly due to a better control of heating processes, necessary for the production of both these new adhesives as well as bronze and iron. The research project described above proves the importance and value of an interdisciplinary approach, not only to reveal new information on pottery function but also to better understand the exploitation of natural substances, such as animal fats, vegetable oils, beeswax, plant resins and tars, that do not leave easily detectable archaeological traces.

## References


**Arnold**, D.E. (1985). *Ceramic Theory and Cultural Process. New Studies in Archaeology*. Cambridge: Cambridge University Press.

**Aveling**, E. and C. Heron (1998). Identification of Birch Bark Tar at the Mesolithic Site of Star Carr. Ancient *Biomolecules* **2,1**: 69-80.

**Badler**, V.R., P.R. McGovern and R.H. Michel (1990). Drink and be Merry! Infrared Spectroscopy and Ancient Near Eastern wine. *MASCA Research Papers in Science and Archaeology* 7: 25-36.

**Barnard**, H., S.H. Ambrose, D.E. Beehr, M.D. Forster, R.E. Lanehart, R.E. Parr, M.E. Malainey, M. Rider, C. Solazzo and R.M Yohe II (2007). Mixed Results of Seven Methods for Organic Residue Analysis Applied to One Vessel with the Residue of a Known Foodstuff. *Journal of Archaeological Science* **34**: 28-37.

**Binder**, D., G. Bourgeois, F. Benoist and C. Vitry (1990). Identification de brai de bouleau (Betula) dans le Néolithique de Giribaldi (Nice, France) par la spectrométrie de masse. *Revue d'archéométrie* **14**: 37-42.

**Blanchet**, J.C. (1992). *Les pirogues néolithiques de Bercy*. Exposition à la Mairie de Paris du XIIème arrondissement.

**Bosquet**, D., N. Dubois, J. Jadin and M. Regert (2001). Identification de brai de bouleau sur quatre vases du site rubané de Fexhe-le-Haut-Clocher 'Podrî l'Cortri'. *Notae Praehistoricae* **21**: 119-127

**Bourgeois**, G. and P. Gouin (1995). Résultats d'une analyse de traces organiques fossiles dans une 'faisselle' harapéenne. *Paléorient* **21,1**: 125-128.

**Charters**, S., R.P. Evershed and L.J. Goad (1993). Identification of an Adhesive used to Repair a Roman Jar. *Archaeometry* **35,1**: 91-101

**Charters**, S., R.P. Evershed, P.W. Blinkhorn and V. Denham (1995). Evidence for the Mixing of Fats and Waxes in Archaeological Ceramics. *Archaeometry* **37**: 113-127.

**Colombini**, M.P., F. Modugno and E. Ribechini (2005). Direct Exposure Electron Ionization Mass Spectrometry and Gas Chromatography/Mass Spectrometry Techniques to Study Organic Coatings on Archaeological Amphorae. *Journal of Mass Spectrometry* **40**: 675-687.

**Condamin**, J., F. Formenti, M.O. Metais, M. Michel and P. Blond (1976). The Application of Gas Chromatography to the Tracing of Oil in Ancient Amphorae. *Archaeometry* **18,2**: 195-201.

**Condamin**, J. and F. Formenti (1978). Détection du contenu d'amphores antiques (huiles, vins). Etude méthodologique. *Revue d'archéométrie* **2**: 43-58.

**Copley**, M.S., R. Berstan, S.N. Dudd, V. Straker, S. Payne and R.P. Evershed (2005a). Dairying in Antiquity I. Evidence from Absorbed Lipid Residues dating to the British Iron Age. *Journal of Archaeological Science* **32**: 485-503.

**Copley**, M.S., R. Berstan, S.N. Dudd, V. Straker, S. Payne and R.P. Evershed (2005b). Dairying in Antiquity I. Evidence from Absorbed Lipid Residues dating to the British Iron Age. *Journal of Archaeological Science* **32**: 505-521.

**Copley**, M.S., R. Berstan, S.N. Dudd, V. Straker, S. Payne and R.P. Evershed (2005c). Dairying in Antiquity III. Evidence from Absorbed Lipid Residues dating to the British Iron Age. *Journal of Archaeological Science* **32**: 523-546.

**Cotte**, J. and C. Cotte (1917. *La caverne de l'Adaouste. Première annexe. Analyses de résidus organiques de l'époque néolithique.* Paris: Extrait des Bulletins et Mémoires de la Société d'Anthropologie de Paris, Séance du 15 mai 1917.

**Craig**, O.E., J. Mulville, M.P. Pearson, R. Sokol, K. Gelsthorpe, R. Stacey and M.J. Collins (2000). Detecting Milk Proteins in Ancient Pots. *Nature* **408**: 312.

**Craig**, O.E. (2002). The Development of Dairying in Europe. Potential Evidence from Food Residues Ceramics. *Documenta Praehistorica* **29**: 97-107.

**Craig**, O.E., J. Chapman, A. Figler, P. Patay, G. Taylor and M.J. Collins (2003). 'Milk Jugs' and other Myths of the Copper Age of Central Europe. *European Journal of Archaeology* **6,3**: 251-265.

**Diallo**, B., M. Vanhaelen and O.P. Gosselain (1995). Plant Constituents involved in Coating Practices among traditional African Potters. *Experientia* **51,1**: 95-97.

**Dietsch**, M.F. (1997). *Milieux humides pré- et protohistoriques dans le bassin parisien: l'étude des diaspores.* Thèse Université Paris X.

**Dudd**, S.N. and R.P. Evershed (1998). Direct Demonstration of Milk as Element of Archaeological Economies. *Science* **282**: 1478-1481.

**Dudd**, S.N., M. Regert, R.P. Evershed (1998). Assessing Microbial Contributions to Absorbed Acyl Lipids During Laboratory Degradations of Fats and Oils and Pure Triacylglycerols Absorbed in Ceramic Potsherds. Organic *Geochemistry* **29**: 1345-1354.

**Dudd**, S.N., R.P. Evershed and A.M. Gibson (1999). Evidence for Varying Patterns of Exploitation of Animal Products in Different Prehistoric Pottery Traditions Based on Lipids Preserved in Surface and Absorbed Residues. *Journal of Archaeological Science* **26**: 1473-1482.

**Duplaix-Rata** A. (1997). Les dépôts alimentaires carbonisés. In P. Pétrequin (ed.). *Les sites littoraux néolithiques de Clairvaux-les-Lacs et de Chalain (Jura) III. Chalain station 3, 3200-2900 av. J.-C., volume 2, sous la direction de Pierre Pétrequin.* Paris: Editions de la Maison des Sciences de l'Homme, pp. 733-745.

**Eglinton**, G. and G.A. Logan (1991). Molecular Preservation. *Philosophical Transactions of the Royal Society B* **333,1268**: 315-328.

**Evans**, K. and C. Heron (1993). Glue, Disinfectant and Chewing Gum. Natural Products Chemistry in Archaeology. *Chemistry & Industry* **12**: 446-449.

**Evershed**, R.P., C. Heron and L.J. Goad (1990). Analysis of Organic Residues of Archaeological Origin by High-temperature Gas Chromatography/Mass Spectrometry. *Analyst* **115,10**: 1339-1342.

**Evershed**, R.P., C. Heron, S. Charters and L.J. Goad (1991). Epicuticular Wax Components in Potsherds as Chemical Indicators of Leafy Vegetables in Ancient Diets. *Antiquity* **65**: 540-544.

**Evershed**, R.P., C. Heron, S. Charters and L.J. Goad (1992a). The Survival of Food Residues, New Methods of Analysis, Interpretation and Application. *Proceedings of the British Academy* **77**: 187-208.

**Evershed**, R.P., S. Charters, C. Heron and L.J. Goad (1992b). Chemical Analysis of Organic Residues in Ancient Pottery. Methodological Guidelines and Applications. In R. White and H. Page (eds.). *Organic Residues in Archaeology. Their Identification and Analysis.* London: UKIC Archaeology Section, pp. 11-25.

**Evershed**, R.P. and N. Tuross (1996). Proteinaceous Material from Potsherds and Associated Soils. *Journal of Archaeological Science* **23**: 429-436.

**Evershed**, R.P., H.R. Mottram, S.N. Dudd, S. Charters, A.W. Stott, G.J. Lawrence, A,M, Gibson, A. Conner, P.W. Blinkhorn and V. Reeves (1997a). New Criteria for the Identification of Animal Fats Preserved in Archaeological Pottery. *Naturwissenschaften* **84**: 402-406.

**Evershed**, R.P., S.J. Vaughan, S.N. Dudd and J.S. Soles (1997b). Fuel for Thought? Beeswax in Lamps and Conical Cups from Late Minoan Crete. *Antiquity* **71**: 979-985.

**Evershed**, R.P., S.N. Dudd, M.S. Copley, R. Berstan, A.W. Stott, H. Mottram, S.A. Buckley and Z. Grossman (2002). Chemistry of Archaeological Animal Fats. *Accounts of Chemical Research* **35**: 660-668.

**Evershed**, R.P., S.N. Dudd, V.R. Anderson-Stojanovic and E.R. Gebhard (2003). New Chemical Evidence for the Use of Combed Ware Pottery Vessels as Beehives in Ancient Greece. *Journal of Archaeological Science* **30**: 1-12.

**Gardin**, J.C. (1979). *Une archéologie théorique.* Paris: Hachette Littérature.

**Garnier**, N., C. Cren-Olivé, C. Rolando and M. Regert (2002). Characterization of Archaeological Beeswax by Electron Ionization and Electrospray Ionization Mass Spectrometry. *Analytical Chemistry* **74**: 4868-4877.

**Garnier**, N., P. Richardin, V. Cheynier and M. Regert (2003). Characterization of Thermally Assisted Hydrolysis and Methylation Products of Polyphenols from Modern and Archaeological Vine Derivatives using Gas Chromatography-Mass Spectrometry. *Analytica Chimica Acta* **493**: 137-157.

**Grünberg**, J.M., H. Graetsch, U. Baumer and J. Koller (1999). Untersuchung der mittelpaläolithischen 'Hartzreste' von Königsaue. Ldkr. Aschersleben-Stassfurt. *Jahresschrift für mitteldeutsche Vorgeschichte* **81**: 7-38.

**Grünberg**, J.M. (2002). Middle Palaeolithic Birch-bark Pitch. *Antiquity* **76**: 15-16.

**Guash-Jané**, M.R., M. Ibern-Gomèz, C. Andrès-Lacueva, O. Jáuregui and R.M. Lamuela-Raventós (2004). Liquid Chromatography with Mass Spectrometry in Tandem Mode Applied for the Identification of Wine Markers in Residues from Ancient Egyptian Vessels. *Analytical Chemistry* **76**: 1672-1677.

**Harmeyer**, B., M. Maggetti and J. Weiss (1995). La céramique. In W.E. Stöckli, U. Niffeler and E. Gross-Klee (eds.). *La Suisse du Paléolithique à l'aube du Moyen-Age Néolithique.* Bâle: Société suisse de préhistoire et d'archéologie, pp. 178-183.

**Hayek**, E.W.H., P. Krenmayr, H. Lohninger, U. Jordis, W. Moche and F. Sauter (1990). Identification of Archaeological and Recent Wood Tar Pitches using Gas Chromatography/Mass Spectrometry and Pattern Recognition. *Analytical Chemistry* **62**: 2038-2043.

**Hayek**, E.W.H., P. Krenmayr, H. Lonhinger, U. Jordis, F. Sauter and W. Moche (1991). GC/MS and Chemometrics in Archaeology. Investigation of Glue on Two Copper-age Arrowheads. *Fresenius Journal of Analytical Chemistry* **340,3**: 153-156.

**Heintzel**, C. (1880). Urnenhartz, Fettgehalt der Urnen, eine Goldmünze und Gletscherspuren. *Zeitschrift für Ethnologie* 12: 375-378.

**Heintzel**, C. (1881). Urnenhartz aus dem Urnenfelde von Borstel bei Stendal. *Zeitschrift für Ethnologie* **13**: 241-242.

**Heron**, C. and R.P. Evershed (1993). The Analysis of Organic Residues and the Study of Pottery Use. *Archaeological Method and Theory* **5**: 247-284.

**Heron**, C., N. Nemcek, K.M. Bonfield, D. Dixon and B.S. Ottaway (1994). The Chemistry of Neolithic Beeswax. *Naturwissenschaften* **81**: 266-269.

**Maksoud**, S.A., M.N. El Hadidi and W.M. Amer (1994). Beer from the Early Dynasties (3500-3400 cal. B.C.) of Upper Egypt detected by Archaeochemical Methods. *Vegetation History and Archaeobotany* **3,4**: 219-224.

**McGovern**, P. (1998) Wine for Eternity. How Molecular Archaeologists identified the Contents of Vessels found in the Tomb of an Egyptian King. *Archaeology* **51,4**: 28-32.

**McGovern**, P., D.L. Glusker, L.J. Exner and M.M. Voigt (1996). Neolithic Resinated Wine. *Nature* **381**: 480-481.

**Orton**, C., P. Tyers and A. Vince (1993). *Pottery in Archaeology.* Cambridge: Cambridge University Press.

**Oudemans**, T.F.M. and J.J. Boon (1991). Molecular Archaeology. Analysis of Charred (Food) Remains from Prehistoric Pottery by Pyrolysis-Gas Chromatography/Mass Spectrometry. *Journal of Analytical and Applied Pyrolysis* **20**: 197-227.

**Pétrequin**, P. (ed.) (1997). *Les sites littoraux néolithiques de Clairvaux-les-Lacs et de Chalain (Jura) III. Chalain station 3, 3200-2900 av. J.-C., volume 2, sous la direction de Pierre Pétrequin.* Paris: Maison des sciences de l'homme.

**Philp**, R.P. and J.N. Oung (1988). Biomarkers. Occurrence, Utility, and Detection. *Analytical Chemistry* **60,15**: 887A-896A.

**Regert**, M. (1996). *Les composés organiques en préhistoire: nouvelles approches analytiques.* Thèse Université Paris X, p. 351.

**Regert**, M. and C. Rolando (1996). Archéologie des résidus organiques: de la chimie analytique à l'archéologie, un état de la question. *Techne* **3**: 118-128.

**Regert**, M., J.M. Delacotte, M. Menu, P. Pétrequin and C. Rolando (1998a). Identification of Neolithic Hafting Adhesives from Two Lake Dwellings at Chalain (Jura, France). *Ancient Biomolecules* **2**: 81-96.

**Regert**, M., H.A. Bland, S.N. Dudd, P.F. van Bergen and R.P. Evershed (1998b). Free and Bound Fatty Acid Oxidation Products in Archaeological Ceramic Vessels. *Proceedings of the Royal Society London B* **265**: 2027-2032.

**Regert**, M., S.N. Dudd, P. Pétrequin and R.P. Evershed (1999). Fonction des céramiques et alimentation au Néolithique final sur les sites de Chalain: de nouvelles voies d'étude fondées sur l'analyse chimique des résidus organiques conservés dans les poteries. *Revue d'Archéométrie* **23**: 91-99.

**Regert**, M., N. Garnier, D. Binder and P. Pétrequin (2000). Les adhésifs néolithiques: quels matériaux utilisés, quelles techniques de production dans quel contexte social? In *Actes des XXèmes Rencontres Internationales d'Archéologie et d'Histoire d'Antibes 'Arts du feu et productions artisanales'. Octobre 1999.* Juan-les-Pins: Editions APDCA, pp. 586-604.

**Regert**, M., S. Colinart, L. Degrand and O. Decavallas (2001a). Chemical Alteration and Use of Beeswax through Time. Accelerated Ageing Tests and Analysis of Archaeological Samples from Various Environmental Contexts. *Archaeometry* **43,4**: 549-569

**Regert**, M., S.N. Dudd, P. van Bergen, P. Pétrequin and R.P. Evershed (2001b). Investigations of Solvent Extractable Lipids and Insoluble Polymeric Components: Organic Residues in Neolithic Ceramic Vessels from Chalain (Jura, France). *British Archaeological Reports S* **939**: 78-90.

**Regert**, M. and C. Rolando (2002). Identification of Archaeological Adhesives Using Direct Inlet Electron Ionization Mass Spectrometry. *Analytical Chemistry* **74**: 965-975.

**Regert**, M., S. Vacher, C. Moulherat and O. Decavallas (2003a). Study of Adhesive Production and Pottery

Function during Iron Age at the site of Grand Aunay (Sarthe, France). *Archaeometry* **45**: 101-120.

**Regert**, M., N. Garnier, O. Decavallas, C. Cren-Olivé and C. Rolando (2003b). Structural Characterization of Lipid Constituents from Natural Substances Preserved in Archaeological Environments. *Measurement Science and Technology* **14**: 1620-1630.

**Regert**, M. (2004). Investigating the History of Prehistoric Glues through Gas Chromatography-Mass Spectrometry. *Journal of Separation Science* **27**: 244-254.

**Regert**, M., J. Langlois and S. Colinart (2005). Characterisation of Wax Works of Art by Gas Chromatographic Procedures. *Journal of Chromatography A* **1091**: 124-136.

**Regert**, M., V. Alexandre, N. Thomas and A. Lattuati-Derieux (2006). Molecular Characterisation of Birch Bark Tar by Headspace Solid-phase Microextraction Gas Chromatography-Mass Spectrometry. A New Way for Identifiying Archaeological Glues. *Journal of Chromatography A* **1101**: 245-253.

**Rice**, P.M. (1987). *Pottery Analysis. A Sourcebook.* Chicago-London: The University of Chicago Press.

**Scalarone**, D.J. van der Horst, J.J. Boon and O. Chiantore (2003). Direct-temperature Mass Spectrometric Detection of Volatile Terpenoids and Natural Terpenoid Polymers in Fresh and Artificially Aged Resins. *Journal of Mass Spectrometry* **38**: 607-617.

**Schiffer**, M.B. (1990). The Influence of Surface Treatment on Heating Effectiveness of Ceramic Vessels. *Journal of Archaeological Science* **17,4**: 373-382.

**Sigaut**, F. (1991). Un couteau ne sert pas à couper mais en coupant. Structure, fonctionnement et fonction dans l'analyse des objets. In P. Pétrequin, P. Fluzin, J. Thiriot and P. Benoit (eds.). *25 ans d'études technologiques en préhistoire. XIè Rencontres Internationales d'Archéologie et d'Histoire d'Antibes*. Juan-les-Pins: Editions APDCA, pp. 21-34.

**Skibo**, J.M. (1992). *Pottery Function. A Use-alteration Perspective.* New York: Plenum Press.

**Urem-Kotsou**, D., B. Stern, C. Heron and K. Kotsakis (2002). Birch-bark Tar at Neolithic Makriyalos, Greece. *Antiquity* **76**: 962-967.

**Vogt**, E. (1949). The Birch Bark as a Source of Raw Material during the Stone Age. *Proceedings of the Prehistoric Society* **5**: 50-51.

# CHAPTER SEVEN
# Fatty Acid Analysis of Archaeological Residues: Procedures and Possibilities

**M.E. Malainey**

Mary Malainey, Associate Professor; Department of Anthropology, Brandon University; 270-18th Street; Brandon, Manitoba, R7A 6A9; Canada. This research was made possible by the generous support of R.P. Pryzbylski, the Department of Human Nutritional Sciences (formerly Foods and Nutrition) and G. Monks, Department of Anthropology, University of Manitoba. M. Quigg, R. Buss, F. Suffridge, A. Malof and others provided reference material and experimental residues. The Social Sciences and Humanities Research Council of Canada provided funding in the form of a Standard Research Grant and Post-Doctoral and Doctoral Fellowships as well as an UM-SSHRC Research Grant. The Canada Foundation for Innovation (CFI) New Opportunities and the CFI-Canada Research Chair Infrastructure Funds as well as the Canada Research Chair Program and the Brandon University Research Council currently support this research.

Archaeological food residues extracted from areas of fat accumulation in artifacts can be characterized on the basis of relative fatty acid composition. Compositions of ancient residues are compared to experimental residues subjected to periods of oven storage, which simulates the effects of oxidative decomposition over time. Levels of medium and very long chain saturated fatty acids, C18:0 and C18:1 isomers indicate the fat content of the material of origin and probable presence of animal or plant material. This technique performs well in blind tests of decomposed residues of previously unknown foods and identification criteria remain valid over time.

Gas chromatography is an effective and efficient method of examining fatty acids in the form of methyl ester derivatives. Instruments are widely available and relatively inexpensive to obtain and operate. Gas chromatography has long been used to determine the fatty acid composition of archaeological residues. While absolute identifications are not possible, a wide range of archaeological residues can be rapidly categorized. Fatty acid compositions of cooking residues change over time; but these variations can be modeled. In particular, decreases in the relative amounts of C18:1 isomers in decomposing residues strongly correlate with logarithmic curves; the functions can then be used to extrapolate further change. Better than expected preservation of monounsaturated and, occasionally, polyunsaturated fatty acids is observed in the cooking residues of certain plant and plant and meat combinations, likely due to the presence of antioxidants.

With careful selection of archaeological samples for residue analysis, good preservation of residues is possible. The effects of microbes found in parkland, prairie and forest soils appear to be mediated by the reduced availability of oxygen in a burial environment. With a reference collection of decomposed foodstuffs from the region, one can establish possible origins of the residues and eliminate others. If desired, other methods of analysis can be employed to provide more precise identifications by confirming the fatty acid identifications and targeting molecules that can serve as biomarkers.

## Previous Research

The major constituents of fats and oils are fatty acids that usually occur in nature as triacylglycerides, consisting of three fatty acids attached to a glycerol molecule by ester-linkages. Their insolubility in water and relative abundance compared to other classes of lipids, such as sterols and waxes, make fatty acids suitable for residue analysis. Since employed by Condamin et al. (1976), gas chromatography has been used extensively to analyze the fatty acid component of archaeological residues (cf. Chapter 5).

The composition of uncooked plants and animals provides important baseline information, but it is not possible to compare modern uncooked plants and animals with highly degraded archaeological residues of prepared foodstuffs. Unsaturated fatty acids, which are found widely in fish and plants, decompose more readily than saturated fatty acids, sterols or waxes. In the course of decomposition, simple addition reactions might occur changing double into single bonds (Solomons 1980), or peroxidation might lead to the formation of a variety of volatile and non-volatile products which continue to degrade (Frankel 1991). Peroxidation occurs most readily in polyunsaturated fatty acids.

Attempts have been made to identify archaeological residues using criteria that discriminate uncooked foods (Marchbanks 1989; Skibo 1992; Loy 1994). Marchbanks' (1989) percent of saturated fatty acids (%S) criteria has been applied to residues from a variety of materials including pottery, stone tools and burned rocks (Marchbanks 1989; Marchbanks and Quigg 1990; Collins et al. 1990). Skibo (1992, 89) could not apply the %S technique and used two ratios of fatty acids: C18:0/C16:0 and C18:1/C16:0. He reported that it was possible to link the uncooked foods with residues extracted from modern cooking pots used to prepare one type of food. However, the ratios could not identify food mixtures. The utility of these ratios did not extend to residues extracted from archaeological potsherds because the ratios of the major fatty acids in the residue changed with decomposition (Skibo 1992, 97). Loy (1994) proposed the use of a Saturation Index (SI), determined by the ratio:

$$SI = 1 - \left[ \frac{(C18:1 + C18:2)}{C12:0 + C14:0 + C16:0 + C18:0)} \right]$$

He admitted that the poorly understood decompositional changes to the suite of fatty acids make it difficult to develop criteria for distinguishing animal and plant fatty acid profiles in archaeological residues. The major drawback of the distinguishing ratios proposed by Marchbanks (1989), Skibo (1992) and Loy (1994) is they have never been empirically tested. The proposed ratios are based on criteria that discriminate food classes on the basis of their original fatty acid composition. The resistance of these criteria to the effects of decompositional changes has not been demonstrated. Skibo (1992) found that his fatty acid ratio criteria could not be used to identify highly decomposed archaeological samples.

In order to identify a fatty acid ratio unaffected by degradation processes, Patrick et al. (1985) simulated the long-term decomposition of one sample and monitored the resulting changes. An experimental cooking residue of seal was prepared and degraded in order to identify a stable fatty acid ratio. Patrick et al. (1985) found that the ratio of two C18:1 isomers, oleic and vaccenic acid, did not change with decomposition and this fatty acid ratio was used to identify an archaeological vessel residue as seal. While the fatty acid composition of uncooked foods must be known, Patrick et al. (1985) showed that the effects of cooking and decomposition over long periods of time on the fatty acids must also be understood.

## Development of the Identification Criteria

As the first stage in developing the identification criteria, fatty acid compositions of more than 130 uncooked native food plants and animals from Western Canada were determined using gas chromatography (Malainey 1997; Malainey et al. 1999a). When the fatty acid compositions of modern food plants and animals were subject to cluster and principal component analyses, the resultant groupings generally corresponded to divisions that exist in nature (Table 1). Clear differences in the fatty acid composition of large mammal fat, large herbivore meat, fish, plant roots, greens and berries/seeds/nuts were detected, but the fatty acid composition of meat from medium-sized mammals resembled berries/seeds/nuts.

| Cluster A | Subcluster | | | |
|---|---|---|---|---|
| | I | II | III | IV |
| Type | Mammal fat and marrow | Large herbivore meat | Fish | Fish |
| C16:0 | 19.90 | 19.39 | 16.07 | 14.10 |
| C18:0 | 7.06 | 20.35 | 3.87 | 2.78 |
| C18:1 | 56.77 | 35.79 | 18.28 | 31.96 |
| C18:2 | 7.01 | 8.93 | 2.91 | 4.04 |
| C18:3 | 0.68 | 2.61 | 4.39 | 3.83 |
| VLCS | 0.16 | 0.32 | 0.23 | 0.15 |
| VLCU | 0.77 | 4.29 | 39.92 | 24.11 |

| Cluster B | Subcluster | | | | | |
|---|---|---|---|---|---|---|
| | V | VI | VII | VIII | IX | X |
| Type | Berries and nuts | Mixed | Seeds and berries | Roots | Seeds | Mixed |
| C16:0 | 3.75 | 12.06 | 7.48 | 19.98 | 7.52 | 10.33 |
| C18:0 | 1.47 | 2.36 | 2.58 | 2.59 | 3.55 | 2.43 |
| C18:1 | 51.14 | 35.29 | 29.12 | 6.55 | 10.02 | 15.62 |
| C18:2 | 41.44 | 35.83 | 54.69 | 48.74 | 64.14 | 39.24 |
| C18:3 | 1.05 | 3.66 | 1.51 | 7.24 | 5.49 | 19.77 |
| VLCS | 0.76 | 4.46 | 2.98 | 8.50 | 5.19 | 3.73 |
| VLCU | 0.25 | 2.70 | 1.00 | 2.23 | 0.99 | 2.65 |

| Cluster C | Subcluster | | | | |
|---|---|---|---|---|---|
| | XI | XII | XIII | XIV | XV |
| Type | Greens | Berries | Roots | Greens | Roots |
| C16:0 | 18.71 | 3.47 | 22.68 | 24.19 | 18.71 |
| C18:0 | 2.48 | 1.34 | 3.15 | 3.66 | 5.94 |
| C18:1 | 5.03 | 14.95 | 12.12 | 4.05 | 3.34 |
| C18:2 | 18.82 | 29.08 | 26.24 | 16.15 | 15.61 |
| C18:3 | 35.08 | 39.75 | 9.64 | 17.88 | 3.42 |
| VLCS | 6.77 | 9.10 | 15.32 | 18.68 | 43.36 |
| VLCU | 1.13 | 0.95 | 2.06 | 0.72 | 1.10 |

Table 1 (this and previous page): Summary of average fatty acid compositions of modern food groups generated by hierarchical cluster analysis. VLCS = very long chain (C20, C22 and C24) saturated fatty acids, VLCU = very long chain (C20, C22 and C24) unsaturated fatty acids.

Samples in cluster A, the large mammal and fish cluster had elevated levels of C16:0 and C18:1 (Table 1). Divisions within this cluster stemmed from the very high level of C18:1 isomers in fat, high levels of C18:0 in bison and deer meat and high levels of very long chain unsaturated fatty acids (VLCU) in fish. Differences in the fatty acid composition of plant roots, greens and berries/seeds/nuts reflect the amounts of C18:2 and C18:3ω3 present. The berry, seed, nut and small mammal meat samples appearing in cluster B have very high levels of C18:2, ranging from 35% to 64% (Table 1). Samples in subclusters V, VI and VII have levels of C18:1 isomers from 29% to 51%. Plant roots, plant greens and some berries appear in cluster C. All cluster C samples have moderately high levels of C18:2; except for the berries in subcluster XII, levels of C16:0 are also elevated. Higher levels of C18:3ω3 and/or very long chain saturated fatty acids (VLCS) are also common except in the roots which form subcluster XV.

The effects of cooking and degradation over time on fatty acid compositions were also examined. Nineteen modern residues of plants and animals from the plains, parkland and forests of Western Canada were prepared by cooking samples of meats, fish and plants, alone or combined, in replica vessels over an open fire (Malainey 1997; Malainey et al. 1999b). After four days at room temperature, the vessels were broken and a set of sherds analyzed to determine changes after a short term of decomposition. A second set of sherds remained at room temperature for 80 days, and then placed in an oven at 75°C for a period of 30 days in order to simulate the processes of long term decomposition. The relative percentages were calculated on the basis of the ten fatty acids that were regularly found in Precontact period vessel residues from Western Canada: C12:0, C14:0, C15:0, C16:0, C16:1, C17:0, C18:0, C18:1ω9, C18:1ω11 and C18:2. Observed changes in fatty acid composition of the experimental cooking residues enabled the development of a method for identifying the archaeological residues (Table 2).

It was determined that levels of medium chain fatty acids (the sum of C12:0, C14:0 and C15:0), C18:0 and C18:1 isomers in the sample could be used to distinguish degraded experimental cooking residues (Malainey 1997; Malainey et al. 1999b). These fatty acids are suitable for the identification criteria because saturated fatty acids are stable and monounsaturated fatty acids degrade very slowly as compared to polyunsaturated fatty acids (deMan 1992). Furthermore, when principal component analysis is applied to the total relative fatty acid composition, the groupings generated by multivariate analysis strongly coincide with identifications made under the criteria (Malainey et al. 1999c, Figure 10). Higher levels of medium chain fatty acids, combined with low levels of C18:0 and C18:1 isomers, were detected in the decomposed experimental residues of plants, such as roots, greens and most berries. High levels of C18:0 indicated the presence of large herbivores. Moderate levels of C18:1 isomers, with low levels of C18:0, indicated the presence of either fish or foods similar in composition to corn. High levels of C18:1 isomers with low levels of C18:0 were found in residues of beaver or foods of similar fatty acid composition. The criteria for identifying six types of residues were established experimentally; the seventh type, plant with large herbivore, was inferred (Table 2).

These criteria were applied to residues extracted from more than 200 pottery cooking vessels from 18 plains, parkland and southern boreal forest sites in Western Canada (Malainey 1997; Malainey et al. 1999c; 2001b). The identifications were consistent with the evidence from faunal and tool assemblages for each site. Settlement and subsistence patterns proposed for Aboriginal hunter-gatherer peoples who occupied the area prior to European contact were also supported (Malainey et al. 2001b).

| Identification | Medium chain | C18:0 | C18:1 isomers |
|---|---|---|---|
| Large herbivore | ≤ 15% | ≥ 27.5% | ≤ 15% |
| Large herbivore with plant or Bone marrow | Low | ≥ 25% | 15% ≤ $X$ ≤ 25% |
| Plant with large herbivore | ≥ 15% | ≥ 25% | No data |
| Beaver | Low | Low | ≥ 25% |
| Fish or Corn | Low | ≤ 25% | 15% ≤ $X$ ≤ 27.5% |
| Fish or Corn with plant | ≥ 15% | ≤ 25% | 15% ≤ $X$ ≤ 27.5% |
| Plant (except corn) | ≥ 10% | ≤ 27.5% | ≤ 15% |

Table 2: Criteria for the identification of archaeological residues based on the decomposition patterns of experimental cooking residues prepared in pottery vessels.

Work has continued to expand, refine and test the identification criteria (Malainey et al. 2000a; 2000b; 2000c; 2001a; Quigg et al. 2001). The reference collection now includes several food plants and animals from the Southern Great Plains. The validity of applying the criteria to cooking residues extracted from burned rocks used for stone boiling and grilling foods has been demonstrated (Quigg et al. 2001, 290-294). This enabled the study of Late Archaic subsistence patterns at the Lino site in South Texas, despite the virtual absence of vertebral faunal and burned macrobotanical remains (Quigg et al. 2001, 295-301). Comparisons of residue identifications with faunal, macrobotanical and other cultural remains recovered from other sites in Texas, New Mexico, Wyoming and Arizona, dating from the Late Paleoindian-Early Archaic to the Late Prehistoric Periods, strongly supports the soundness of applying the technique to burned rocks (Malainey 2004).

The identification criteria for archaeological residues in Table 2 were based on foodstuffs exclusively from the plains, parkland and southern boreal forest of Western Canada. In order to make the identifications more applicable to materials from the Central and Southern Great Plains and adjacent regions, data from the analysis of food plants and animals from these regions were incorporated into these criteria.

The identification of residues with high levels of C18:0 as 'Large herbivore' was found to be valid throughout the Great Plains and adjacent areas. The flesh of deer, bison, cow and moose produce these residues. Levels of C18:0 were found to increase as animals became fat-depleted over winter (Perrin 2002). Levels of C18:1 isomers in large herbivore residues were found to be a reliable indicator of the degree of meat fattiness. Very lean meat had the lowest levels of C18:1 isomers and highest C18:0; relative amounts of C18:0 were somewhat lower in fat meat. Conversely, high levels of C18:0 combined with elevated levels of C18:1 isomers was found in meat with higher levels of fat, such as in well-marbled cuts. In addition to large herbivore meat, the decomposed cooking residues of javelina meat (*Tayassu tajacu*) and tropical oils from plants such as sotol (*Dasylirion wheeleri*) are also known to have high levels of C18:0. They must be considered as potential sources of the residue if it is possible that they could have been utilized. The decomposed cooking residues of yucca root produce residues with elevated levels of medium chain fatty acids and C18:0. Under the criteria, they would be identified as 'Plant with large herbivore'.

The original category 'Plant' was renamed 'Low fat content plant' to more accurately reflect the types of foods known to fall under this category. Plant roots, greens and berries from Western Canada produced residues with high levels of medium chain fatty acids. Further south, jicama tuber (*Pachyrhizus erosus*), onion bulbs, yopan leaves (*Ilex vomitoria*), buffalo gourd (*Cucurbita foetidissima*) and biscuit root (*Lomatium* species) produce similar residues. The levels of C18:1 isomers in the decomposed cooking residues of fleshy fruit of prickly pear (*Opuntia engelmannii*) and Spanish dagger (*Yucca treculeana*) is slightly higher; these are referred to as 'Medium-low fat plants'.

The original category 'Fish or Corn' is now referred to as 'Medium fat content foods'. In addition to fish and corn, the decomposed cooking residues of mesquite beans (*Prosopis glandulosa*), cholla (*Opuntia* species), certain snails (*Rabdotus* species), box terrapin (*Terrapene* species), and the fat-depleted meat of late winter elk (*Cervus elaphus*) produce similar residues. Medium fat content food residues of plants were found to often have elevated levels of medium chain fatty acids or very long chain saturated fatty acids.

In Western Canada, the decomposed cooking residues of fatty meat of medium-size mammals, such as beaver (*Castor canadensis*) was found to have levels of C18:1 isomers in excess of 25% with much lower levels of C18:0 and medium-chain fatty acids. The rendered fat of mammals such as bear, known as bear grease or bear oil, had even higher levels of C18:1 isomers. In the Central and Southern Plains and adjacent regions, a wide variety of seeds and nuts are available and were exploited by humans in the past. Levels of C18:1 isomers in the decomposed cooking residues of some of these foods far exceeded 25%. The original category 'Beaver' has been subdivided to better reflect these findings. From lower to highest levels of C18:1 isomers, the divisions are 'Moderate-high', 'High', 'Very high' and 'Extremely high fat content foods'. The fatty meat of beaver and the seeds of Texas ebony (*Pithecellobium ebano*) are examples of moderate-high fat content foods. Olive oil, mescal seeds (*Agave perryi*) and bear meat are examples of high or very high fat content foods. Fresh bear (*Ursus americanus*) grease and the decomposed cooking residue of piñon (*Pinus edulis*) are examples of extremely high fat content foods. Again, the residues of plant origin are more likely to have elevated levels of medium chain fatty acids or very long chain saturated fatty acids.

In general, elevated levels of C18:0 are associated with the presence of large herbivores, but javalina and tropical seed oils must be considered as possible sources if they were locally available. The relative amount of C18:1 isomers in the residue indicate the fat content of the material of origin. Medium and very long chain saturates facilitate the discrimination between foods of plant origin and those of animal origin. It must be understood that the identifications given do not necessarily mean that those particular foods were actually prepared because different foods of similar fatty acid composition and lipid content will produce similar residues. It is possible only to say that the material of origin for the residue was similar in composition to the foods indicated.

## Testing the Validity of Identification Criteria

Fatty acids decompose over time; this is the biggest criticism levied against their use as a means of identifying archaeological residues. While the fact they degrade is indisputable, clear decomposition trends emerged through the study of oven stored experimental cooking residues and led to the development of the identification criteria outlined in Table 2 (Malainey 1997; Malainey et al. 1999b). The next task undertaken was to test the validity of the criteria and the degree to which fatty acid degradation in cooking residues is predictable. This research was of particular importance to assessing criterion that included C18:1 isomers, as these monounsaturated fatty acids continue to slowly degrade over time. Over long periods of time, levels of C18:1 isomers continue to drop and relative amount of the other fatty acids in the residue increase. The goal was to determine how well decomposing cooking residues from different parts of a vessel conform to the various criteria and at what point are they no longer applicable. The possible effect of burial environment, in particular, microbial action in different types of soil, was also investigated.

In order to assess the validity of the criteria, fatty acid decomposition patterns of cooking residues were traced over extended periods of oven or soil storage. A variety of foods were boiled in clay cylinders after which the cylinders were cut into tiles. In one set of experiments, these tiles were stored at 75°C. Samples were taken periodically over a 68 day period and changes in the relative fatty acid composition were monitored. In another set of experiments, clay tiles were buried in soils from the plains, parkland and southern boreal forest of Manitoba stored at room temperature. Samples were taken periodically over a period of ten months. The relative fatty acid composition of the residue extracted from buried tiles was compared to that of controls after ten months, one month of oven storage and two months of oven storage.

## Experimental Procedures

Residues were prepared by cooking foods in clay cylinders (approximately 15 cm outside diameter x 15 cm tall) made of medium coarse raku clay and fired in an annealing oven. The cylinders were placed inside glass beakers after which about 100 g of each food type and one liter distilled water was added. This was brought to the boil over a hot plate. The total cooking time was approximately 2 hours, 1.5 hours at a simmer. After cooling at least one hour, the clay cylinder was removed and allowed to dry and stored at room temperature. The cylinder was cut into tiles of approximately 2.5 cm square and labeled. Tiles from the two highest, the two center and the two bottom rows were designated 'upper', 'middle' and 'lower', respectively.

After being subjected to the designated periods of oven, soil and room temperature storage, lipids were extracted from the tiles using a variation of the method developed by Folch et al. (1957). Possible surface contaminants were removed by grinding exterior surfaces off with a Dremel tool fitted with a silicon carbide bit. The sample was crushed with a hammer mortar and pestle and extracted twice in 40 ml chloroform-methanol (2:1, v/v) using ultrasonication (2 x 10 min). Solids were removed by filtering the solvent mixture into a separatory funnel. The lipid/solvent filtrate was washed with 22 ml of ultrapure water. Once separation into two phases was complete, the lower chloroform-lipid phase was transferred to a round-bottomed flask and the chloroform removed by rotary evaporation. Any remaining water was removed by evaporation with 2 ml benzene and 2 ml chloroform-methanol (2:1, v/v) was used to transfer the dry total lipid extract into a screw-top glass vial with a Teflon-lined cap. The vial was flushed with nitrogen and stored at -20°C.

Depending on its apparent concentration, 100-600 µl of total lipid extract (TLE) solution was placed in a screw-top test tube and dried in a heating block under nitrogen after which 1 ml of a 0.01 mg/ml solution of C21:0 (heneicosanoic acid) was added as an internal standard. Fatty acid methyl esters (FAMEs) were prepared by treating the dry lipid and internal standard with 6 ml of 0.5 N anhydrous hydrochloric acid in methanol (60 min at 65-70°C). Fatty acids that occur in the sample as di- or triacylglycerides are detached from the glycerol molecule (saponification) and converted to methyl esters. After cooling to room temperature, 4 ml of ultrapure water was added and the FAMEs were recovered with 3 ml petroleum ether and transferred into a vial. Solvents were removed by heat under nitrogen, the FAMEs dissolved in 75 µl iso-octane and transferred into a GC vial with a conical glass insert.

Analysis was performed on a Hewlett-Packard 5890 gas chromatograph fitted with a flame ionization detector connected to a personal computer. Samples were separated using a DB-23 fused silica capillary column (30 m x 0.25 mm I.D.; J&W Scientific; Folsom, CA). An autosampler injected a 1-2 µl sample using a split injection system with the ratio set between 1:20 and 1:80, depending on sample concentration. Hydrogen was used as the carrier gas at a linear velocity of 40 cm/sec. Column temperature was programmed from 155°C to 215°C at 2°C/min; lower and upper temperatures were held for 4 and 5 minutes, respectively. Chromatogram peaks were integrated using ChromPerfect software and identified through comparisons with known compounds (NuCheck Prep; Elysian, MN). Solvents and chemicals were checked for purity by running a blank sample. Traces of contamination were subtracted from sample chromatograms. Relative percentage composition was calculated by dividing the integrated peak area of each fatty acid by the total area of fatty acids present. Using this protocol, fatty acids are detectable to the nanogram ($10^{-9}$ g) level.

## Oven Storage of Cooking Residues

For samples undergoing oven storage only, fatty acid compositions of residues extracted from the upper, middle and lower portions of the clay cylinder were determined at the beginning of the oven storage experiment (time 0). Nine other tiles from each portion were placed in an oven at 75°C. For the first 20 days, one tile from each portion was sampled every four days; they were then sampled every 12 days until the end of the experiment after 68 days.

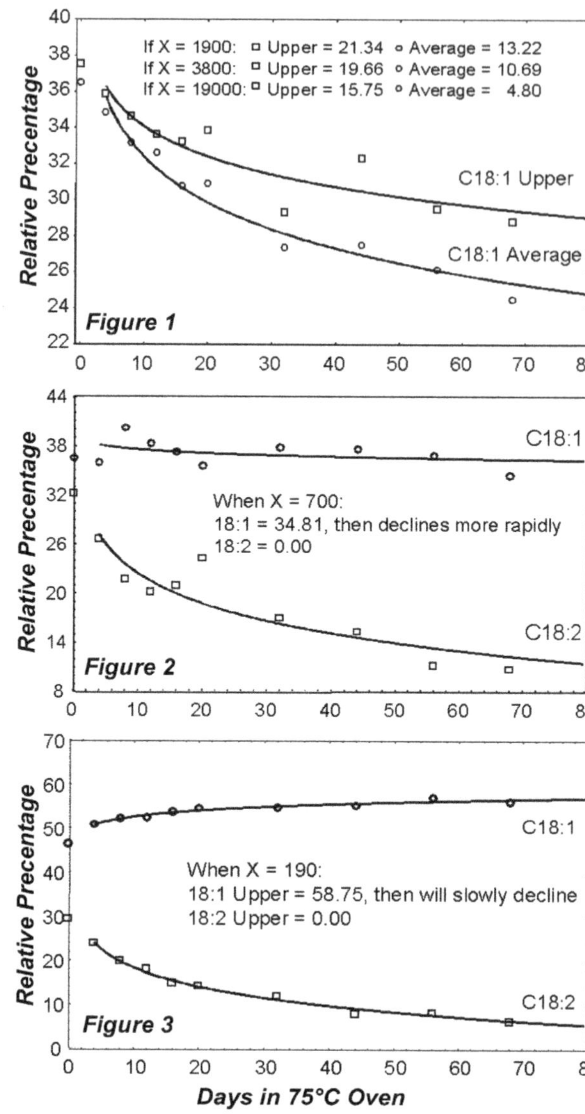


Figure 1: Plot of changes in relative amounts of C18:1 isomers in the cooking residue of smoked trout after 68 days oven storage. The logarithmic decay curve for residues from the upper portion of the clay cylinder is C18:1 isomers = 39.67 - 5.59*log10(x) + e. The overall average for residues from all areas is C18:1 Average = 40.82 - 8.412*log10(x) + e.

Figure 2: Plot of changes in amounts of C18:1 isomers and C18:2 in the cooking residue of deer and Spanish dagger after 68 days oven storage. The logarithmic decay curve for relative amounts of C18:1 isomers in residues from the upper portion of the clay cylinder is C18:1 Upper = 38.95 - 1.46*log10(x) + e. Levels of C18:2 in residues from the same area is C18:2 Upper = 34.57 - 12.15*log10(x) + e.

Figure 3 (previous page): Plot of changes in amounts of C18:1 isomers and C18:2 in the cooking residue of mescal seeds after 68 days oven storage. The logarithmic decay curve for relative amounts of C18:1 isomers in the residues from the upper portion of the clay cylinder is C18:1 Upper = 48.22 + 4.62*log10(x) + e. Levels of C18:2 in residues from the same area is C18:2 Upper = 33.12 - 14.54*log10(x) + e.

Oven storage is widely used by food scientists to accelerate oxidative decomposition of fatty acids. It has been shown that the rate of oxidative degradation doubles with every 10°C increase in temperature (Labuza 1971). One day of storage at 65°C is the equivalent of one month of storage at room temperature (Malcomson et al. 1994). At 75°C, the rate of oxidative degradation is 192 times faster than at a site with an average yearly temperature of 0°C. Oven storage at 75°C for 68 days produces oxidative degradation approximating 36.3 years at 0°C and 1900 days (5.2 years) of oven storage would simulate approximately 1000 years of oxidative degradation at 0°C. To understand the effects of the loss of monounsaturated fatty acids, relative percentages of C18:1 isomers and, where relevant, C18:2, in the residues were plotted from 'time 0' to 'day 68'. Graphs for smoked trout, the combination of deer meat and Spanish dagger fruit, and mescal, fitted with logarithmic decay curves are presented in Figures 1, 2 and 3.

Smoked trout (107 g) was cut into pieces of 10-15 g and placed inside a clay cylinder that was itself set in a glass beaker. One liter of distilled water was added and the contents were brought to the boil on a hot plate and allowed to simmer. The total cooking time was two hours. The clay cylinder was removed after the broth cooled and stored at room temperature for approximately six months. Small amounts of polyunsaturated fatty acids were still detectable in the residue at time 0 and after four days of oven storage. These were omitted prior to the calculation of the relative fatty acid composition to make time 0 and day 4 compositions comparable with the rest of the data set. At time 0, C18:1 isomer levels in residues from the upper, middle and lower portions of the clay cylinder ranged from 35.5% to 37.50% (Figure 1). Despite initial similarities, relative fatty acid compositions of the residues quickly diverged. After 68 days of oven storage, the level of C18:1 isomers from the upper portion was 28%, compared to 24% for the cylinder average.

According to the identification criteria, decomposed cooking residues of freshwater fish have C18:1 isomer levels between 15% and 27.5%. Using the logarithmic decay curves in Figure 1 to predict the amounts of C18:1 isomers, only residues from the upper portion can be correctly identified for extended periods of time. If oven storage continued for 1900 and 3800 days (10.4 years), approximating 1000 and 2000 years oxidative degradation at 0°C, C18:1 isomer levels of 19.66% and 21.34% will occur in residues from the upper portion. The relative percentage of C18:1 isomers in smoked trout residues from the upper portion would conform to the identification criteria until 25,600 days oven storage, approximating almost 13,500 years oxidative decomposition at 0°C. The identification criteria can not be successfully applied using average C18:1 isomer levels for the cylinder after a mere 1167 days oven storage, simulating just over 600 years oxidative decomposition.

Similar results were observed in the decomposed residues of another freshwater fish, pickerel. After 1900 days of oven storage, the level of C18:1 isomers is predicted to be 23.95%; after 3800 days, it would be 22.05%. After 19,000 days of oven storage, approximating 10,000 years oxidative decomposition, C18:1 isomer levels would still conform to the identification criteria for fish. Cooking residues with the highest levels of C18:1 isomers most often occur in the upper portion; although residues from the lower portion can sometimes be equally well or better preserved. Average levels of C18:1 isomers drop below 15% after less than 3000 years of oxidative decomposition.

Another cooking residue was prepared by combining the meat of a larger herbivore, 110 g deer, with 70 g chopped Spanish dagger fruit, which alone produces a medium-low fat content residue. Changes in the levels of C18:1 isomers and C18:2 in the residues extracted from the upper portion of the clay cylinder are presented in Figure 2. A notable feature of this residue is that amounts of C18:2 were maintained at levels over 10% after 68 days oven storage. By using logarithmic decay curves, C18:2 is predicted to completely disappear only after 700 days of oven storage. The relative amounts of C18:1 isomers in the residue changed only slightly over 68 days of decomposition. It is predicted that after 700 days of oven storage, the relative amount of C18:1 isomers in the residue will exceed 34%. Once C18:2 is completely lost, it is possible that the relative amounts of C18:1 isomers may decline more rapidly.

For the preparation of another residue, 33 g mescal seeds were ground and prepared as described above. Levels of C18:1 isomers and C18:2 observed in the cooking residue over 68 days of oven decomposition are shown in Figure 3. At time 0, levels of C18:2 isomers formed 30% of the relative fatty acid composition of the residue from the upper portion of the clay cylinder; after 68 days of oven storage, it dropped to 6%. With the loss of this polyunsaturated fatty acid, the relative amount of C18:1 isomers increased from about 46% to 56%. Levels of C18:2 are predicted to reach zero after 190 days oven storage; at that time, C18:1 isomers will form almost 59% of the relative fatty acid composition of the residue and the relative amount of C18:1 isomers will begin to decline. The maintenance of C18:2 after an extended period of oven storage was also observed in the cooking residue of Texas ebony seeds, a moderate-high fat content plant. After ten months of storage at room temperature and two months of oven storage, levels of C18:2 remained at about 4% and C18:1 isomers formed more than 30% of the relative fatty acid composition of the residue.

Clearly the upper portion of the vessels in which foods were prepared (the boil line) should be targeted for residue analysis. Initial levels of C18:1 isomers are the highest, and the rate of decomposition the slowest, in residues recovered from this region. Where present, C18:2 can be preserved in the upper portion of clay cylinders exposed to extended periods of oven storage, possibly several hundred days in the case of the deer and Spanish dagger residue. Residues from the middle and lower portions are more likely to have lower levels of monounsaturated fatty acids. Fatty acid compositions of these residues conform to the identification criteria for a shorter period of oxidative degradation, the equivalent of several hundred to a few thousand years.

## Soil Storage of Cooking Residues

Average air temperature in the plains and parkland of Western Canada ranges from 1.5 to 5.0°C (Hare and Hay 1979). During the winter the frost penetrates deep into the soil. The harsh winter conditions in Manitoba made an outdoor experiment impractical as microbial activity would drop significantly. In order to accelerate the effects of microbial decomposition, the storage was conducted at room temperature. Very dark gray sandy loam was collected from an uncultivated grassland environment on a terrace of the Pembina Hills Escarpment in southern Manitoba (Michalyna et al. 1988). Chernozemic parkland soil was collected from an aspen-oak grove on the grounds of a University of Manitoba Research Station south of the City of Winnipeg (Michalyna et al. 1975). Sandy forest soil was collected from the southern boreal forest near Nopiming Provincial Park in east-central Manitoba. While a professional survey has not been conducted in this particular area, the soil has the characteristics of a well-drained Minimal Podzol developed on fine sand and is quite acidic (Smith and Ehrlich 1967).

A 5 cm thick basal layer of soil was placed in clear plastic storage containers with lids and a capacity of 11.4 l. Seven clay tiles measuring about 2.5 cm square of one type of residue were laid in a single layer on the soil; at least 5 cm apart. The tiles were covered with a second 5 cm thick layer of soil. Tiles from the upper portion of the clay cylinder were placed in parkland soil; tiles from the middle portion of the clay cylinder were placed in prairie grassland soil; tiles from the lower portion of the clay cylinder were placed in the sandy forest soil. For each type of soil, residues containing lower amounts of fat were placed in one container; residues containing higher amount of fat were placed in a separate container. The containers were closed with their lids so that the soil remained moist. Three tiles of each type of residue remained exposed to the atmosphere at room temperature to serve as controls.

One tile of each residue type was collected from each type of soil after 60, 120, 180 and 240 days. At that time, 100 ml of water was sprinkled evenly on the surface to maintain soil dampness. Over the course of the storage experiment, insect life was observed and seeds in soil sprouted. After 300 days, the three remaining tiles were removed from the soil and the three control tiles were collected. Residues from one tile stored in each type of soil and one control tile were selected to show the effects of 10 months of storage. The remaining soil-stored and control tiles were immediately placed in an oven at 75°C. One tile stored in each type of soil and control tiles from the upper, middle and lower cylinders for each type of residue were stored for one month; another set underwent two months of oven storage. This was to simulate the effects of long-term decomposition on a residue modified by burial in different types of soil.

The residue for bison, a large herbivore, was prepared by boiling 112 g of meat as described above. Tiles from the upper portion of the clay cylinder, which included the area of fat accumulation, were placed in parkland soil. Compared to other areas of the clay cylinder, levels of C18:1 isomers were highest in these residues. Levels of C18:1 isomers in residues buried in parkland soil for ten months prior to one and two months of oven storage were higher than those in residues extracted from unburied control tiles (Figure 4). After a month of oven storage, C18:1 isomers formed about 48% of the total fatty acid composition of the residues; whereas the level of C18:1 isomers was only 40% in the residue extracted from the control. Similarly, after two months of oven storage C18:1 isomers represented 39% of the total fatty acid composition of the residue extracted from the tile buried in parkland soil for 10 months compared to 35% in the control. While the difference is less profound after two months of oven storage, storage in parkland soil for ten months appears to have enhanced the preservation of these monounsaturated fatty acids.

Tiles from the middle part of the clay cylinder with bison cooking residue were stored in prairie grassland soil for ten months. Residues extracted from these tiles were compared to those from unburied controls. Levels of C18:1 isomers in residues from the tiles that were buried were higher than the controls. At about 38%, levels of C18:1 isomers are virtually identical in residues from buried tiles after one and two months of oven storage. This is higher than the C18:1 isomer levels observed in both control sherds, which were an unexpectedly low 26% after one month oven storage and about 32% after two months oven storage. Levels of C18:2 are somewhat higher in this residue which depresses the relative amount of C18:1 isomers. The exact placement on the clay cylinder may have added to the variation as the middle portion of the cylinder, as with the other sections, was cut into upper and lower rows.

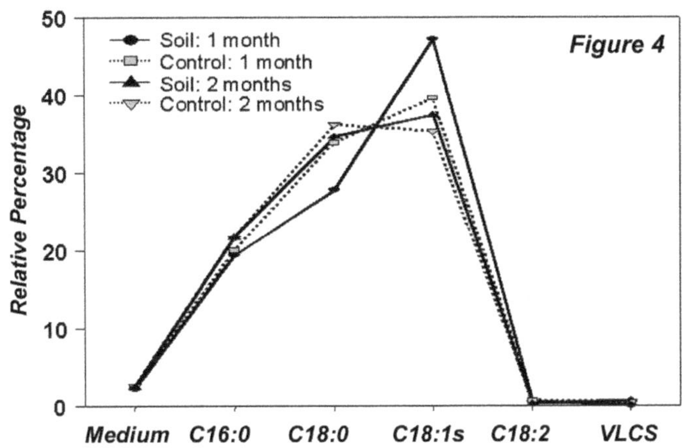


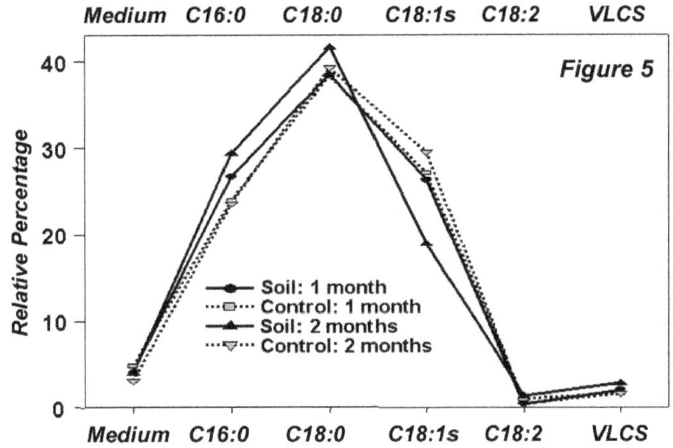


Figure 4 (top): Relative fatty acid compositions of bison cooking residues from the upper portion of the clay cylinder stored in parkland soil compared to controls after one and two months of oven storage at 75°C. VLCS = very long chain (C20, C22 and C24) saturated fatty acids.

Figure 5 (bottom): Relative fatty acid compositions of bison cooking residues from the middle portion of the clay cylinder stored in prairie soil compared to controls after one and two months of oven storage at 75°C. VLCS = very long chain (C20, C22 and C24) saturated fatty acids.

Tiles from the lowest portion of the bison residue clay cylinder were buried in forest soil and residues extracted were compared to controls (Figure 5). After one month of oven storage, levels of C18:1 isomers in residues extracted from both the control and the tile buried in forest soil were almost identical, about 29% and 29.5%, respectively. Levels of C18:1 isomers in residues extracted from tiles after two months of oven storage diverged widely. At about 30%, the level in the control exceeded all others; the amount in the residue from the buried tile was unexpectedly low, about 19%. Again, it is possible that the exact placements of the tiles on the clay cylinder may have played a role in these results.

Overall, microbial activity in parkland, prairie grassland and forest soil does not appear to adversely affect C18:1 isomer preservation in bison cooking residues. After oven storage, the degree of preservation in residues extracted from tiles buried for ten months is often equal to or better than in unburied controls. The exact placement of the tile on the clay cylinder may account for some of the variation observed.

A residue of pike was prepared by boiling 200 g of fish as described above on two consecutive days. The preservation of C18:1 isomers in the cooking residue of the pike, a freshwater fish, were similar in that the levels in residues buried in the different types of soil were generally equal to or higher than amounts in residues from unburied control tiles. After ten months of oven storage, levels of C18:1 isomers in the residue extracted from the tile buried in parkland soil exceeded 40%; the amount measured in the residue from the control tile was 33% (Figure 6). The same level of C18:1 isomers (33%) was found in the control tile after two months of oven storage; residue from the tile buried in parkland soil was only slightly lower, 31%. Levels of C18:1 isomers in residues extracted from tiles buried in forest soil were virtually identical to that of controls after both ten-months and two-months oven storage (Figure 7). A greater range of variation was observed in the levels of C16:1; it was higher in the buried tile after ten months and higher in the control after two months oven storage.

One notable exception to the pattern of equal or better preservation of C18:1 isomers was in the cooking residue of Texas ebony seeds. It was prepared by boiling 95 g of seeds that were ground immediately prior to cooking in the manner described above. Tiles from the lowest portion of the clay cylinder were stored in forest soil. The relative fatty acid compositions of residues extracted from these tiles are presented in Figure 8. After ten months of storage in soil, C18:1 levels in the residue from the buried tile was about 40%; in the unburied control, C18:1 isomer levels exceeded 50%. After two months of oven storage, levels of C18:1 isomers in the control were about 45%; in the residue from the tile buried in forest soil, levels were about 29%. The differences in preservation are so profound that further investigation is required to find an explanation for the variation.

In most residues, ten months of burial in any type of soil produced little or no difference in the preservation of C18:1 isomers. It is possible that in some cases, the burial environment was protective in that the residues were less susceptible to oxidative degradation. This could be verified experimentally by burying tiles with exactly the same degrees of residue accumulation in the different soils and comparing them to unburied controls.

## Sample Selection and Handling Guidelines

Ideally, archaeological samples for fatty acid analysis should be selected in the field. To prevent the introduction of contaminants, samples should only be handled with clean tools and gloved hands. It is preferable to examine unwashed artifacts, but it is possible to extract lipid residues from samples washed in clean water. In all cases, samples should be selected from the area of fat accumulation. With pottery, the area of fat accumulation varies with how the vessel was used (Charters et al. 1995). On the Northern Plains, where most vessels were cooking pots used for boiling, sherds from the neck or shoulder area are most suitable. Vessel morphology, paste characteristics, decoration and use-alterations, such as the location of soot and carbonized residues, provide clues as to the function of a vessel if it is not known (Henrickson and McDonald 1983; Hally 1983; Smith 1983; Skibo 1992).

With thermally altered rock, it is important to identify rocks, if possible which surfaces of those rocks, are most likely to have absorbed food residues. It is also necessary to assess in which rock would residues be best preserved. Burned rock features and cooking pits are widely distributed in the archaeological record in several parts of the United States (Ellis 1997). On the basis of ethnographic research, a variety of foods were likely prepared in pits using hot rocks as heat reservoirs.

Wandsnider (1997) reported that differences in food preparation relate to fat, protein and carbohydrate content. Foods prepared in the ovens may have been dry-roasted or water may have been added so that the food was steamed. The construction of earth ovens in terms of the physical arrangement of hot rocks, insulating material, and the food within ovens is also known to have varied (Ellis 1997). On the Northern Plains, there is ethnographic evidence that some pits were lined with hide and served as receptacles for stone boiling; some of the pits in the Southern Plains may have served similar functions (Quigg et al. 2001). Features resembling platforms or beds of burned rock may have been heated then used to broil, sear or parch foods (Ellis 1997). If rocks were exposed to secondary heating after the introduction of food residues, changes in the fatty acid decomposition due to thermal degradation may lead to erroneous identifications.

In the case of earth ovens, there is ethnographic evidence that insulating material served to absorb fat and other cooking juices lost by roasted animals (Wandsnider 1997). Thick layers of insulation between the food and the heat source would prevent residues from reaching the rocks. In this case, soil from the pit wall at the apparent food layer should be targeted and a natural control from the same depth below the surface should also be collected so that the soil lipids can be assessed.

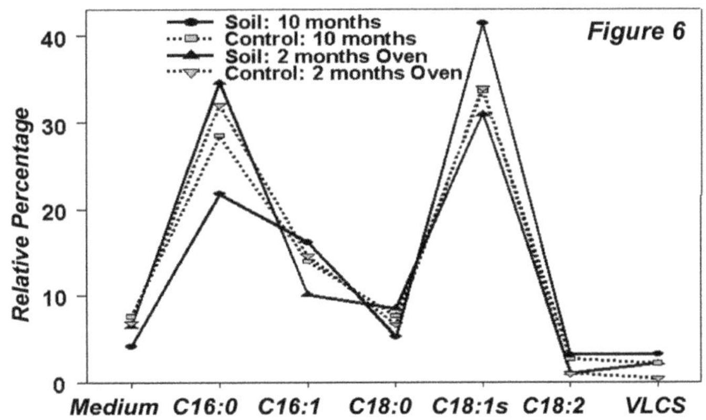


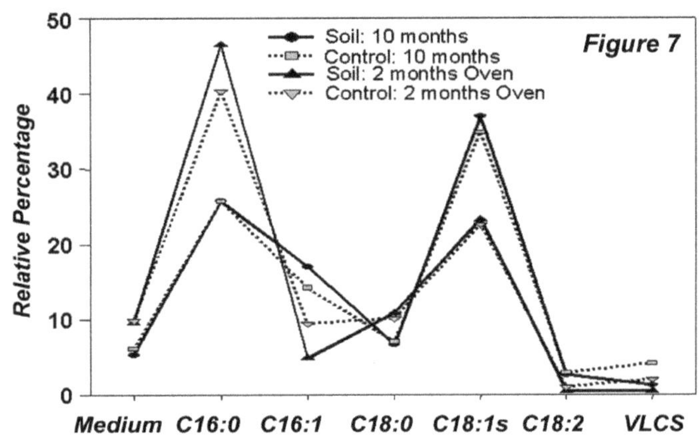


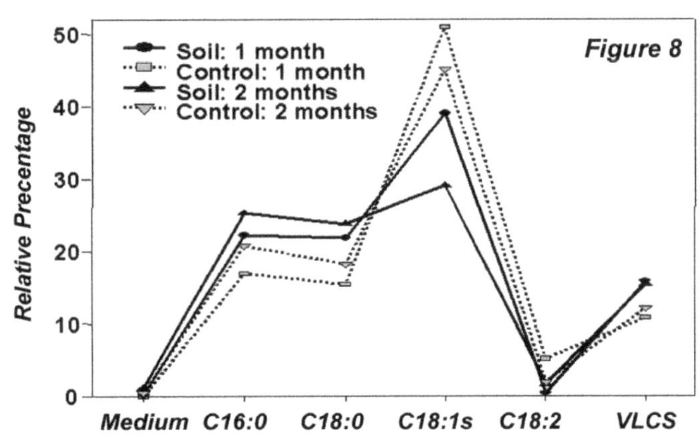


Figure 6 (top): Relative fatty acid compositions of pike cooking residues from the upper portion of the clay cylinder stored in parkland soil compared to controls after ten months and after two months of oven storage at 75°C. VLCS = very long chain (C20, C22 and C24) saturated fatty acids.

Figure 7 (middle): Relative fatty acid compositions of pike cooking residues from the lower portion of the clay cylinder stored in forest soil compared to controls after ten months and after two months of oven storage at 75°C. VLCS = very long chain (C20, C22 and C24) saturated fatty acids.

Figure 8 (bottom): Relative fatty acid compositions of Texas ebony cooking residues from the lower portion of the clay cylinder stored in forest soil compared to controls after one and two months of oven storage at 75°C. VLCS = very long chain (C20, C22 and C24) saturated fatty acids.

## Conclusion

Fatty acid analysis is a relatively rapid, inexpensive and accessible method for obtaining a characterization of an archaeological residue. It is especially valuable when dealing with the material remains of cultures from periods for which there is no written history, such as those in North America prior to European contact. In these situations, a wide range of plant and animal foods may have been utilized. Under these circumstances, general characterizations of residues from many vessels may be far more beneficial than proving that a specific food was prepared in a certain vessel.

While fatty acid analysis alone cannot prove that a particular food was cooked, it is useful for showing that certain types of foods were not. This is particularly useful for delineating high fat content foods, low fat content foods and large herbivore products. It provides an independent line of evidence that can should used with other types of information, from artifact recoveries, faunal and archaeobotanical remains and other types of residue analyses, to paint a more complete picture of the activities of the site inhabitants. If desired or necessary, the total lipid extract can undergo further analyses, such as the identification of biomarkers or establishing stable isotope ratios of carbon and nitrogen (Barnard et al. 2007), which will refine the residue identification.

## References


**Barnard**, H., S.H. Ambrose, D.E. Beehr, M.D. Forster, R.E. Lanehart, R.E. Parr, M.E. Malainey, M. Rider, C. Solazzo and R.M Yohe II (2007). Mixed Results of Seven Methods for Organic Residue Analysis Applied to One Vessel with the Residue of a Known Foodstuff. *Journal of Archaeological Science* **34**: 28-37.

**Charters**, S., R.P Evershed, P.W. Blinkhorn and V. Denham (1995). Evidence for the mixing of fats and waxes in archaeological ceramics. *Archaeometry* **37,1**: 113-127.

**Collins**, M.B., B. Ellis and C. Dodt-Ellis (1990). *Excavations at the Camp Pearl Wheat Site (41KR243). An Early Archaic Campsite on Town Creek, Kerr County, Texas. Studies in Archaeology 6*. Austin: Texas Archaeological Research Laboratory, The University of Texas at Austin.

**Condamin**, J., F. Formenti, M.O. Metais, M. Michel and P. Blond (1976). The Application of Gas Chromatography to the Tracing of Oil in Ancient Amphorae. *Archaeometry* **18,2**: 195-201.

**deMan**, J.M. (1992). Chemical and Physical Properties of Fatty Acids. In C.K. Chow (ed.). *Fatty Acids in Foods and their Health Implications*. New York: Marcel Dekker, pp. 17-39.

**Ellis**, L.W. (1997). Hot Rock Technology. In S.L. Black, L.W. Ellis, D.G. Creel and G.T. Goode (eds.). *Hot rock Cooking on the Greater Edwards Plateau. Volume 1. Studies in Archaeology 22*. Austin: Texas Archeological Research Laboratory, The University of Texas at Austin, pp. 79-139.

**Evershed**, R.P., C. Heron and L.J. Goad (1990). Analysis of Organic Residues of Archaeological Origin by High Temperature Gas Chromatography and Gas Chromatography-Mass Spectroscopy. *Analyst* **115**: 1339-1342.

**Evershed**, R.P., C. Heron, S. Charters and L.J. Goad (1992). The Survival of Food Residues. New Methods of Analysis, Interpretation and Application. *Proceedings of the British Academy* **77**: 187-208.

**Folch**, J., M. Lees and G.H. Sloane-Stanley (1957). A Simple Method for the Isolation and Purification of Lipid Extracts from Brain Tissue. *Journal of Biological Chemistry* **191**: 833.

**Frankel**, E.N. (1991). Recent Advances in Lipid Oxidation. *Journal of the Science of Food and Agriculture* **54**: 465-511.

**Frankel**, E.N. (1993). In Search of Better Methods to Evaluate Natural Antioxidants and Oxidative Stability in Food Lipids. *Trends in Food Science and Technology* **4**: 220-225.

**Hare**, F.K. and J.E. Hay (1974). The Climate of Canada and Alaska. In R.A. Bryson and F.K. Hare (eds.). *Climates of North America. World Survey of Climatology Volume 11*. Amsterdam: Elsevier, pp. 49-192.

**Hally**, D.J. (1983). Use alteration of Pottery Vessel Surfaces. An Important Source of Evidence of the Identification of Vessel Function. *North American Archaeologist* **4**: 3-26.

**Labuza**, T.P. (1971). Kinetics and lipid oxidation of foods. *CRC Critical Reviews in Food Technology* **1**: 355-372.

**Loy**, T. (1994). Residue Analysis of Artifacts and Burned Rock from the Mustang Branch and Barton Sites (41HY209 and 41HY202). In R.A. Ricklis and M.B. Collins (eds.). *Archaic and Late Prehistoric Human Ecology in the Middle Onion Creek Valley, Hays County, Texas. Volume 2. Topical Studies. Studies in Archeology 19*. Austin: Texas Archaeological Research Laboratory, The University of Texas at Austin, pp. 607-627.

**Malainey**, M.E. (1997). *The Reconstruction and Testing of Subsistence and Settlement Strategies for the Plains, Parkland and Southern Boreal Forest*. University of Manitoba Ph.D. thesis.

**Malainey**, M.E. (2004). *Assessing the Fatty Acid Analysis of Residues Extracted from Thermally-Altered Rock and Groundstone*. Paper presented at the 62nd Annual Plains Anthropological Conference, Billing, Montana, October 2004.

**Malainey**, M.E., K.L. Malisza, R. Przybylski and G. Monks (2001a). *The Key to Identifying Archaeological Fatty Acid Residues*. Paper presented at the 34th Annual

Meeting of the Canadian Archaeological Association, Banff, Alberta, May 2001.

**Malainey**, M.E., R. Przybylski and G. Monks (2000a). *The Identification of Archaeological Residues using Gas Chromatography and Applications to Archaeological Problems in Canada, United States and Africa.* Paper presented at The 11th Annual Workshops in Archaeometry, State University of New York at Buffalo, February 2000.

**Malainey**, M.E., R. Przybylski and G. Monks (2000b). *Refining and Testing the Criteria for Identifying Archaeological Lipid Residues using Gas Chromatography.* Paper presented at the 33rd Annual Meeting of the Canadian Archaeological Association, Ottawa, May 2000.

**Malainey**, M.E., R. Przybylski and G. Monks (2000c). *Developing a General Method for Identifying Archaeological Lipid Residues on the Basis of Fatty Acid Composition.* Paper presented at the Joint Midwest Archaeological and Plains Anthropological Conference, Minneapolis, Minnesota, November 2000.

**Malainey**, M.E., R. Przybylski and B.L. Sherriff (1999a). The Fatty Acid Composition of Native Food Plants and Animals of Western Canada. *Journal of Archaeological Science* **26,1**: 83-94.

**Malainey**, M.E., R. Przybylski and B.L. Sherriff. (1999b). The Effects of Thermal and Oxidative Decomposition on th Fatty Acid Composition of Food Plants and Animals of Western Canada. Implications for the Identification of Archaeological Vessel Residues. *Journal of Archaeological Science* **26,1**: 95-103.

**Malainey**, M.E., R. Przybylski and B.L. Sherriff (1999c). Identifying the Former Contents of Late Precontact Period Pottery Vessels from Western Canada using Gas Chromatography. *Journal of Archaeological Science* **26,4**: 425-438.

**Malainey**, M.E., R. Przybylski and B.L. Sherriff (2001b). One Person's Food. How and Why Fish Avoidance May Affect the Settlement and Subsistence Patterns of Hunter-Gatherers. *American Antiquity* **66,1**: 141-161.

**Malcomson**, L.J., M. Vaisey-Genser, R. Przybylski and N.A.M. Eskin (1994). Sensory Stability of Canola Oil. Present States of Shelf Life Studies. *Journal of the American Oil Chemists' Society* **71**: 435-440.

**Marchbanks**, M.L. (1989). *Lipid Analysis in Archaeology. An Initial Study of Ceramics and Subsistence at the George C. Davis Site.* The University of Texas at Austin M.A. thesis,

**Marchbanks**, M.L. and J.M. Quigg (1990). Appendix G. Organic Residue and Phytolith Analysis. In D.K. Boyd, J.T. Abbott, W.A. Bryan, C.M. Garvey, S.A. Tomka and R.C. Fields (eds.). *Phase II Investigations at Prehistoric and Rock Art Sites, Justiceburg Reservoir, Garza and Kent Counties, Texas, Volume II. Reports of Investigations Number 71.* Austin: Prewitt and Associates, Inc., pp. 496-519.

**Michalyna**, W., W. Gardiner and G. Podolsky (1975). *Soils of the Winnipeg Region Study Area.* Winnipeg: Municipal Planning Branch, Manitoba Department of Municipal Affairs.

**Michalyna**, W., W. Gardiner and E. St. Jacques (1988). *Soils of the Rural Municipalities of Grey, Dufferin, Roland, Thompson and part of Stanley. Report D60.* Winnipeg: Manitoba Department of Agriculture.

**Patrick**, M., A.J. de Konig and A.B. Smith (1985). Gas Liquid Chromatographic Analysis of Fatty Acids in Food Residues from Ceramics Found in the Southwestern Cape, South Africa. *Archaeometry* **27,2**: 231-236.

**Perrin**, J.A. (2002). *The Effects of Seasonal Fat Depletion of Large Herbivores on Archaeological Lipid Residues.* Brandon University Undergraduate Thesis.

**Quigg**, J.M., M.E. Malainey, R. Przybylski and G. Monks (2001). No Bones about it. Using Lipid Analysis of Burned Rock and Groundstone Residues to examine Late Archaic Subsistence Practices in South Texas. *Plains Anthropologist* **46,177**: 283-303.

**Skibo**, J.M. (1992). *Pottery Function. A Use-Alteration Perspective.* New York: Plenum Press.

**Solomons**, T.W.G. (1980). *Organic Chemistry.* Toronto: John Wiley & Sons.

**Smith**, R.E. and W.A. Ehrlich (1967). *Soils of the Lac Du Bonnet Area. Soils Report Number 15.* Winnipeg: Manitoba Soil Survey, Manitoba Department of Agriculture.

**Wandsnider**, L. (1997). The Roasted and the Boiled. Food Composition and Heat Treatment with Special Emphasis on Pit-hearth Cooking. *Journal of Anthropological Archaeology* **16**: 1-48.

# CHAPTER EIGHT
# Organic Residue Analysis and the Decomposition of Fatty Acids in Ancient Potsherds
## J.W. Eerkens

Jelmer Eerkens, Associate Professor; Department of Anthropology, University of California, Davis; One Shields Avenue; Davis, CA 95616; USA; <jweerkens@ucdavis.edu>.

Determining the function of prehistoric artifacts has long been an important avenue of archaeological inquiry (for instance Bennett 1943, 1944; Clark 1939; Linton 1944; Smith 1910 or Steward and Setzler 1938). Although archaeology has moved beyond the 'functionalist' theoretical paradigm, so influential in the 1930's through 1950's (Trigger 1989; Willey and Sabloff 1980), where every artifact was understood to confer some adaptive advantage to the people who used it, determining the use of artifacts continues to play an important role in reconstructing prehistoric behavior. In archaeology today, determining the function of an object is rarely an end product. Instead, the function of an artifact is usually used as one data set to help inform on other behavior, such as the organization of technology, the division of labor, gender relations and issues concerning diet, among other topics.

Organic residue analysis reflects one line of investigation that archaeologists have employed to attempt to deduce function of artifacts, especially pots (Charters et al. 1997; Copley et al. 2005; Deal and Silk 1988; Eerkens 2002, 2005; Evershed et al. 1997, 2003; Heron et al. 1991; Malainey 1999c; Morton and Schwarcz 2004; Mottram et al. 1999; Reber and Evershed 2004; Skibo 1992; Stott et al. 1999) which are the subject of this chapter, though pipes (Rafferty 2002), hunting weapons (Craig and Collins 2002; Fullager and Jones 2004; Pearsall et al. 2004; Rots and Williamson 2004; Wadley et al. 2004) and cooking stones (Quigg et al. 2001, Buonasera 2005) have also been examined. There are other approaches to help reconstruct the function of ancient pots, including engineering analyses (Arnold 1985; Bronitsky and Hamer 1986; Brown 1989; Feathers 1989; Juhl 1995; Linton 1944; Rice 1987; Rye 1976; Skibo et al. 1989; Smith 1985), use wear studies (Beck et al. 2002; Halley 1983; Rice 1987; Shiffer 1989; Skibo 1992) and ethnographic analogy (Costin 2000; Hegmon 2000; Henrickson and McDonald 1983), but these methods are not often definitive. They usually provide only hypotheses about the types of foods that may have been cooked or stored in a pot. Organic residue analysis has the potential to be more precise about the foods that were prepared or stored within a pot, hence the intense interest by archaeologists in developing this method over the last decades. Indeed, if the rapidly expanding literature is any measure, the expected future payoffs from this field are high. However, as several of the chapters in this volume attest, residue analysis is still in its infancy and there is much to be learned and fine-tuned, particularly on the methodology of the final interpretation of the biochemical findings.

### Organic Residue Analysis

Although a range of organic compounds have been recovered from archaeological potsherds including amino acids, waxes, and cholesterol, fatty acids have been the principal class of compounds targeted for analysis in archaeological studies. This is due in part to their ease of extraction from potsherds and the widespread availability of the instruments needed to detect and quantify their presence. However, the main reason fatty acids have been targeted is undoubtedly the stability of these biomolecules over long periods of time (Christie 1989; Evershed 1993). Relative to DNA, proteins or carbohydrates, lipids (including fatty acids) are relatively resistant to decomposition and degradation.

That fatty acids are often present in ancient sherds, occasionally in very high quantities, has been amply demonstrated by archaeologists and chemists. Fatty acids are particularly prevalent in the interior walls of pots, especially near the neck and rim and occupy small vugs or open spaces within the ceramic fabric. What is less clear is that such residues actually represent the unmodified, or little modified, byproducts of ancient foods cooked or stored in the vessels as several alternatives exist. First, fatty acids are produced by nearly every organism, from bacteria to mammals, and are therefore present in virtually every environment on earth and can simply be native to the clays that people use to make pots. Second, fatty acids could represent post-depositional contamination by, for example, bacteria that are consuming other food residues such as proteins within the sherds, or free fatty acids leaching into sherds from the surrounding soil. Third, the fatty acids that we find may only represent a small fraction of what remains after decomposition due to processes such as oxidation and hydrolysis. Finally, fatty acids may be entirely the product of laboratory contamination.

We can probably rule out several of these possibilities. The first possibility, native contamination, is unlikely. The exposure of fatty acids to high temperatures, which promotes oxidation, leads to decomposition. Fatty acids are quite stable in temperatures below 200°C, but rapidly oxidize between 200-250°C (De Souza et al. 2004; Frankel 1980, 1987, 2005; Santos et al. 2002), with polyunsaturated and monounsaturated fats degrading at slightly lower temperatures than saturated fats. This is much lower than the minimum temperature required to fire a pot, 500-800°C. Thus, any native fatty acids in a clay are extremely unlikely to survive firing. As well, it is higher than the temperature achieved in most pre-industrial ceramic cooking methods. Thus, we can be fairly confident that pots start out clean of fatty acid s and that the act of cooking will not degrade them. Indeed, experiments by Johnson et al. (1988) suggest that firing of clay tiles to 400-600°C in both oxidizing and non-oxidizing environments essentially removes all hydrocarbons and fatty acids, though some carbon remains in the form of inorganic compounds and pure carbon (such as coal).

Similarly, the fourth possibility, that fatty acids are merely the product of laboratory contamination, is also dismissible. Most laboratories run blanks and other controls to evaluate the influence of such contamination. Low levels of fatty acid contamination are in most cases unavoidable due to their ubiquity in the environment. However, archaeological sherds often contain concentrations of fatty acids that are a level of magnitude or greater than the blanks, indicating that most are native to the sherds themselves. In any case, these are the sherds that archaeologists should be including in their interpretations.

As well, it has been demonstrated that sherds buried in archaeological sediments are not contaminated by the influx of fatty acids from nearby soils. Tests examining the fatty acid profiles of sherds and the immediately surrounding soil show that the two are quite different (Deal and Silk 1988; Heron et al. 1991). This is likely due to the fact that fatty acids, like all lipids, are water insoluble which would also help keep water out of the walls of pots infused with lipids or coated with residues or carbonized remains. Many ethnographic studies suggest people either coat cooking pots with, or simply soak them in, lipid-rich products (Arnold 1985, 140). These activities help to prevent water from the interior from leaking through the pot. As well, experimental studies suggest that water rich in organic matter penetrates the walls of porous pots and deposit residues there (Skibo 1992, 151), presumably as the water evaporates leaving behind organic materials. Such residues are not removed by washing. These findings suggest that the residues in the interior walls most likely represent the application of such lipid-rich mixtures, or the primarily remains of the first several uses of a pot. After a pot has become infused with organic residues, water no longer leaks through and there is little room for the accumulation of additional residues.

I am unaware of studies examining the potential role of bacteria in contributing to the pool of fatty acids recovered from archaeological sherds. Bacteria produce the same types of fatty acids as most plants and animals, especially the more common saturated and monounsaturated fats typically found in potsherds. This is a concern that needs to be addressed by future research beyond the scope of this chapter. Instead, I will here examine the final possibility mentioned above, that is, the potential effects of decomposition of fatty acids in ancient sherds. Specifically, I aim to examine how differential decomposition between organic compounds affects our interpretation of fatty acid profiles.

## Decomposition: Food Sciences Perspective

While fatty acids are relatively resistant to decomposition when compared to many other biomolecules, they still degrade when exposed to oxygen and water, processes that will be accelerated by higher temperatures.[1] The processes of decomposition have been of much interest to food scientists. Many of the foul tastes and smells associated with spoiled food are the byproducts of fatty acid degradation. As long-chained compounds break down, they form short-chained and often volatile (airborne) aromatic ones. Evolution has predisposed humans to recognize these compounds as foul, representing foods to be avoided, because various toxic-compound-producing bacteria live on rancid foods. These processes, then, have attracted much research by food chemists to understand what exactly happens when foods, and the fatty acids within them, decompose.

Research in the food sciences has demonstrated that fatty acid decomposition is an extremely complex process that can produce a diverse range of organic compounds depending on environment (Frankel 2005; Fritsch and Deatherage 1956; Hudlicky 1990). For example, saturated fatty acids can oxidize to produce several shorter derivative compounds, each of which may further decompose into other unstable and short-lived compounds, which may again decompose into yet other compounds. Indeed, the decomposition process for many isolated fatty acids is still not completely understood by food scientists, much less for whole foods.

In any case, the relevance of such decomposition research to archaeology is unclear. Food scientists are interested primarily in the short-term (months) decomposition products while archaeologists are, of course, interested in the long-term outcomes (centuries

[1] In fact, a prominent food chemist at the University of California, Davis expressed amazement when I told him that I was studying fatty acids over 500 years old. He was surprised that any fatty acids could survive that long.

to millennia). In most cases, archaeologists will not be interested in all the intermediary biomolecules, only the final and relatively stable endproducts. To date, this does not seem to be an arena that has attracted much attention by archaeologists.

## Decomposition: Archaeological Perspective

In linking organic residues to particular types of foods, there are basically two methods by which residue studies in archaeology have progressed. The first involves the presence of distinct biomarkers (Evershed et al. 1999). Biomarkers are hypothesized to be synthesized only by particular species or genera, hence their presence in organic residues is indicative of the cooking or storage of that plant or animal in the past. Claims for biomarkers include certain ω-3 fatty acids for fish and erucic acid (C22:1) for plants in the mustard family. Unfortunately, such biomarker lipids are rare in nature, and in fact many plants and animals are producing small amounts of compounds that are considered biomarkers for other species. Thus, although plants in the mustard family produce significantly higher densities of erucic acid, other plants also produce this molecule in small amounts.

The second method is more generalized and assumes that different plants and animals produce different quantities of fatty acids. For example, certain plant families might produce higher relative quantities of long-chained fatty acids than others, or certain categories of food, such as nuts, have higher quantities of saturated relative to unsaturated fatty acids. By this logic, the ratios of different compounds will be different for different plant and animal groups, rather than only the presence or absence of specific organic compounds as above.

Much work has been carried out in both archaeology (Eerkens 2005; Evershed et al. 1997; Malainey et al. 1999a; Mottram et al. 1999; Skibo 1992) and food science to show that modern fresh foods do indeed have different and recognizable fatty acid profiles. This is hardly surprising, for different plants and animals eat different foods and metabolize and store energy in different ways. Similarly, different parts of a plant (root, leaves, seeds) serve different biological functions and will be made up of varying quantities of organic compounds that help to serve those functions. For archaeologists, the assumption is that these profiles remain relatively unchanged over time. To determine what types of foods were processed or stored within a pot, the assumption is that one need only extract the organic residues, determine the relative amounts of different fatty acids, and then compare this profile to some fatty acid database of different foods.[2]

## Dealing with Decomposition

Decomposition of fatty acids will have differing effects on these two methods. In the first method, decomposition of biomarkers below the detection threshold of whatever equipment is being used (GC/MS, HPLC) will simply lead to the inability to identify ancient foods. Either the biomarkers are there, and the pot function can be identified, or not. In the second method, decomposition is potentially more problematic. If all fatty acids decompose at the same rate, their relative percentages will stay the same, and it is a simple matter to calculate the percentage of each fatty acid which will stay constant over time. If, on the other hand, fatty acids decompose at different rates, the relative percentages will constantly change over time.

Unfortunately, it is the latter case that is the rule. Unsaturated fats oxidize faster than saturated ones,. As well, longer- and shorter-chained compounds (those greater than 18 and less than 14 carbon atoms) oxidize more quickly than medium-chained compounds (with 14 to 18 carbon atoms). Unfortunately, the precise relative rate at which fats decompose depends on a number of factors including temperature, the availability of oxygen and water, and the original relative densities of different compounds. and the rate increasing over ten times for each double bond present. For example, deMan (1992) estimated that the rate of oxidation between C18:0 (stearic acid), C18:1 (oleic acid), C18:2 (linoleic acid), and C18:3 (linolenic acid) at 100°C is 1:100:1200:2500, but these ratios are likely to change under different conditions such as higher or lower temperatures. As well, longer-chain compounds oxidize more quickly than shorter-chained compounds. Reconstructing all these parameters for archaeological pot sherds is extremely difficult, if not impossible. It also suggests that the approximation of long-term decomposition using artificial means, such as high temperature or exposure to oxygen, may not necessarily replicate natural decomposition. Additional research is necessary to determine the accuracy of such methods.

2 In essence, this process is similar to provenance analysis of lithic and ceramic artifacts using techniques such as X-ray fluorescence (XRF) or instrumental neutron activation analysis (INAA). In the organic residue case, the relative quantity of different fatty acids are the analytical analogue for parts-per-million quantities of elements, such as rubidium, strontium, etc.

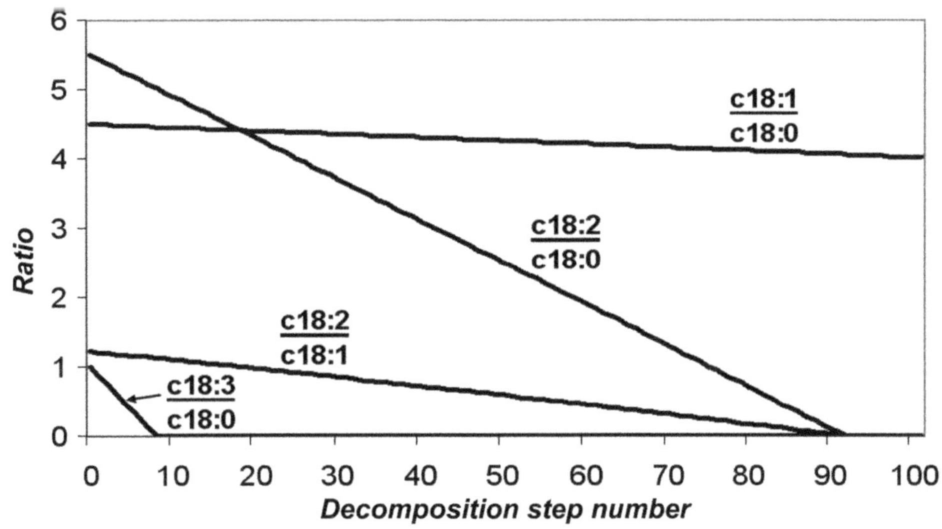


Figure 1: Simulated decomposition for four fatty acid ratios over 100 decomposition steps (based on deMan 1992).

The upshot is that the identification of any archaeological residues relying on the relative percentage of different fatty acids to one another is potentially problematic, unless those fatty acids decompose at similar rates or we can estimate the precise age of a pot, know the relative rate of decomposition between different fatty acids, and determine how much decomposition has taken place. The latter is unlikely given the complex nature of decomposition (Frankel 2005). In other words, if some food resource, such as maize, has high quantities of C18:1, it is not enough to simply detect a high quantity of C18:1 in an archaeological sherd to identify the reside as deriving from maize. A high quantity of C18:1 could mean that C18:1 was present in large amounts in the original residue (potentially maize), but it could also mean that C18:1 was originally present in low quantities, but that several other high-quantity fatty acids have decomposed, leaving mainly C18:1 after such long-term decomposition.

To demonstrate this problem, Figure 1 shows the simulated ratios of several fatty acids relative to one another over time, using the relative decomposition rate provided by DeMan (1992). The starting values represent the approximate concentration of the four different fatty acids in rabbit meat. Each unit along the X-axis (time) represents a 'decomposition step.' This is an arbitrary unit that is proportional to real time, but will accelerate or decelerate depending on environmental factors such as temperature, humidity and the presence of oxygen. As seen, the ratios have different slopes indicating quicker or slower change, and none of the four ratios is constant over time. In particular, ratios involving polyunsaturated fatty acids quickly approach zero as these molecules degrade.

Not surprisingly, the recovery of polyunsaturated fatty acids is rare in archaeological sherds. All of this demonstrates that the use of fatty acid ratios to identify ancient residues is problematic (Skibo 1992), again, unless those fatty acids decompose at similar rates. The use of advanced statistical methods such as principal components analysis (PCA) or cluster analysis appears not to resolve this issue (Malainey 1999c).

## Fatty Acid Ratios

To deal with this decomposition issue, it is necessary for archaeologists to develop fatty acid ratios that are relatively constant despite decomposition, in other words, to use the ratios of fatty acids that decompose at similar rates. This would include isomers of the same fatty acid, for example, C18:1ω9 and C18:1ω7, two monounsaturated fats with the double bond located at different positions along the carbon chain, or fatty acids with identical number of double bonds and of similar length, for example, C18:0 and C16:0 or C18:1 and C16:1. Unfortunately, compounds that degrade at similar rates tend to be related and serve similar biological functions in plants and animals. As a result, they tend to be produced in similar amounts within a plant or animal, and the ratios of these compounds may be similar across different species. In other words, the use of fatty acids that decompose at similar rates may not provide great discriminatory power for determining pot function. I have shown elsewhere that such fatty acid ratios are very similar across different plant species, but also that they do vary systematically by food product (Eerkens 2005). Thus, the leaves and greens of plants have ratios other than seeds and nuts, which are again different than roots and bulbs. Similarly, terrestrial mammals are different than fish, which are again different than plant products.

Experimental data generated by Malainey et al. (1999b) serve to show that such ratios remain relatively stable in spite of overall fatty acid degradation. They boiled various food products in small earthenware pots, broke the pots and separated the sherds into three groups. They extracted residues from the first group immediately, subjected the second group to 100°C temperatures for four days, before extracting residues, and subjected the last group to 100°C temperatures for approximately 30 days. Figure 2 plots two fatty acid ratios, C16:0/C18:0 and C16:1/C18:1 for three different food products (bison, catfish and greens) across these three time frames. As seen, although there is some variability, the fatty acid ratios stay relatively constant, particularly the latter ratio. The major exception is C16:0/C18:0 for greens which shows marked change in its ratio over time.

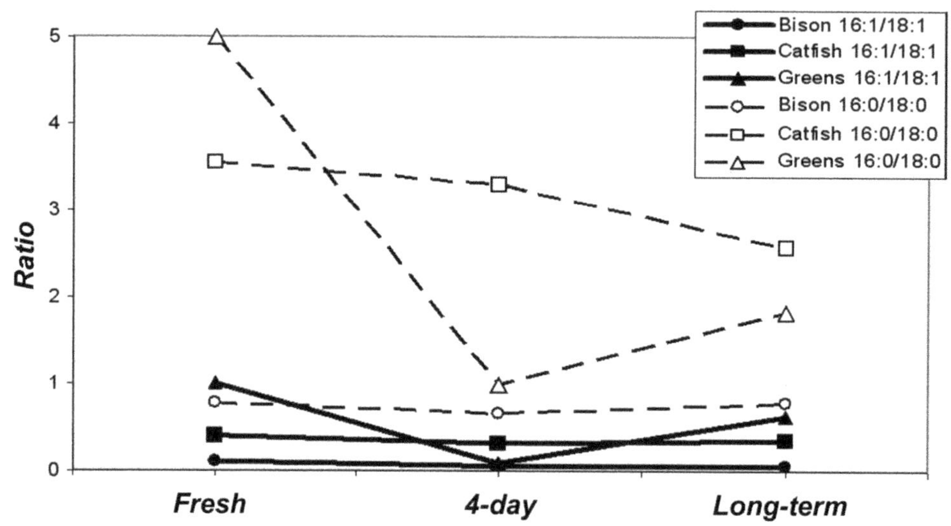


Figure 2: Induced decomposition for three foods and two fatty acid ratios (based on Malainey et al. 1999b).

Although this approach using fatty acid ratios cannot make fine-scale distinctions between different specific foods in degraded residues, it can make some general divisions for different food classes. Using data on boiled foods generated by Malainey et al. (1999a), I was able to derive four ratios that were useful in discriminating five different food classes (Eerkens 2005), including terrestrial mammals, fish, seeds and berries, roots, and greens. These four ratios are C12:0/C14:0, C16:0/C18:0, C16:1/C18:1 and (C15:0 + C17:0)/C18:0, consisting of eight different fatty acids that are commonly encountered in archaeological residues.

Figures 3 and 4 plot these ratios and show the separation between the general food classes discussed above. Ellipses represent subjective estimations of the range of values for each food group, not mathematically-defined confidence intervals. As shown in these figures, there is some overlap in the ellipses. This makes it difficult to assign a definitive function to individual sherds. I would argue that the best use of such classification schemes is to analyze multiple sherds from a site or region and evaluate which food groups seem to be best represented among the samples. Correlating these assignments by pot shape, burial context and decoration motifs may assist further in the reconstruction of pot function.

## Conclusions

Two approaches have characterized most archaeological residue studies involving fatty acids. The first consists of locating specific biomarkers distinctive of particular species, genera, or families of plants or animals. While useful in some cases, fatty acid biomarkers are unfortunately rare in nature. As a result this approach has seen only limited application. The second approach has been to use the ratios of more common fatty acids to define particular classes of foods. The problem with this approach is that different fatty acids decompose at different rates over due to oxidation and hydrolysis. As a result such ratios are not stable over time, unless care is taken to only rely on ratios of fatty acids that decompose at similar rates.

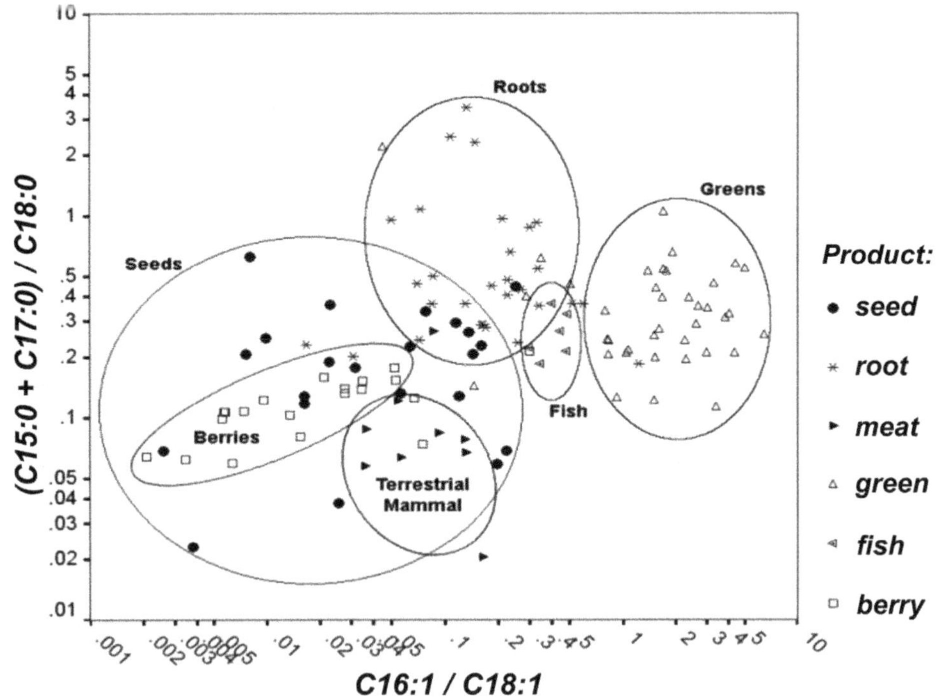


Figure 3: Biplot of two conservative fatty acid ratios for modern food products (ellipses represent subjective estimations of the range of values for each food group).

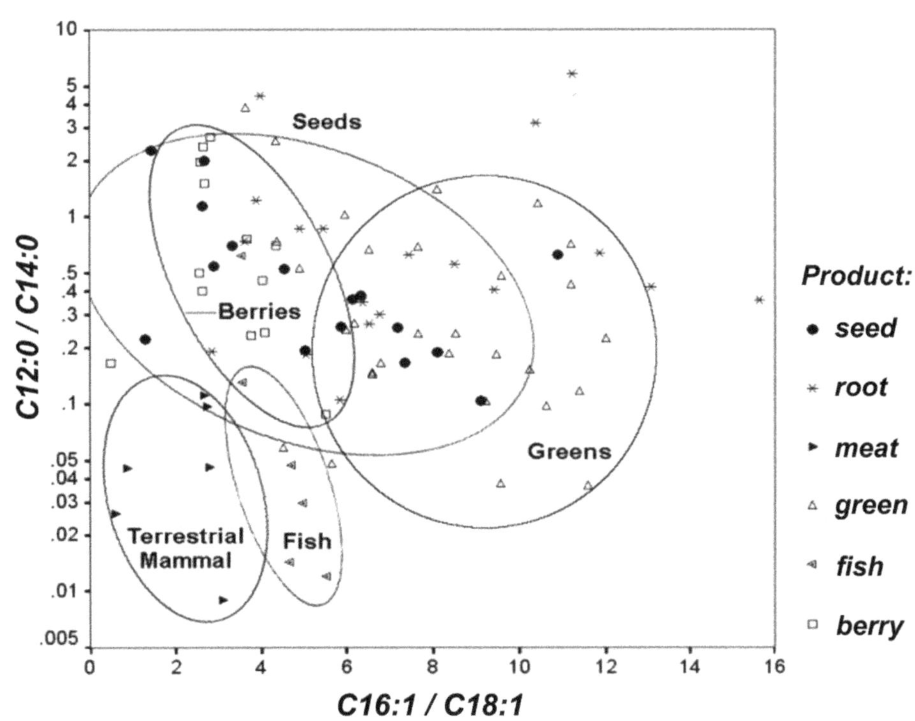


Figure 4: Biplot of two additional conservative fatty acid ratios for modern foods. Ellipses represent subjective estimations of the range of values for each food group (after Eerkens 2005).

In this chapter I argue that a small set of four ratios comprising eight common fatty acids regularly recovered in archaeological studies are relatively stable over time. Preliminary studies suggest that these four ratios can be used to separate fatty acid residues into five to six very general food classes which can help classify archaeological samples (Eerkens 2005). While this level of detail may not allow archaeologists to answer all of the questions they have about pots, or other residue containing artifacts, they can be useful as a starting point for generating basic data about the function of artifacts from archaeological settings. Further, I argue that it is not possible to assign individual sherds specific functions, but that groups of sherds can be evaluated based on their overall fatty acid makeup. Thus, despite the promise of such residue analyses to identify specific functions of artifacts, as discussed in the opening paragraphs, decomposition and other confounding problems probably relegate such analyses more to the role of producing hypotheses or supporting information about the function of artifacts.

In this respect, it would be worthwhile to augment fatty acid analysis with the extraction and evaluation of other organic residues, such as longer-chained lipids, waxes, carbohydrates or amino acids, and to use a range of analytical and derivatization techniques, such as stable isotope analysis and the isolation of isomers of the same fatty acid. Some in the field perform such multi-pronged approaches, but the majority (myself included) do not, favoring only one method for various reasons. The identification of additional organic compounds and new analytical techniques could serve to further subdivide some of the ellipses plotted in Figures 3 and 4 into smaller and more specific food categories.

## References


**Arnold**, D.E. (1985). *Ceramic Theory and Cultural Process*. Canbridge: Cambridge University Press.

**Beck**, M.E., J.M. Skibo, D.J. Hally and P. Yang (2002). Sample Selection for Ceramic Use-alteration Analysis. The Effects of Abrasion on Soot. *Journal of Archaeological Method and Theory* **29**: 1-15.

**Bennett**, J.W. (1943). Recent Developments in the Functional Interpretation of Archaeological Data. *American Antiquity* **9**: 208-219.

**Bennett**, J.W. (1944). The Interaction of Culture and Environment in the Smaller Societies. *American Anthropologist* **46**: 461-478.

**Bronitsky**, G. and R. Hamer (1986). Experiments in Ceramic Technology. The Effects of Various Tempering Materials on Impact and Thermal-shock Resistance. *American Antiquity* **51**:89-101.

**Brown**, J.A. (1989). The Beginnings of Pottery as an Economic Process. In S.E. van der Leeuw and R. Torrence (eds.). *What's New? A Closer Look at the Process of Innovation*. London: Unwin Hyman, pp. 203-224.

**Charters**, S., R.P. Evershed, A. Quye, P.W. Blinkhorn, and V. Reeves (1997). Simulation Experiments for Determining the Use of Ancient Pottery Vessels: the Behavior of Epicuticular Leaf Wax during Boiling of a Leafy Vegetable. *Journal of Archaeological Science* **24**:1-7.

**Christie**, W.W. (1989). *Gas Chromatography and Lipids*. Ayr (Scotland): Oily Press.

**Clark**, J.G.D. (1939). *Archaeology and Society*. London: Methuen.

**Craig**, O.E. and M.J. Collins (2002). The Removal of Protein from Mineral Surfaces: Implications for Residue Analysis of Archaeological Materials. *Journal of Archaeological Science* **29**: 1077-1082.

**Copley**, M.S., R. Berstan, S.N. Dudd, V. Straker, S. Payne and R.P. Evershed (2005). Dairying in Antiquity I. Evidence from Absorbed Lipid Residues dating to the British Iron Age. *Journal of Archaeological Science* **32**: 485-503.

**Costin**, C.L. (2000). The Use of Ethnoarchaeology for the Archaeological Study of Ceramic Production. *Journal of Archaeological Method and Theory* **7**: 377-403.

**DeMan**, J.M. (1992). Chemical and Physical Properties of Fatty Acids. In C.K. Chow (ed.). *Fatty Acids in Foods and their Health Implications*. New York: Marcel Dekker Inc., pp. 17-45.

**De Souza**, G.A., O.J.C. Santos, M.M. Conceição, D.M.C. Silva and S. Prasad (2004). A Thermoanalytic and Kinetic Study of Sunflower Oil. *Brazilian Journal of Chemical Engineering* **21**: 265-273.

**Deal**, M., and P. Silk (1988). Absorption Residues and Vessel Function. A Case Study from the Maine-Maritimes Region. In C.C. Kolb and L.M. Lackey (eds.). *A Pot for All Reasons. Ceramic Ecology Revisited.* Philadelphia: Laboratory of Anthropology, Temple University, pp. 105-125.

**Eerkens**, J.W. (2002). The Preservation and Identification of Piñon Resins by GC-MS in Pottery from the Western Great Basin. *Archaeometry* **44**: 95-105.

**Eerkens**, J.W. (2005). GC-MS Analysis and Fatty Acid Ratios of Archaeological Potsherds from the Western Great Basin of North America. *Archaeometry* **47, 1**: 83-102.

**Evershed**, R.P. (1993). Biomolecular Archaeology and Lipids. *World Archaeology* **25**: 74-93.

**Evershed**, R.P., S.N. Dudd, S. Charters, H. Mottram, A.W. Scott, A. Raven, P.F. van Bergen and H.A. Bland

(1999). Lipids as Carriers of Anthropologenic Signals from Prehistory. *Philosophical Transactions of the Royal Society of London B* **354**: 19–31.

**Evershed**, R.P., S.N. Dudd, V.R. Anderson-Stojanovic and E.R. Gebhard (2003). New Chemical Evidence for the Use of Combed Ware Pottery Vessels as Beehives in Ancient Greece. *Journal of Archaeological Science* **30**: 1-12.

**Evershed**, R.P., H.R. Mottram, S.N. Dudd, S. Charters, A.W. Stott and G.J. Lawrence (1997). New Criteria for the Identification of Animal Fats preserved in Archaeological Pottery. *Naturwissenschaften* **84**: 402-406.

**Feathers**, J.K. (1989). Effects of Temper on Strength of Ceramics: Response to Bronitsky and Hamer. *American Antiquity* **54**: 579-588.

**Frankel**, E.N. (1980). Lipid Oxidation. *Progress in Lipids Research* **19**: 1-22.

**Frankel**, E.N. (1987). Secondary Products of Lipid Oxidation. *Chemistry and Physics of Lipids* **44**: 73-85.

**Frankel**, E.N. (2005). *Lipid Oxidation* (second edition). Dundee: Oily Press.

**Fritsch**, C.W. and F.E. Deatherage (1956). A Study of the Volatile Compounds Produced by the Autoxidation of Methyl Oleate, Oleic Acid, and Cis-9-octadecene. *Journal of the American Oil Chemists Society* **33**: 109-113.

**Fullager**, R. and R. Jones (2004). Usewear and Residue Analysis of Stone Artefacts from the Enclosed Chamber, Rocky Cape, Tasmania. *Archaeology in Oceania* **39**: 79-93.

**Hally**, D.J. (1983). Use Alteration of Pottery Vessel Surfaces. An Important Source of Evidence for the Identification of Vessel Function. *North American Archaeologist* **4**: 3-26.

**Hegmon**, M. (2000). Advances in Ceramic Ethnoarchaeology. *Journal of Archaeological Method and Theory* **7**: 129-137.

**Henrickson**, E.F. and M.A. McDonald (1983). Ceramic Form and Function: an Ethnographic Search and an Archaeological Application. *American Anthropologist* **85**: 630-643.

**Heron**, C., R.P. Evershed and L.J. Goad (1991). Effects of Migration of Soil Lipids on Organic Residues Associated with Buried Potsherds. *Journal of Archaeological Science* **18**: 641-59.

**Hudlicky**, M. (1990). *Oxidations in Organic Chemistry*. Washington D.C.: American Chemical Society.

**Johnson**, J.S., J. Clark, S. Miller-Antonio, D. Robins, M.B. Schiffer and J.M. Skibo (1999). Effects of Firing Temperature on the Fate of Naturally occurring Organic Matter in Clays. *Journal of Archaeological Science* **15**: 403-414.

**Juhl**, K. (1995). *The Relation Between Vessel Form and Vessel Function: A Methodological Study*. Stavanger: Arkeologisk Museum.

**Linton**, R. (1944). North American Cooking Pots. *American Antiquity* **9**. 369-380.

**Malainey**, M.E., R. Przybylski and B.L. Sherriff (1999a). The Fatty Acid Composition of Native Food Plants and Animals of Western Canada. *Journal of Archaeological Science* **26**: 83-94.

**Malainey**, M.E., R. Przybylski and B.L. Sherriff (1999b). The Effects of Thermal and Oxidative Degradation on the Fatty Acid Composition of Food Plants and Animals of Western Canada. Implications for the Identification of Archaeological Vessel Residues. *Journal of Archaeological Science* **26**: 95-103.

**Malainey**, M.E., R. Przybylski and B.L. Sherriff (1999c). Identifying the Former Contents of Late Precontact Period Pottery Vessels from Western Canada using Gas Chromatography. *Journal of Archaeological Science* **26**: 425-438.

**Morton**, J.D. and H.P. Schwarcz (2004). Palaeodietary Implications from Stable Isotopic Analysis of Residues on Prehistoric Ontario Ceramics. *Journal of Archaeological Science* **31**: 503-517.

**Mottram**, H.R., S.N. Dudd, G.J. Lawrence, A.W. Stott and R.P. Evershed (1999). New Chromatographic, Mass Spectrometric and Stable Isotope Approaches to the Classification of Degraded Animal Fats Preserved in Archaeological Pottery. *Journal of Chromatography A* **833**: 209-221.

**Pearsall**, D.M., K. Chandler-Ezell and J.A. Zeidler (2004). Maize in Ancient Ecuador. Results of Residue Analysis of Stone Tools from the Real Alto Site. *Journal of Archaeological Science* **31**: 423-442.

**Quigg**, J.M., M.E. Malainey, R. Przyylski and G. Monks (2001). No Bones About it. Using Lipid Analysis of Burned Rock and Groundstone Residues to Examine Late Archaic Subsistence Practices in South Texas. *Plains Anthropologist* **46**: 283-303.

**Rafferty**, S.M. (2002). Identification of Nicotine by Gas Chromatography/Mass Spectroscopy Analysis of Smoking Pipe Residue. *Journal of Archaeological Science* **29**: 897-907.

**Reber**, E.A. and R.P. Evershed (2004). Identification of Maize in Absorbed Organic Residues. A Cautionary Tale. *Journal of Archaeological Science* **31**: 399-410.

**Rice**, P.M. (1987). *Pottery Analysis: a Sourcebook*. Chicago: University of Chicago Press.

**Rots**, V., and B.S. Williamson (2004). Microwear and Residue Analyses in Perspective: the Contribution of Ethnoarchaeological Evidence. *Journal of Archaeological Science* **31**: 1287-1299.

**Rye**, O.S. (1976). Keeping your Temper under Control: Materials and the Manufacture of Papuan Pottery. *Archaeology and Physical Anthropology in Oceania* **11**: 106-137.

**Santos**, J.C.O., A.V. Santos, A.G. Souza, S. Prasad and I.M.G. Santos (2002). Thermal Stability and Kinetic Study on Thermal Decomposition of Commercial Edible Oils by Thermogravimetry. *Journal of Food Science* **67, 4**: 1393-1398.

**Schiffer**, M.B. (1989). A Provisional Theory of Ceramic Abrasion. *American Anthropologist* **91**: 101-115.
**Skibo**, J.M. (1992). *Pottery Function: A Use-Alteration Perspective*. New York: Plenum Press.
**Skibo**, J.M., M.B. Schiffer and K.C. Reid (1989). Organic-tempered Pottery: an Experimental Study. *American Antiquity* **54**: 122-143.
**Smith**, H.I. (1910). *The Prehistoric Ethnology of a Kentucky Site. Anthropological Papers 6(2)*. New York, American Museum of Natural History.
**Smith**, M.F. Jr. (1985). Towards an Economic Interpretation of Ceramics. Relating Vessel Size and Shape to Use. In B.A. Nelson (ed.). *Decoding Prehistoric Ceramics*. Carbondale: Southern Illinois University Press, pp. 254-309.
**Steward**, J.H. and F.M. Setzler (1938). Function and Configuration in Archaeology. *American Antiquity* **4**: 4-10.
**Stott**, A.W., R.P. Evershed, S. Jim, V. Jones, J.M. Rogers and N. Tuross (1999). Cholesterol as a New Source of Palaeodietary Information. Experimental Approaches and Archaeological Applications. *Journal of Archaeological Science* **26**: 705-716.
**Trigger**, B. (1989). *A History of Archaeological Thought*. Cambridge: Cambridge University Press.
**Wadley**, L., M. Lombard and B. Williamson (2004). The First Residue Analysis Blind Tests: Results and Lessons Learnt. *Journal of Archaeological Science* **31**: 1491-1501.
**Willey**, G.R. and J.A. Sabloff (1980). *A History of American Archaeology* (second edition). San Francisco: Freeman.

# CHAPTER NINE
# A Comparative Study of Extractable Lipids in the Sherds and Surface Residual Crusts of Ceramic Vessels from Neolithic and Roman Iron Age Settlements in the Netherlands

T.F.M. Oudemans and J.J. Boon

Tania Oudemans; Institute for Atomic and Molecular Physics (AMOLF); P.O.-Box 41883; 1009 DB Amsterdam; the Netherlands; and Faculty of Archaeology; University of Leiden; P.O.-Box 9515; 2300 RA Leiden; the Netherlands; <info@taniaoudemans.com>; and Jaap Boon; Institute for Atomic and Molecular Physics (AMOLF);; P.O.-Box 41883; 1009 DB Amsterdam; the Netherlands. This work is part of the research program of the Faculty for Archaeology of the University of Leiden and Program 49 (BIOMSL) of the Foundation for Fundamental Research on Matter (FOM), a subsidiary of the Dutch Organization for Scientific Research (NWO). We gratefully acknowledge both the FOM and the Faculty for Archaeology, Leiden for financial support of T.F.M. Oudemans. In addition, we would like to thank, T.J. ten Anscher, L.P. Louwe Kooijmans, L.L. Therkorn and the late A.A. Abbink for supplying samples and contextual information about the various sites. We also gratefully acknowledge the extensive technical support and the access to laboratory facilities provided by R.P. Evershed and co-workers. Finally, we would like to thank S.V. Tsygankova, for advice on our graphics, and C.C. Bakels, for critical reading of the manuscript.

## Introduction: Lipid Analysis in Ceramic Studies

Pottery assemblages are a rich and durable source of information for the study of the daily behavior of people in the past. In order to assess the value of the information obtained from these assemblages, the use of the ancient vessels is an essential prerequisite. The identification of organic remains of ancient vessel contents can enable the retrieval of information about original vessel use. Since the 1970s, the study of organic residues has shown the preservation of many organic compounds in association with ceramics (Craig et al. 2000; Evershed et al. 1999; Evershed et al. 1992; Heron and Evershed 1993; McGovern et al. 1996; Mills and White 1987; Oudemans and Boon 1996; Oudemans et al. 2005; Pastorova et al. 1993; Regert and Rolando 2002; Rottländer and Schlichtherle 1979; Rottländer and Schlichtherle 1980).

The study of organic residues has focused primarily on fatty materials. Lipids are favored for organic residue studies due to their easy retrieval with solvent extraction and the continuous development of analytical techniques such as GC, GC/MS and gas chromatography isotope ratio mass spectrometry. Lipids also have obvious potential as diagnostic markers for the original vessel use due to their chemical stability (Eglinton and Logan 1991). In contrast to proteins and carbohydrates, lipids possess only a limited number of reactive sites resulting in relatively high resistance to thermal degradation during heating (Davídek et al. 1990, 169). In addition, the aliphatic nature of lipids results in low water solubility and thus enhances the immobilization of the molecular debris considered crucial to long term preservation at a molecular level (Eglinton and Logan 1991). Post-depositional exchange of lipids between residues and their surrounding soil has been shown to be very limited (Heron et al. 1991; Oudemans and Boon 1991; Oudemans, 2006).

## Introduction: Types of Residues

In a few rare cases, lipids have been preserved as solidified or liquid substances in sealed vessels (Gibson and Evans 1985; Shedrinski et al. 1991), but most frequently lipids have survived in visible crusts adhering to the interior or exterior surface of a vessel (Hill and Evans 1988; Oudemans and Boon 1991; 1996; Oudemans et al. 2005; 2007; Oudemans and Erhardt 1996; Patrick et al. 1985; Regert and Rolando 2002; Rottländer and Schlichtherle 1979) or absorbed within the ceramic matrix of the vessels (Charters et al. 1995; Charters et al. 1993; Condamin et al. 1979; Dudd et al. 1998; Evershed et al. 1994; Evershed et al. 1990; Evershed et al. 1997; Gianno et al. 1990; Heron et al. 1991; Mottram et al. 1999; Passi et al. 1981; Regert et al. 1998).

The relative suitability of different types of residues for the identification of original vessel content has been discussed by a number of investigators. Although substances in sealed vessels can be in relatively good condition, their sparseness makes them less suitable for systematic study of vessel use. Absorbed lipids may occur more frequently than visible surface residues (Evershed et al. 1991), and have been claimed to be easier to identify due to their better preservation (Rottländer 1990). On the other hand, some researchers detected lipids in surface residues while none were found in the adjacent sherd (Needham and Evans 1987; Regert et al. 2001). A number of additional methodological advantages have been formulated for the study of surface residues (Oudemans and Boon 1991; Oudemans et al. in press). In short, the study of surface residues makes it possible to sample only a limited number of use phases, while absorbed residues represent the accumulated deposits of multiple use-phases in addition to possible post-firing sealing agents. Extractions of absorbed residues may also include such sealing agents, complicating interpretation even more. Post-firing surface sealing with organic mixtures, the 'seasoning' of the vessel, is common amongst traditional potters and is performed with a variety of materials including common foodstuffs such as milk, oil and various starch-rich foods (Rice 1987, 163-164), as well as less edible materials such as beeswax, various resins and other plant materials (Arnold 1985, 139-140; Diallo et al. 1995; Kobayashi 1994). Stern et al. (2000) confirm that fatty acids extracted from Bronze Age Canaanite amphorae show that the jars were used to hold a lipid product, but that it was impossible to distinguish single use and multiple use. An additional reason to use surface residues in cooking vessels is the relatively higher degree of thermal degradation that has likely taken place in absorbed residues. Absorbed residues have usually been exposed to more severe heating regimes (both in temperature and in time) than residues situated on the interior surface of the vessel. Although numerous quantitative studies have been performed on lipids preserved in different residue types, no quantitative comparison of lipids extracts was ever published.

## Introduction: Aims

In this study the extractable lipids of different types of residues are quantitatively analyzed using corrected flame ionization detector (FID) response factors for each compound. Comparisons are made to increase our knowledge of the differences in lipid chemistry between charred and non-charred surface residues; between surface residues and the lipids absorbed in the underlying ceramic material and between charred surface residues from the Roman Iron Age and the Neolithic. In order to facilitate the comparison of the lipid profiles, three operational parameters (the saturation index, the hydrolysis index and the odd carbon number fatty acid index) are defined. The main purpose of this chapter is to address the potential variation in lipid preservation in different sample materials and to discuss the possible biomolecular origin of the extracted lipids.

## Experimental: Sample Material and Treatment

Organic residues from five different prehistoric contexts in the Netherlands were studied (Table 1). The main focus of this study was a ceramic assemblage recovered from an indigenous settlement at Uitgeest-Groot Dorregeest dating back to the Roman Iron Age (Abbink 1985; 1999). Both charred and non-charred residues were chosen for analysis. Non-charred surface residues from this settlement can appear as cream-colored crusts adhering to the interior vessel wall, or as red-brown films or dripping patterns on the interior or exterior vessel wall (Table 1). Surface residues were sampled as well as the ceramic fabric of the vessel directly underneath the surface residue. In one case (sample 34-0-12 from Uitgeest-Groot Dorregeest) three longitudinal sections of the vessel wall were sampled and lipids from the interior (S3), middle (S2) and exterior (S1) section of the vessel wall were extracted separately.

Charred surface residues of different age were collected to study the effect of burial time on the preservation of lipids. Residues from the Roman Iron Age settlements Schagen-Muggenburg (Abbink 1999; Therkorn 2004), Uitgeest-Groot Dorregeest and Uitgeesterbroekpolder 54 (Reyers 1985; Therkorn 2004) and from the Neolithic sites NO-Polder 14 (ten Anscher 2000/2001) and Hazendonk (Louwe Kooijmans 1974; 1976) were collected. All ceramic assemblages had roughly comparable burial conditions in peaty soil interspersed with sand and clay layers.

Most ceramics were washed in tap water, dried and stored in plastic bags for different lengths of time (up to 20 years). Ceramics from NO-polder 14 were treated specifically for organic residue sampling: directly after recovery from the field, pottery was wrapped in aluminum foil and stored at -20°C. Surface residues (about 5-10 mg) were scraped from the ceramics with a solvent cleaned scalpel, after removal of the upper 0.5 mm of the residue. Ceramic samples (about 2 g) were cut out of the vessel with a solvent cleaned scalpel, after removal of any surface residue and an additional 1 mm of ceramic. Samples were crushed in an agate mortar and stored in glass vials. Samples were prepared according to Evershed et al. (1990). In short, an internal standard (IS = 20 µg n-heptadecane) was added to each weighed sample, prior to extraction by solvent washing (10 ml chloroform/methanol, 2:1 v/v, 30 min ultrasonication). After centrifuging, the supernatant was dried in a round-bottomed flask by rotary evaporation at 50°C (in vacuum). A small amount (100 µl) of the solvent was added to transfer the total lipid extract (TLE) into a vial. One fifth (20 µl) of this extract was transferred into a second screw-topped vial and silylated with 25 µl N,O-bis(trimethylsilyl)trifluoroacetamide (BSTFA) containing 1% of trimethylchlorosilane (TMCS) and heated at 60°C for 10 min directly prior to analysis. All analytical grade solvents were distilled before use.

| Site | Period* | Sample number** | Appearance | N [%] | C [%] | H [%] | Total organic content [%] | C/N | C/H |
|---|---|---|---|---|---|---|---|---|---|
| Uitgeest-Groot Dorregeest | RIA | 34-0-30 | Char | 5.93 | 41.16 | 3.72 | 50.81 | 6.94 | 11.06 |
| Uitgeest-Groot Dorregeest | RIA | 35-7-28 | Cream colored | 0.18 | 3.52 | 1.20 | 4.90 | 19.56 | 2.93 |
| | | 35-7-28 S | Ceramic | | | | | | |
| Uitgeest-Groot Dorregeest | RIA | 34-0-12 | Char | 3.55 | 21.91 | 1.55 | 27.01 | 6.17 | 14.14 |
| | | 34-0-12 S3 | Ceramic | | | | | | |
| | | 34-0-12 S2 | Ceramic | | | | | | |
| | | 34-0-12 S1 | Ceramic | | | | | | |
| Uitgeest-Groot Dorregeest | RIA | 8-1 | Red brown | 0.79 | 6.55 | 1.25 | 8.59 | 8.29 | 5.24 |
| | | 8-1 S | Ceramic | | | | | | |
| Uitgeest-Groot Dorregeest | RIA | 14-6-4.4 | Char | 5.51 | 60.13 | 4.12 | 69.76 | 10.91 | 14.59 |
| | | 14-6-4.4 S | Ceramic | | | | | | |
| Uitgeest-Groot Dorregeest | RIA | 14-6-4.3c | Char | 4.10 | 42.46 | 2.37 | 48.93 | 10.36 | 17.92 |
| | | 14-6-4.3c S | Ceramic | | | | | | |
| Uitgeest-Groot Dorregeest | RIA | 14-6-4.2b | Char | 4.97 | 29.19 | 3.20 | 37.36 | 5.87 | 9.12 |
| | | 14-6-4.2b S | Ceramic | | | | | | |
| Schagen-Muggenburg | RIA | 79-1-1 | Char | 7.08 | 49.63 | 4.08 | 60.79 | 7.01 | 12.16 |
| Uitgeest 54 | RIA | 226-48 | Char | 7.78 | 40.04 | 4.14 | 51.96 | 5.15 | 9.67 |
| Uitgeest 54 | RIA | 320-17 | Char | 5.14 | 51.69 | 3.92 | 60.75 | 10.06 | 13.19 |
| Hazendonk | Neo | 32.740 | Char | 6.18 | 43.30 | 3.13 | 52.61 | 7.01 | 13.83 |
| Hazendonk | Neo | 33.781 | Char | 4.95 | 55.36 | 1.69 | 62.00 | 11.18 | 32.76 |
| NO-Polder 14 | Neo | 6745 | Char | 4.71 | 52.68 | 3.37 | 60.76 | 11.18 | 15.63 |
| NO-Polder 14 | Neo | 7054 | Char | 3.02 | 43.54 | 2.25 | 48.81 | 14.42 | 19.35 |

* RIA = Roman Iron Age; Neo = Neolithic

** R = surface residues S = ceramic material from vessel wall

Table 1: Overview of the archaeological samples discussed in this chapter.

## Experimental: Instrumentation

The amounts of carbon, hydrogen and nitrogen (CHN analysis) were determined for all surface residues in order to get a rough indication of the overall organic composition of the sample. Elemental composition was performed after samples were dried, weighted and analyzed twice using a Carlo Erba 1500 CHN analyzer. Results were referenced in weight percentages using N-phenyl-acetamide or acetanilide ($C_8H_9NO$) as a standard to determine relative detector response. C/N and C/H ratios are directly calculated from their weight percentages (not on a molar basis).

The analytical GC work was on a Hewlett-Packard 5890A gas chromatograph equipped with a FID and a Hewlett-Packard 3396A computing integrator and plotter. On-column injection was used to introduce samples into a 60 cm x 0.32 mm inner diameter retention gap of de-activated fused silica, connected to the analytical column, a polyamide clad analytical column of 12 m x 0.22 mm coated with a BP1 stationary phase (OV-1 equivalent, 0.1 μm film thickness), via a stainless-steel union with an inner diameter of 0.8 mm (SGE). The GC oven was programmed from 50°C (2 min isothermal hold after injection) to 350°C at a rate of 10°C/min, after which the temperature was maintained isothermal for 15 min. Helium was used as carrier gas at a constant column head pressure of 1.7 atm. The GC/MS was performed using a similar column in a Pye Unicam 204 GC linked to a VG 7070H double-focusing magnetic sector mass spectrometer. The MS was operated in the EI+ mode (70 eV) with a source temperature of approximately 300°C, an acceleration voltage of 4 kV. The effluent was scanned over the range *m/z* 40-700 in a total cycle time of 3 s. The date acquisition and processing was performed on a Finnigan INCOS 2300 data system.

## Experimental: Quantification

The quantitative data were derived from the peak areas measured using a GC with a FID. The peak areas were corrected for compound specific response with using the effective carbon number (ECN) per compound (Table 2), calculated according to Kaiser (1969, 99-103). The contribution of ester bonds was considered to be equal to the sum of an alcohol and a ketone group, being 0.55 for the 1- and 3-position and 0.35 for the 2-position in the acylglycerols (Ackman 1964). The ECN of unsaturated free fatty acids and monoacylglycerols (MAGs) is decreased with 0.1 per double bond (Scanlon and Willis 1985). Because saturated and unsaturated forms of diacylglycerols (DAGs) and triacylglycerols (TAGs) co-elute under current conditions, the effect of double bonds of acylglycerols (varying from 0.6 % for $D_{40:2}$ to 1.1 % for $T_{54:6}$) were neglected. The contribution of trimethylsilyl-groups (TMS) to the ECN of acids (3.0 for the -$CO_2$-TMS) and alcohols (3.69 for the H-C-O-TMS) were defined according to Scanlon and Willis (1985). Primary and secondary silylated alcohols were assumed to have the same contribution. The ECN of cholesterol was calculated at 29.19, assuming that cyclic C-atoms are comparable to aliphatic C-atoms, minus one double bond equivalent per closed ring, and that the TMS derivative of the 3β-hydroxyl group in cholesterol is comparable to the same group in an alcohol. One double bond in the 5-position was included in the ECN calculation of cholesterol.

The relative molar response factor $F(R\ molar)_i$ (Equation 1) expresses the relative amount of a component i necessary to obtain the same response (in area measured) as the IS (Kaiser 1969, 99-103; Scanlon and Willis 1985) and is defined as:

$$F(Rmolar)_i = \frac{ECN_{is}}{ECN_i} \qquad [1]$$

where $ECN_{is}$ is the calculated ECN for the IS (17.00 for heptadecane) and $ECN_i$ is the calculated ECN for compound i. Therefore the amount of every compound i present in the total sample $A_i$ can now be calculated and expressed in mol:

$$A_i = A_{is} \cdot F(Rmolar)_i \cdot \frac{X_i}{X_{is}} \qquad [2]$$

where $A_{is}$ is the known amount of IS added to the total sample (in mol), $X_i$ is the measured relative peak area for compound i(in percent), and $X_{is}$ is the measured relative peak area for the IS (in percent). In order to calculate the composition of samples before derivatization, the normalized weight percentage $WP_i$ of the original compounds is calculated according to:

$$WP_i = \frac{A_i \cdot MW_{i\,(underivatized)}}{\sum_{i=1}^{n}\left(A_i \cdot MW_{i\,(underivatized)}\right)} \cdot 100\% \qquad [3]$$

where $MW_{i\ (underivatized)}$ is the molecular weight of compound i in underivatized form (in mg), and n is the total number of compounds in the sample.

| Compound class | Mass peak (*m/z*) | Diagnostic fragment ions | Range** |
|---|---|---|---|
| Fatty acids* | $M^{+}$. | | e/s: C8-C30 |
| | $[M-15]^{+}$ | $[M-CH_3]^{+}$ | e/us: C16:1, C18:1, |
| | *m/z* 73 | $[Si(Me)_3]^{+}$ | C18:2, C20:1, C22:1 |
| | *m/z* 75 | $[HO=Si(Me)_3^{+}$ | o/s: C9-19, C23, C25 |
| | *m/z* 117 | $[Si(Me)_3OCO]^{+}$ | |
| Monoacylglycerols * | $[M-15]^{+}$ | $[M-CH_3]^{+}$ | M14:0, M16:0, M16:1 |
| | *m/z* 129 | $[(Me)_3Si-O=CH-CH=CH_2]^{+}$ | M18:0, M18:1 |
| 1-monoacyl | $[M-103]^{+}$ | $[M-(CH_2-O-Si(Me)_3)]^{+}$ | |
| 2-monoacyl | *m/z* 218 | $[(Me)_3SiO-CH=CH-CH_2-OSi(Me)_3]^{+}$ | |
| Diacylglycerols * | $[M-15]^{+}$ | $[M-CH_3]^{+}$ | e: D26-D36 |
| | *m/z* 129 | $[(Me)_3Si-O=CH-CH=CH_2]^{+}$ | o: D29-D35 |
| | $[M-RCOO]^{+}$ | | |
| | $[M-(RCOO+1)]^{+}$ | $[M-RCOOH]^{+}$ | |
| | $[M-(RCOO+14)]^{+}$ | $[M-RCOOCH_2]^{+}$ | |
| | $[RCO]^{+}$ | acyl fragment ion | |
| | $[RCO+74]^{+}$ | $[RCOO-CH_2-CH(OH)CH_2]^{+}$ | |
| | $[RCO+128]^{+}$ | $[RCOO-CH_2-CH(O-C(CH-CH_2)-OCH_2]^{+}$ | |
| Triacylglycerols | $[M-RCOO]^{+}$ | | e: T40-T45 |
| | $[M-(RCOO+1)]^{+}$ | $[M-RCOOH]^{+}$ | o: T43-T53 |
| | $[M-(RCOO+14)]^{+}$ | $[M-RCOOCH_2]^{+}$ | |
| | $[RCO]^{+}$ | acyl fragment ion | |
| | $[RCO+74]^{+}$ | | |
| | $[RCO+128]^{+}$ | | |
| Cholesterol * | $M^{+}$. | | |
| | $[M-15]^{+}$ | $[M-CH_3]^{+}$ | |
| | *m/z* 129 | $[(Me)_3Si-O=CH-CH_2=CH]^{+}$ | |
| | $[M-129]^{+}$ | $[M-(Me)_3Si-O=CH-CH_2=CH]^{+}$ | |
| Alcohols * | $M^{+}$. | | e: C12-C18, C24-C32 |
| | $[M-15]^{+}$ | $[M-CH_3]^{+}$ | o: C15 |
| | *m/z* 103 | $[(Me)_3Si-O-CH_2]^{+}$ | |
| | *m/z* 75 | | |
| Elementary sulphur | *m/z* 64, 128, 256 | $S_2$, $S_4$, $S_8$ | |
| Other steroids * | *m/z* 215, 257 | | Stanols |
| Alkanes | *m/z* 57, 71, 85 | | C15-C32 |
| Phthalate esters | *m/z* 149 | | dibutyl, dimethyl |
| Squalene | *m/z* 410 | $M^{+}$. | |

* detected in silylated form
** e/s = even numbered, saturated; o/s = odd numbered, saturated; e/us = even numbered, unsaturated

Table 2: Compounds detected by GC/MS.

The total lipid yield (TLY, Equation 4) of the extraction procedure is defined (in mg/g), according to:

$$TLY = \frac{\sum_{i=1}^{n}\left(A_i \cdot MW_{i\ (underivatized)}\right)}{W_s} \qquad [4]$$

where $W_s$ is the amount of sample used for extraction in gram. This calculation is based on the assumptions that the extraction is equivalent per lipid species and that a 100% extraction of the added IS is achieved.

In order to facilitate the comparison of lipid profiles, three operational parameters are defined that represent major aspects of lipid preservation and degradation. These indices are based on the relative total weight percentages per compound as calculated in Equation 3. The saturation index $I_{sat}$ of the free fatty acids serves to express the proportion of saturated even carbon number fatty acids in the extract. The saturation index $I_{sat}$ is defined as:

$$I_{sat} = \frac{\sum(WP_i\ saturated\ even\ FA)}{\sum(WP_i\ all\ even\ FA)} \qquad [5]$$

This value is a tentative measure for the degree of polymerization that has occurred in the sample as a result of oxidation or heating under anoxic circumstances. The hydrolysis index $I_{hydr}$ represents the proportion of even carbon number free fatty acids relative to all even carbon number acyl fragments in TAGs or free fatty acids. The hydrolysis index $I_{hydr}$ is defined as:

$$I_{hydr} = \frac{\sum(WP_i\ even\ FA)}{\sum(WP_i\ even\ FA\ and\ all\ TAG)} \qquad [6]$$

This parameter provides a measure for the degree of hydrolysis that has taken place in a sample. Acyl fragments can be hydrolyzed by microbial activity (enzymatic hydrolysis) and under alkaline or acidic conditions (chemical hydrolysis). The odd carbon number fatty acid index $I_{o/e}$ corresponds to the proportion of odd carbon number free fatty acids to the total fatty acid abundance and is defined as:

$$I_{o/e} = \frac{\sum(WP_i\ odd\ FA)}{\sum(WP_i\ all\ FA)} \qquad [7]$$

Because the fatty acids C15:0 and C17:0 are the major contributors to the total weight in the numerator of this index, the $I_{o/e}$ can be interpreted as a reflection of the amount of bacterial material in the sample.

## Results: CHN Analysis

Elemental CHN analysis (Table 1) shows a distinct difference in total organic content between the charred residues (27-70%) and the non-charred residue (4-9%). The non-charred residues consist primarily of inorganic compounds and contain hardly more organic material than the ceramic material, which has a total organic content of 4.7% (Oudemans et al. in press). Although the charred residues from Uitgeest-Groot Dorregeest showed more variation than charred residues from other sites, they are roughly comparable in overall chemical contents (Table 1). There is a considerable variation in elemental composition of the charred residues. The C/H ratios vary from 9.12-32.76, indicating a less aliphatic and more condensed nature of the material as the ratio increases. The C/N ratios vary from 5.15-14.42, indicating a decrease in the amount of nitrogen present in the material as the ratio increases.

## Results: Qualitative Lipid Analysis

The compounds identified by GC/MS are summarized in Table 2 and further illustrated in Figures 1 and 2. The identity of the compounds in the TLEs were deduced largely from their EI+ mass spectra or from their TMS-derivatives using the characteristic ions (Table 2), given by Odham and Stenhagen (1972a; b) and Waller et al. (1972; 1980). Although isomers of C18:1 and diacylglycerols (DAGs) were detected, no isomer specific identification can be given under the analytical conditions employed. Isomers of C15:0 and C17:0 (normal-, iso- and anteiso-) and monoacylglycerols (1-, and 2- forms) were identified in some of the samples, but have been summarized in the quantitative results. The EI+ mass spectra of TAGs display such weak molecular ions ($M^+$) and fragment ions ($[M-18]^+$) that they are of little diagnostic value. The total carbon number of the TAGs was therefore established by comparison of the retention times with those of authentic compounds. Fragment ions representing the loss of one acyl moiety give information about the nature of diacyl fragments. The ions representing the acyl fragments give an indication of the ratio of acyl moieties present in the intact TAGs (Figure 3). All TAGs of a given total carbon number co-elute on the stationary phase as employed in this study. Hence, the identification of all TAGs is limited to molecular species.

Due to the high complexity of the mixtures analyzed, it was not always possible to identify all the peaks produced by high temperature gas chromatography (HTCG). Some of the minor components (including alkanes, wax esters and some steroids) could not be fully identified due to low signal-to-noise ratios in the GC/MS analyses, or the absence of diagnostic ions in the mass spectra. The total measured peak area of all identified compounds (including the IS) in the HTGC varied between 51-100% (Table 3), with an average of 84% for surface residues (although in two samples only 53% was identified), and 69% for ceramic samples (although two samples contained no identifiable lipids and one sample only 5% that could be identified).

## Results: Quantitative Lipid Analysis

A first assessment of a sample is made by calculation of the TLY based on all compounds identified by GC with an ECN which could be calculated (Equation 4, Tables 3 and 4). Phthalate esters were not included because they were considered contamination. TLYs show considerable variation between samples (Table 3), but general trends are visible. First, surface residues always yield more lipids per gram sample than the ceramic directly adjacent to it (20 to 1000 times higher). Surface residues from Uitgeest-Groot Dorregeest produce TLYs between 0.47-27.52 mg/g, while the adjacent ceramic samples yield TLYs between 0.00-0.16 mg/g. Second, most charred surface residues (9 of the 12 samples) produce lipid yields 5-50 times higher than non-charred surface residues (averaging 1.71 mg/g). Finally, charred residues from different excavations vary considerably in lipid yield. Those from Schagen-Muggenburg and Uitgeest 54 gave relatively high TLYs (between 43.43-139.56 mg/g) while those from Neolithic sites exhibit lower yields (between 1.77-19.59 mg/g) with an average TLY comparable to that from Uitgeest-Groot Dorregeest.

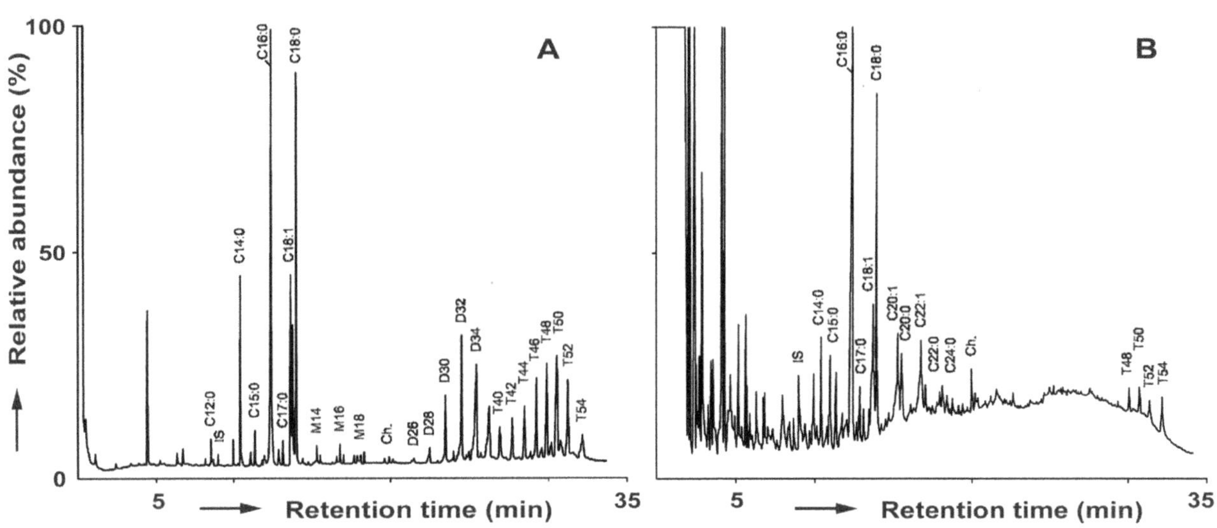


Figure 1: High temperature gas chromatograms of lipids from charred surface residues from sample 14-6-4.2b R from the Roman site Uitgeest-Groot Dorregeest (A) and sample 33.781 from the Neolithic site Hazendonk (B). IS = internal standard; M = monoacylglycerol; D = diacylglycerol; T = triacylglycerol, Ch = cholesterol and Ph = phthalate ester (numbers represent the total number of carbon atoms in the acyl moieties of lipids).

In order to provide a quantitative assessment of the highly complex lipid extracts, the normalized weight percentages of each identified underivatized lipid were calculated (Equation 3, Table 4). The chromatogram of charred residue 14-6-4.2b, from Uitgeest-Groot Dorregeest, shows several classes of compounds including free fatty acids, MAGs, DAGs, TAGs, cholesterol and phthalate esters (Figure 1A). Although some variation can be seen between the charred residues from Uitgeest-Groot Dorregeest, most samples were found to be of comparable lipid composition (Figure 2C R). However, the lipid composition of non-charred surface residues from Uitgeest-Groot Dorregeest appeared significantly different. These lipid profiles showed no odd carbon number fatty acids and relatively low percentages of free fatty acids as can be seen in the chromatogram of cream-colored residue 35-7-28 and red-brown residue 8-1 (Figures 2A R and 2B R).

Absorbed residues from Uitgeest-Groot Dorregeest yielded lower proportions of acyl lipids, and relatively more odd carbon number fatty acids than those of the surface residues (Figure 2, Table 4). Vessels with non-charred residues yielded little or no absorbed lipids from the ceramic matrix (Figures 2A S and 2B S). The comparison between lipid extracts from vessel walls and their directly adjacent surface residues was performed on six vessels, of which four preserved charred and two preserved non-charred residues. The profile of the lipids absorbed in the vessel walls does not necessarily reflect that of the solid residue situated on the vessel surface. Only in two cases (14-6-4.3c and 34-0-12) very similar profiles were observed (Figures 2C R and S). All vessels with charred surface residues contained absorbed lipids. In sample 34-0-12 the middle section contained no identifiable lipids, while both interior and exterior produced low TLYs (Table 4). The interior section showed a lipid profile similar to the surface residue directly adjacent. The charred surface residues from the Neolithic sites rendered lipid traces that contained a relatively high percentage of free fatty acids, including long-chain fatty acids, up to 24 carbon atoms, no detectable MAGs and DAGs and higher proportions of odd carbon number fatty acids.

| Site | Sample number | Appearance | Total lipid yield (mg/g) | | Identified peak area [%] | I sat [5]* | I hyd [6]* | I o/e [7]* |
|---|---|---|---|---|---|---|---|---|
| | | | Corrected [4]* | Assuming linear response | | | | |
| Uitgeest-Groot Dorregeest | 34-0-30 | Char | 27.52 | 25.07 | 98% | 0.78 | 0.51 | 0.07 |
| Uitgeest-Groot Dorregeest | 35-7-28 | Cream | 1.32 | 2.05 | 53% | 1.00 | 0.39 | 0.00 |
| | 35-7-28 S | Ceramic | 0.01 | 0.03 | 5% | - | 1.00 | 0.00 |
| Uitgeest-Groot Dorregeest | 34-0-12 | Char | 0.47 | 0.58 | 91% | 0.80 | 1.00 | 0.05 |
| | 34-0-12 S3 | Ceramic | 0.02 | 0.04 | 62% | 1.00 | 1.00 | 0.21 |
| | 34-0-12 S2 | Ceramic | - | - | - | - | - | - |
| | 34-0-12 S1 | Ceramic | 0.02 | 0.02 | 92% | 0.98 | 1.00 | 0.00 |
| Uitgeest-Groot Dorregeest | 8-1 | Red brown | 2.10 | 2.30 | 100% | 1.00 | 0.10 | 0.00 |
| | 8-1 S | Ceramic | - | - | - | - | - | - |
| Uitgeest-Groot Dorregeest | 14-6-4.4 | Char | 14.77 | 14.42 | 99% | 0.70 | 0.78 | 0.07 |
| | 14-6-4.4 S | Ceramic | 0.01 | 0.03 | 85% | 0.85 | 1.00 | 0.11 |
| Uitgeest-Groot Dorregeest | 14-6-4.3c | Char | 4.71 | 4.99 | 92% | 0.89 | 0.82 | 0.10 |
| | 14-6-4.3c S | Ceramic | 0.16 | 0.18 | 100% | 0.92 | 0.76 | 0.09 |
| Uitgeest-Groot Dorregeest | 14-6-4.2b | Char | 9.97 | 9.13 | 96% | 0.82 | 0.54 | 0.08 |
| | 14-6-4.2b S | Ceramic | 0.04 | 0.16 | 71% | 0.92 | 1.00 | 0.18 |
| Schagen-Muggernburg | 79-1-1 | Char | 139.56 | 132.42 | 94% | 0.61 | 0.39 | 0.10 |
| Uitgeest 54 | 226-48 | Char | 52.48 | 53.70 | 74% | 1.00 | 0.48 | 0.20 |
| Uitgeest 54 | 320-17 | Char | 43.43 | 42.66 | 86% | 0.96 | 0.43 | 0.15 |
| Hazendonk | 32.740 | Char | 19.59 | 22.06 | 83% | 0.85 | 0.80 | 0.22 |
| Hazendonk | 33.781 | Char | 7.38 | 7.79 | 74% | 0.74 | 0.90 | 0.12 |
| NO-Polder 14 | 6745 | Char | 11.86 | 13.96 | 84% | 1.00 | 0.43 | 0.33 |
| NO-Polder 14 | 7054 | Char | 1.77 | 2.80 | 53% | 1.00 | 0.53 | 0.64 |

Table 3: Total lipid yield and preservation indices. *) Numbers refer to equations in the text.

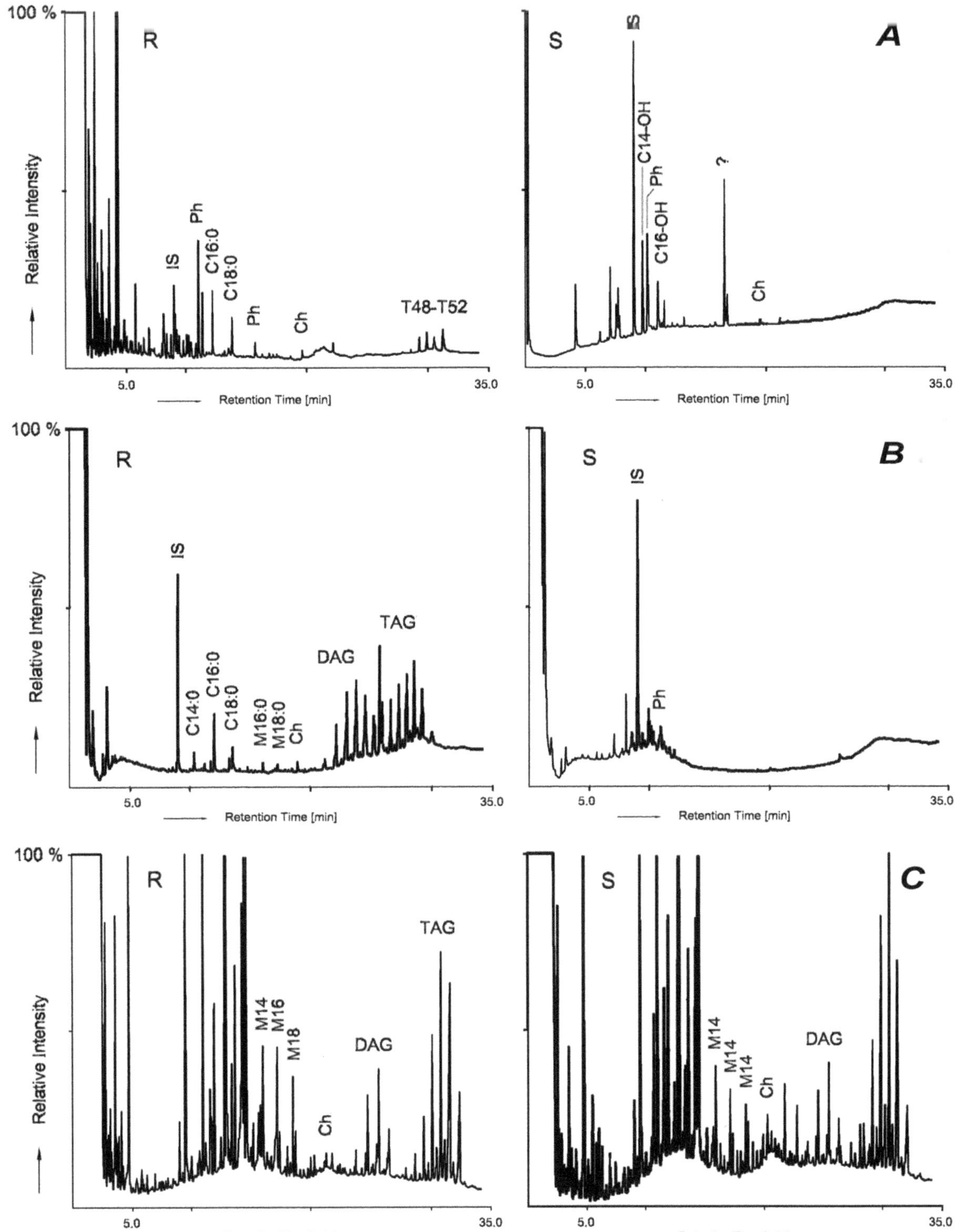


Figure 2: Comparison between lipids from surface residues (R) and lipids absorbed in sherds (S) in cream colored crust 34-7-28 (A), brown residue 8-1 (B) and charred residue 14-6-4.3c (C). IS = internal standard; M = monoacylglycerols; DAG = diacylglycerols; TAG = triacylglycerols, Ch = cholesterol and Ph = phthalate ester (numbers represent the total number of carbon atoms in the acyl moieties of lipids).

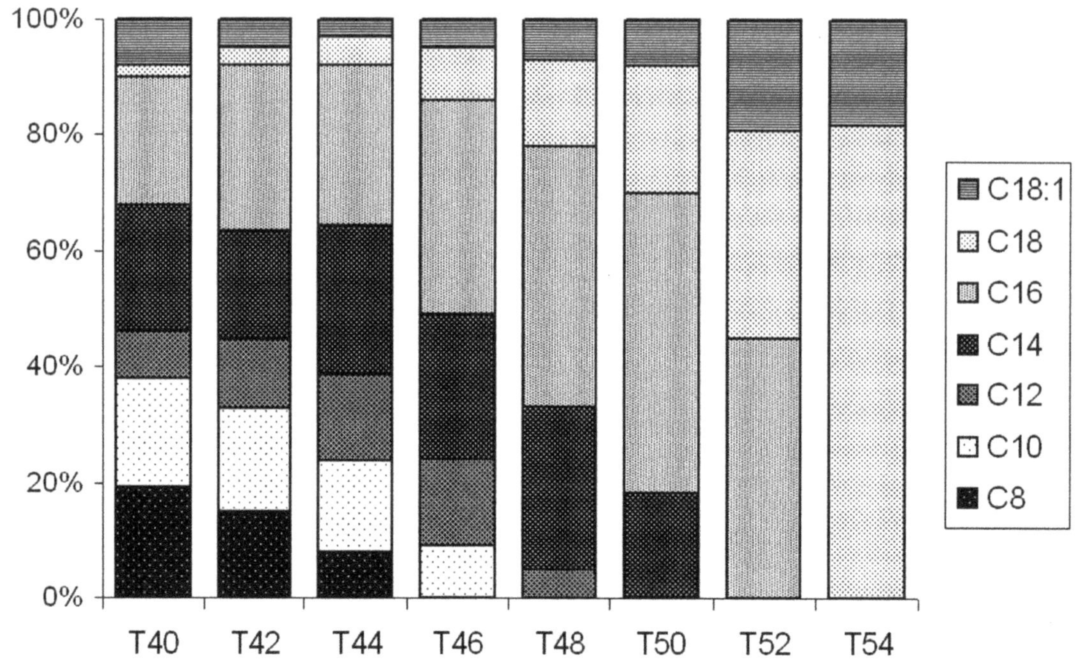


Figure 3: Relative composition of acyl fragments in mass spectra of intact even carbon number TAGs of charred surface residue 34-0-30 from Uitgeest-Groot Dorregeest. The percentages are relative numbers based on the intensities of R-$COO^+$ fragments in the EI+ mass spectra of each TAG. Assuming that fragmentation is equivalent for all acyl chains, these figures can be seen as representing the acyl composition of intact TAGs.

## Discussion: Lipid Quantification

Most earlier quantitative lipid studies have been based on the assumption that all compounds exhibit similar responses in the FID of the GC. Although this assumption is valid when closely related compounds are being investigated, differences in response of the FID may be observed when compounds show widely varying chemical properties (Kaiser 1969, 99-103). In this study, rather than assuming equivalent responses for all components of the TLE, corrected response factors for each compound were calculated in order to enhance quantitative precision. Although differences between such TLYs and the traditional uncorrected TLYs are shown to be considerable (10-20%), especially when dealing with low overall yields (Table 3), they are not in the same order of magnitude as the differences between TLYs calculated for lipid extracts originating from different excavations (or even between different kinds of residues within one excavation). For instance, the ceramic material from Uitgeest-Groot Dorregeest shows uncorrected lipids yields between 0.02-0.18 mg/g, while lamps and dripping dishes from the medieval site at Raunds in the UK frequently contained yields between 0.1-1.0 mg/g (Charters et al. 1993; Evershed et al. 1999; Evershed et al. 1991), and amphorae from the Late Bronze Age in the Western Isles of Scotland contained between 0.025-0.3 mg/g lipid (Craig et al. 2005). These large differences completely overshadow the smaller differences due to correction of the FID response factors.

The approximation of equivalent response is probably sufficiently precise to allow general comparisons of extractable lipids in soil and potsherds (Heron et al. 1991), or comparisons between excavations, and probably sufficiently precise to demonstrate rough differences in concentration of lipids accumulated in different parts of vessels (Charters et al. 1993). However, the use of corrected FID response factors for each compound is especially relevant when comparing relative lipid compositions. Discrepancies of ± 10-15% for various compounds can be seen (Table 4). When quantification of each compound to microgram precision is needed for comparisons with published lipid compositions of reference materials, or detailed comparison between lipid profiles, correction is highly desirable.

## Discussion: Chemotaxonomic Markers

The suitability of lipids as chemotaxonomic markers, or biomarkers, depends on their diagnostic value and their capacity for survival during long-term burial. Although even carbon number free fatty acids, with 4-24 carbon atoms, MAGs, DAGs and TAGs occur commonly in plant and animal fats (Hillditch & Williams 1964, 6-25), not all different compound classes are equally suitable as taxonomic markers. In this study only TAGs, sterols and free fatty acids were used as diagnostic chemotaxonomic markers. MAGs and DAGs are excluded because their origin is too ambiguous. They can be part of the original prehistoric lipid profile, be formed during hydrolysis of the original lipids or be the result of microbial activity. Although the same is true for free fatty acids, they can be diagnostic in specific cases.

TAGs are not produced by micro-organisms and therefore highly diagnostic for the plant or animal origin of the residue. Due to their insolubility in water, TAGs are not likely to leach out of their original depositional matrix and are unlikely to be exchanged with the surrounding soil. Oils, on the other hand, will undergo a 'drying process', a combination of cross-linking, polymerization and oxidation, if sufficient di- or tri-unsaturated acyl fragments are present in combination with oxygen (Mills and White 1987, 30-32). This effect will lead to a selective preservation of saturated extractable TAGs. A second chemical change that commonly occurs in TAGs is the chemical or enzymatic hydrolysis of the ester moieties leading to an overall loss of TAGs (Evershed et al. 1995a). Additionally, long-chain carboxylic acids in free fatty acids or TAGs can undergo condensation (though ketonic decarboxylation) when exposed to temperatures around 400ºC in the presence of calcium salts (Evershed et al. 1995b; Raven et al. 1997). These condensation processes cause the formation of long-chain ketones in cooking vessels and a loss of TAGs. In short, even when the lipid profile contains adequate amounts of TAGs, they may not exactly reflect the original TAG composition.

Sterols are an important minor class of lipids with diagnostic value in organic residue studies (Evershed et al. 1992). Sterols are diagnostic for animal (cholesterol) or plant (sitosterol and campesterol) products. Sterols have low solubility in water and are not easily damaged by heating (damage will occur around 280-300ºC), but are relatively easily oxidized into fats and oils (Davídek et al. 1990, 204, 216). The interpretation of cholesterol as an indicator for animal products must be made with caution because of the possibility of post-excavation contamination with cholesterol through handling of the potsherds. Some oxidation products of cholesterol have been detected in Saxon oil (Evershed et al. 1992). Microbial reduction of 5-sterols (like cholesterol) to 5α(H)- and 5β(H)-stanols occurs commonly under anaerobic conditions in the intestines of humans and animals and during diagenesis in sediments (Mackenzie et al. 1982). This process may also take place in the context of the original residue.

Although free fatty acids are abundantly present in most organic residues, they also illustrate most clearly the difficulty in assigning degraded lipids to a specific source. Extracted fatty acids may result from a mixture of different sources, including the remains of the original vessel content as well as the secondary products of microbial activity. In addition, a wide range of degradative pathways exists for free fatty acids, causing the overall free fatty acid composition to become an unreliable chemotaxonomic indicator. First, selective degradation of unsaturated fatty acids can occur as a result of oxidation or autoxidation in fresh materials (Davídek et al. 1990, 201-204). Additional condensation processes take place during heating or cooking of lipids (Malainey et al. 1999). When heated up to 270-300°C with limited access to oxygen, unsaturated lipids (primarily the polyunsaturated fatty acids typical for plant oils) will form cyclic hydrocarbons or acyclic polymers (Davídek et al. 1990, 195). Long-chain carboxylic acids can undergo condensation (though ketonic decarboxylation) when exposed to temperatures around 400ºC in the presence of calcium salts (Evershed et al. 1995b; Raven et al. 1997). Together, all these network forming processes are likely to be responsible for the formation of non-extractable aliphatic structures of which the fragments (alkanes and alkenes) were detected in pyrolysates of some surface residues and most sherd samples (Oudemans and Boon 1991). Anoxic conditions and temperatures up to 300°C may well have been present in the ceramic wall of vessels during cooking, and in some of the surface residues during severe charring. Stern and co-workers confirmed the hypothesis that hard to extract fatty acids may indeed be present in the ceramic matrix, bound as cross-linked macromolecules (Stern et al. 2000).

Secondly, selective loss of short-chain fatty acids can occur as a result of enzymatic and non-enzymatic hydrolysis of acyl lipids during the use of the vessel as well as after deposition. The enhanced volatility and water solubility of short-chain fatty acids may also result in a selective loss. Enzymatic degradation of intact fatty acids by micro-organisms through β-oxidation can also play a role in this process through loss of one or more pairs of C atoms from the acyl chain (Leninger 1977). This effect is commonly observed in bog bodies and buried fats such as bog butter (Evershed 1992; Thornton et al. 1970). Third, alkaline environments enhance the transformation of free fatty acids to salts of fatty acids and can produce salts of various nature. Transformation of fatty acids into insoluble salts occurs commonly in fresh fat buried in the ground, for instance during the formation of adipocere (mortuary wax) which consists mainly of fatty acids and their calcium salts (Eglinton and Logan 1991). Some of these salts are relatively soluble in water and can cause fatty acids to leach out of their original matrix, while others, such as calcium and magnesium salts, are virtually insoluble in either water or organic solvents. Although this prevents leaching out, it also prevents extraction during analysis, resulting in deviant lipid profiles. Some researchers have warned for this phenomenon (Condamin et al. 1979; Rottländer and

Schlichtherle 1979). This conversion was shown to occur under arid conditions in an Ancient Egyptian sealed stone vessel. Of the oil probably stored in this vessel only a mixture of salts of long-chain fatty acids remained (Shedrinski et al. 1991). Stern and co-workers undertook to extract such salts using an acidic extraction of ceramic samples, but only released very low amounts of the 'recalcitrant' fatty acids. The researchers concluded the fatty acids were not salts but bound as cross-linked macromolecules (Stern et al. 2000).

Other known degradative pathways that commonly occur in fatty materials buried in the ground are the formation of hydroxy fatty acids formed through the hydration of double bonds in adipoceres (Den Dooren de Jong 1961; Evershed 1991; 1992), and the formation of isomers of mono-unsaturated fatty acids observed in bog bodies (Evershed 1991; 1992). No hydroxy fatty acids were detected in the lipids extracted in this study, so this pathway was obviously not active. Although the latter pathway may have taken place, isomers of C18:1 were not separated in this analysis, so no conclusions can be drawn about this process.

## Discussion: Lipid Preservation and Degradation

Some significant aspects of lipid preservation and degradation can be studied using the operational parameters defined in this chapter (Table 4). The saturation index $I_{sat}$ (Equation 5, Table 3) expresses the proportion of saturated even carbon number fatty acids in the residue and is a tentative measure for the degree of polymerization that has occurred in the sample as a result of thermal or oxidative degradation. Contrary to expectations, no correlation could be found between the saturation index (measuring the amount of polymerization in fatty acids) and the C/H ratio (a measure of the overall condensation in residue). This suggests that oxidation without heating plays a prominent additional role in the degree of saturation of the extractable lipids.

The hydrolysis index $I_{hydr}$ (Equation 6, Table 3) provides a measure for the degree of hydrolysis that has taken place in a sample. Acyl fragments can be hydrolyzed by microbial activity (enzymatic hydrolysis), under alkaline or acidic conditions or as a result of heating in the presence of water (non-enzymatic hydrolysis). It must be kept in mind that, under alkaline conditions, free fatty acids may be present in the form of insoluble salts, which excludes them from extraction. However, under acidic conditions free fatty acids will be preserved in their free form in the original matrix (Eglinton and Logan 1991), unless subsequent degradation pathways have effected their preservation (such as the selective loss of short-chain or unsaturated fatty acids). It was noted that the hydrolysis index does not appear to be correlated to TLY in this study. This could indicate that the degree of hydrolysis does not determine the overall lipid preservation and that lipids are commonly preserved even after hydrolysis.

The odd carbon number fatty acid index $I_{o/e}$ (Equation 7, Table 3) corresponds to the proportion of odd carbon number free fatty acids to the total abundance of free fatty acids. In the extracts under investigation C15:0 and C17:0 are the major contributors to the total weight in the numerator of this index. Since these fatty acids are primarily formed during bacterial growth, the $I_{o/e}$ can be interpreted as a reflection of the relative amount of bacterial matter (directly or indirectly) contributed to the sample. Bacterial matter can be incorporated into the residue as part of a ruminant milk fat, during the original vessel use, as ruminant milk fat contains odd carbon number TAGs that can produce odd carbon number fatty acids after hydrolysis (Breckenridge and Kuksis 1967; Murata 1977). However, bacterial matter can also be incorporated into the residue in an indirect way during post-depositional bacterial degradation. The presence of the typical combination of n-, iso- and anteiso- isomers of C15:0 and C17:0 is diagnostic for bacterial growth (Nes and Nes 1980, 135; Shaw 1974). When odd carbon number TAGs are absent and lipid hydrolysis is not complete, a high $I_{o/e}$ should be considered indicative of post-depositional bacterial degradation.

In the charred residues a rough positive correlation exists between the C/N ratio and the C/H ratio, indicating that an increase in condensation goes hand in hand with a decrease in the amount of nitrogen present in the material. This is consistent with the conclusions from a combined FTIR/NMR study of the solid fraction of surface residues (Oudemans et al. 2007). Severe heating (over 250ºC) over a longer period of time (over 2 hours) was shown to create progressively condensed materials with high C/H ratios (13-16), high overall organic contents (57-67%) and relatively few remaining biomolecular characteristics, such as nitrogen containing compounds or compounds with lipid characteristics. Because part of the charred residues under investigation fall within these parameters, low TLYs were expected from these residues (Table 1). However, the data show clearly that the highest amounts of lipid are extracted from charred residues with both a high C/H and a high C/N, thus necessitating modification of the above model of condensation. Although a certain amount of condensation is obviously desirable for preservation, the charred residue needs to have a nitrogen component in order to yield lipids. This correlation suggests that the presence of lipids is either determined by the presence of original biomaterials containing protein (meat, fish, high fat content seeds) or that the preservation of lipids is determined by the presence of nitrogen containing compounds contributing to the formation of the charred residue. The last effect could be caused by Maillard reactions (chemical reactions between amino acids and reducing sugars, usually requiring the addition of heat), known to produce highly insoluble materials of strongly refractory nature (non-enzymatic browning).

## Discussion: Possible Origin of Lipids

It is obvious from the above that the total extractable lipid composition in archaeological materials cannot be diagnostic unless careful consideration is given to all possible degradation mechanisms involved. A plant origin is hard to assign to any of the extracts studied. Although the relative proportion of unsaturated fatty acids is relatively low for plants, the presence of very long chain free fatty acids, with 26-30 carbon atoms, known to be the hydrolytic degradation products of wax esters (Kollattukudy 1976, 12), combined with the presence of long chain alcohols (such as occurred in sample 34-0-12 S3 from Uitgeest-Groot Dorregeest) suggests that this residue probably, at least partly, originated from plant material.

The presence of cholesterol leads to the tentative conclusion that several residues (34-0-30, 35-7-28, 34-0-12, 34-0-12.S3, 14-6-4.3c, 14-6-4.2b, 79-1-1, 320-17, 32740, 33781 and 6745) were, at least partly, of animal origin. The diagnostic use of cholesterol must be treated with caution as this compound also occurs in the surface lipids of human skin. If squalene is detected in the same extract this must be considered contaminated during preparation as the isoprenoid unsaturated hydrocarbon squalene also occurs in human skin fats. No squalene was detected in the extracts under consideration.

TAGs with an odd number of carbons in their acyl chains are know to occur in milk fats from cow milk (Murata 1977). Such TAGs were detected in five charred residues (34-0-30, 14-6-4.3c, 14-6-4.3c S, 14-6-4.2b and 79-1-1). Short-chain fatty acids, with 4-12 carbon atoms, are reported to be characteristic for dairy products (Breckenridge and Kuksis 1968; Hilditch and Williams 1964, 144-145), but are absent in the extracts under consideration. Their absence can easily be caused by selective evaporation during heating, or selective leaching into the surrounding soil during burial. The presence of short-chain fatty acid moieties in some intact TAGs, as illustrated for residue 34-0-4 in Figure 3, is rather significant in this respect. A high abundance of components with 30-46 carbon atoms among the even carbon number TAGs was shown to be correlated to the presence of degraded ruminant milk fats in ceramics from the Iron Age and Roman period site in Stanwick in the UK (Dudd and Evershed 1998). A comparable TAG distribution pattern is clearly visible in five charred residues in Figure 4 (34-0-30, 14-6-4.3c, 14-6-4.3c S, 14-6-4.2b and 79-1-1). The additional presence of cholesterol leads to the conclusion that the lipids in these charred residues are, at least partially, derived from ruminant milk fats. TAG distribution patterns without a high abundance of components with 40-46 carbon atoms, lacking odd carbon number TAGs, but containing cholesterol, can be interpreted as, at least partially, derived from animal depot fats (Figure 4). Non-charred residue 35-7-28 and charred residues 320-17, 6745, 32.740 and 33.781 fall within this category.

Odd carbon number free fatty acids are primarily formed during bacterial growth and indicate the relative amount of bacterial matter that has become part of the residue. As described above, bacterial matter can be incorporated in the residue as part of ruminant milk fat. Bacterial matter incorporated in the residue as a result of post-depositional bacterial degradation in the soil, is shown in charred residues from Uitgeest 54 and NO-Polder 14. No odd carbon number TAGs were present in these residues and lipid hydrolysis is incomplete (average $I_{hydr}$ = 0.48 for NO-Polder 14 and average $I_{hydr}$ = 0.45 for Uitgeest 54). In these residues bacterial growth took place during post-depositional degradation. Charred residues from Hazendonk show a similar pattern, but due to the higher degree of hydrolysis (average $I_{hydr}$ = 0.85) the origin of the odd carbon number fatty acids could not be ascribed to post-depositional bacterial degradation with any degree of certainty.

## Discussion: Lipids from Surface Residues

Charred surface residues from Uitgeest-Groot Dorregeest exhibited an average $I_{hydr}$ = 0.73, with charred residue 34-0-30 being less hydrolyzed then average and charred residue 34-0-12 completely hydrolyzed. The low $I_{o/e}$ with an average value of 0.07, indicates the presence of some odd carbon number free fatty acids, but not higher then 7% of the corrected TLY. The low $I_{sat}$ with an average of 0.08, indicates limited bacterial activity and a relatively well preserved lipid profile. All charred residues except 34-0-12 contained extractable lipids derived from ruminant milk fats. Comparison of these results with those obtained from non-charred residues suggests a difference in original material or in mode of formation.

The non-charred surface residues from Uitgeest-Groot Dorregeest are completely saturated (average $I_{sat}$ = 1.0), indicating a greater exposure to oxidizing conditions. The effects of hydrolysis are very limited in these residues (average $I_{hydr}$ = 0.25), resulting in well-preserved TAG profiles completely without odd carbon number TAGs. Non-enzymatic hydrolysis of lipids is greatly enhanced by heating in the presence of water (Davídek et al. 1990, 186), which would suggest that these vessels were not used for cooking fatty substances in water. The complete absence of odd carbon number fatty acids shows that bacterial growth has occurred to a very limited extent, suggesting the formation of a denatured material prior to deposition in the soil, possibly as a result of polymerization or network formation. These two residues are visually and chemically different, but it is possible that they were both regularly exposed to the air during their formation. The vessel containing residue 35-7-28 may have been used for storage or transport of solid materials, while residue 8-1 may have been applied as decoration prior to the use of the vessel. It is clear that these vessels were not used as cooking vessels.

## Discussion: Lipids Absorbed in Ceramics

The ceramic samples of two vessels containing non-charred residues did yield extremely low concentrations of extractable lipids (<0.01 mg/g) that might easily be dismissed as blanks (Figure 2). The ceramic material 8-1 S contained only traces of a few unidentified aliphatic compounds and contaminants (phthalates), indicating a lack of absorption of lipids into the vessel during use. The ceramic material of 35-7-28 S contained contaminants and some even carbon number alcohols, with 12-18 carbon atoms, that are, as yet, unexplained.

The four vessels containing charred residues all yielded extractable lipids from the ceramic material directly adjacent to the residues. In one case (14-6-4.3c S) a lipid profile was obtained from the ceramic material that was almost identical to that from the charred residue, suggesting the same origin (Figure 2C). In the other sample pairs (34-0-12, 14-6-4.2b and 14-6-4.4) extractable lipid profiles from surface residues and ceramic samples were quite different. The most significant difference is an increased saturation (average $I_{sat}$ = 0.92), combined with a complete hydrolysis (average $I_{hydr}$ = 1.0), and an increased odd carbon number fatty acid index (average $I_{o/e}$ = 0.17). Lipids extracted from charred surface residues are obviously better preserved than those extracted from the directly adjacent ceramic material of the vessel. This difference in preservation is most likely the result of a combination of chemical mechanisms. Most important, absorbed lipids will have been submitted to a more extreme thermal regime due to higher temperatures inside the ceramic vessel wall and repetitive cooking phases. This more extensive thermal exposure may have caused both the complete hydrolysis of lipids, due to heating in the presence of water, and the high degree of saturation, due to heat induced polymerization. The increase of $I_{o/e}$ is hard to explain due to the complete hydrolysis. In the case of 14-6-4.2 S the extractable lipids may have resulted from complete hydrolysis of a milk fat residue (just like its adjacent residue). In the case of 14-6-4.4 and 34-0-12 an increased bacterial influence could be the cause. Although this explanation seems counter-intuitive, because absorption of lipids in the ceramic matrix of the vessel wall would be expected to reduce external influences, the refractory nature of the charred material obviously better prevents bacterial degradation. Evens (1990) and Oudemans and Boon (1991) propose that the preservation of lipids in the charred matrix may be enhanced by means of micro-encapsulation of small amounts of lipids during the formation of the charred residue, although the mechanisms of encapsulation are still unknown.

## Discussion: Chars from Other Sites

Extractable lipid profiles from Neolithic charred residues differ from those from the Roman period in that they often lack MAGs and DAGs, and that no odd carbon number TAGs are preserved. All the charred residues from the Neolithic contain cholesterol and present even carbon number TAG distribution patterns that can be interpreted as originating from animal depot fats (Figure 4). Hydrolysis (average $I_{hydr}$ = 0.67) and saturation (average $I_{sat}$ = 0.90) are comparable to those of charred residues from the Roman period, with an average $I_{hydr}$ = 0.71 and an average $I_{sat}$ = 0.94. The only profound difference is an increase in the $I_{o/e}$ of the Neolithic charred residues (average $I_{o/e}$ = 0.33) compared to charred residues from the Roman period (average $I_{o/e}$ = 0.12). Because of the absence of other indicators for milk fats, this increase is interpreted as a higher degree of bacterial growth resulting from a longer period of burial in the ground.

Some combinations of index values appear to be typical for a specific settlement. For example, settlement NO-Polder 14 shows a high saturation index ($I_{sat}$ = 1.0) combined with relatively low degree of hydrolysis ($I_{hydr}$ = 0.5), while Hazendonk shows a low degree of saturation ($I_{sat}$ = 0.8) and a much more extensive hydrolysis ($I_{hydr}$ = 0.9). Although both sites date to the Neolithic, the index values of NO-Polder 14 resemble those of the native Roman site Uitgeest 54 more closely than those of Hazendonk. These site-specific effects may be the result of local preserving conditions at the different sites, although the number of samples studied is too small to draw firm conclusion.

## Discussion: Sampling Issues

Quantitative comparison of extractable lipids of surface residues and absorbed residues shows an apparent greater degree of preservation of extractable lipids in surface residues than in the directly adjacent ceramic fabric. Surface residues are a more attractive target material for identification of original vessel contents. An argument can be made for the combined study of both surface residues and absorbed residues. The mechanisms responsible for the formation of surface residues and absorbed residues are clearly dependant on vessel use. Data presented in this study show the virtual absence of absorbed extractable lipids underneath non-charred surface residues, indicating that some kinds of vessel use lead to the accumulation of surface residues and little or no absorbed extractable lipids (such as decoration of vessels or their use as serving dishes or storage and transport vessel of dry goods). Others uses may cause the accumulation of absorbed lipids but produce little or no surface residues (such as storage or transport of oily or fatty liquids). Another argument for dual studies of surface residues and absorbed residues is the possibility of detecting multiple use-phases in one vessel. For these reasons, ideally, samples of both surface residues and absorbed organics should be studied in order to enhance the overall understanding of vessel use based on organic residue analysis.

## Conclusions

The quantitative study of the extractable lipid composition (including fatty acids, MAGs, DAGs, TAGs, sterols and long-chain alcohols), shows an apparently greater degree of preservation of lipids in surface residues than in the directly adjacent ceramic fabric of the vessel. Not only is the TLY per gram sample much higher in surface residues (especially in charred surface residues), but the amount of intact acyl lipids (as expressed in the hydrolysis index) and the amount of unsaturated fatty acids are also much higher in surface residues. This difference in preservation is proposed to be the result of a more severe thermal regime inside the vessel wall and the highly refractory nature of charred surface residues. This observation may have important consequences for sampling strategies in archaeological organic residue analysis.

Lipid extracts of charred and non-charred surface residues are very different in composition. Charred surface residues show varying degrees of condensation (C/H ratios between 9-33), indicating more or less severe thermal degradation due to the heating of organic materials. In spite of this thermal degradation, charred surface residues show the highest yields of extractable lipids per gram sample. Non-charred residues show many characteristics (low overall organic contents, a lower degree of hydrolysis, little or no bacterial degradation and little and no absorption of lipids into the directly adjacent ceramic vessel) that suggest a very different kind of vessel use. Most likely these organic residues are the result of a use with longer period of exposure to oxygen without severe heating. Non-charred residues may be the result of decoration with organic materials, or the residue of solids stored or transported in the vessel.

Lipids from charred surface residues from two Neolithic sites (ca. 5000 years old) were compared to charred residue from three native Roman settlements (ca. 1800-2000 years old). Although Neolithic charred residues showed comparable lipid yields, their lipid profiles contained a relatively higher proportion of material of bacterial origin. In spite of indications for site-specific degradation, this phenomenon is proposed to be the result of ongoing low-level microbial degradation.

An important focus for future work in archaeological organic residue analysis should be the mechanisms responsible for the deposition and accumulation of residues in vessels. Our results show that surface residues are a more attractive sample material for the identification of the original vessel content. Some types of vessel-use were shown to lead to the accumulation of surface residues and little or no absorbed lipids. In theory, others may lead to the accumulation of absorbed lipids without forming any surface residue. For this reason, ideally, sample of both surface residues and absorbed organics, preferably taken from the same vessel, should be studied.

| | | Uitgeest-Groot Dorregeest | | | | | | | |
|---|---|---|---|---|---|---|---|---|---|
| | | 34-0-30 *) | | | 35-7-28 | | 34-0-12 | | |
| | | R | | | R | S | R | S3 | S1 |
| | | Charred | | | Cream-colored | | Charred | | |
| TLY [mg/g]: | | 27.52 | | | 1.32 | 0.01 | 0.47 | 0.02 | 0.02 |
| Lipid | ECN | Wpi [%] | Δ [%] | Xi [%] | Wpi [%] | Wpi [%] | Wpi [%] | Wpi [%] | Wpi [%] |
| **Free fatty acids** | | | | | | | | | |
| C12:0 | 14.00 | 0.63 | 6.6 | 0.68 | | | | | |
| C13:0 | 15.00 | | | | | | | | |
| C14:0 | 16.00 | 3.78 | 6.8 | 4.12 | | | 4.39 | 9.04 | 24.57 |
| C15:0 | 17.00 | 0.50 | 6.9 | 0.55 | | | | 7.93 | |
| C15:0 | 17.00 | 0.75 | 6.9 | 0.82 | | | | 5.05 | |
| C16:1 | 17.90 | | | | | | 1.45 | | |
| C16:0 | 18.00 | 15.84 | 7.0 | 17.39 | 25.15 | | 29.56 | 21.64 | 49.38 |
| C17:0 | 19.00 | 0.65 | 7.1 | 0.71 | | | 4.08 | | |
| C17:0 | 19.00 | 0.52 | 7.1 | 0.57 | | | | 2.74 | |
| C18:2 | 19.80 | 0.24 | 7.6 | 0.27 | | | | | |
| C18:1 | 19.90 | 8.56 | 7.4 | 9.38 | | | 16.78 | | 2.30 |
| C18:0 | 20.00 | 9.58 | 7.2 | 10.48 | 13.20 | | 35.35 | 23.80 | 23.75 |
| C19:0 | 21.00 | 0.09 | 7.3 | 0.10 | | | | | |
| C20:1 | 21.90 | | | | | | | | |
| C20:0 | 22.00 | 0.45 | 7.3 | 0.49 | | | | | |
| C22:1 | 23.90 | | | | | | | | |
| C22:0 | 24.00 | 0.20 | 7.4 | 0.21 | | | | | |
| C23:0 | 25.00 | 0.28 | 7.6 | 0.31 | | | | | |
| C24:0 | 26.00 | 0.26 | 7.5 | 0.29 | | | 1.56 | 3.96 | |
| C26:0 | 28.00 | 0.13 | 7.6 | 0.14 | | | | 6.10 | |
| C28:0 | 30.00 | | | | | | | 5.26 | |
| C30:0 | 32.00 | | | | | | | 3.44 | |
| **Alcohols** | | | | | | | | | |
| C12-OH | 14.69 | | | | | 14.70 | | | |
| C14-OH | 16.69 | | | | | 59.65 | | | |
| C15-OH | 17.69 | | | | | | | | |
| C16-OH | 18.69 | | | | | 17.97 | | | |
| C18-OH | 20.69 | | | | | 7.68 | | | |
| C24-i-OH | 26.69 | | | | | | | 3.84 | |
| C26-i-OH | 28.69 | | | | | | | | |
| C28-OH | 30.69 | | | | | | | | |
| C30-OH | 32.69 | | | | | | | | |
| C32-OH | 34.69 | | | | | | | | |

Table 4a: Lipid composition extracts from surface (R) and absorbed (S) residues 34-0-30, 35-7-27 and 34-0-12 expressed in relative weight percentages $WP_i$ (in %). ECN is the calculated effective carbon number per compound. *) For sample 34-0-30 R the corrected $WP_i$ and the uncorrected $X_i$ are given per detected compound. The difference is expressed by Δ in percent relative to $WP_i$.

| | | Uitgeest-Groot Dorregeest | | | | | | | |
|---|---|---|---|---|---|---|---|---|---|
| | | 34-0-30 *) | | | 35-7-28 | | 34-0-12 | | |
| | | R | | | R | S | R | S3 | S1 |
| | | Charred | | | Cream-colored | | Charred | | |
| TLY [mg/g]: | | 27.52 | | | 1.32 | 0.01 | 0.47 | 0.02 | 0.02 |
| Lipid | ECN | Wpi [%] | Δ [%] | Xi [%] | Wpi [%] | Wpi [%] | Wpi [%] | Wpi [%] | Wpi [%] |
| **MAG** | | | | | | | | | |
| M14:0 | 20.93 | 0.14 | 5.5 | 0.15 | | | | | |
| M16:0 | 22.93 | 0.45 | 5.8 | 0.49 | | | | 1.32 | |
| M18:1 | 24.83 | 0.49 | 5.2 | 0.53 | | | | | |
| M18:0 | 24.93 | | | | | | 1.76 | 1.97 | |
| **DAG** | | | | | | | | | |
| D28 | 30.69 | 0.82 | -8.8 | 0.77 | | | | | |
| D29 | 31.69 | | | | | | | | |
| D30 | 32.69 | 2.27 | -7.9 | 2.13 | | | | | |
| D31 | 33.69 | 0.27 | -7.4 | 0.26 | | | | | |
| D32 | 34.69 | 4.79 | -5.4 | 24.62 | | | | | |
| D33 | 35.69 | 0.30 | -6.7 | 0.29 | | | | | |
| D34 | 36.69 | 5.55 | -5.6 | 5.35 | | | | | |
| D35 | 37.69 | 0.12 | -5.9 | 0.11 | | | | | |
| D36 | 38.69 | 4.58 | -5.7 | 4.41 | | | | | |
| **TAG** | | | | | | | | | |
| T40 | 38.45 | 2.66 | -15.7 | 2.29 | | | | | |
| T42 | 40.45 | 2.44 | -14.7 | 2.12 | | | | | |
| T43 | 41.45 | | | | | | | | |
| T44 | 42.45 | 2.72 | -13.8 | 2.39 | | | | | |
| T45 | 43.45 | 1.02 | -13.4 | 0.90 | | | | | |
| T46 | 44.45 | 3.52 | -13.0 | 3.12 | | | | | |
| T47 | 45.45 | 0.42 | -12.6 | 0.38 | | | | | |
| T48 | 46.45 | 4.79 | -12.3 | 4.29 | 8.59 | | | | |
| T49 | 47.45 | 1.18 | -11.6 | 1.06 | | | | | |
| T50 | 48.45 | 7.19 | -11.5 | 6.45 | 14.19 | | | | |
| T51 | 49.45 | 1.71 | -11.2 | 1.55 | | | | | |
| T52 | 50.45 | 6.86 | -10.9 | 6.23 | 6.16 | | | | |
| T53 | 51.45 | | | | | | | | |
| T54 | 52.45 | 3.05 | -10.3 | 2.80 | 29.94 | | | | |
| **Other** | | | | | | | | | |
| C9-diacid | 13.00 | | | | | | | | |
| Cholesterol | 29.19 | 0.20 | 15.1 | 0.23 | 2.77 | | 5.07 | 3.92 | |
| **Total** | | **100** | | **100** | **100** | **100** | **100** | **100** | **100** |

Table 4b: Lipid composition extracts from surface (R) and absorbed (S) residues 34-0-30, 35-7-27 and 34-0-12 expressed in corrected relative weight percentages $WP_i$ (in %). The numbers indicating the different DAGs and TAGs represent the total number of carbon atoms in the acyl fragments. ECN is the calculated effective carbon number per compound. *) For sample 34-0-30 R the corrected $WP_i$ and the uncorrected $X_i$ are given per detected compound. The difference is expressed by Δ in percent relative to $WP_i$.

| | | Uitgeest-Groot Dorregeest | | | | | | |
|---|---|---|---|---|---|---|---|---|
| | | 8-1 | 14-6-4.4 | | 14-6-4.3c | | 14-6-4.2b | |
| | | R | R | S | R | S | R | S |
| | | Brown | Charred | | Charred | | Charred | |
| TLY [mg/g]: | | 2.10 | 14.77 | 0.01 | 4.71 | 0.16 | 9.97 | 0.04 |
| Lipid | ECN | Wpi [%] | Wpi [%] | Wpi [%] | Wpi [%] | Wpi [%] | Wpi [%] | Wpi [%] |
| **Free fatty acids** | | | | | | | | |
| C12:0 | 14.00 | | 0.29 | 1.99 | 0.50 | 0.67 | 0.46 | |
| C13:0 | 15.00 | | | | | 0.60 | 0.07 | |
| C14:0 | 16.00 | 1.00 | 5.34 | 12.10 | 4.35 | 5.47 | 2.76 | 13.22 |
| C15:0 | 17.00 | | 0.72 | | 1.54 | | 0.39 | |
| C15:0 | 17.00 | | 1.21 | 5.35 | 1.71 | 2.39 | 0.65 | 14.95 |
| C16:1 | 17.90 | | 0.44 | 4.23 | 0.46 | | | |
| C16:0 | 18.00 | 3.55 | 27.69 | 40.36 | 26.20 | 27.80 | 12.11 | 37.74 |
| C17:0 | 19.00 | | 1.46 | | 1.68 | 1.32 | 0.54 | |
| C17:0 | 19.00 | | 1.37 | 3.98 | 1.84 | 1.73 | 0.48 | |
| C18:2 | 19.80 | | | | | | | |
| C18:1 | 19.90 | | 19.63 | 9.10 | 7.04 | 5.25 | 6.52 | 7.13 |
| C18:0 | 20.00 | 2.06 | 12.81 | 21.94 | 30.82 | 24.56 | 12.48 | 26.96 |
| C19:0 | 21.00 | | | | 0.34 | | | |
| C20:1 | 21.90 | | | | | | | |
| C20:0 | 22.00 | | 1.61 | | 0.70 | 1.22 | 0.38 | |
| C22:1 | 23.90 | | | | | | | |
| C22:0 | 24.00 | | | | | | 0.43 | |
| C23:0 | 25.00 | | | | | | 0.60 | |
| C24:0 | 26.00 | | | | 0.36 | 0.50 | 0.49 | |
| C26:0 | 28.00 | | | | | | 0.30 | |
| C28:0 | 30.00 | | | | | | | |
| C30:0 | 32.00 | | | | | | | |
| **Alcohols** | | | | | | | | |
| C12-OH | 14.69 | | | | | | | |
| C14-OH | 16.69 | | | | | | | |
| C15-OH | 17.69 | | | | | | | |
| C16-OH | 18.69 | | | | | | | |
| C18-OH | 20.69 | | | | | | | |
| C24-i-OH | 26.69 | | | | | | | |
| C26-i-OH | 28.69 | | | | | 0.17 | | |
| C28-OH | 30.69 | | | | | 0.23 | | |
| C30-OH | 32.69 | | | | | 0.70 | | |
| C32-OH | 34.69 | | | | | 0.73 | | |

Table 4c: Lipid composition extracts from surface (R) and absorbed (S) residues 8-1, 14-6-4.4, 16-6-4.3c and 14-6-4.2b expressed in corrected relative weight percentages $WP_i$ (in %). ECN is the calculated effective carbon number per compound.

| | | Uitgeest-Groot Dorregeest | | | | | | |
|---|---|---|---|---|---|---|---|---|
| | | 8-1 | 14-6-4.4 | | 14-6-4.3c | | 14-6-4.2b | |
| | | R | R | S | R | S | R | S |
| | | Brown | Charred | | Charred | | Charred | |
| TLY [mg/g]: | | 2.10 | 14.77 | 0.01 | 4.71 | 0.16 | 9.97 | 0.04 |
| Lipid | ECN | Wpi [%] | Wpi [%] | Wpi [%] | Wpi [%] | Wpi [%] | Wpi [%] | Wpi [%] |
| **MAG** | | | | | | | | |
| M14:0 | 20.93 | | | | | | 0.66 | |
| M16:0 | 22.93 | | 0.29 | | 1.29 | 1.18 | 1.07 | |
| M18:1 | 24.83 | | | | | | | |
| M18:0 | 24.93 | | | | 1.05 | 0.83 | 0.94 | |
| **DAG** | | | | | | | | |
| D28 | 30.69 | | | | 0.19 | | 0.90 | |
| D29 | 31.69 | | | | | | 0.07 | |
| D30 | 32.69 | 2.91 | 1.10 | | 0.39 | 0.51 | 3.65 | |
| D31 | 33.69 | | | | | | 0.88 | |
| D32 | 34.69 | 8.81 | 2.82 | | 0.99 | 1.32 | 7.45 | |
| D33 | 35.69 | | | | | | 1.32 | |
| D34 | 36.69 | 11.56 | 3.39 | | 1.54 | 1.45 | 8.40 | |
| D35 | 37.69 | | | | | | | |
| D36 | 38.69 | 11.19 | 0.53 | | 0.98 | 0.63 | 4.76 | |
| **TAG** | | | | | | | | |
| T40 | 38.45 | 7.17 | | | | | 0.26 | |
| T42 | 40.45 | 7.22 | 0.59 | | 0.18 | 0.23 | 0.91 | |
| T43 | 41.45 | | | | | | 0.19 | |
| T44 | 42.45 | 6.48 | 1.11 | | 0.38 | 0.55 | 1.60 | |
| T45 | 43.45 | | | | 0.07 | | | |
| T46 | 44.45 | 7.51 | 2.09 | | 0.82 | 1.57 | 2.85 | |
| T47 | 45.45 | | | | 0.26 | 0.68 | 1.13 | |
| T48 | 46.45 | 9.81 | 4.12 | | 1.78 | 3.28 | 4.52 | |
| T49 | 47.45 | | | | 0.54 | 0.88 | 1.80 | |
| T50 | 48.45 | 13.36 | 6.63 | | 3.52 | 4.98 | 6.55 | |
| T51 | 49.45 | | | | 1.15 | 1.01 | 1.85 | |
| T52 | 50.45 | 7.38 | 4.77 | | 4.36 | 4.65 | 6.15 | |
| T53 | 51.45 | | | | 0.34 | | 0.82 | |
| T54 | 52.45 | | | | 2.46 | 2.46 | 2.02 | |
| **Other** | | | | | | | | |
| C9-diacid | 13.00 | | | | | | | |
| Cholesterol | 29.19 | | | 0.95 | 0.20 | 0.47 | 0.62 | |
| **Total** | | **100** | **100** | **100** | **100** | **100** | **100** | **100** |

Table 4d: Lipid composition extracts from surface (R) and absorbed (S) residues 8-1, 14-6-4.4, 16-6-4.3c and 14-6-4.2b expressed in corrected relative weight percentages $WP_i$ (in %). The numbers indicating the different DAGs and TAGs represent the total number of carbon atoms in the acyl fragments. ECN is the calculated effective carbon number per compound.

| | | Sch | Utg54 | | Haz | | P14 | |
|---|---|---|---|---|---|---|---|---|
| | | 79-1 | 226 | 320 | 32 | 33 | 6745 | 7054 |
| | | R | R | R | R | R | R | R |
| | | Charred | Charred | Charred | Charred | Charred | Charred | Charred |
| TLY [mg/g]: | | 139.56 | 52.48 | 43.43 | 19.59 | 7.38 | 11.86 | 1.77 |
| Lipid | ECN | Wpi [%] | Wpi [%] | Wpi [%] | Wpi [%] | Wpi [%] | Wpi [%] | Wpi [%] |
| **Free fatty acids** | | | | | | | | |
| C12:0 | 14.00 | 0.30 | 1.77 | 1.33 | 2.32 | 9.53 | | |
| C13:0 | 15.00 | | | | | | | |
| C14:0 | 16.00 | 2.01 | 5.26 | 4.48 | 3.15 | 4.17 | | 14.21 |
| C15:0 | 17.00 | | | | 6.32 | 3.14 | | |
| C15:0 | 17.00 | 2.30 | 5.99 | 4.30 | 1.37 | 1.10 | 11.89 | 22.80 |
| C16:1 | 17.90 | | | | | 0.54 | | |
| C16:0 | 18.00 | 9.01 | 16.04 | 14.29 | 27.56 | 27.44 | 22.24 | 14.40 |
| C17:0 | 19.00 | | | | 4.46 | 3.15 | | |
| C17:0 | 19.00 | 0.63 | | | | 1.13 | | |
| C18:2 | 19.80 | | | | | | | |
| C18:1 | 19.90 | 12.00 | | 1.31 | 6.22 | 12.78 | | |
| C18:0 | 20.00 | 7.42 | 8.74 | 9.46 | 13.51 | 12.43 | 14.05 | 7.08 |
| C19:0 | 21.00 | | | | | | | |
| C20:1 | 21.90 | | | | 2.13 | 5.25 | | |
| C20:0 | 22.00 | | | | 1.77 | 4.64 | | |
| C22:1 | 23.90 | | | | | 3.10 | | |
| C22:0 | 24.00 | | | | | 2.10 | | |
| C23:0 | 25.00 | | | | | | | |
| C24:0 | 26.00 | | | | | | | |
| C26:0 | 28.00 | | | | | | | |
| C28:0 | 30.00 | | | | | | | |
| C30:0 | 32.00 | | | | | | | |
| **Alcohols** | | | | | | | | |
| C12-OH | 14.69 | | | | | | | |
| C14-OH | 16.69 | | | | | | | |
| C15-OH | 17.69 | | | | | | | |
| C16-OH | 18.69 | | | | | | | |
| C18-OH | 20.69 | | | | | | | |
| C24-i-OH | 26.69 | | | | | | | |
| C26-i-OH | 28.69 | | | | | | | |
| C28-OH | 30.69 | | | | | | | |
| C30-OH | 32.69 | | | | | | | |
| C32-OH | 34.69 | | | | | | | |

Table 4e: Lipid composition extracts from surface residues from Schagen-Muggenburg (Sch), Uitgeesterbroekpolder 54 (Utg54), Hazendonk (Haz) and NO-Polder 14 (P14) in corrected relative weight percentages $WP_i$ (in %). ECN is the calculated effective carbon number per compound.

| | | Sch | Utg54 | | Haz | | P14 | |
|---|---|---|---|---|---|---|---|---|
| | | 79-1 | 226 | 320 | 32 | 33 | 6745 | 7054 |
| | | R | R | R | R | R | R | R |
| | | Charred | Charred | Charred | Charred | Charred | Charred | Charred |
| TLY [mg/g]: | | 139.56 | 52.48 | 43.43 | 19.59 | 7.38 | 11.86 | 1.77 |
| Lipid | ECN | Wpi [%] | Wpi [%] | Wpi [%] | Wpi [%] | Wpi [%] | Wpi [%] | Wpi [%] |
| **MAG** | | | | | | | | |
| M14:0 | 20.93 | 0.57 | | | | | | |
| M16:0 | 22.93 | 0.97 | 1.92 | | | | | |
| M18:1 | 24.83 | 0.52 | | | 15.63 | | | |
| M18:0 | 24.93 | 0.90 | | | | | | |
| **DAG** | | | | | | | | |
| D28 | 30.69 | 1.50 | 7.85 | 5.24 | | | | |
| D29 | 31.69 | | | | | | | |
| D30 | 32.69 | 3.26 | 4.17 | 3.48 | | | | |
| D31 | 33.69 | 0.64 | | | | | | |
| D32 | 34.69 | 6.56 | 6.95 | 7.08 | | | | |
| D33 | 35.69 | 0.69 | | | | | | |
| D34 | 36.69 | 2.46 | 8.25 | 7.62 | | | | |
| D35 | 37.69 | | | | | | | |
| D36 | 38.69 | | | | | | | |
| **TAG** | | | | | | | | |
| T40 | 38.45 | 0.63 | | | | | | |
| T42 | 40.45 | 1.64 | | 2.93 | | | | |
| T43 | 41.45 | | | | | | | |
| T44 | 42.45 | 2.47 | 4.15 | 1.66 | | | | |
| T45 | 43.45 | 0.44 | | | | | | |
| T46 | 44.45 | 4.18 | 4.13 | 3.20 | | | | |
| T47 | 45.45 | 0.78 | | | | | | |
| T48 | 46.45 | 8.09 | 7.14 | 7.55 | 2.07 | 1.12 | 6.28 | 5.43 |
| T49 | 47.45 | 1.70 | | | | | | |
| T50 | 48.45 | 12.14 | 3.81 | 10.21 | 4.55 | 1.73 | 13.00 | 9.20 |
| T51 | 49.45 | | | | | | | |
| T52 | 50.45 | 11.27 | 10.84 | 10.21 | 1.79 | 1.93 | 4.53 | |
| T53 | 51.45 | | | | | | | |
| T54 | 52.45 | 4.69 | 2.98 | 4.09 | 4.96 | 3.12 | 23.65 | 17.03 |
| **Other** | | | | | | | | |
| C9-diacid | 13.00 | | | | | | | 9.84 |
| Cholesterol | 29.19 | 0.29 | | 1.55 | 2.22 | 1.60 | 4.41 | |
| **Total** | | **100** | **100** | **100** | **100** | **100** | **100** | **100** |

Table 4f: Lipid composition extracts from surface residues from Schagen-Muggenburg (Sch), Uitgeesterbroekpolder 54 (Utg54), Hazendonk (Haz) and NO-Polder 14 (P14) in corrected relative weight percentages $WP_i$ (in %). The numbers indicating the different DAGs and TAGs represent the total number of carbon atoms in the acyl fragments. ECN is the calculated effective carbon number per compound.

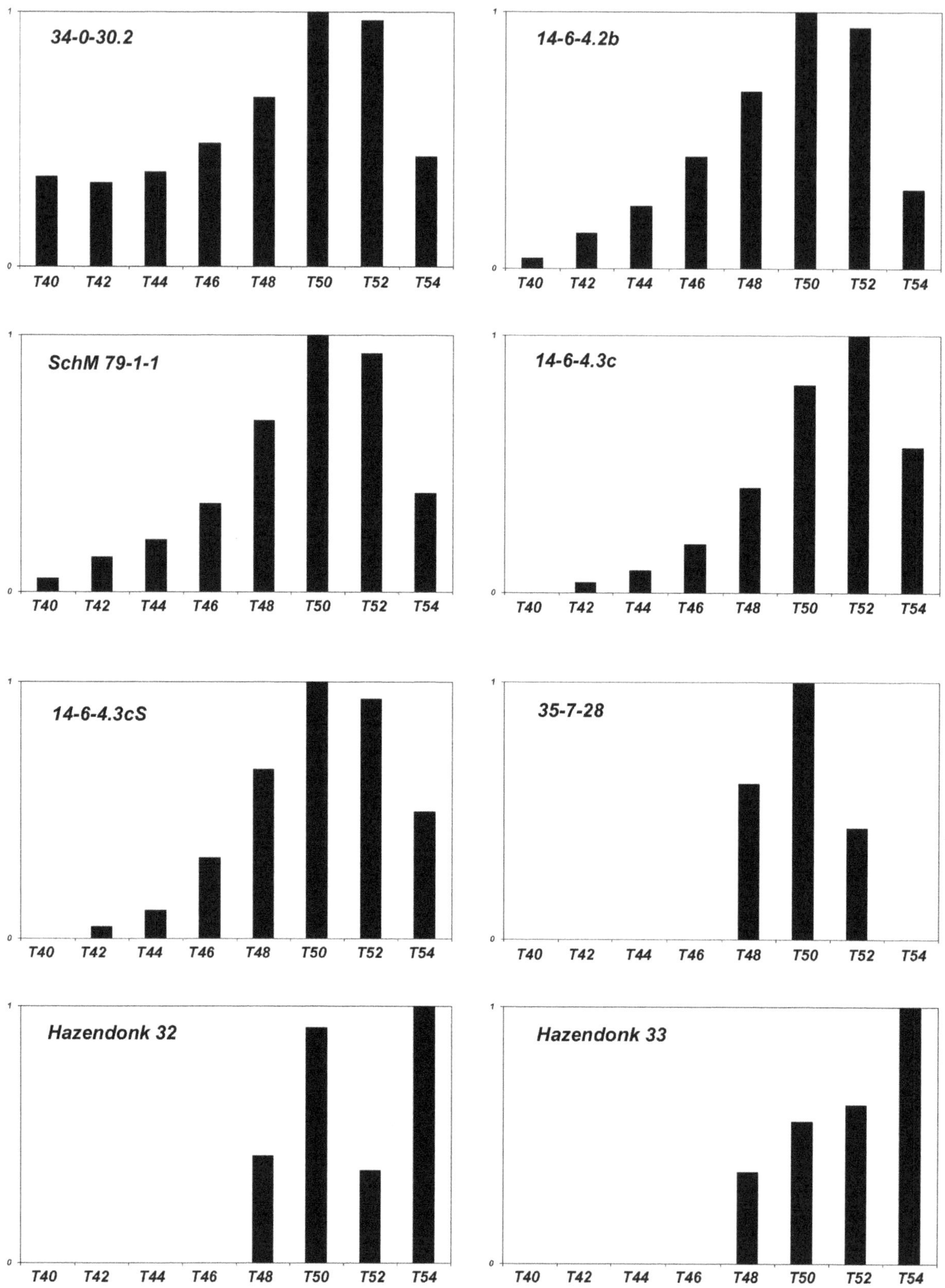
34-0-30.2
14-6-4.2b
SchM 79-1-1
14-6-4.3c
14-6-4.3cS
35-7-28
Hazendonk 32
Hazendonk 33
1
0
T40 T42 T44 T46 T48 T50 T52 T54

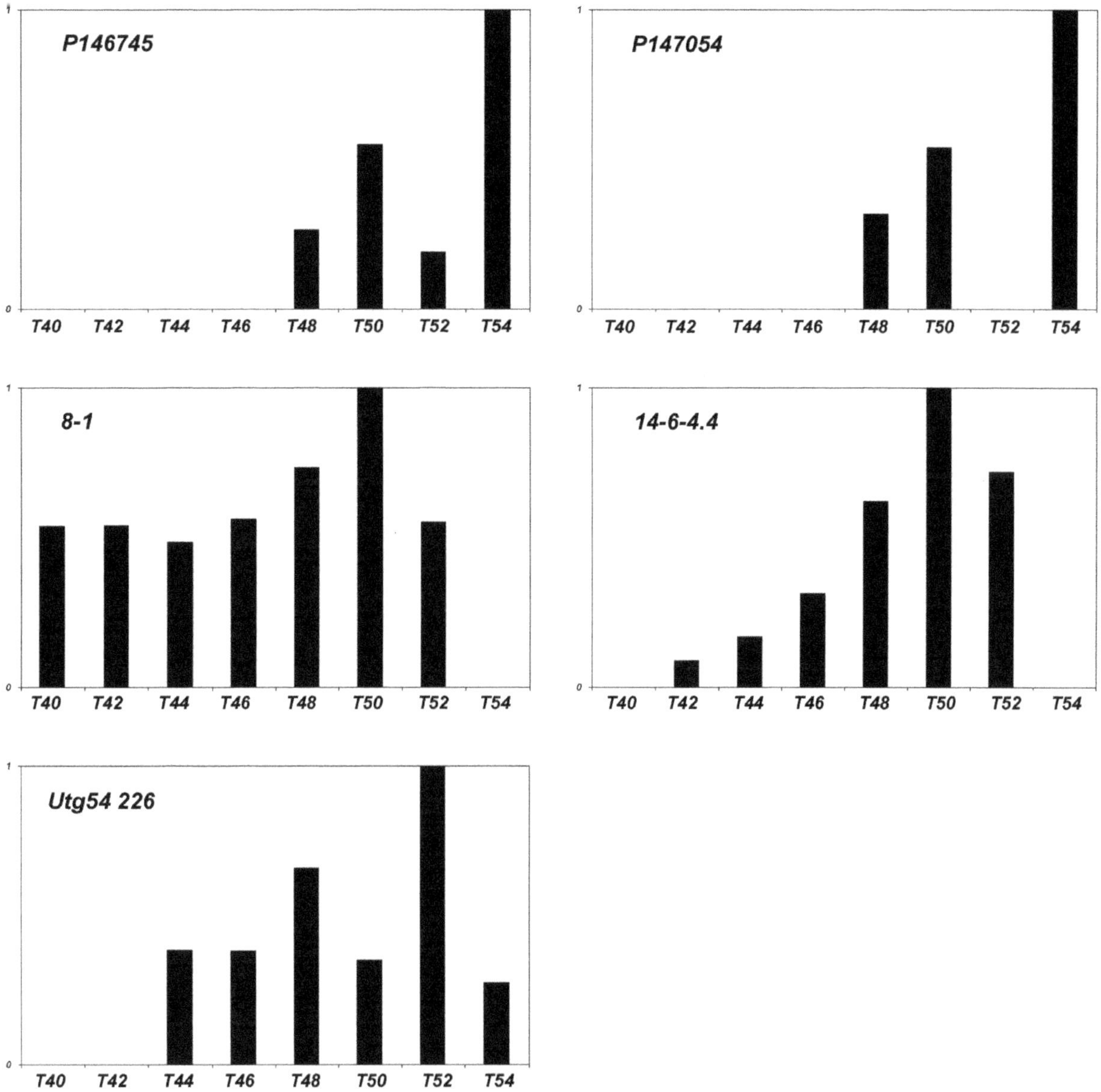


Figure 4 (on this and the previous page): Histograms of the relative abundance (%) of even carbon number TAGs (T40-T54) in total lipid extracts from ceramics from Neolithic and Roman Iron Age settlements in the Netherlands. Histogram 34-0-30.2 through 14-6-4.3C S (on the previous page) show lipid distributions associated with degraded ruminant milk fats. This identification is supported by the presence of odd carbon number TAGs and cholesterol. Histogram 35-7-28 through P147054 (on this and the previous page) show lipid distributions associated with degraded animal depot fats. This identification is supported by the presence of cholesterol. Histogram 8-1 through Utg54 226 (above) show lipid distributions from other origins. This identification is supported by the absence of both cholesterol and odd-carbon TAGs.

## References

**Abbink**, A.A. (1985). Onderzoek naar Handgevormd Aardewerk uit de Romeinse Tijd uit de Nederzetting Uitgeest Groot-Dorregeest. De Relaties tussen Vorm, Functie en Technologie. In W.J.H. Willems (ed.). *Jaarverslag over 1983 van de Rijksdienst voor Oudheidkundig Bodemonderzoek.* Amersfoort-Rijswijk: Rijksdienst voor Oudheidkundig Bodemonderzoek: pp. 27-28 (in Dutch).

**Abbink**, A.A. (1999). *Make it and Break it. Cycles of Pottery. A study of the Technology, Form, Function and Use of Pottery from Settlements at Uitgeest-Groot Dorregeest and Schagen-Muggenburg 1, Roman period,*

*North-Holland, the Netherlands*. Leiden: Leiden University Press.

**Ackman**, R.G. (1964). Fundamental Groups in the Responce of Flame Ionisation Detectors to Oxygenated Aliphatic Hydrocarbons. *Journal of Gas Chromatography* **2**: 173-179.

**Arnold**, D.E. (1985). *Ceramic Theory and Cultural Process*. Cambridge: Cambridge University Press.

**Breckenridge**, W.C. and A. Kuksis (1967). Molecular Weight Distribution of Milk Fat Triglycerides from Seven Species. *Journal of Lipid Research* **8**: 473-478.

**Breckenridge**, W.C. and A. Kuksis (1968). Specific Distribution of Short-chain Fatty Acids in Molecular Distillates of Bovine Milk Fat. *Journal of Lipid Research* **9**: 388-393.

**Charters**, S., R.P. Evershed, P.W. Blinkhorn and V. Denham (1995). Evidence for the Mixing of Fats and Waxes in Archaeological Ceramics. *Archaeometry* **37**: 113-127.

**Charters**, S., R.P. Evershed, L.J. Goad, A. Leyden, P.W. Blinkhorn and V. Denham (1993). Quantification and Distribution of Lipid in Archaeological Ceramics. Implications for Sampling Potsherds for Organic Residue Analysis and the Classification of Vessel Use. *Archaeometry* **35,2**: 211-223.

**Condamin**, J., F. Formenti, M.O. Metais and P. Blond (1979). The Application of Gas Chromatography to the Tracing of Oil in Ancient Amphorae. *Archaeometry* **18**: 195-201.

**Craig**, O.E., J. Mulville, M. Parker Pearson, R. Sokol, K. Gelsthorpe, R. Stacey and M. Collins (2000). Detecting Milk Proteins in Ancient Pots. *Nature* **408**: 312.

**Craig**, O.E., G. Taylor, J. Mulville, M.J. Collins and M. Parker Pearson (2005). The Identification of Prehistoric Dairying Activities in the Western Isles of Scotland. An Integrated Biomolecular Approach. *Journal of Archaeological Science* **32**: 91-103.

**Davídek**, J., J. Velísek and J. Pokorný (1990) *Chemical Changes during Food Processing. Volume 21*. Amsterdam: Elsevier Science Publishers.

**Den Dooren de Jong**, L.E. (1961). On the Formation of Adipocere from Fats. Contribution to the Microbiology of Systems Containing Two Liquid Phases. *Antonie van Leeuwenhoek Journal of Microbiology and Serology* **27**: 337-361.

**Diallo**, B., M. Vanhaelen and O.P. Gosselain (1995). Plant Constituents Involved in Coating Practices among Traditional African Potters. *Experientia. Interdisciplinary Journal for Life Sciences* **51, 1**: 95-97.

**Dudd**, S.N. and R.P. Evershed (1998). Direct Demonstration of Milk as Element of Archaeological Economies. *Science* **282**: 1478-1481.

**Dudd**, S.N., M. Regert and R.P. Evershed (1998). Assessing Microbial Lipid Contributions during Laboratory Degradations of Fats and Oils and Pure Triacylglycerols Absorbed in Ceramic Potsherds. *Organic Geochemistry* **29,5-7**: 1345-1354.

**Eglinton**, G. and G.A. Logan (1991). Molecular *Preservation. Philosophical Transactions of the Royal Society of London B* **333**: 315-328.

**Evans**, J. (1990). Come back King Alfred, All is Forgiven! In P.E. McGovern (ed.). *Organic Contents of Ancient Vessels. Materials Analysis and Archaeological Investigation. Volume 7*. Philadelphia: The University Museum of Archaeology and Anthropology, pp. 7-9.

**Evershed**, R.P. (1991). Bog Body Lipid Taphonomy. In B. Ottaway (ed.). *Archaeological Sciences 1989. Proceedings of a Conference on the Application of Scientific Techniques to Archaeology. Bradford, September 1989. Volume 9*. Oxford: Oxbow Books, pp. 352-361.

**Evershed**, R.P. (1992). Chemical Composition of a Bog Body Adipocere. *Archaeometry* **34**: 253-265.

**Evershed**, R.P., K.I. Arnot, J. Collister, G. Eglinton and S. Charters (1994). Application of Isotope Ratio Monitoring Gas Chromatography-Mass Spectrometry to the Analysis of Organic Residues of Archaeological Origin. *Analyst* **119, 5**: 909-914.

**Evershed**, R.P., S. Charters and A. Quye (1995a). Interpreting Lipid Residues in Archaeological Ceramics. Preliminary Results from Laboratory Simulations of Vessel Use and Burial. *Materials Research Symposia Proceedings* **352**: 85-95.

**Evershed**, R.P., S.N. Dudd, S. Charters, H. Mottram, A.W. Stott, A.M. Raven, P.F. van Bergen and H. Bland (1999). Lipids as Carriers of Anthropogenic Signals from Prehistory. *Philosophical Transactions of the Royal Society of London B* **354**: 13-31.

**Evershed**, R.P., C. Heron, S. Charters and L.J. Goad (1992). The Survival of Food Residues. New Methods of Analysis, Interpretation and Application. *Proceedings of the British Academy* **77**: 187-208.

**Evershed**, R.P., C. Heron and L.J. Goad (1990). Analysis of Organic Residues of Archaeological Origin by High-temperature Gas Chromatography and Gas Chromatography-Mass Spectrometry. *Analyst* **115**: 1339-1342.

**Evershed**, R.P., C. Heron and L.J. Goad (1991). Epicuticular Wax Components Preserved in Potsherds as Chemical Indicators of Leafy Vegetables in Ancient Diets. *Antiquity* **65**: 540-544.

**Evershed**, R.P., A.W. Stott, A.M. Raven, S.N. Dudd, S. Charters and A. Leyden (1995b). Formation of Long-chain Ketones in Ancient Pottery Vessels by Pyrolysis of Acyl Lipids. *Tetrahedron Letters* **36,48**: 8875-8878.

**Evershed**, R.P., S.J. Vaughan, S.N. Dudd and J.S. Soles (1997). Fuel for Thought? Beeswax in Lamps and Conical Cups from Late Minoan Crete. *Antiquity* **71**: 979-985.

**Gianno**, R., D. Erhardt, D.W. von Endt, W. Hopwood and M.T. Baker (1990). Archaeological Resins from Shipwrecks off the Coasts of Saipan and Thailand. In P.E. Mc Govern (ed.). *Organic Contents of Ancient Vessels. Materials Analysis and Archaeological*

*Investigation. Volume 7*. Philadelphia: University of Pennsylvania, pp. 59-67.
**Gibson**, J. and J. Evans (1985). Some Eighteenth-Century Pharmaceutical Vessels from London. *Post-Medieval Archaeology* **19**: 151-157.
**Heron**, C. and R.P. Evershed (1993). The Analysis of Organic Residues and the Study of Pottery Use. In M. Schiffer (ed.). *Archaeological Method and Theory 5*. Tucson: University of Arizona Press, pp. 247-287.
**Heron**, C., R.P. Evershed and L.J. Goad (1991). Effects of Migration of Soil Lipids on Organic Residues Associated with Buried Potsherds. *Journal of Archaeological Science* **18**: 641-659.
**Hill**, H.E. and J. Evans (1988). Vegeculture of Solomon Islands Prehistory from Pottery Residues. In J.O. Tate (ed.). *Science and Archaeology. Proceedings of a Conference on the Application of Scientific Techniques to Archaeology. Glasgow, September 1987. British Archaeological Reports British Series 196*. Oxford: BAR Publishing, pp. 449-458.
**Hillditch**, T.P. and P.N. Williams, P.N. (1964). *The Chemical Constitution of Natural Fats*. London: Chapman and Hall.
**Kaiser**, R. (1969). *Gas Phase Chromatography. Volume III*. Washington: Buttersworths.
**Kobayashi**, M. (1994). Use-Alteration Analysis of Kalinga Pottery. Interior Carbon Deposits of Cooking Pots. In J.M. Skibo (ed.). *Kalinga Ethnoarchaeology. Expanding Archaeological Method and Theory*. Washington D.C., Smithsonian Institution: pp. 127-168.
**Kollattukudy**, P.E. (ed.) (1976). *Chemistry and Biochemistry of Natural Waxes*. Amsterdam: Elsevier Science Publishing Company.
**Leninger**, A.L. (1977). *Biochemistry*. New York: Worth Publishers.
**Louwe Kooijmans**, L.P. (1974). *The Rhine/Meuse Delta. Four Studies on its Prehistoric Occupation and Holocene Geology. Analecta Praehistorica Leidensia. Volume 7*. Leiden: Brill.
**Louwe Kooijmans**, L.P. (1976). Hazendonk (gem. Molenaarsgraaf). In H. Sarfatij (ed.). *Archeologische Kroniek van Zuid-Holland over 1975, Holland 8*. Amersfoort-Lelystad: Rijksdienst voor het Oudheidkundig Bodemonderzoek: pp. 263-265 (in Dutch).
**Mackenzie**, A.S., S.C. Brassell, G. Eglinton and J.R. Maxwell (1982). Chemical Fossils. The Geological Fate of Sterols. *Science* **217**: 491-504.
**Malainey**, M.E., P. Przybylski and B.L. Sherriff (1999). The Effects of Thermal and Oxidative Degradation on the Fatty Acids Composition of Food Plants and Animals of Western Canada. Implications for the Identification of Archaeological Vessel Residues. *Journal of Archaeological Science* **26**: 95-103.
**McGovern**, P.E., D. Glusker, L.J. Exner and M.M. Voigt (1996). Neolithic Resinated Wine. *Nature* **381**: 480-481.
**Mills**, J.S. and R. White (1987). *The Organic Chemistry of Museum Objects*. London: Butterworths.
**Mottram**, H.R., S.N. Dudd, G.J. Lawrence, A.W. Stott and R.P. Evershed (1999). New Chromatographic, Mass Spectrometric and Stable Isotope Approaches to the Classification of Degraded Animal Fats Preserved in Archaeological Pottery. *Journal of Chromatography A* **833**: 209-221.
**Murata**, T. (1977). Analysis of Triglycerides by Gas Chromatography/Chemical Ionisation Mass Spectrometry. *Analytical Chemistry* **49**: 2209-2213.
**Needham**, S. and J. Evans (1987). Honey and Dripping. Neolithic Food Residues from Runneymede Bridge. *Oxford Journal of Archaeology* **6**: 21-28.
**Nes**, W.R. and W.D. Nes (1980). *Lipids in Evolution*. New York: Plenum Press.
**Odham**, G. and E Stenhagen (1972a). Fatty Acids (Chapter 8). In G.R. Waller (ed.). *Biochemical Applications of Mass Spectrometry*. New York: Wiley-Interscience, pp. 211-228.
**Odham**, G. and E. Stenhagen (1972b) Complex Lipids (Chapter 9). In G.R. Waller (ed.). *Biochemical Applications of Mass Spectrometry*. New York: Wiley-Interscience, pp. 229-249.
**Oudemans**, T.F.M. (2006). *Molecular Studies of Organic Residues in Ancient Vessel*. Leiden University Ph.D. dissertation.
**Oudemans**, T.F.M. and J.J. Boon (1991). Molecular Archaeology. Analysis of Charred (Food) Remains from Prehistoric Pottery by Pyrolysis-Gas Chromatography/Mass Spectrometry. *Journal of Analytical and Applied Pyrolysis* **20**: 197-227.
**Oudemans**, T.F.M. and J.J. Boon (1996). Traces of Ancient Vessel Use. Investigating Prehistoric Usage of Four Pot Types by Organic Residue Analysis Using Pyrolysis Mass Spectrometry. *Analecta Praehistorica Leidensia* **26**: 221-234.
**Oudemans**, T.F.M., J.J. Boon, J.J. and R.E. Botto (in press). FTIR and Solid-State $^{13}C$ CP/MAS NMR Spectroscopy of Charred and Non-charred Solid Organic Residues Preserved in Roman Iron Age Vessels from the Netherlands. *Archaeometry*.
**Oudemans**, T.F.M., G.B. Eijkel and J.J. Boon (2007). Identifying Biomolecular Origins of Solid Organic Residues Preserved on Iron Age Pottery using DTMS and MVA. *Journal of Archaeological Science* **34**: 173-193.
**Oudemans**, T.F.M., G.B. Eijkel and J.J. Boon (2005). DTMS and DTMS/MS Study of Solid Organic Residues Preserved on Ancient Vessels. In E. Burke (ed.). *The 33rd International Symposium on Archaeometry, 22-26 April 2002*. Amsterdam: Vrije Universiteit, pp. 501-505.
**Oudemans**, T.F.M. and D. Erhardt (1996). Organic Residue Analysis in Ceramic Studies. Implications for Conservation Treatment and Collections Management. In P. Smith (ed.). *Archaeological Conservation and its Consequences*. London-Copenhagen: The International

Institute for Conservation of Historic and Artistic Works, pp. 137-142.
**Passi**, S., M.C. Rothshild-Boros, P. Fasella, M. Nazzaro-Porro and D. Whitehouse (1981). An Application of High Performance Liquid Chromatography to the Analysis of Lipids in Archaeological Samples. *Journal of Lipid Research* **22**: 778-784.
**Pastorova**, I., T.F.M. Oudemans and J.J. Boon (1993). Experimental Polysaccharide Chars and their 'Fingerprints' in Archaeological Food Residues. *Journal of Analytical and Applied Pyrolysis* **25**: 63-75.
**Patrick**, M., A.J. de Koning and A.B. Smith (1985). Gas Liquid Chromatographic Analysis of Fatty Acids in Food Residues from Ceramics found in the Southwestern Cape, South Africa. *Archaeometry* **27**: 231-236.
**Raven**, A.M., P.F. van Bergen, A.W. Stott, S.N. Dudd and R.P. Evershed (1997). Formation of Long-chain Ketones in Archaeological Pottery Vessels by Pyrolysis of Acyl Lipids. *Journal of Analytical and Applied Pyrolysis* **40-41**: 267-285.
**Regert**, M., H.A. Bland, S.N. Dudd, P.F. van Bergen and R.P. Evershed (1998). Free and Bound Fatty Acid Oxidation Products in Archeological Ceramic Vessel. *Philosophical Transactions of the Royal Society of London B* **265,1409**: 2027-2032.
**Regert**, M., S. Colinart, L. Degrand and O. Decavallas (2001). Chemical Alteration and Use of Beeswax through Time. Accelerated Aging Tests and Analysis of Archaeological Samples from Various Environmental Contexts. *Archaeometry* **43, 4**: 549-569.
**Regert**, M. and C. Rolando (2002). Identification of Archaeological Adhesives using Direct Inlet Electron Ionization Mass Spectrometry. *Analytical Chemistry* **74, 5**: 965-975.
**Reyers**, O. (1985) *Eindrapport Uitgeest 54. Project Oer-IJ Estuarium.* Amsterdam: Universiteit van Amsterdam (in Dutch).
**Rice**, P.M. (1987). *Pottery Analysis. A Sourcebook.* Chicago: Chicago University Press.
**Rottländer**, R.C.A. (1990). Lipid Analysis in the Identification of Vessel Contents. In P.E. McGovern (ed.). *Organic Contents of Ancient Vessels. Materials Analysis and Archaeological Investigation, MASCA.* Philadelphia: University of Pennsylvania, pp. 37-40.
**Rottländer**, R.C.A. and H. Schlichtherle (1979). Food Identification of Samples from Archaeological Sites. *Archaeo-Physika* **10**: 260-267.
**Rottländer**, R.C.A. and H. Schlichtherle (1980). Gefäßinhalte. Eine kurz kommentierte Bilbliographie. *Archaeo-Physika* **7**: 61-70.
**Scanlon**, J.T. and D.E. Willis (1985). Calculation of Flame Ionisation Detector Relative Response Factors using the Effective Carbon Number Concept. *Journal of Chromatographic Science* **23**: 333-340.
**Shaw**, N. (1974). Lipid Composition as a Guide to Classification of Bacteria. In D. Perlman (ed.). *Advances in Applied Microbiology 17.* New York: Academic Press, pp. 63-109.
**Shedrinski**, A.M., R.E. Stone and N.S. Baer (1991). Pyrolysis Gas Chromatographic Studies on Egyptian Archaeological Specimens. Organic Patinas on the 'Three Princesses' Gold Vessels. *Journal of Analytical and Applied Pyrolysis* **20**: 229-238.
**Stern**, B., C. Heron, M. Serpico and J. Bourriau (2000). A Comparison of Methods for Establishing Fatty Acid Concentration Gradients Across Potsherds. A Case Study Using Late Bronze Age Canaanite Amphorae. *Archaeometry* **42,2**: 399-414.
**Ten Anscher**, T.J. (2000/2001). Huisplattegronden uit de late Swifterbantcultuur op P14 gem. Noordoostpolder. *Archeologie* **10**: 80-85 (in Dutch).
**Therkorn**, L.L. (2004). *Landscaping the Powers of Darkness and Light. 600 BC-350 AD Settlement Concerns of Noord-Holland in Wider Perspective.* Amsterdam: Universiteit van Amsterdam.
**Thornton**, N.D., E.D. Morgan and F.S.C. Celoria (1970). The Composition of Bog Butter. *Science and Archaeology*, **2/3**: 20-25.
**Waller**, G.R. (1972). *Biochemical Applications of Mass Spectrometry.* New York: Wiley-Interscience.
**Waller**, G.R. and O.C. Dermer (1980). *Biochemical Applications of Mass Spectrometry. First Supplementary Volume.* New York: John Wiley and Sons.

# CHAPTER TEN
# Patterns of Subsistence Change During the Final Neolithic in the Primorye Region of the Russian Far East as Revealed by Fatty Acid Residue Analysis
## J. Cassidy

Jim Cassidy, Ph.D.; Natural Resources and Environmental Affairs; Marine Corps Air Ground Combat Center (MCAGCC); Twentynine Palms, CA 92277; USA.

Global climatic instability dating to approximately 3700 years ago appears to correlate with the expansion of Northeast Asian 'Bronze Age' cultures and the introduction of core-periphery interaction spheres into the frontier areas of the Primorye Region of the Russian Far East. These processes appear to have led to the demise of long-standing traditions based on hunting-gathering-fishing and the introduction of new patterns of social organization and food production. The appearance of new stylistic patterns associated with the Margarita Culture on the northern littoral of the Sea of Japan seems to represent a transitional phase between Stone Age and Early Iron Age cultural expressions in the region. The degree to which the expansion of food procurement, in the form of millet and barley cultivation, as well as competing influences from northern pastoral societies, played a role in these processes is presently a matter of debate. The extremely poor preservation of organic materials in the archaeological record has hampered the exploration of these questions. In an attempt to gain a more direct understanding of subsistence practices and potential food production represented in deposits dating to this period of rapid culture change a total of 18 Neolithic and 48 Margarita Period ceramic sherds were subjected to fatty acid residue analysis. This procedure revealed a number of similarities and differences related to subsistence practices, as well as patterns of seasonality.

Starting in the 1950's, the field of anthropological archaeology strongly emphasized ecologically precepts in the formation of culture theory. These ideas placed heavy emphasis on biological models that address group behavior, such as optimal foraging theory (Jochim 1998; Smith and Winterhalder 1992). The methodology employed requires the empirical measurement of interaction between human populations and their environment. This especially applies to the analysis of subsistence practices and settlement strategies as adaptive mechanisms that account for processes of culture change (Sabloff and Ashmore 2001). In recent years, modifications to the principles of cultural ecology have incorporated concepts pertaining to human biological evolution that place greater emphasis on individual survival and reproductive fitness (Kelly 1995). This has come to be known as 'behavioral ecology'. Under this theoretical paradigm, it is assumed that evolutionary stable states may modulate the rate of culture change only as long as stress between human populations and the environment remains at a relatively low level. In the face of catastrophic and destabilizing events, such as rapid and severe climatic oscillations or pandemic disease, the level of decision making shifts to that of individual agents who are motivated by more immediate goals, such as personal survival. A potential reward of such efforts may be the accumulation of personal wealth and power, thereby leading to enhanced reproductive success. Cultural ecology assumes that innovations will take place slowly and variability will be minimized. Behavioral ecology assumes that innovations will occur rapidly, especially under conditions of significant stress, and that cultures will change in a punctuated manner. While the scale of analysis provided by cultural ecology allows us to examine behavior at group level, a greater emphasis on personal agency provides enhanced insights into dynamic processes of change reflected in human behavior. Unfortunately, the fine-grained scale required for the empirical measurement of agency-based behavior is seldom available in the inherently coarse-grained scale of the archaeological record.

In spite of these considerations, the distinctions between cultural and behavioral ecology do provide the opportunity for heuristically useful analytical contrasts. Where archaeological deposits suggest limited variation and incremental processes of innovation over extended periods of time, it may be assumed that environmental stress remained at a relatively stable level. Where the archaeological record reflects increasing variability and significant innovations over relatively short periods of time, it may be assumed that environmental stresses were active, which resulted in agency-based processes (Maschner 2000). When addressing such concepts as cultural variability and rates of innovation, it must be recognized that archaeological interpretations remain largely contingent upon the scale of data collection and analysis. Initial efforts of investigation are by nature constrained to relatively coarse-grained data collection procedures and the descriptive formation of artifact typologies that may permit the analytic construction of archaeological cultures. As our knowledge of specific archaeological cultures accumulates it is possible to refine the scale of analysis and establish more rigorous analytic objectives.

Data recovery efforts in the Russian Far East initially concentrated on the construction of relatively coarse-grained historical-descriptive models and the definition of archaeological cultures (Olkadnikov 1964). As data has accumulated over the past 50 years, these models have gradually gained finer-grained resolution and begin to shed light on regional variability in the prehistoric record. As a result of these activities, the present level of data resolution has begun to permit the exploration of ecologically based theoretical perspectives. The idea of a Bronze Age invokes a number of very specific theoretical implications among Western archaeologists centering on the development of social complexity. Among the most commonly cited components of social complexity are domesticated plants and animals, surplus food production, storage facilities, the technological ability to fire materials at high temperatures, social stratification by work, gender or status, political, religious, and craft specialization, extensive exchange of raw materials and exotic goods, production of prestige goods, hierarchically organized village clusters, fortified structures and warfare (Barnes 1993; Shelach 1999). Any hypothesis pertaining to Bronze Age cultures should specifically test for the presence or absence of these items.

## The Bronze Age in the Primorye Region

Over the past 50 years more than a thousand archaeological sites have been discovered throughout the Primorye Region in the Russian Far East, which occupies the coastal area along the northern littoral of Sea of Japan (Figure 1). Examination of the artifact assemblages from surface collections and excavations has resulted in the definition of a number of archaeological cultures that have been assigned to the technological stages of Paleolithic, Early Neolithic, Late Neolithic, Bronze Age, and Early Iron Age.

As a result of the discovery of ground stone replicas of bronze spear points, ground stone daggers, occasional metal objects and associated plain-ware ceramic vessels found in archaeological contexts in Primorye in the 1960's, a Bronze Age component spanning the period between 4000 and 2700 years before present was proposed. During the decades that followed, many new sites have been discovered and excavated. These discoveries have resulted in the refinement of typological distinctions between these Bronze Age cultures that include the Senii Gai Culture of the interior Khanka Plain, the Lidovka Culture that arguably extended 800 km along the east coast and into the western interior along the southern flanks of the Sihote-Alin Mountain Range, and the Margarita Culture along the coast of eastern Primorye.

From an ecological perspective, it is fairly evident that the Bronze Age Cultures of Primorye occupied fairly distinct zones, as did the previous Early and Late Neolithic periods. The Senii Gai Culture is situated in the foothills surrounding Khanka Lake and appears to be the most heavily influenced by the spread of developing complex societies of Northeast Asia. The east coast along the margin of the Sihote-Alin Mountains appears to be the nexus of the overlapping Margarita and Lidovka Cultures that shared a number of common traits. While they also shared some generalized chronological traits found in the interior, they appear to represent a distinctly separate course of development. In fact, Dyakov (1989, 210) argues that the Margarita Culture represents an early phase of the Lidovka Culture. However, an examination of available radiocarbon dates does not support this idea but reveals a pattern of cultural succession commencing with the Margarita Culture, followed by the Senii Gai Culture and finally the Lidovka Culture (Figure 2).

In fact, it appears that a gap of 400 years separates the Margarita Culture from the Lidovka Cultural, with an intervening period in which the Senii Gai Culture could develop and spread. This time period, commencing around 3700 years ago, coincides with the sudden onset of a severe cold period that was initially very humid but later became cold and dry before returning to warmer and wetter conditions approximately 3300 years ago (Besednov and Vostretsov 1997, 124; Vaganov 1998). This severely cold interval was of global proportions (Steig 1999), and appears to correspond to the last available date of the Late Neolithic Zaisanovka Culture on the south coast and the onset of the Margarita Culture on the east coast. It seems reasonable to assume that such a climatic anomaly resulted in significant ecological disruptions of marine and terrestrial animal and plant communities as well as the related subsistence practices of the people throughout the region. It has been suggested that these events coincided with the spread of the interaction sphere of the northern pastoral groups to the east coast and the displacement of Zaisanovka populations southward.

It is obvious that the classification 'Bronze Age' in the Primorye Region represents a generalized chronological period rather than a specific cultural manifestations. The division in three distinct cultures and the absence of this cultural expression on the south coast, reveals the presence of considerable variety in the form of ecological adaptations and subsistence practices. Furthermore, the interplay between indigenous populations and influences penetrating the region from the north and west appears highly dynamic and complex.

Figure 1: Map of Northeast Asia showing the position of the Primorye Region (after Aikens and Rhee 1992, 2).

## Zera Lake

In 1995 a joint Russian-American research project started with the intent of investigating the succession of archaeological cultures in the Primorye Region from an ecological perspective. Initial activities included investigations into the paleogeography and paleoecology of the region (Kononenko and Cassidy 1998). As the project progressed the subject of investigation was narrowed to focus on ecological processes that surrounded the terminal Neolithic period and the onset of the Bronze Age (Cassidy and Kononenko 2001; 2002). These activities included numerous site visits and an analysis of regional settlement patterns.

Sites in the Glazkovka drainage system and nearby Zera Lake, located towards the south of Kit Inlet (Figure 3), were selected for excavation during the 1998-2001 field seasons. These comprised terminal Late Neolithic Zaisnovka Culture deposits and early Bronze Age

Margarita Culture sites (Kluev and Yanshino 1997). Pollen samples were obtained from the sites, as well as non-cultural river and lake profiles, for the purpose of reconstructing the paleoenvironment and paleoclimate. Post-depositional preservation of organic remains in archaeological sites in the Primorye Region is extremely poor as a result of highly acidic soils, intense annual hydrologic saturations and freeze-thaw cycles. In spite of the wet screening of all midden deposits through a 3 mm (1/8th inch) mesh and the flotation of all deposits from hearths, minimal amounts of organic material were recovered from these sites. This greatly limited our ability to test ecological hypotheses relating to subsistence activities and seasonality. The second most common abundant material recovered were ceramic sherds and we decided to process a sample of these sherds for fatty acid residues for the purpose of diet reconstruction. This chapter provides a review of the results of these research procedures.

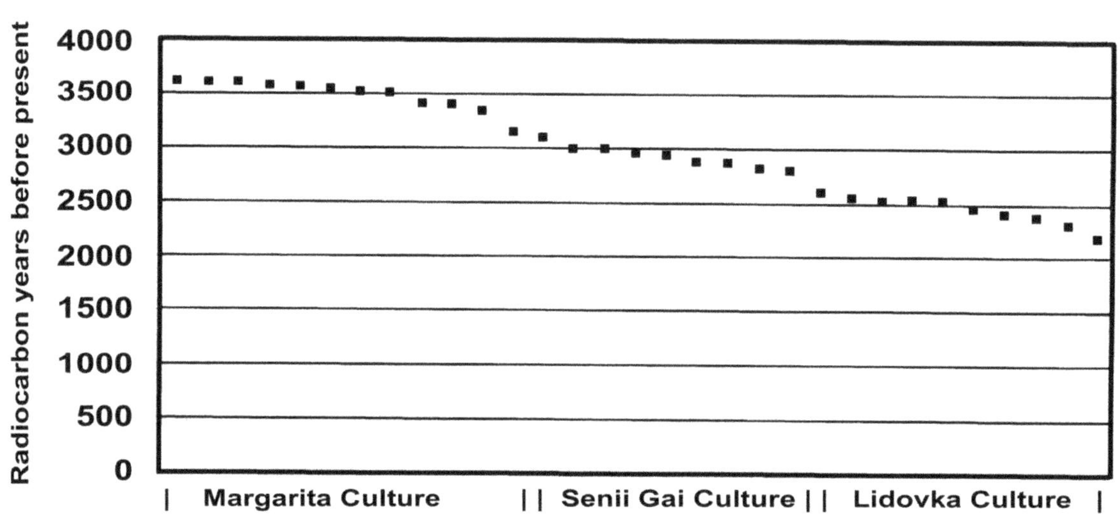


Figure 2: Chronological relationship of the Bronze Age Cultures in the Primorye Region.

Between 3800 and 3500 years ago climatic conditions appear to have become colder in the Primorye Region (Korotkii et al. 1988). Simultaneously, precipitation levels declined to approximately 700 mm per year. It was during this climatic oscillation that the transition from the Neolithic Zaisnovka Period and the Bronze Age is thought to have taken place on the east coast of Primorye (Brodyanski 1995; Dykov 1989; Garkovik 1983). Some preliminary observations can be made regarding the occupation of Margarita Culture sites as a result of this research project. First, there is no evidence of agricultural activities. Second, there is no evidence of any metallurgical activities, including knowledge of technologies for the firing of objects in non-oxidizing environments or at high temperatures. Third, there are no ornaments or other objects, such as ground spear points and knives, that resemble metal objects. Finally, there are no artistic or iconographic objects that would suggest the onset of class distinctions or hierarchical social organization. Thus, the Margarita Culture may be more appropriately classified as a Final Neolithic phase, rather than a Bronze Age culture.

Optimal foraging theory suggests that where highly ranked targeted subsistence resources are abundant and predictable, lower ranked species will only be pursued on an encounter basis, if at all. Thus, if the exploitation aggregated seasonal migrations of fish and birds were the focus of occupations at Zera Lake, then the pursuit of dispersed larger mammals would result in a lowered cost-benefit ratio for the expenditure of finite human labor and be counterproductive.

Carbonized nutshell fragments recovered from our sites were visually identified as Manchurian walnuts, hazelnuts, acorns and pine nuts. All of these have been commonly identified in Neolithic and later deposits throughout the Russian Far East. These nuts served as a reliable and abundant source of carbohydrates, fats and proteins. The nuts ripen in September but can be stored and used year-round and their presence in a site cannot be used as an indicator of the season of occupation. The majority of shell fragments were found at Zera-3. In addition to shell fragments, nut-processing tools, such as pitted anvilstones and pestles, are commonly found in Primorye archaeological sites. A few of these tools were recovered from Zera-3.

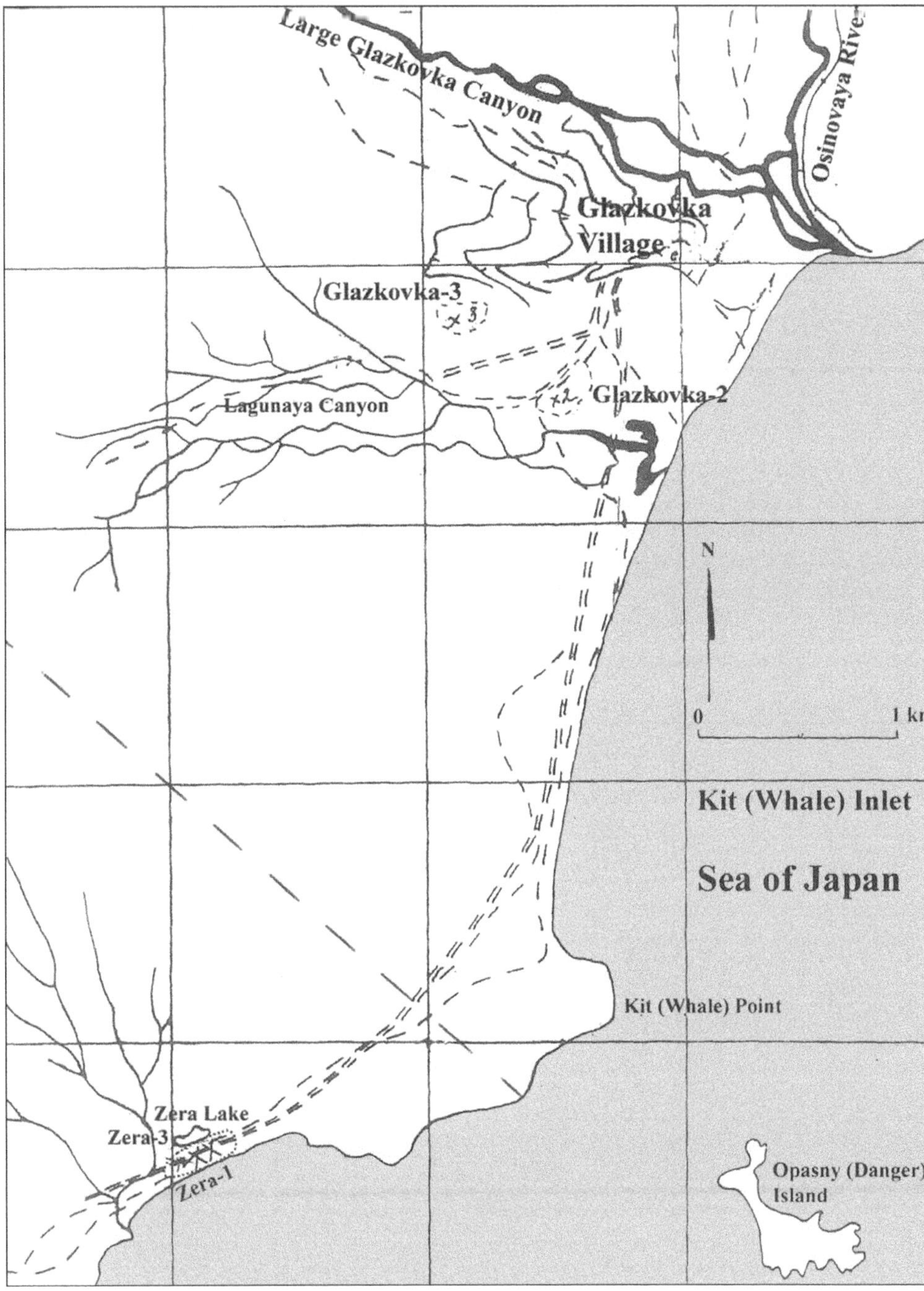


Figure 3: Map showing the locations of Zera Lake and Glazkovka.

## Macrobotanical Analysis

One explicit goal of our research project was to ascertain the relationship of the Margarita Culture with indigenous Neolithic cultures of east coast Primorye and the expanding interaction spheres of agricultural and pastoral societies to the northwest. A primary indicator of such a link would be the presence of cultigens such as millet or barley at the sites. Samples were taken from all identified hearths and cooking pits and subjected to flotation procedures to identify macrobotanical elements in the middens. Samples taken from hearths and cooking pits were separated into one liter increments, for volume control, and sun dried for at least 48 hours. Each sample was then processed through a flotation device to separate the macrobotanical remains.

Elena Sergusheva (2001), research specialist at the Institute of History, Archaeology and Ethnography in Vladivostok, microscopically examined the macrobotanical samples. The flotation samples from the Zera-1A and Zera-1D units dating to the Zaisanovka Culture yielded almost no identifiable macrobotanical results, the samples from the Zera-3A and Zera-3B dating to the Margarita Culture were more productive. Nutshells were recovered from virtually all dwellings at Zera-3 and included Manchurian walnuts, hazelnuts and acorns. Wild grape and possible fruit husks were recovered from Zera-3A, legume and grass seeds were found in Dwellings 1 and 5, dating to the Margarita Culture. The most interesting of the macrobotanical finds was a total of four seeds from levels 4, 5 and 6 of Dwelling 1 that were identified as either Setaria sp. (Broomcorn millet) or Echinochloa sp. (Foxtail millet). Although three of these seeds were considered to be too fragmentary or deformed to determine if they were wild or cultivated varieties, one was classified as a cultigen. Gary Crawford, University of Toronto at Mississauga, Canada, also inspected these macrobotanical samples and agreed that cultivated forms of foxtail and broomcorn millet were probably present (personal communication 2004, Figure 4).

The concentration of these seeds in one of the Margarita Period dwellings and their absence in all other deposits is provocative. If these seeds were of the wild variety, they should have been dispersed in all of the dwellings. The concentration of all four possible millet seeds in a single hearth in Dwelling 1 suggests their purposive presence and a relationship with agricultural practices, either through trade or transport by Margarita Culture populations. The question of whether cultigens were associated with the Margarita Culture cannot be definitively answered, but such an association now appears more likely. Ceramic and obsidian evidence strongly suggests that ongoing interaction existed between the inhabitants of interior Primorye and the Margarita Culture population. It has now been established that people of the interior practiced low level production of millet and barley during the Zaisanovka Period and that their dependence on these resources increased over time (Kluev et al. 2002; Vostretsov et al. 2003).

It seems reasonable to assume that the Margarita Culture people not only had knowledge of the existence of cultigens among the interior populations, but also the means of acquiring these products through social interaction and systems of exchange. This does not imply that the Margarita Culture people were actively engaged in the cultivation of plants at Zera Lake. The absence of agricultural tools as well as the seasonality of the occupation of the site appears to argue against such a conclusion. However, if the occupation of Dwelling 1 took place after the summer harvest in the interior, it is not unreasonable to expect that small quantities of grain may have been transported to the site as a component of the diet while in residence at Zera Lake.

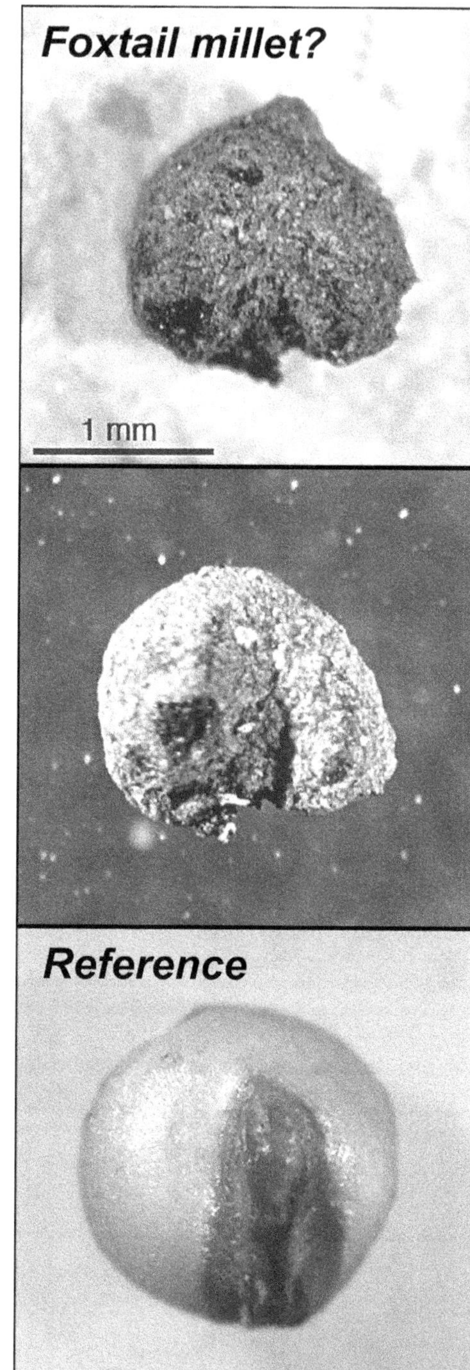


Figure 4: Archaeological macrobotanical find from Zera-3A (top and middle) compared to a modern seed of cultivated Foxtail millet (courtesy of Gary Crawford).

## Fatty Acid Ceramic Residue Analysis

The invention of pottery often coincides with the transition of highly mobile hunter-gatherers to a more settled and culturally conditioned lifestyle (Bettinger et al. 1994). As a result, populations became tethered to restricted geographic regions and annually conditioned seasonal cycles. The durable nature of fired clay has placed it among the most abundant and useful artifacts available for the reconstruction of prehistoric cultural systems. Ceramics are especially informative in the study of when, why and how people first came to intensively exploit specific subsistence resources (Boserup 1965; Morrison 1994).

In the field of ceramic analysis, one direct method of examining vessel function is fatty acid residue analysis (Condamin et al. 1976; Barnard et al. 2007). It is now known that the fatty acids released in the cooking process are absorbed into the walls of ceramic vessels (Skibo 1992). Through the use of combined gas chromatography mass spectroscopy equipment (GC/MS) the chemical signatures of the lipid components of specific categories of cooked plants and animals can be identified (Eerkens 2001; Malainey 1997; Malainey et al. 1998; Malainey et al. 1999). It is known that thermal, oxidative and chemical post-depositional processes degrade fatty acid residues in archaeological contexts, but that the ratios of lipid components remain relatively constant. Through the comparison of archaeological samples with modern experimentally derived reference collections we can identify the types of foods that were cooked in prehistoric clay vessels to the family level. This process is comprehensively discussed in several publications (Condamin et al. 1976; Eerkens 2001; Evershed 1993; Malainey 1997; Malainey et al. 1998; Malainey et al. 1999; Skibo 1992), and will only be briefly summarized here.

Hand made ceramic vessels fired at relatively low temperatures, such as the ones found in Neolithic deposits in Primorye, tend to contain numerous air pockets, called 'vugs', in their walls. When these vessels are employed for preparing or storing food, lipids may liquefy and get absorbed into the vugs where they displace the airpocket within the walls. Upon cooling, many of the lipids will solidify. It appears that the first materials cooked in a pot tend to saturate the walls and prevent the absorbsion of subsequent lipids (Fankhauser 1997). The extraction of the lipids from ceramic sherds, in the laboratory, is a relatively simple and minimally destructive process (cf. Chapter 5).

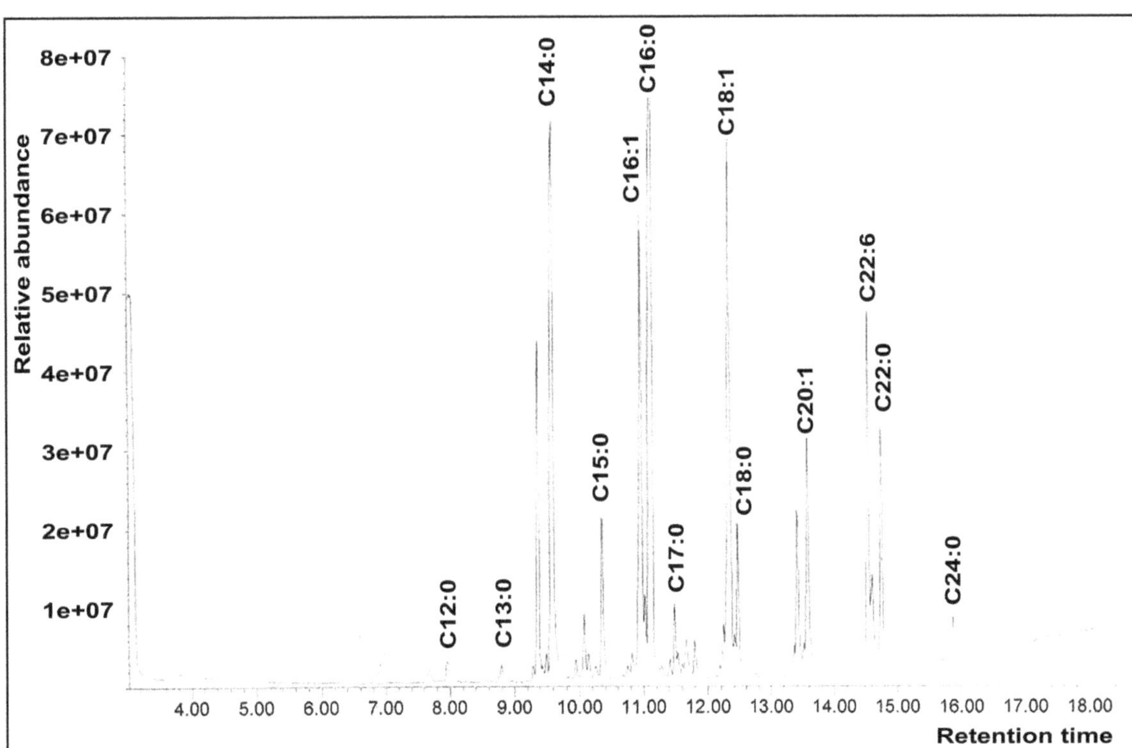


Figure 5: Chromatogram of an experimental residue of wild Pacific Salmon (Table 1).

Approximately one square centimeter of the sherd was taken for analysis. The outer millimeter was ground off all sides to remove most external contaminants. The remainder of the sherd was pulverized, with a mortar and pestle, and a 400 mg sample was placed into a test tube. The powder was then suspended in a chloroform-methanol (2:1, v/v) solvent, gently agitated and placed in a centrifuge to draw down the remaining ceramic fraction. The supernatant, containing the dissolved lipids, was drawn off with a pipette and placed into a second test tube. This was placed in a vacuum centrifuge to remove the solvents, and the remaining dried lipids were derivatized by adding methanolic HCl and heating the sample to 60°C for one hour. The result was dried in a vacuum centrifuge and taken up in 50 μl hexane. About 1 μl of this was injected splitless into a 5890 Hewlett Packard Gas Chromatograph coupled to a 5790 Hewlett Packard Mass Spectrometer. The GC-oven was heated from 100°C to 325°C at a rate of 12°C/min. Different lipids are transformed to a gaseous state at different temperatures and pass through a flame ionization detector, measuring the volume of gas over time. This process results in a chromatogram (Figure 4). Data on the archaeological compounds obtained by GC/MS were analyzed using the National Institute of Standards and Technology (NIST) 1998 Mass Spectral Library software and quantified with the linked Automated Mass Deconvolution and Identification System (AMDIS) software. The raw scores obtained from this process were converted to percentages to permit a comparison between archaeological and experimental samples (Table 1).

| Wild Pacific Salmon | | | | | |
|---|---|---|---|---|---|
| **Compound** | **Retention time** | **Amount** | **Confidence** | **Conversion** | **Systematic name** |
| C12:0 | 7.97 | 0.22% | 67.70 | 0.92 | Dodecanoic |
| C13:0 | 8.81 | 0.09% | 76.60 | 0.37 | Tridecanoic |
| C14:0 | 9.61 | 4.02% | 36.20 | 16.72 | Tetradecanoic |
| C15:0 | 10.37 | 1.02% | 73.50 | 4.24 | Pentadecenoic |
| C16:0 | 10.79 | 0.02% | 46.40 | 0.08 | Hexadecanoic |
| C16:1 | 10.97 | 3.28% | 65.70 | 13.64 | Hexadecanoic |
| C16:0 | 11.14 | 5.85% | 79.40 | 24.34 | Hexadecanoic |
| C17:0 | 11.55 | 0.07% | 65.30 | 0.29 | Heptadecanoic |
| C17:0 | 11.56 | 0.05% | 71.60 | 0.21 | Heptadecanoic |
| C17:0 | 11.62 | 0.01% | 61.50 | 0.04 | Heptadecanoic |
| C18:2 | 12.3 | 0.10% | 16.30 | 0.42 | Octadecadienoic |
| C18:1 | 12.34 | 3.53% | 10.10 | 14.68 | Octadecenoic |
| C18:0 | 12.48 | 0.84% | 74.80 | 3.49 | Octadecanoic |
| C20:4 | 13.38 | 0.15% | 62.90 | 0.62 | Eicosatetraenoic |
| C20:1 | 13.59 | 1.40% | 77.70 | 5.82 | Eicosenoic |
| C20:0 | 13.75 | 0.01% | 94.30 | 0.04 | Eicosanoic |
| C22:6 | 14.54 | 2.36% | 73.60 | 9.82 | Docosahexaenoic |
| C22:0 | 14.9 | 1.00% | 80.40 | 4.16 | Docosanoic |

Table 1: Quantification of GC/MS results after processing through the NIST and AMDIS databases (Figure 5).

Although long carbon chains are relatively stable, double bonds tend to oxidize over time. Thus, while poly-unsaturated fatty acids may often be present in experimental samples, they are relatively rare in archaeological samples, which predominately contain saturated and mono-unsaturated fatty acids. The ratios of these are available for comparison with experimental samples to determine food groups represented in the archaeological sample. Some of data on experimental samples were obtained from publications by Malainey et al. (1997; 1998; 1999). Additional experimental samples were analyzed by Jelmer Eerkens (2001) and me. These were then compared with the 18 archaeological samples from Zera-1 and 48 archaeological samples from Zera-3.

The identification of specific food groups is based on the fact that individual animals and plants contain different ratios and types of lipids that make each of them more or less unique. Some animals and plants also contain more or less specific biomarkers that make them more easily identifiable; birds and ruminant animals being examples (Eerkens 2001, 112-115), as well as saltwater versus freshwater fish (Eerkens 2001, 101). On the other hand, some other food groups are so similar in their lipid contents that it is difficult to tell them apart, such as seeds and nuts (Eerkens 2001, 114). Where biomarkers are present it is sometimes possible to identify specific plants or animals. However, given the variability of post-depositional degradation of archaeological samples and the similarity of some lipid compound ratios (Table 2), the most conservative approach is to identify specific food groups rather than individual species (Eerkens 2001, 93-95). In addition to the need to destroy a small portion of a ceramic sherd there are a number of other problems that should be acknowledged. Given the low firing temperatures used to produce these relatively porous ceramic vessels, and the amount of time they have been exposed to destructive post-depositional processes, the ceramic sherds obtained from archaeological excavations are often highly fragmented and non-diagnostic, and the fatty acids may also have undergone a substantial amount of degradation (Schiffer and Skibo 1987). When possible, only sherds that show visual traces of cooking through soot buildup or carbonaceous residues should be selected for analysis.

| Ratios | Mammal meat and fat | Fish | Bird | Roots | Greens | Seeds and nuts | Berries |
|---|---|---|---|---|---|---|---|
| (C12:0 + C14:0) / (C20:0 + C22:0) | >3.5 | >5 | >3 | <2 | <2 | <2 | <0.8 |
| (C16:0 + C16:1) / (C18:0 + C18:1) | <1 | <1.5 | >0.5 and <0.8 | >1.5 | >2 | <0.5 | <0.5 |
| C16:1 / C18:1 | <0.4 | >0.2 and <0.7 | <0.3 | <1 | >1 | <0.1 or >0.1 and <0.3 | <0.2 |
| C16:0 / C18:0 | <4 | >3 | <3 | >3 | >3 | variable*) | <4 |

Table 2: Distinguishing food types by fatty acid ratios (after Eerkens 2001, 106). *) Some seeds and nuts have values well over 4, making this a worthwhile ratio.

The 48 sherds from Zera-3 Margarita Period levels and the 18 sherds from Zera-1 Zaisanovka Period levels showed evidence of food preparation and were assessed for fatty acid residues to ascertain the types of foods cooked. The results of the fatty acid residue analysis of these ceramic sherds revealed a number of subsistence related patterns. First, the presence of a high ratio of berries among the Zaisanovka Period samples suggests a warm period of occupation, when these fruits are known to ripen. This was further supported by the absence of nuts, which are known to ripen in September. Second, the presence of nuts in the Margarita Period samples, along with more than doubling of the presence of greens, such as legumes that are available during the monsoon season of late August and early September, suggests a late summer and fall occupation. These results are interpreted to support the hypothesis that variation in site location and dwelling type reflects a shift in seasonal occupancy. Although the lipid ratios of nuts and seeds are almost identical, the volume of nutshells found in the sites compared to the cultivated millet seeds may argue for a more likely interpretation of nuts in the fatty acid residue results. It should be kept in mind that nuts can usually be consumed without cooking, while it is necessary to boil seeds to release their nutrient contents. Interestingly, the consumption of each of the highly ranked resources of birds, greens and fish far outweigh the consumption of nuts and seeds (Figure 6). This suggests that nuts and seeds did not represent a highly ranked subsistence resource and were merely used to supplement the diet. The frequent placement of Zaisanovka Period sites on coastal beaches, along with occasional fish bones at the sites, has led to a general assumption that this represented a primary focus on saltwater fishing activities (Andreeva et al. 1987; Popov and Yesner 2002; Vostretsov 1998). In contrast, the Margarita Period sites are usually located a short distance from the coast, adjacent to freshwater streams. This pattern suggested that the transition between the two periods coincided with a shift from saltwater fishing to freshwater fishing.

Some species of plants and animals occasionally possess properties that permit a more specific level of identification, such as freshwater fish with a relatively high level of the C18:1 while saltwater fish appears to have a relatively high C24:1 content. These differences permitted the potential to evaluate the hypothesized shift from saltwater to freshwater fishing between the two culture periods (Figure 7).

Surprisingly, there was little difference between the percentages of freshwater fish consumed by both groups. Further, the quantity of freshwater fish consumed by the Zaisanovka Period population was almost double that of saltwater fish, the reverse of what was expected. The most significant difference was that the Margarita Period group at Zera Lake consumed approximately twice the percentage of saltwater fish compared to the Zaisanovka Period population.

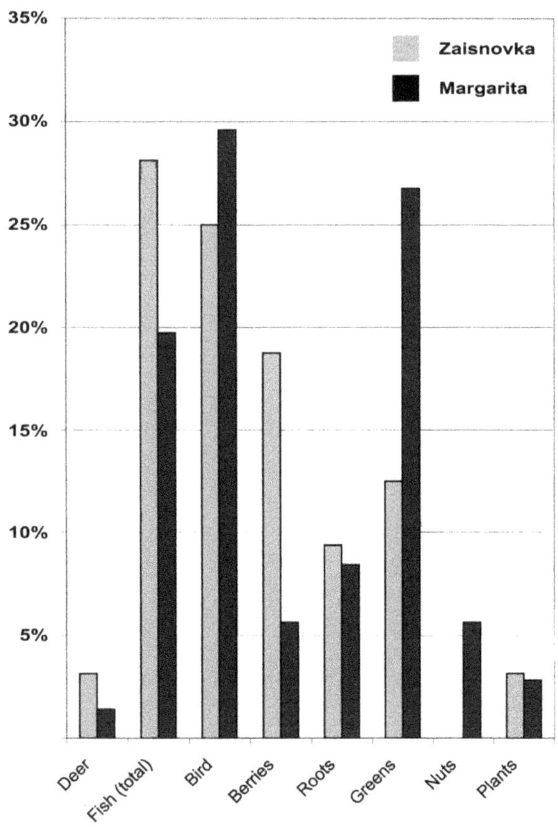


Figure 6: Comparison of percentages of GC/MS derived food groups between the Zaisnovka and the Margarita Periods.

Like everywhere else in the world, the Primorye Region experienced a significant cold and dry climatic anomaly around 3600 years ago (Besednov and Vostretsov 1997; Steig 1999; Vostretsov 1998). It was during this time period that the salmon productivity in the north Pacific reached optimum proportions (Chatters et al. 1995). Although salmon spawn in freshwater habitats, they spend their lives in saltwater environments and possess a similar fatty acid composition to those of saltwater fish (Figure 4). Thus, the increase in saltwater fish consumption during the Margarita Period at Zera Lake correlates directly with a significant drop in sea surface temperatures and the increase in abundance and predictability of salmon spawning events in the rivers and near-coastal lakes of the Primorye Region.

The presence of small mammals and birds in the diet, suggested by a few isolated bone fragments, was also supported by the residue results. In contrast to our expectations, the quantity of birds consumed was statistically equal to, or slightly larger than, the amount of fish during both periods of occupation. Landscape analysis revealed that Zera Lake was substantially larger during the Middle Holocene occupations and may have served as a stopover during seasonal bird migration. Thus, humans may have chosen to settle in this location as much for the presence of migratory waterfowl as for the availability of fish (Berthold 2001; Laptev 1999).

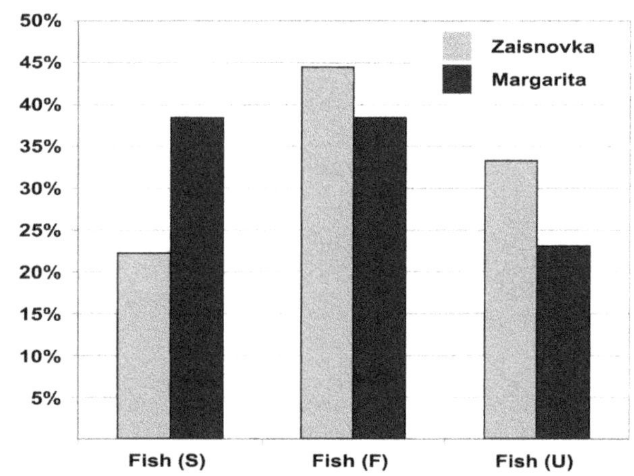


Figure 7: Comparison of saltwater (S) to freshwater (F) fish (U = undifferentiated) identified through GC/MS analysis between the Ziasnovka and the Magarita Periods.

Another property of the lipid profiles that proved revealing were odd number versus even number fatty acid chains, which are related to the types of meat consumed. These suggested that the animals hunted were mostly ruminants, such as deer (Malainey 1997). In both sample groups the numbers of sherds indicating meat were substantially fewer than those associated with fish or birds. This disparity was especially apparent among the Margarita Period samples. These results support the earlier proposal that non-gregarious (solitary) large mammals were not systematically hunted for subsistence and were only taken on an opportunistic basis.

## Conclusions

The post-depositional environment in Primorye Period sites creates numerous challenges to the reconstruction of diets among the prehistoric population. However, through the application of a number of analytic procedures, a clearer picture of subsistence practices of the Zaisanovka and Margarita Period populations at Zera Lake is now beginning to emerge.

The wet screening and flotation analysis of midden and hearth soils has revealed the presence of small quantities of bird and small mammal bones in the deposits at Zera Lake. This is especially true of the excavations conducted at Zera-3. The absence of larger and thicker mammal bones suggests that these were not hunted during the seasonal occupations of either site. While nutshell fragments were recovered from the deposits at both sites, the majority were associated with Zera-3. Nut-processing tools, such as pestles and anvilstones, were present in the Margarita Period stone tool assemblage. As nuts ripen in September, their presence in higher numbers in the Margarita Period deposits suggests a late summer or fall occupation.

Analysis of the macrobotanical remains derived from flotation revealed probable millet seeds from the hearth in Dwelling 1 at Zera-3A. Although it cannot presently be concluded whether these seeds are of wild or of domesticated varieties, their concentration in only one dwelling suggests a degree of intentionality in their acquisition. The lack of agriculturally related tools in the Margarita Period deposits does not necessarily imply that the population was unaware of domesticated plants. A fall occupation would have been beyond the planting season for these types of cultigens, and the seeds may have been grown elsewhere earlier in the summer and then transported to Zera Lake for consumption during the fall when the salmon spawning migrations take place. That these people maintained ongoing contacts with agriculturalists in the interior zone has been well established, so they must at least have been aware of the existence and value of cultivated crops. Further support for a shift in seasonality between the two periods was derived from fatty acid residue analysis. The Zaisanovka Period samples revealed a higher ratio of summer berries and virtually no nut residues. In contrast, the Margarita Period samples revealed high quantities of nuts and greens, abundant in the fall, but only a small amount of berries, which ripen during the summer months.

Although no fish bones were found in the deposits of either site, the results from the GC/MS analysis firmly established that both freshwater and saltwater fish were a primary focal resource exploited by both groups. Further, the consumption of saltwater fish by the Margarita Period population was double that during the Zaisanovka Period occupation. This appears to correlate directly with a substantial increase in salmon spawning migrations during the fall months in the streams and lakes along the east coast of Primorye.

A final subsistence element of significance revealed by lipid analysis was the large quantities of birds that were consumed. In both the Zaisanovka and Margarita Period occupations the percentage of birds cooked in the pots was equal to, or greater than, the quantities of fish. Zera Lake forms a strategic location along the path of migratory waterfowl, both now and apparently in the past. From the volume of birds consumed in these sites, it appears that these were an abundant and reliable source of food that was targeted as a primary resource.

While it appears that both the Zaisanovka and Margarita Period population engaged in periodic contact with people of the interior, they do not appear to have actively engaged in the cultivation of plants or the domestication of animals while in residence at Zera Lake. The subsistence regime revealed provides a clear picture of hunter-gatherer-fisher foraging activities. These activities appear to be seasonally oriented, with the Zaisanovka Period population focusing on summer resources and the Margarita Period people taking more advantage of fall resources.

The more substantial nature of the Margarita Period semi-subterranean dwellings, compared to the Zaisanovka Period light surface structures, implies a greater need for protection from the elements. However, the smaller size of the Margarita Period site and the more specialized nature of the tool assemblage suggest a short-term logistical hunting-fishing camp of foragers, rather than a long-term base camp for gatherers. This implies that abundant and predictable aggregated resources were hunted and processed for storage at Zera-3 and were then transported elsewhere for consumption during the cold winter months. This hypothesis is supported by the presence of large smoking pits associated with the dwellings at Zera-3 and the absence of features associated with long-term food storage.

## References

**Andreeva**, Z.V., A.B. Garkovik, N.A. Kononenko and I.S. Zhushchikhovskaya (1987). *Valentin-Peresheek.* Moscow: Nauka (in Russian).

**Barnard**, H., S.H. Ambrose, D.E. Beehr, M.D. Forster, R.E. Lanehart, R.E. Parr, M.E. Malainey, M. Rider, C. Solazzo and R.M Yohe II (2007). Mixed Results of Seven Methods for Organic Residue Analysis Applied to One Vessel with the Residue of a Known Foodstuff. *Journal of Archaeological Science* **34**: 28-37.

**Barnes**, G.L. (1993). *China, Korea and Japan. The Rise of Civilization in East Asia.* London: Thames and Hudson.

**Berthold**, P. (2001). *Bird Migration. A General Survey.* Oxford: Oxford University Press.

**Besednov**, L.N. and Y.E. Vostretsov (1997). Marine Fishing and Sea Mammal Hunting During the Early and Middle Holocene. *News of Pacific Ocean Scientific Research from the Center of Fishing Economies* **122**:108-130 (in Russian).

**Bettinger**, R.L. (1991. *Hunter-Gatherers, Archaeological and Evolutionary Theory*. New York: Plenum Press.

**Bettinger**, R.L., D.B. Madsen and R.G. Elston (1994. Prehistoric Settlement Categories and Settlement Systems in the Alashan Desert of Inner Mongolia, PRC. *Journal of Anthropological Archaeology* **13**: 74-101.

**Boserup**, E. (1965). *The Conditions of Agricultural Growth*. Chicago: Aldine.

**Brodyanski**, D.L. (1995). *Neolithic and Paleometal of Southern Primorye*. Novosibirsk: Far Eastern State University (in Russian).

**Cassidy**, J. and N.A. Kononenko (2001). *Cultural Entanglements and Transformation During the Bronze Age in the Russian Far East*. Paper presented at the annual meetings of the Society for American Archaeology, New Orleans, 21 April 2001.

**Cassidy**, J. and N.A. Kononenko (2002). Cultural Transformations of Prehistoric Societies in the Primorye Region of the Russian Far East During the Middle Holocene. In N. Khristoforovo, N. Kononenko and M. Glassow (eds.). *Man in the Coastal Zone. Experience of Centuries*. Vladivostok: TINRO-Center, pp. 141-148.

**Chatters**, J.C., V.L. Butler, M.J. Scott, D.M. Anderson and D.A. Neitzel (1995). A Paleoscience Approach to Estimating the Effects of Climate Warming on Salmonid Fisheries of the Columbia River Basin. In R.J. Beamish (ed.). *Climate Change and Northern Fish Populations*. Ottawa: National Research Council of Canada, pp. 357-372.

**Glassow**, M.A. (ed.). *Materials of the UNESCO International Conference*. Kamchatka: Petropovlovsk, pp. 141-148.

**Condamin**, J., F. Formenti, M.O. Metais, M. Michel and P. Blond (1976). The Application of Gas Chromatography to the Tracing of Oil in Ancient Amphorae. *Archaeometry* **18**: 195-201.

**Dyakov**, V.I. (1989). *Primorye in the Bronze Age*. Vladivostok: Far Eastern University Press (in Russian).

**Eerkens**, J.W. (2001). *The Origins of Pottery among Late Prehistoric Hunter-Gatherers in California and the Western Great Basin*. University of California Santa Barbara Ph.D. thesis.

**Fankhauser**, B.L. (1996). Amino Acid Analysis of Food Residues in Pottery. A Field and Laboratory Study. *Archaeology in Ocenia* **32**: 131-140.

**Garkovik**, A.V. (1983). *The Materials of Prehistoric Archaeology of the South of the Far East U.S.S.R. and his Neighboring Territory*. Vladivostok: Dalnauka (in Russian).

**Jochim**, M.A. (1998). *A Hunter-Gatherer Landscape. Southwest Germany in the Late Paleolithic and Mesolithic*. New York: Plenum Press.

**Kelly**, R.L. (1995). *The Foraging Spectrum*. Washington D.C.: Smithsonian Institution Press.

**Kluev**, N.A. and O.V. Yaushina (1997). New Material of Paleo-metal Period in Primorye. Glazkovka-2 Site. In: *Questions in Archaeology, History and Ethnology of the Far East. Journal of the Institute of History, Archaeology and Ethnography of the People of the Russian Far East*. Vladivostok: Dalnauka (in Russian).

**Kluev**, N.A., E.A. Sergusheva and N.B. Verkhovskaya (2002). On Agriculture at the Final Neolithic in the Primorye (on the Material of the Settlement of Novocelishe-4). *Traditional Cultures of Eastern Asia* **4**: 102-126 (in Russian).

**Kononenko**, N.A. and J. Cassidy (1997). Subsistence Activity in the Context of Changing Environments During the Pleistocene-Holocene Boundary in the Zerkalnaya River Valley of the Russian Far East. *Annual Proceedings of the Society for California Archaeology* **12**: 57-61.

**Korotkii**, A.M., S.P. Pletnev, V.S. Pushkar, T.A. Grebennikova, N.G. Razhigaeva, E.D. Sakhebgareeva and L.M. Mokhova (1988). *The Development of Natural Environment of the USSR Far East Southern Part (Late Pleistocene-Holocene)*. Moscow: Nauka (in Russian).

**Laptev**, A. (1999). *Taiga at the Ocean. Red Book of Endangered Species of Primorye*. Vladivostok: Russian Island Almanac.

**Maschner**, H.D.G. (2000). Catastrophic Change and Regional Interaction. The Southern Bering Sea in a Dynamic World System. In M. Appelt, J. Berglund and H.C. Gullov (eds.). *Identities and Cultural Contacts in the Arctic. Proceedings from a Conference at the Danish National Museum, Copenhagen*. Copenhagen: The Danish National Museum and Danish Polar Center, pp. 252-264.

**Malainey**, M.E. (1997). *The Reconstruction and Testing of Subsistence and Settlement Strategies for the Plains, Parkland, and Southern Boreal Forest*. University of Manitoba Ph.D. thesis.

**Malainey**, M.E., R. Przybyiski and B.L. Sherriff (1998). The Fatty Acid Composition of Native Plants and Animals of Western Canada. *Journal of Archaeological Science* **26**: 83-94.

**Malainey**, M.E., R. Przybyiski and B.L. Sherriff (1999). The Effects of Thermal and Oxidative Degradation on the Fatty Acid Composition of Food Plants and Animals of Western Canada. Implications for the Identification of Archaeological Vessel Residues. *Journal of Archaeological Science* **26**: 95-103.

**Morrison**, K.D. (1994). The Intensification of Production. Archaeological Approaches. *Journal of Archaeological Method and Theory* **1**: 111-159.

**Okladnikov**, A.P. (1964). Siberia in the Stone Age. In: *Materials on the Ancient History of Siberia*. Ulan-Ude (in Russian).

**Popov**, A.N. and D.R. Yesner (2002). *Early Maritime Adaptation on the Southern Coast of the Far East of Russia in Ancient Times*. Paper presented at the Beyond the Steppe and the Sown, Integrating Local and Global Visions Conference at the University of Chicago, 4 May 2002.

**Sabloff**, J.A. and W. Ashmore (2001). An Aspect of Archaeology's Recent Past and Its Relevance in the New

Millennium. In G.M. Feinman and T.D. Price (eds.). *Archaeology at the Millennium. A Sourcebook*. New York: Kluwer Academic/Plenum Publishers, pp. 11-32.
**Schiffer**, M.B. and J.M. Skibo (1987. Theory and Experiment in the Study of Technological Change. *Current Anthropology* **28**: 595-622.
**Sergusheva**, E. (2001). *Preliminary Report. Identification of the Archaeobotanical Collection from Zera-3A, Zera-3B, Zera-1D, Zera-1A*. Unpublished report.
**Shelach**, G. (1999). *Leadership Strategies, Economic Activity, and Interregional Interaction. Social Complexity in Northeast China*. New York: Kluwer Academic/Plenum Publishers.
**Skibo**, J.M. (1992). *Pottery Function. A Use-Alteration Perspective*. New York: Plenum Press.
**Smith**, E.A. and B. Winterhalder (1992). *Evolutionary Ecology and Human Behavior*. New York: Aldine De Gruyter.
**Steig**, E.J. (1999). Mid-Holocene Climate Change. *Science* **286**: 1485-1487.
**Vostretsov**, Y. (1998). *The First Fishers in Peter the Great Bay. Environment and Ancient Man in Boisman Bay*. Vladivostok: Nauka (in Russian).
**Vostretsov**, Y.E., H. Obata, E.A. Sergusheva, K. Masayaki and M. Kazuo (2003). New Data on the Early Stage of Agriculture in Primorye. The Neolithic Complex of Krounovka-1 Settlement. In A.P. Derevianko (ed.). *Problems of Archaeology and Paleoecology of Northern, Eastern and Central Asia*. Novosibirsk: Institute of Archaeology and Ethnography, pp. 158-160 (in Russian).

# CHAPTER ELEVEN
# Using Residue Analysis to Confirm Trade Connections at Pella, Jordan
## H.A. Hoekman-Sites

Hanneke Hoekman-Sites; Department of Anthropology, Florida State University; 1847 West Tennessee Street; Tallahassee, FL 32306; USA; <hah04@garnet.acns.fsu.edu>. This research was done while I was a student at the College of Wooster (Ohio) as part of the college's Senior Independent Study program. Many thanks are due to my faculty advisors P. Nick Kardulias and Paul Edmiston, whose guidance and support helped me to complete this project successfully. Several others made valuable contributions to this project, including Robert H. Smith, Ruth Beeston, William Parkinson, Marilyn Loveless, Walter Zurko and Brian Sites. Thanks also to Wooster's Copeland Fund for financial support.

Residues on ceramic samples from 43 separate vessels from Pella, Jordan, were analyzed for the presence of residues, focusing specifically on identifying biomarkers in the samples, in order to investigate trade during the Late Byzantine and Medieval Periods. The results provide information about diet and trade in this city that was an important part of the complex commercial network at the western terminus of Asian caravan routes. The ceramic samples were soaked in dichloromethane to extract the residues, which were then analyzed using gas chromatography-mass spectrometry (GC/MS). The resulting spectra were compared to the National Institute of Standards and Technology (NIST) 1998 Mass Spectral Library to find matching compounds. From the identity of individual compounds, the contents of the original pottery vessels were tentatively inferred.

The site of Pella in Jordan was excavated by a joint team from the University of Sydney (Australia) and the College of Wooster (Ohio, USA) starting in the late 1960s. The goal of this study was to learn more about the population of ancient Pella by uncovering specific organic substances, both local and imported, used in daily life. These substances may be foodstuffs or other compounds like perfumes or body oils. The study described in this chapter focused on ceramic pieces from the Late Byzantine and Medieval Periods, although Pella was occupied long before these times.

Pella was destroyed by the Hasmonean leader Alexander Jannaeus in 83/2 BCE (Smith 1981, 48). During the Roman Period, Pella was rebuilt after Pompey liberated it from Jannaeus (Smith 1973, 39). By the 6th century CE, towards the end of the Byzantine Period, Pella was at the height of its prosperity and size (Smith and Day 1989, 7-9). After the earthquake in 747 CE, Pella was never again a major city. However, it seems that several small villages have continued sporadic occupation of the site, from medieval herders to today.

Previous archaeological evidence, primarily based on ceramic styles, suggests that Pella was a major commercial center during the Byzantine Period (4th-7th centuries CE), trading with North Africa, Cyprus and Phocaea (Smith and Day 1989, 103). The evidence supports the view of Pella as a bustling trade center during the Byzantine Period, suggesting trade connections with China. Specific plant species were identified to support this: *Schisandra chinensis* (wu wei zi) and possibly *Camellia sinensis* (theaceae). In the Medieval Period (8th-13th centuries CE),[1] Pella had lost its position as a long-distance trade center. Only local goods were found on medieval sherds: *Micromeria fruticosa* subsp. barbata (tasnanesi), *Phoenix dactylifera* (date palm), and possibly *Pimpinella anisum* (aniseed).

### Trade at Pella

Pella is uniquely located on the edge of the Arabian Desert at a crossroad of cultures with Europe to the north, the Mediterranean to the west, China to the east and the Arabian Peninsula and Africa to the south (Figure 1). This location made it a major commercial center during many periods of its history. Recent archaeological research has shown trade connections with Cyprus, Rhodes, Egypt, Asia Minor, Palestine, Syria and North Africa (Smith 1981).

Trade has been modeled in many ways, including Central Place Theory (Christaller 1966), World System's Theory (Gunder Frank 1993; Kardulias 1999) and Peer Polity Interaction (Renfrew and Cherry 1986). Pella was positioned at the edge of the Byzantine Empire, far from its core (Byzantium/Constantinople, modern day Istanbul). However, it was not functioning as a typical peripheral location, but instead as a major market center. The idea of peer polity interaction applied to empire-level organization has been dubbed Dual Empire Interaction (Libby 2003). Illustrated by the interaction between Monte Alban and Teotihuacan, Dual Empire Interaction occurs when two neighboring elite groups develop and maintain their positions of power through the trade of exotic goods back and forth. However, Pella was an important trade location on the fringe of an

[1] The Byzantine and Medieval Periods are seperated by the Umayyad Khalifate (611-750 CE) which had its capital in Damascus (Syria).

empire, or between several bordering empires, rather than a powerful center. Borderland markets are market centers located on the fringes of major city-states of empires. States and their rulers are especially interested in such borderland markets, and often this type of market is under state control (Berdan 1985, 354). Their interest stems from the fact that borderland markets promote the flow of exotic goods to the ruling class. These goods, which are not locally available, help secure the power of the ruling class through their scarcity. Along with acting as a buffer between the empire and the lands outside of its control, Pella also functioned as a borderland trade center. Many of the exotic goods entering the empire from the east came through Pella, or another of the Decapolis cities. The Decapolis, a confederation of cities on the northern Transjordanian Plain, was formed to promote commerce by maintaining political stability (Smith 1981, 48-51).

The formation of the Decapolis illustrates how trade can contribute to the development of larger governmental groupings (Earle 1997). The logic behind the Decapolis is clear: the instability caused by its location on the fringe of the Roman Empire, and later the Byzantine Empire, needed to be negated by local stability in order to promote commerce. Pella joined the Decapolis after the Romans subdued the Hasmonean Revolt that had disrupted much of the Levant. During its time as part of the Decapolis, Pella enjoyed a relatively secure political situation which led to its most prosperous time in the 6th century CE.

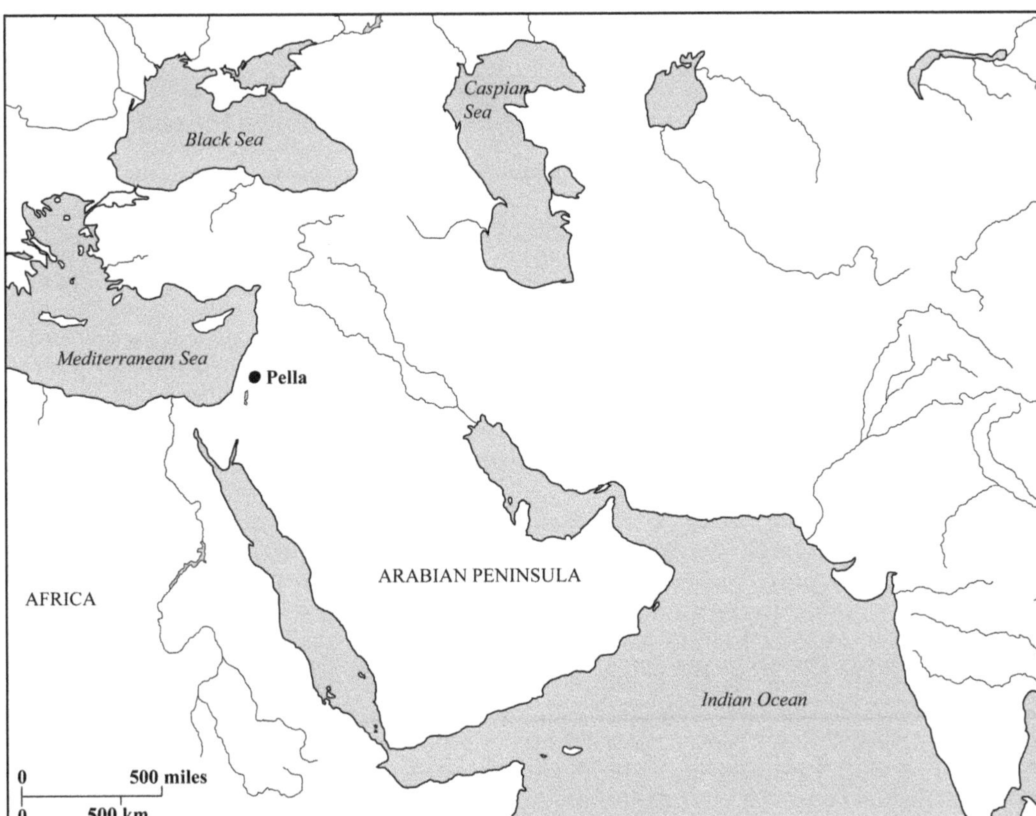


Figure 1: Map showing the location of Pella.

Pella and the other cities in the Decapolis functioned as borderland market exchange centers. Market exchange involves a central location where goods are exchanged for other goods (Renfrew 1984, 93). Pella most likely employed a market exchange system because no storage facilities have been found, although a redistribution system is also possible. In his description of market exchange, Renfrew (1984) notes that it involves a social relationship between people exchanging goods and people controlling the central market location. Those who control the central market location provide a secure location for trade usually in exchange for a tax of some kind.

Borderland markets are of special interest to elites because they increase their access to exotic goods, which help secure their positions of power. D'Altroy and Earle (1985) make a distinction between staple finance and wealth finance. Based on their research in South America, they state that staple finance involves the transport of bulk products over short distances while wealth finance involves the transport of specialty products over long distances. The products being transported to Pella from long distances were most likely specialty items instead of staple crops. The flow of these exotic specialty goods increased the importance of Pella to the large empire governments that controlled the area. Borderland markets can be recognized in the archaeological record in two ways: through their location on the edge of major states and through the presence of exotic trade goods at the site.

During the Byzantine Period, Pella was located on the edge of the Byzantine Empire. An initial review of the ceramics stylistically suggests exchange with other areas in the empire, showing Pella to be a prosperous trade center (Smith and Day 1989, 103-109). During the Medieval Period there were no major states surrounding Pella and there is no ceramic evidence suggesting exchange with other areas. Using organic residue analysis, I will examine ceramics for evidence of exotic trade goods to confirm Pella's status as a borderland market during the Byzantine Period.

## Theory behind Derivatization

Archaeological materials, including organic residues in ceramic vessels, are now commonly investigated with chemical methods. The choice of method depends on the sample and the expected compounds. Visible surface residues assumed to contain only one compound (Michel et al. 1993; McGovern et al. 1996) are treated differently than invisible residues of unknown content extracted from a ceramic matrix (Evershed et al. 1990; Eerkens 2002). GC/MS was chosen for this study because of its ability to separate and identify a mixture of individual chemical residues. This technique primarily separates compounds based on their boiling point, thus all analytes must be volatile. A solvent containing the compound mixture (the analytes) is injected into a heated injection port, where the analytes change into gas and are transported into the column by the carrier gas. In the cool column, the analytes change back into liquid and are deposited onto the column walls. As the column temperature is slowly raised, each analyte reaches its boiling point, changes into a gas, and is swept by the carrier gas through the column where analytes are further separated by polarity. At the end of the column, the analytes enter a detector. The detector first takes the gaseous analyte molecules and ionizes them by electron impact. The ions then pass through the mass selector, for this study a quadrupole mass spectrometer: four parallel metal rods that send the ions on complex trajectories using both a constant and an oscillating voltage. Only ions with a certain mass-to-charge ratio (m/z) are allowed to pass to the detector at a certain time, with the specific m/z changing as the oscillating voltage varies. Comparing the masses of the analyte and its fragments with a library of known compounds, the identity of the molecule can be determined with reasonable certainty (cf. Chapter 5).

GC/MS is not only extremely sensitive, which is crucial to a project involving the detection of trace amounts of materials, but can also yield quantitative information, which is important when comparing amounts of different compounds in a single residue. In order to aid the detection of compounds otherwise not detectable by GC/MS, derivatization by trimethylsiylation was performed on all samples. Trimethylsiylation is a reaction which replaces a terminal hydrogen in an -OH, $-NH_2$, or -COOH group to a trimethylsilyl group. The derivatization agent used for this reaction is N,O-bis(trimethylsilyl)-trifluoroacetamide (BSTFA). The resonance forms for BSTFA can be seen in Figure 2, although the main form is the first one, which will be used in the reaction mechanisms (Pierce 1968, 69-71). The exact mechanism of derivatization is unknown; two possibilities can be seen in Figure 2.

Mechanism 1, a nonconcerted reaction illustrated with an amide, begins with the electron pair from the amide [2] forming a bond with the derivatization agent [1]. Both the oxygen and nitrogen of [1] have a trimethylsilyl constituent. The oxygen is the favored reaction center because it is more electron withdrawing than the nitrogen and a more stable amide results from the oxygen-centered reaction (Pierce 1968, 71). An intermediate [3] is formed in the nonconcerted reaction. An electron pair from the oxygen bond with a hydrogen from the nitrogen, thus breaking off the amide with a trimethylsilyl group, forming [6]. The derivatization agent loses one trimethylsilyl group, forming [4]. [4] converts into the more stable amide [5] through the tautomerism reaction.

Mechanism 2, a concerted reaction illustrated with an alcohol, begins with the electron pair from the alcohol [7] attacking the trimethylsilyl group on the derivatization agent's oxygen [1] while the electrons from oxygen-silyl bond in [1] move onto the oxygen. This result is the formation of [8] and [9]. An electron pair from the negatively charged oxygen on [9] attacks the hydrogen on [8], which [8] is prepared to get rid of because of the positive charge on its oxygen. The end result is the same as mechanism 1, with the derivatization forming [4] and the alcohol gaining a trimethylsilyl group to form [10].

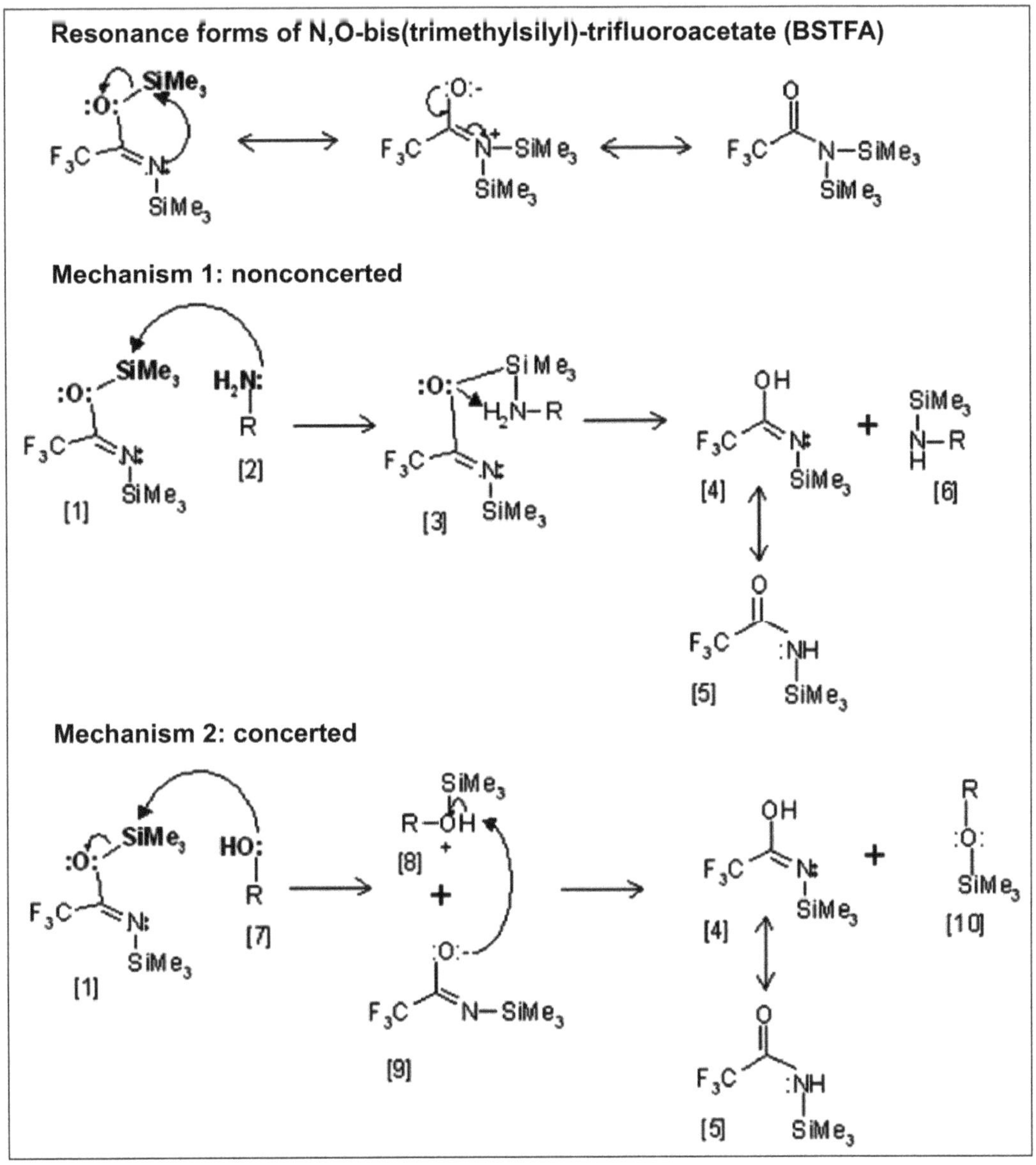


Figure 2: Trimethylsilation reactions using N,O-bis(trimethylsilyl)-trifuoroacetamide (BSTFA) as the derivatizing agent. Mechanism 1 shows how the reaction proceeds if it is nonconcerted, with [3] as the intermediate form. Mechanism 2 shows how the reaction proceeds if it is concerted. The exact reaction mechanism, concerted or nonconcerted, is not known. Mechanism 1 shows the derivatization of an amide, mechanism 2 shows the derivatization of an alcohol. The reaction also works with organic acids (R-COOH).

## Sample Selection and Preparation

Forty-three ceramic samples were selected from the Pella Collection at the College of Wooster for their provenience and variation. The pieces were all from the Byzantine and Medieval Periods. The Byzantine pieces primarily came from the excavation of the Odeum, a section of the Civic Complex which was used as a dumping ground by the 6th century CE (Smith and Day 1989, 8). The Medieval Period pieces came from Tomb 7, which was originally a Byzantine tomb. After its roof collapsed, around 1100 CE, Medieval Period ceramics from a domestic settlement uphill silted in (Smith 1973, 193). A few Byzantine pieces came from Tomb 7's original use. Most pieces had no glaze or paint of any type, a few had decorated exteriors. Samples were broken off the ceramic vessels using a clean wrench.

The samples were ground into a fine powder using a IKA A11 Basic Mill and stored in a sterile glass vial with a Teflon lined lid. A second glass vial with 15 ml dichloromethane ($CH_2Cl_2$, Aldrich 32,399-3) was prepared for each sample after which 1.5 g pottery powder was transferred into these. The vials were shaken vigorously and placed on a shaking table for 24 h. The samples were analyzed in batches of eight for practical reasons. Along with seven samples from Pella, one blank was run with each batch to control for contamination. Two newly made ceramic vessels were used as blanks. As a control 30 µl 0.5% dodecane ($CH_3$-$(CH_2)_{10}$-$CH_3$) in dichloromethane was added to each blank. Glass and Teflon were used exclusively because of the inert nature of those materials. Other containers, especially those of plastic, will leach molecules during the extraction process. The resulting signal from the container may overwhelm the sample signal, making it impossible to detect and identify the ancient residues. As water was used to wash the samples after excavation dichloromethane was chosen as the extraction solvent. Water, a polar liquid, will have dissolved salts and polar compounds and these were assumed lost. Thus a solvent that would extract the remaining non-polar residues was chosen. In previous work (Hoekman 2002), dichloromethane was used for initial extraction, followed by methanol. Because the second methanol extraction resulted in no signal, identical to a clean solvent run, it can be assumed that dichloromethane is an effective solvent to extract all residues.

After each vial was agitated, the sample was transferred into centrifuge tubes and each vial was rinsed with dichloromethane. The rinse was poured into the same centrifuge tube. Each sample was then centrifuged, 5500 rpm for 12 min, and transferred into a sterile vial by pipette. The solvent was evaporated by a stream of nitrogen or by simply placing the open vial in a fume hood overnight. Evaporation in the hood became the preferred method because water condensed in the vials treated with nitrogen. Once the samples were evaporated down to less than 2 ml, they were transferred into GC sample vials after which dichloromethane was added to raise the volume to 2 ml.

As most fatty acids are not able to travel through a GC/MS, the fatty acids in the samples were changed into fatty acid methyl esters (FAMEs) by trimethylsiylation. Derivatization was achieved by adding 50 µl BSTFA, BSTFA (with 1% trimethylchlorosilane, Fluka 15238) to each sample. The vials were then heated in an oven at 60°C for 2 hours and allowed to cool on the bench.

## Procedures and Criteria

About 50 µl sample was first used to acquire an infrared (IR) spectrum after which the samples were run on the GC/MS. Before running samples, a clean baseline was achieved by running pure solvent. After this 1 µl of each sample was injected into the injection port set at a split rate of 1:3. The CG-column was a fused-silica analytical column (Agilent 19091S-433), which was held at 55°C, followed by a ramp from 55°C to 300°C at 15°C per minute. A 6 min solvent delay was used to allow the solvent and excess derivatizing agent time to elute from the column before the mass detector began measurement. Once the oven reached 300°C, this temperature was held for 15 min to allow any remaining compounds to exit the column.

Previous work on this GC/MS instrument did show that it can detect trace amounts of chemicals, down to 0.25 pg ($2.5\times10^{-13}$ g) under ideal circumstances. As the background level of the chromatogram was low, on average about 2000 abundance counts, the criterion for signal was set at an abundance of 15,000 over the baseline. This threshold resulted in roughly 50% of the individual peaks in the chromatogram being included in the analysis, the rest being discarded as noise. Some samples also had an area of heavily overlapping peaks occurring between 15 and 20 min. These peaks appeared to be predominantly hydrocarbons, coming off the GC-column after comparable retention times because of their structural similarity. In these areas, only the most prominent peaks could be identified.

## IR and GC/MS Results

IR provides a spectrum for the extracted residues as a whole, it provides no spatial information. The IR results for the samples in this study did not yield valuable information because the amounts of derivatizing agent dominated all spectra.

The GC/MS chromatograms were analyzed using ChemStation Enhanced Data Analysis software with the NIST 1998 Mass Spectral Library installed. For all peaks with an abundance higher than 15,000 the software suggested possible identities based on the ion fragment pattern compared to a library of 107,829 compounds. The likelihood of each suggested identity was considered and the best identity was chosen based on the presence of matching ion fragment peaks in the spectra. The chemical identity was also ranked on a scale of 1-5, with 5 being the most certain. Certainty was judged based on how well the mass spectrum of the sample matched with the spectrum of the chemical suggested from the library. Retention time, abundance, molecular weight, possible identity and certainty factor were recorded in a separate table for each peak in all chromatograms.

One blank sample was run for every seven archaeological samples. The purpose of the blanks was to detect any contaminants, including all chemicals that were extracted from the ceramic matrix. Chemicals that are found in the blanks are not from archaeological residues. Thus, the peaks that were detected in the blanks were not considered part of the chemical signature for the ancient samples.

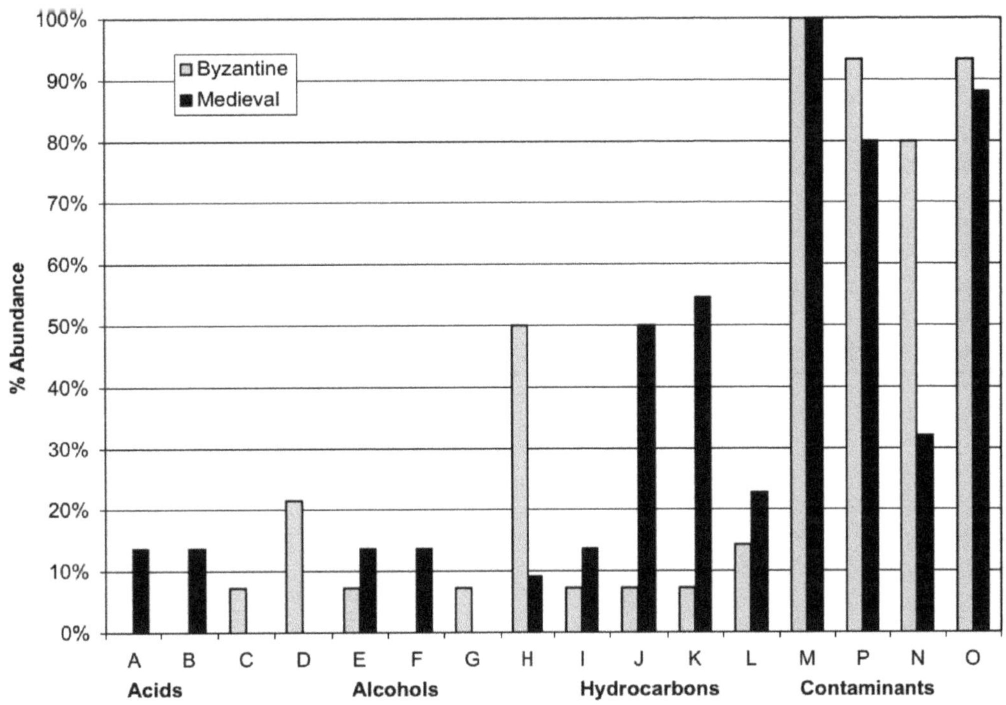


Figure 3: Specific residues present in Byzantine and Medieval Period samples (Table 1).

| |
|---|
| **Acids** |
| A:propanedioic acid diethyl ester (diethyl malonate) |
| B: hexanoic acid |
| C: dehydroabietic acid |
| D: 2,3-dihydroxy-propyl ester octadecanoic acid |
| **Alcohols** |
| E: phenol |
| F: 2-phenoxyethanol |
| G: 1-(hydroxy)-2-(hydroxy)-benzene |
| H: 1-dodecanol |
| **Hydrocarbons** |
| I: hexadecane |
| J: octadecane |
| K: nonadecane |
| L: heneicosane |
| **Contaminants** |
| M: octamethyl-trisiloxane |
| N: bis(2-methylpropyl) ester 1,2-benzenedicarboxylic acid |
| O: dibutyl phthalate |
| P: unknown |

Table 1: Chemicals identified in residues (Figure 3).

## Interpretation of the Results

A comparison of molecules present in Byzantine and Medieval Period samples was performed (Figure 3). Table 1 lists molecules by type, with four organic acids followed by four alcohols followed by four hydrocarbons and finally three contaminants and an unknown. Both Byzantine and Medieval Period residues contain organic acids, but medieval samples included propanedioic acid diethyl ester [A] and hexanoic acid [B], which were not present in any of the Byzantine samples, while the Byzantine samples included dehydroabietic acid [C] and 2,3-dihydroxypropyl ester octadecanoic acid [D], which were not present in any of the medieval samples.

Fatty acids signal the presence of oil or fat (Stern et al. 2000). Animal fats can sometimes be distinguished from plant oils as a larger percentage of animal fatty acids have an even number of carbon atoms compared to fatty acids of plant origin (Condamin et al. 1976, 199; Persaud 2003). Animal fatty acids also have fewer double bonds between carbon atoms than those of plant origin. Mono-unsaturated fatty acids oxidize over time into dicarboxylic acid (Regert et al. 1998). Neither unsaturated fatty acids nor dicarboxylic acids were found in the Pella residues. Alcohols were present in both Byzantine and Medieval Period samples to differing degrees. Dodecanol [H] is fairly common in Byzantine samples. One remarkable trend is that there are no long-chain hydrocarbons present in the Byzantine samples from the Odeum. This could be due to the immersion that the Odeum samples endured, although it could be explained in other ways as well. Long-chain hydrocarbons are indicative of plant remains. The presence of long-chain hydrocarbons in the Byzantine samples from Tomb 7 make it likely that the lack of hydrocarbons in the other Byzantine samples is due more to their specific context than to their time period or original contents. All samples contained contaminants, possibly from the ceramic itself (octamethyl-trisiloxane) or from the column. Phthalate esters were present in all archaeological samples, presumably from the plastic bags in which they were stored since excavation: phthalates are man-made molecules added to most plastics. Two examples are shown in Figure 3, bis(2-methylpropyl) ester 1,2-benzenedicarboxylic acid [N] and dibutyl phthalate [O]. Although the origin of bis(2-methylpropyl) ester 1,2-benzenedicarboxylic acid is unclear, it must be assumed a modern contaminant because it was present in all of the blanks.

Modern sources of contamination are a major concern for any archaeologist wishing to utilize chemical analytical techniques. The presence of contaminants in a sample can easily mask the presence of smaller residue peaks. Contamination can be introduced from many sources: the hands of the excavator, the storage facility and the laboratory equipment, to name but a few. The major identifiable source of contamination for the Pella samples was the plastic bags in which the samples were stored. While the constant presence of water probably washed away many residues from the samples recovered from the Odeum, the constant presence of volatile molecules from their plastics surroundings built up a sizable deposit of phthalates in them. Slight changes in our archaeological methods could prevent much of the contamination introduced during excavation and sample storage. Ideally, when a sample is found that is to be tested using chemical methods, it should not be washed. Instead the sample should be carefully deposited into a storage container along with some of the surrounding soil. Instead of plastic containers, glass, Teflon or metal receptacles should be considered instead, provided those can be properly sealed. Paint cans are sometimes used for forensic sampling as they are airtight and free of contamination.

## Specific Plant Identification

Plants containing the chemicals identified in the residues were found by searching 'Dr. Duke's Phytochemical and Ethnobotanical Database' (Duke 1996), an online database funded by the Agricultural Research Service (USDA) and maintained by Dr. James Duke, a well-known ethnobotanist who has published many significant books on the subject.[2] When a chemical name is entered into the database search, a list of plants containing that chemical is generated along with the part of the plant in which the chemical is concentrated. This database is currently the best tool available for identifying plant residues based on their chemical composition on a large scale. A list of possible plants for each chemical was compiled using the native habitats of the plants as a method for weeding out bad matches (Mabberley 1997). Since each residue sample contained several chemicals, a plant lists for all chemicals was compiled, with plants appearing several times getting extra emphasis and a list of possible plant species could be compiled for each sample. While these general lists provide interesting suggestions for further research, they are not conclusive.

Another method to identify plant species in archaeological residues is by looking specifically for biomarkers and ignoring other chemicals. Biomarkers are chemicals that clearly signify the presence of one genus or even a specific species of plant (see Eerkens 2002). Chemicals identified by GC/MS were entered into the Ethnobotanical Database (Duke 1996). Some chemicals (such as phenol) are present in many types of plants while others (such as propanedioic acid) have only a few possible plant sources. Some of the chemicals identified in the residues, along with the number of plant species that contain each chemical from the Ethnobotanical Database, are presented in Table 2. When trying to identify plant species based on biomarkers, the focus will be on the chemicals that resulted in the smallest number of plant species. Chemicals that are present in few plant species, biomarkers, make it possible to identify specific plant species from chemical residues. The number of plant species containing specific chemicals identified in the Pella residues can be seen in Table 2. Plant species identifications can be seen in Table 3 for chemicals that are only present in a few plants (biomarkers).

Propanedioic acid, dehydroabietic acid, 2,3-dihydroxypropyl ester octadecanoic acid and 3-methoxy-4-hydroxy-benzaldehyde are chemicals that have been identified in a single plant species. Propanedioic acid is found in *Micromeria fruticosa* subsp. barbata (tasnanesi), a plant whose leaves are used for tea (Mabberley 1997). It grows wild in Israel and Turkey (Fleisher and Fleisher 1991; Kinmer et al. 1993), in the

[2] Including Handbook of Legumes of World Economic Importance (1981), CRC Handbook of Medicinal Herbs (1985), Medicinal Plants of China (1985), Field Guide to Medicinal Plants (2000), Handbook of Phytochemical Constituents of GRAS Herbs and Other Economic Plants (1992) and Handbook of Edible Weeds (2001).

same region as Pella. Dehydroabietic acid is found in *Pinus sylvestris* (Scots pine), a plant used for medicines and scented oils (Mabberley 1997). It is native to temperate areas in Eurasia, specifically in Scotland and Norway (Mabberley 1997). The Scots pine was probably not native to Pella, because of its climate, but this chemical could also indicate the presence of another species from the genus Pinus, native to the Mediterranean. *Schisandra chinesis* (wu wei zi), which contains 2,3-dihydroxypropyl ester octadecanoic acid, is a berry that was used for food (Thompson 1982) as well as medicine (Gunther 1959). Wu wei zi, native to China, would have to be traded from the Far East to Pella. *Phoenix dactylifera* (date palm), which contains 3-methoxy-4-hydroxy-benzaldehyde, produces dates, a major source of food in North Africa and the Middle East (Mabberley 1997). A chemical that can be found in two plant species is 2,6,10,14-tetramethyl-pentadecane. The first species is *Camellia sinensis* (theaceae), a Chinese plant whose leaves are used for tea (Mabberley 1997). The second species is *Pimpinella animum* (aniseed), a flavoring for many foods, drinks and used in medicines (Mabberley 1997). *Pimpinella animum* is a native plant in Greece and along the east coast of the Mediterranean to Egypt (Mabberley 1997).

| Chemical compound | Number of plant species | Number of Byzantine samples | Number of Medieval samples |
|---|---|---|---|
| Pentanoic acid | 4 | 1 | 0 |
| Propanedioic acid diethyl ester | 1 | 0 | 3 |
| Hexanoic acid | 11 | 0 | 3 |
| Dehydroabietic acid | 1 | 1 | 0 |
| 2,3-dihydroxypropyl ester octadecanoic acid | 1 | 3 | 0 |
| Phenol | 36 | 1 | 3 |
| Dodecanol | 6 | 7 | 2 |
| Pentadecane | 7 | 0 | 2 |
| Nonadecane | 6 | 1 | 12 |
| 3-methoxy-4-hydroxy-benzaldehyde | 1 | 0 | 2 |
| 2,6,10,14-tetramethyl-pentadecane | 2 | 1 | 6 |

Table 2: Number of plant species identified for key molecules found in Byzantine and Medieval Period residues.

Using Table 2, the biomarkers identified can be divided into two groups: Byzantine and Medieval. Biomarkers that are present in Byzantine samples are dehydroabietic acid, 2,3-dihydroxypropyl ester octadecanoic acid and 2,6,10,14-tetramethyl-pentadecane. Dehydroabietic acid is indicative of *Pinus sylvestris* (Scots pine); 2,3-dihydroxypropyl ester octadecanoic acid is indicative of *Schisandra chinesis* (wu wei zi); 2,6,10,14-tetramethyl-pentadecane is indicative of either *Camellia sinensis* (theaceae) or *Pimpinella animum* (aniseed). All three of these biomarkers point to species that are not native to Pella or the Decapolis region, suggesting contacts between Pella and regions to the north and east. Although the Pinus residue could point to local species, the presence of *Schisandra chinesis* clearly points to long-distance trade. The biomarkers not present in the Byzantine samples are propanedioic acid and 3-methoxy-4-hydroxy-benzaldehyde, which are indicative of local goods. It appears that the 'intense international trade' proposed by R.H. Smith (Smith and Day 1989, 103) was not exaggerated. The evidence suggests that during the Byzantine Period, Pella had an extensive trade network, exchanging goods with China.

Biomarkers that are present in medieval samples are propanedioic acid, 3-methoxy-4-hydroxy-benzaldehyde and 2,6,10,14-tetramethyl-pentadecane. Propanedioic acid is indicative of *Micromeria fruticosa* subsp. barbata (tasnanesi); 3-methoxy-4-hydroxy-benzaldehyde is indicative of *Phoenix dactylifera* (date palm); 2,6,10,14-tetramethyl-pentadecane is indicative of either *Camellia sinensis* (theaceae) or *Pimpinella animum* (aniseed). All three of these biomarkers point to local or regional species. The biomarkers not present in the medieval samples are dehydroabietic acid and 2,3-dihydroxypropyl ester octadecanoic acid, which are indicative of long-distance trade. The evidence suggests that by the Medieval Period, Pella was no longer an international trade center.

| Chemica compound | Plant species | Common name | Use | Habitat |
|---|---|---|---|---|
| Propanedioic acid diethyl ester | *Micromeria fruticosa* subsp. barbata | tasnanesi | tea (leaves)[1] | Israel, Turkey[2, 3] |
| 2,6,10,14-tetramethyl-pentadecane | *Camellia sinensis* | theaceae | tea (leaves)[1] | China[1] |
| | *Pimpinella animum* | aniseed | spice, flavor, medicine (oil)[1] | Greece to Egypt[1] |
| Dehydroabietic acid | *Pinus sylvestris* | Scots pine | medicine (bark, leaves), scented oil (leaves)[5] | Eurasia[1] |
| 2,3-dihydroxypropyl ester octadecanoic acid | *Schisandra chinesis* | wu wei zi | food (fruit), medicine (berries)[1, 6] | China[1] |
| 3-methoxy-4-hydroxy-benzaldehyde | *Phoenix dactylifera* | date palm | food (fruit)[1] | North Africa, Middle East[1] |
| Pentanoic acid | *Glycyrrhiza glabra* | liquorice | flavor (root), medicine (berries)[1, 5] | Mediterranean to Central Asia[1] |
| | *Laurus nobilis* | bay laurel | flavor (leaves), medicine (leaf oil)[4] | Greece[1] |
| | *Pogostemon cablin* | patchouli | scented oil[1] | Indomalasia, New Guinea[1] |
| | *Prunus persica* | peach | food (fruit)[1] | China[1] |

Table 3: Plant species information for a selection of possible biomarkers (references: 1: Mabberley 1997; 2: Fleisher and Fleisher 1991; 3: Kinmer et al. 1993; 4: Thompson 1982; 5: Gunther 1959; 6: Weirsma and Leon 1999).

## Conclusion

The Byzantine ceramic samples contained residue from plants native to China, an area far from the Byzantine Empire. The identification of these exotic residues, along with Pella's known location on the eastern edge of the Byzantine Empire, establishes Pella's status as a borderland market during the Byzantine Period.

Extraction and chemical analysis of residues from pottery sherds discovered at Pella of the Decapolis were conducted to examine trade connections. Sherds were extracted with dichloromethane and the resulting residues were analyzed using GC/MS. The concept of biomarkers was used to identify with a greater degree of certainty specific plant species for the Byzantine and Medieval Periods. Pella was part of the Decapolis, a complex commercial network at the western terminus of Asian caravan routes. Archaeological evidence suggests Pella was a major commercial center during the Byzantine Period, trading with North Africa, Cyprus, and Phocaea (Smith and Day 1989, 103). The evidence upholds the view of Pella as a bustling trade center during the Byzantine Period and suggests additional trade connections with China. Specific plant species were identified to support these trade claims: *Schisandra chinensis* and possibly *Camellia sinensis*. In the Medieval Period, Pella had lost its prestige as a trade center. Only local goods were found associated with medieval sherds: *Micromeria fruticosa*, *Phoenix dactylifera* and possibly *Pimpinella anisum*.

To make the chemical identifications more secure, samples of each chemical should be run on the same GC/MS instrument as was used for the ancient samples. An ideal match would have an identical retention time and mass spectrum.

## References

**Berdan**, F.F. (1985). Markets in the Economy of Aztec Mexico. In S. Plattner (ed.). *Markets and Marketing*. Lanham: University Press of America, pp. 339-367.

**Christaller**, W. (1966). *Central Places in Southern Germany*. Englewood Cliffs: Prentice-Hall (translated by C.W. Baskin).

**Condamin**, J., F. Formenti, M.O. Metais, M. Michel and P. Blond (1976). The Application of Gas Chromatography to the Tracing of Oil in Ancient Amphorae. *Archaeometry* **18,2**: 195-201.

**d'Altroy**, T. and T.K. Earle (1985). Staple Finance, Wealth Finance, and Storage in the Inka Political Economy. *Current Anthropology* **26**: 187-197.

**Duke**, J.A. (1996). Dr. Duke's Phytochemical and Ethnobotanical Databases. Agricultural Research Service. <http://www.ars-grin.gov/duke/> (accessed 25 February 2004).

**Earle**, T.K. (1997). *How Chiefs come to Power. The Political Economy in Prehistory*. Stanford: Stanford University Press.

**Eerkens**, J.W. (2002). The Preservation and Identification of Piñon Resins by GC/MS in Pottery from the Western Great Basin. *Archaeometry* **44,1**: 95-105.

**Evershed**, R.P., C. Heron and L.J. Goad (1990). Analysis of Organic Residues of Archaeological Origin by High-temperature Gas Chromatography and Gas Chromatography-Mass Spectrometry. *Analyst* **115**: 1339-1342.

**Fleisher**, Z. and A. Fleisher (1991). The Essential Oil of *Micromeria fruticosa* (L.) Druce subsp. Barbata (Boiss et. Ky.) P.H. Davis. Aromatic plants of the Holy Land and the Sinai. Part VII. *Journal of Essential Oil Research* **3**: 477-479.

**Gunder Frank**, A. (1993). The Bronze Age World System Cycles. *Current Anthropology* **34**: 383-413.

**Gunther**, R.T. (ed.) (1959). *The Greek Herbal of Dioscorides. Illustrated by a Byzantine A.D. 512*. New York: Hafner Publishing Co. (translated by John Goodyer in 1655, edited and first printed in 1933, reprinted in 1959).

**Hoekman**, H.A. (2002). Analyzing Residues in Bronze Age Ceramics from Pella of the Decapolis. The College of Wooster Junior Independent Study Thesis.

**Kardulias**, P.N. (ed.) (1999). *World Systems Theory in Practice. Leadership, Production, and Exchange*. Lanham: Rowman and Littlefield.

**Kinmer**, N., T. Özek, K.H.C. Baser and G. Tümen (1993). The Essential Oil of Micromeria fruticosa (L.) Druce subsp. Barbata (Boiss. Et Kotschy) P.H. Davis of Turkish Origin. In Ch. Franz, R. Seitz and N. Verlet (eds.). *First World Congress on Medicinal and Aromatic Plants for Human Welfare (WOCMAP). Quality, Phytochemistry, Individual Aspects, Economic Aspects. July 19-25, 1992*. Wageningen: International Society for Horticultural Science.

**Libby**, J. (2003). Exploring Dual Empire Interaction in the Pre-Inca Andean World. The Tiwanaku and Wari states. *Focus 2003/2004*. <http://www.focusanthro.org> (accessed 27 October 2005).

**Mabberley**, D.J. (1997). *The Plant-book. A Portable Dictionary of the Vascular Plants*. Cambridge: Cambridge University Press (2nd edition).

**McGovern**, P.E., D.L. Glusker, L.J. Exner and M.M. Voigt (1996). Neolithic Resinated Wine. *Nature* **381**: 480-481.

**Michel**, R.H., P.E. McGovern and V.R. Balder (1993. First Wine and Beer. Chemical Detection of Ancient Fermented Beverages. *Analytical Chemistry* **65**: 408A-413A.

**Persaud**, D.R. (2003). Fatty Acids Present in Animals. Simon Fraser University (Canada). <http://www.sfu.ca/bisc/bisc-429/lipid.html> (accessed 5 March 2004).

**Pierce**, A.E. (1968). *Silylation of Organic Compounds. A Technique for Gas-phase Analysis*. Rockford: Pierce Chemical Company.

**Regert**, M., H. Bland, S. Dudd, P. van Bergen and R. Evershed (1995). Free and Bound Fatty Acid Oxidation Products in Archaeological Ceramic Vessels. *Proceedings of the Royal Society* **265**: 2027-2032.

**Renfrew**, C. (1984). *Approaches to Social Archaeology*. Cambridge: Harvard University Press.

**Renfrew**, C. and J. F. Cherry (eds.) (1986). *Peer Polity Interaction and Socio-political Change*. Cambridge: Cambridge University Press.

**Smith**, M.E. (2004). The Archaeology of Ancient State Economies. *Annual Review of Anthropology* **33**: 73-102.

**Smith**, R.H. (1973). *Pella of the Decapolis. The 1967 Season of The College of Wooster Expedition to Pella. Volume 1*. Wooster: The College of Wooster.

**Smith**, R.H. (1981). Pella of the Decapolis. An American-Australian Expedition Reconstructs 7,000 Years of History in the Jordan Valley. *Archaeology* **355**: 46-53.

**Smith**, R.H. and L. P. Day (1989). *Pella of the Decapolis. Final Report on The College of Wooster Excavations in Area IX, The Civic Complex, 1979-1985. Volume 2*. Wooster: The College of Wooster.

**Stern**, B., C. Heron, M. Serpico and J. Bourriau (2000). A Comparison of Methods for Establishing Fatty Acid Concentration Gradients across Potsherds. A Case Study using Late Bronze Age Canaanite Amphorae. *Archaeometry* **42**: 399-414.

**Thompson**, D.B. (1982). *Garden Lore of Ancient Athens. Meriden Gravure Company Number 8*. Meriden: American School of Classical Studies at Athens.

**Weirsma**, J.H. and B. Leon (1999). World Economic Plants. A Standard Reference. Boca Raton: CRC Press.

# CHAPTER TWELVE
# The Well-Tempered Pottery Analysis: Residue and Typological Analysis of Potsherds from the Lower Mississippi Valley
## E.A. Reber

Eleanora Reber; Anthropology Department, UNC Wilmington; 601 South College Road; Wilmington, NC 28403; rebere@uncw.edu. This study was made possible by NSF grant # BCS-9980294 (USA). The publication of these results has been a long time coming and particular thanks are due to T.R. Kidder of Washington University at St. Louis and Nathanael Heller of R. Christopher Goodwin & Associations, who waited patiently for results, gave information, feedback and assistance on the regional archaeology. Dr. Nikolaas J. van der Merwe, University of Cape Town and Harvard University, offered unstinting support throughout this project. The laboratory work and interpretations were possible only because of the generosity of Professor Dr. Richard P. Evershed and the Organic Geochemistry Unit, Biogeochemistry Research Centre, School of Chemistry, University of Bristol (UK). Paul Perkins patiently listened to far more about residue analysis than he ever wanted to hear, and gave many helpful suggestions. Any mistakes are, of course, my own

This chapter discusses the relationship between archaeological pottery typology and prehistoric pottery use through absorbed pottery residue analysis. Using residue analysis we can validate some aspects of this relationship that have been previously addressed only through indirect methods, experimental studies or theoretical discussion. Residue analysis provides an additional way of looking at archaeological typology and may be one of the best ways to interpret archaeological pottery use, shedding light on human behavior. This approach requires a thorough grasp of the form of the vessel prior to sampling, which means that a rim sherd is the ideal sampling type. Since absorbed pottery residue analysis is destructive, analysis of form and function using residue analysis has seldom been combined. Given sufficient documentation of the sherds to be sampled, however, the destruction of a relatively small rimsherd (2-10 g) can yield valuable knowledge about the ways in which form and function were linked in ancient pots.

A series of 24 rimsherds and 15 bodysherds from six sites in the Lower Mississippi Valley were submitted for absorbed pottery residue analysis, allowing a preliminary comparison of vessel shape and vessel contents in the Tensas basin of Louisiana. This showed a widely varied diet, which continued through time as maize was introduced to the region, and that variation in vessel form and treatment corresponded to variations in vessel use.

## Archaeological Background

The Lower Mississippi Valley (LMV) is defined as beginning at Cairo, Illinois, where the Ohio River flows into the Mississippi (Kidder 1998). In this study, however, all the sherds are from Louisiana. The chronological sequence for the region is divided into three periods following the Middle Woodland Issaquena period: Baytown, Coles Creek, and Mississippi. Due to the abundant natural resources throughout the LMV, agriculture or horticulture was not extensively used until the end of the Coles Creek period. Particularly in the rich floodplains of the Tensas and Yazoo river basins, foraged resources were more than sufficient to support towns and mound construction activities. The advent of maize into the region, which has been estimated at about 1100 CE, did change the culture noticeably, but in ways unique to this highly varied region, and noticeably different from the effects of maize further north along the Mississippi River (Kidder 1992; Kidder and Fritz, 1993).

Most of the cultures in the Baytown period seem to have been relatively egalitarian. Group burials in or near small mounds, often associated with feasting remains were typical of the Troyville culture. The Troyville mound centers were organized around plazas, but the mounds seem to have been centers of ceremonial activity for the entire group. All members of the group were buried in the mound, and grave goods were rare (Kidder 1998). Towards the end of the Baytown period and the beginning of the Coles Creek, shifts begin to appear in burial patterns, diet and community organization. Despite this, there is considerable continuity between the periods, with Coles Creek centers often built atop Baytown sites. Coles Creek diet seems to have depended more upon starchy seeds, particularly maygrass, than the earlier groups. Despite this, horticulture appears to have been very limited, and subsistence was still based primarily upon foraging. Burials during the Coles Creek period are more concentrated on the individual than the group, and access to mounds is apparently restricted by architectural barriers (Kidder 1998). Kidder (1998) suggests that some members of powerful lineages began to occupy the mound centers and restrict less powerful members of society from mound access. At the same time, mound centers tend to grow in size and decrease in number, though there are still a large number of apparently independent mound centers spread across the floodplain.

Over the course of the Coles Creek period, access to mounds seems to become increasingly restricted. By about 1200 CE, increased centralization, exclusivity and stratification led to the development of the Plaquemine culture. This regional culture operated within the Mississippian cultural sphere, and is distinguished from other Mississippian societies primarily on the basis of pottery. Whereas most Mississippian cultures adopted the use of shell tempered wares, peoples of the Plaquemine culture in the vicinity of Vicksburg and Baton Rouge adopted mixed grog/organic/shell tempered 'Addis' wares, while further south along the coast other Plaquemine peoples maintained the use of grog (crushed pottery) tempered wares until the Contact period. Like other Mississippian cultures Plaquemine communities demonstrate increased political centralization around mound centers.

By the 15th century CE, there were a relatively small number of large mound centers across the landscape, maize horticulture was common and the majority of the non-elite population lived in small farming hamlets that were evenly spread across the landscape. Maize appears in measurable quantity in the archaeological record about 1100 CE (calibrated), during the late Coles Creek culture, after the development of mound centers and chiefdoms in the LMV. In sum, maize adoption in the LMV did not cause increased stratification and centralization so much as it allowed the strengthening of a cultural trend already present in the Coles Creek culture prior to the common use of the plant.

## Description of the Sites

Sherds were sampled from six sites in the Lower Mississippi Valley: Bayou des Familles, Emerson, Morgan Mounds, Osceola, Reno Brake, and St. Gabriel Mound (Table 1, Figure 1). These sites can be grouped into two smaller regions: three from the Tensas basin in northern Louisiana (Reno Brake, Osceola and Emerson) and three from the Gulf Coast region of the state (St. Gabriel Mound, Morgan Mound and Bayou des Familles). Sherds contributed by Dr. T.R. Kidder, then of Tulane University, from the Tensas basin sites included a large percentage of rimsherds, which allowed for a better understanding of the types of vessels sampled. Of these sites, the 20 sherds from the Osceola site made up more than half of the total sherds sampled, and thus represent the largest dataset to date from any site in the Lower Mississippi Valley.

The Reno Brake and Osceola sites are located about 400 m apart, along the abandoned Clark Bayou channel of the Mississippi River in the Tensas River basin of Louisiana. Due to their geographic and chronological proximity, the two sites must have been closely linked. The Reno Brake site consists of four dome-shaped mounds along the southern end of an oxbow lake. Although Reno Brake may have been first occupied during the Middle Woodland Issaquena phase, the majority of the occupation dates to the Troyville culture of Late Woodland Baytown period. The area was occupied throughout the Baytown period, but by the end of this period, the southern end of the lake had apparently either closed or begun to silt, and occupation shifted 400 m to the north, to the Osceola site.

| Site | Body sherds | Rim sherds | Total (% total) | Residue present |
|---|---|---|---|---|
| Bayou des Familles | 4 | 0 | 4 (10%) | 2 |
| Emerson | 1 | 2 | 3 (8%) | 3 |
| Morgan Mounds | 4 | 0 | 4 (10%) | 4 |
| Osceola | 1 | 19 | 20 (51%) | 20 |
| Reno Brake | 1 | 3 | 4 (10%) | 4 |
| St. Gabriel Mound | 4 | 0 | 4 (10%) | 3 |

Table 1: Number of sherds from each site in this study.

Subsistence at Reno Brake seems to have been based on foraging. Wild plant species were dominated by nuts and fleshy fruits, rather than starchy or oily grains. Acorn, pecan nuts, persimmon seeds, and palmetto seeds were among the most commonly found archaeobotanical remains from the period. Some chenopod and amaranth seeds were recovered, but appeared to be from wild populations. The faunal assemblage consisted of 71% large mammal fragments, primarily deer, opposed to only 5% fish and 10% reptile, primarily turtle (Kidder and Fritz 1993). However, fish have been underrepresented due to poor bone preservation at the site. The ubiquity of deer in the assemblage from Reno Brake does agree with evidence of feasting at other Baytown burial sites. Ethnobotanical and faunal analysis suggests that the rich environment around the Reno Brake site was enough to support the local population without recourse to domesticated eastern horticultural complex crops. The four sherds from the Reno Brake site were taken from a 1 x 2 m excavation unit near the crest of Mound A, containing a well-preserved midden.

The occupational shift to the Osceola site corresponded both to the silting of the southern end of the oxbow lake, and to the shift to the Coles Creek culture. The site was occupied throughout the Coles Creek period, until the complete silting of the lake, around 1100 CE (calibrated), caused the abandonment of the site. Osceola was occupied immediately after the abandonment of

Reno Brake, during the Mt. Nebo phase. Mound construction began possibly as early as the Sundown phase (750-850 CE), but was definitely underway by the Saranac phase (850-1000 CE), while reaching its peak during the Balmoral phase (1000-1100 CE) (Kidder and Fritz 1993; Kidder 1998). The end of the Balmoral phase saw the site reach its present shape of six mounds surrounding a plaza, a pattern often associated with Mississippian influence.

Diet at Osceola seems to represent a shift from the Reno Brake site. Fish make up 76% of the bone fragments found during the early Coles Creek phases, while large mammals make up only 10% (Kidder and Fritz, 1993). This is an almost direct reversal of the results from the final period at the Reno Brake site. Fish use seems to decrease in the Middle and Late Coles Creek periods, but is still a very important part of the diet: 36% of the Middle Coles Creek assemblage, and 56% of that from the Late Coles Creek period. Part of this apparent shift in diet may be again due to preservation issues; bone preservation was particularly good in the submound middens at Osceola, most of which date to the early Coles Creek periods that show increased ubiquity in fish remains, and possibly decreased community feasting centered around the mounds.

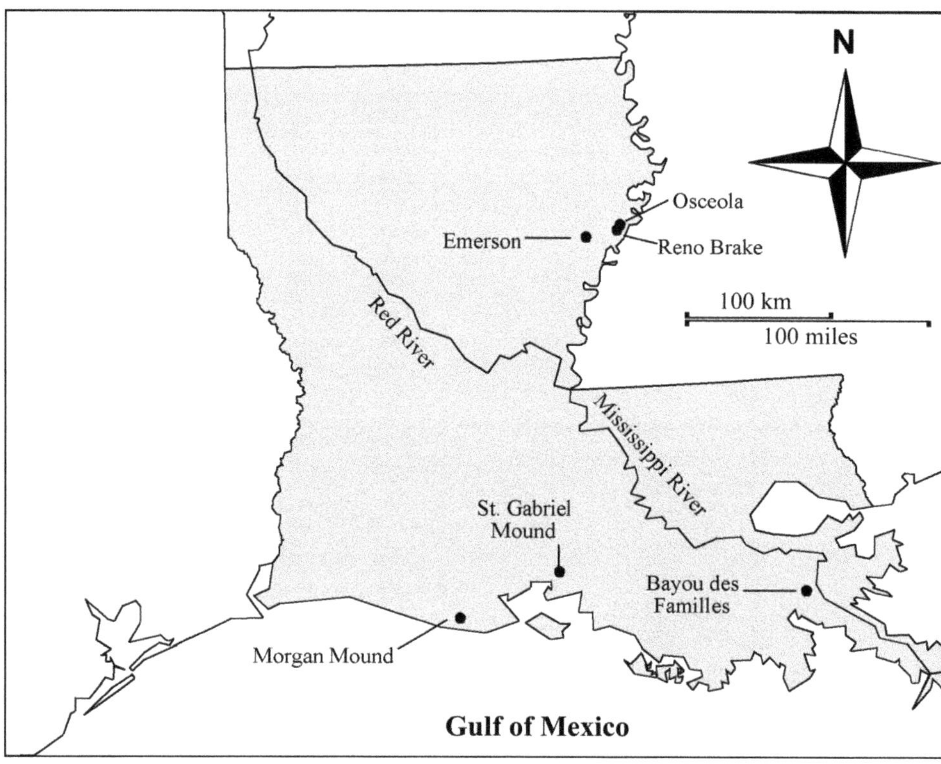


Figure 1: Map showing the position of the sites mentioned in the text.

A shift in diet between the Baytown and Coles Creek cultures is also suggested by the floral assemblage, which shows a greater incidence of starchy seeds at the Osceola site with maygrass, sumpweed, cheno-am, panicoid grass and squash all appearing for the first time in measurable amounts. That said, the most important archaeobotanical remains from the Osceola site are the same as those from the Reno Brake site: acorn, pecan, persimmon and palmetto seeds (Kidder and Fritz 1993). This suggests that the starchy seeds were added to the local diet during the Coles Creek period, which retained its reliance on foraged nuts and fruits. Maize was found only in the Late Coles Creek features at Osceola. In general, maize remains were scattered and sparse; however, a single feature excavated into the Late Coles Creek surface on Mound B contained hundreds of whole, carbonized maize kernels as well as eleven tobacco seeds. The excavators suggest that this feature is representative of ritual activity related to mound construction or use, rather than the common diet of the site occupants (Kidder and Fritz 1993).

The Emerson site is also located in the Tensas River basin, and is a Plaquemine site radiocarbon dated to the middle of the 15$^{th}$ century CE. This site is not a mound center, unlike Osceola or Reno Brake, but rather a small farming hamlet. The site is made up of two contemporary house middens, typifying the increasing political centralization of the Plaquemine culture and the spread of farming hamlets mentioned above (Kidder 1998). Floral remains from the site demonstrate a substantial use of maize, paired with foraging of wild plants and hunting. Emerson exemplifies the Mississippian Plaquemine subsistence pattern of utilizing maize as an important part of the diet, but retaining part of the traditional regional foraging subsistence. Reno Brake, Osceola and Emerson can be viewed as a temporal grouping of culturally related sites, given their close geographical distribution in the Tensas basin.

The Morgan Mounds site is located in the northern Gulf Coastal Plain and on the western edge of the greater Lower Mississippi Valley cultural region. It is an important classic Coles Creek site, comprised originally of four mounds surrounding a plaza. Investigations at the site focused primarily on the largest mound, which was excavated by the Peabody Museum (Harvard University) prior to its destruction in 1986 (Fuller and Fuller 1987). Excavations revealed the remains of a circular structure on the mound summit, a midden on the slope that was probably associated with activities on the mound summit and a submound feature plane with associated midden. Radiocarbon dates from the mound placed occupation between 950-1170 CE (calibrated, $p = 0.93$), which matches well with the ceramic data suggesting a Middle to Late Coles Creek period primary occupation. No subsistence data has been published from the site, although one author confirms large quantities of white-tail deer (*Odocoileus virginianus*) associated with the mound summit and midden on the slope (N. Heller 2005, personal communication).

St. Gabriel Mound comprises a village and single mound at the transition between Coles Creek and Early Plaquemine periods (about 1000-1400 CE), situated on a natural levee of the Mississippi River. The mound was excavated prior to its destruction. Although features associated with the mound summit were poorly preserved, a well-preserved circular structure was exposed in the sub-mound stratum. It appears that the site represented a small village that depended on a mixed diet based on both horticulture and foraging. Burials indicate a relatively egalitarian society, which corresponds to the pattern expected from a Coles Creek site, prior to the increased centralization and social stratification of the later Mississippi Plaquemine period.

Bayou des Familles is a shell midden located in the coastal bayou region of southeastern Louisiana. The site is associated with the Plaquemine culture, with radiocarbon dates of 1300-1500 CE (calibrated). The site is characterized as a small, briefly occupied seasonal habitation, with very little evidence of community planning or segregated space (Kidder 1995). Although shells of the brackish water clam were ubiquitous at the site, a variety of faunal and floral species were also present, including samples of maize.

## Methods

All sherds were cleaned with a drill bit, washed in the extraction solvent, to remove dirt, surface impurities and labels. Glassware and all surfaces that came into contact with samples were also cleaned with the extraction solvent. Each sherd was weighed after cleaning. If the sherd weighed 6 g or less, it was powdered in a mortar and pestle. If the sherd weighed more than 6 g, half to two-thirds was powdered. If the sample weighed less than 12 g, absorbed organic residues were extracted through ultrasonication (2 x 20 min) in a 10 ml mix of chloroform and methanol (2:1, v/v) for every 2-3 g sherd. If the sherd weighed more than 12 g, the large sample size required lipid extraction with a Soxhlet apparatus.[1]

The supernatant was separated from the sample by centrifugation and filtration. The solvents were evaporated under nitrogen and mild heat. An aliquot of this residue was treated with N,O-bis-(trimethylsilyl)-trifluoro-acetamide (BSTFA) with 1% trimethyl-chlorosilane (TMCS) for approximately 10 min at 60°C. The derivatized total lipid extract (TLE) was screened using a Hewlett Packard 5890 gas chromatograph with a fused silica 0.15 mm DB-1 high temperature column. The initial temperature at 50°C was held for 2 min after injection, after which the temperature was raised 10°C/min to 350°C, which was held for 10 min. Following this initial screening, the residues were separated into fatty acid methyl esters (FAMEs) and a trimethylsilylated neutral fraction following the method described by Evershed et al. (1991). The neutral fraction was derivatized using the same procedure as that for the TLE and screened using an identical gas chromatograph, column and temperature program. The FAMEs were

---

[1] Invented by Franz von Soxhlet in 1879, this apparatus is used to separate a soluble compound from a solid. It consists of an extraction chamber, in which the sample is placed, attached to a condenser fed from a container with the extraction solution. This container is heated, causing the extraction solution to evaporate and travel into the condenser. After condensation the solvents drip down on the sample in the extraction chamber. After a while the extraction chamber is emptied, usually by siphon action, back into the heated container. This cycle, during which a portion of the soluble compounds in the sample travels from the extraction chamber into the heated container, is repeated several times. After each cycle the concentration of compounds dissolved in the extraction solution increases, while the solvents in the extraction chamber remain clean (Soxhlet 1879).

screened using the same gas chromatograph and column, but a temperature program set to raise the temperature of the oven from 50-150°C at 15°C/min, 150-250°C at 3°C/min, followed by 10 min at 250°C. This temperature program was also used during the further analysis of the samples.

All samples containing lipids (TLE, neutral and FAME fractions) were analyzed on a gas chromatograph mass spectrometer (GC/MS), either a Carlo Erba Mega Series gas chromatograph interfaced with a Finnigan MAT 4500 mass spectrometer or a Varian TSQ GC/MS with autoinjector. Both instruments had the column and temperature settings as above, operated in electron ionization (EI+) mode and scanned m/z 50-700. Samples with sufficient fatty acids or neutral components underwent compound-specific carbon isotope analysis using a Finnigan MAT delta XL irm-GCMS with CP-Sil-5 column and the same temperature settings at above. Neutral fractions were separated using a CP-Sil 5 column. Each sample was run in triplicate, with a nonadecane ($CH_2$-$(CH_3)_{17}$-$CH_2$) isotopic standard run after each nine runs (three samples). This allowed the monitoring of instrument drift. This procedure allows the identification of maize components in an absorbed residue, as described previously (Reber et al. 2004; Reber and Evershed 2004a).

Interpretation of the residues tended to concentrate first on compound-specific isotopic values, then on the presence of biomarkers, such as sterols or terpenoids, and finally on the amounts of fatty acids. The results hopefully provided a full view of the various components present within the ceramic matrix of each sherd. The TLE was used to determine the approximate degradation undergone by lipids in the residue; sherds with primarily intact triacylglycerides are believed to be less degraded than those with few intact triacylglycerides (Aillaud 2001; Dudd et al. 1998; Hemsley 1997).

## Results

The most obvious result of this study was the excellent preservation of organic residues, which shows promise for future residue studies in the area. All sherds from the Tensas basin contained organic residues, though not all were interpretable. A wide variety of components and biomarkers were identified, confirming that the diet in the LMV was rich and varied.

The sherds from the Tensas basin can be viewed as points on a chronological continuum since the sites are so geographically close to each other. The four residues from the earliest of these sites, Reno Brake, contained similar fatty acid distributions despite their different appearance (Figure 2). There were three rimsherds from Reno Brake; two were plain straight-rimmed jar sherds, and one was a burnished constricted rim jar. Three of the samples contained no identifiable neutral compounds, and could not be categorized. Their isotopic values were $C_3$ in origin.

HRL 195 contained three long-chain alcohols (with 26, 28 and 32 carbon atoms), demonstrating at least a partial plant presence in the residue, given the absence of markers for insect wax. The C16:0 fatty acid in this residue was isotopically enriched in carbon stable isotopes (-20.1‰), which is surprising given the early date of the site. Possible explanations include the presence of very early maize, or the presence of lipid components from a non-maize $C_4$ or CAM plant.[2] Fatty acids originating in a $C_4$ member of the genus Portulaca, which appears in the floral assemblage from the site, seem most likely. The overall similarity of the residues from this site suggests that the four sampled pots had been used to process a similar range of resources. Despite the differences in rim shape and surface treatment, all four vessels seem to have functioned as jars.

Among the 20 sherds sampled from the Osceola site, there were strong similarities among the residues from unslipped plain and burnished jars (Figure 3). Overall fatty acid content seemed somewhat similar to that from Reno Brake. All of the residues appeared to be comprised primarily of plant or fish compounds. Burnished sherds from Osceola had similar fatty acids and residue interpretations as the plain vessels, though not identical. HRL 199 in particular contained a large relative abundance of C16:1 and C22:1, as well as cholesterol and a wide range of other saturated fatty acids (Figure 3). This sherd was interpreted as containing fish, due to the unsaturated fatty acids, meat or animal products, and possibly $C_3$ plants, because of the wide range of long-chain saturated fatty acids. It is

---

2 CAM stands for 'Crassulacean acid metabolism', named after the plant family in which it was first discovered (Cushman 2001). It occurs in more than 20,000 species of plants including succulents (Agavaceae, Cactaceae, etc.), many orchids, some aquatic angiosperms and epiphytes such as *Tillandsia usneoides*. CAM has evolved more than once in these plant families, mostly from hot and dry habitats, as a mechanism to take up $CO_2$ without excessive evaporation when stomata are opened to let in $CO_2$. The chemical reactions resemble the $C_4$ pathway, but are seperated in time rather than space: the $C_4$ metabolism is continuous as long as there is light, while CAM occurs at night. The $CO_2$ is stored in the vacuoles as malate or isocitrate and is used for photosynthesis during the day. This discovery of the phenomenon is traditionally attributed to Benjamin Heyne who reported in 1815 that the leaves of *Cotyledon calycina* were bitter in the morning, but tasteless by noon (Heyne 1815).

unclear whether HRL 199 contained an unusual mixture of components, or the residue was simply better preserved than the other residues in the group and was typical of the Osceola plain and burnished potsherds. This sample was too small to distinguish whether it was from a bowl or jar, and no vessel functional interpretation could be attempted. It is worth mentioning that no study has yet taken place on the effects of burnishing on residue preservation. The decreased number and size of the pores may inhibit the formation of residues, or improve their preservation. Without such a baseline study, it is impossible to determine whether the slight differences between plain and burnished jar residues from Osceola are due to differences in pot function, or systematic variation in residue preservation.

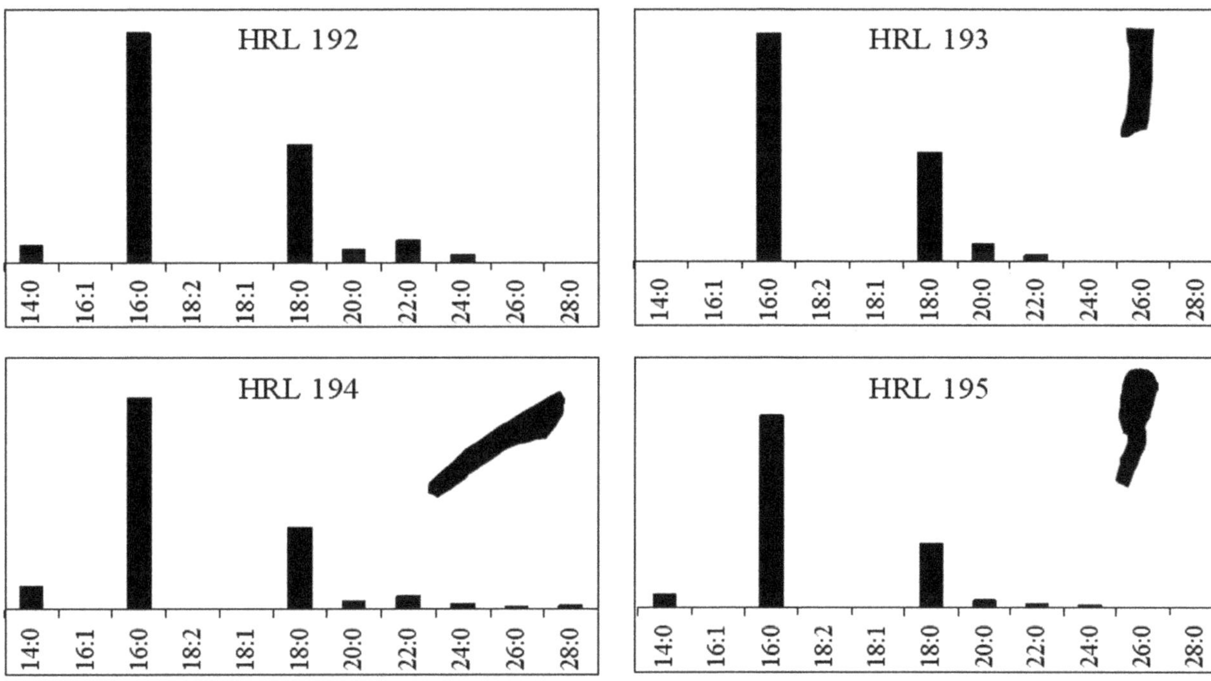


Figure 2: Relative abundances of fatty acids in sherds from Reno Brake, the shapes of their rims are given to the right. The similarity of the fatty acid profiles suggests that the pots were used to process similar resources, despite their stylistic differences.

The only clearly identified bowl sherd from Osceola contained a series of ketones, which were isotopically linked to the fatty acids from the sherd. This pattern suggests that the ketones were formed by pyrolysis of the fatty acids, which only occurs under extremely high temperatures, usually in excess of 400°C (Evershed et al. 1995). Sherds containing similar series of ketones have been interpreted as oil lamps in Europe (cf. Chapter 13). This bowl sherd could represent an oil lamp or fire carrier from the LMV. In any case, the bowl sampled underwent exposure to temperatures hotter than those experienced by jars from the site and hotter than would normally be reached in a cooking vessel, suggesting that the bowl form was linked to a unique high-temperature function.

The relationship between form and function for the sherds from Osceola sherds becomes particularly striking when the relationship between color and function is examined. Although the fatty acids in residues from colored pots are not noticeably distinct from those in the plain and burnished pots (Figure 3), there are unique biomarkers in both. For example, the only black-slipped pot from the Tensas River basin contained the only evidence of maize use in this study. Compound-specific isotope analysis of the neutral component of the residue clearly shows the $C_4$ component in the long-chain alcohol *n*-dotriacontanol (Figure 5). The $C_3$ values for the shorter long-chain alcohols and fatty acids demonstrate that maize and $C_3$ plants were cooked together in this large and unusually-slipped pot. Maize was known to exist in Coles Creek sites in general, and at Osceola in particular, but was interpreted by the excavator as a rare foodstuff used ritually. This interpretation may be borne out by its presence in the black-slipped sherd. Black, in terms of southeastern color symbolism, is often associated with the west, the earth and death and thus has specific ritual importance

(Hall 1997; La Flesche and Bailey 1995). This pattern continues with the only red-slipped sherd from the Tensas river valley, which was also the only sherd containing tree resin. Red, in terms of color symbolism, is often associated with the east, the sun and life (La Flesche and Bailey 1995; Hall, 1997). This sherd, from a smaller vessel than the maize pot, contained two isomers of the compound lupeol (lup-20(29)-en-3β-ol), a triterpenoid known to be a biomarker for tree resins (Charters et al. 1993; Eerkens 2002; Hayek et al. 1990), as well as an unidentified triterpenoid, which also indicates the presence of tree resins. Moreover, the presence of pentacyclic triterpenoids in the residue indicate that the original source of the resin was not a conifer, but rather a non-coniferous species of tree such as birch or persimmon (Mills and White 1977; Regert et al. 2003; Mallavadhani et al. 1998); conifer resins are distinguished by the presence of diterpenoids, which were absent in the red-slipped sherd.

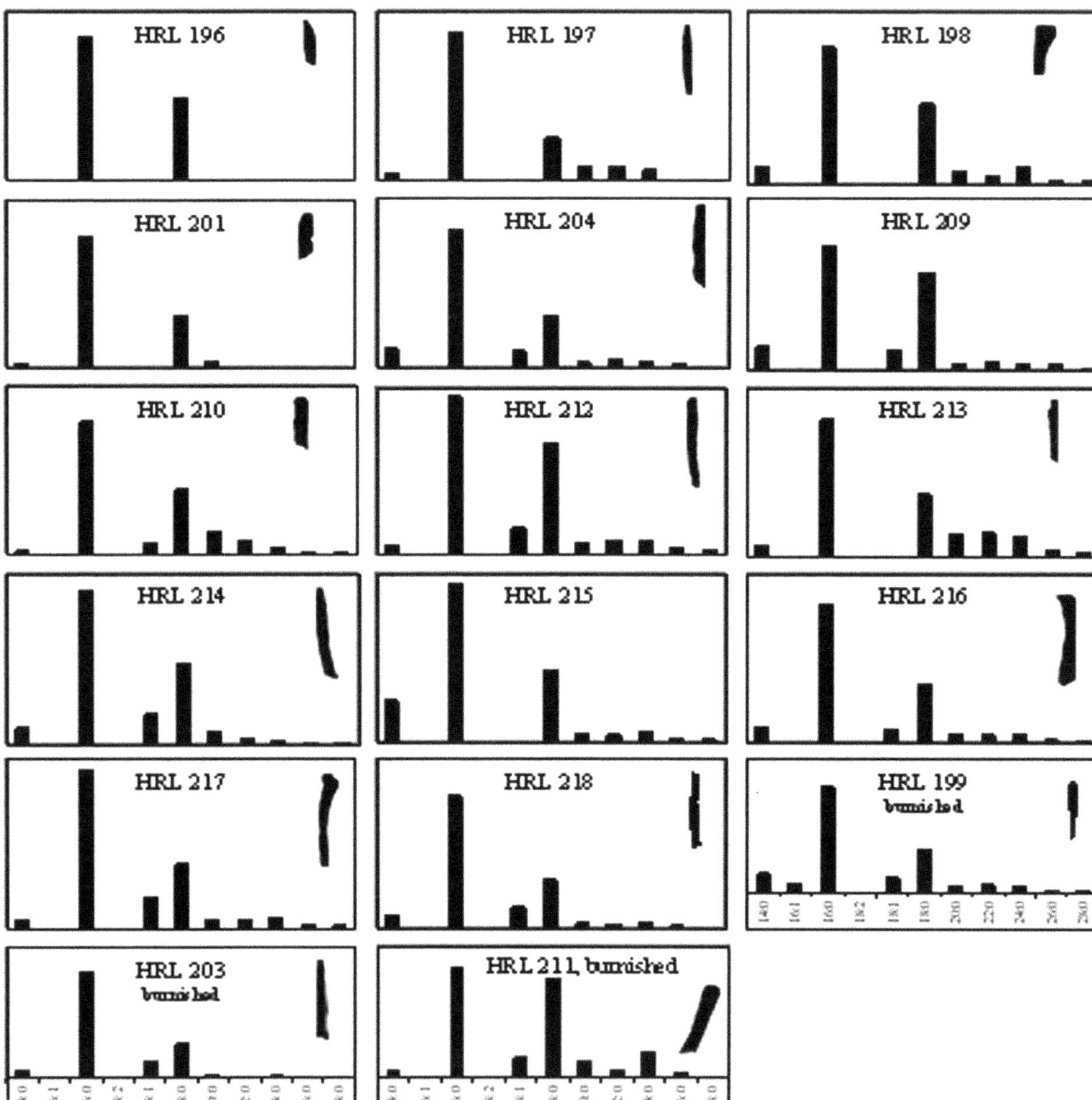


Figure 3: Relative abundance of fatty acids in plain and burnished jars from Osceola, the shapes of their rims are given to the right.

Also present in this residue were cholesterol and a wide range of fatty acids, as shown in Figure 4. The fatty acids could have come from many sources, including the cutin of tree bark (Walton 1990), but cholesterol could have originated only in meat. This data suggests that the pot underwent one of three uses. It may have been sealed with a non-coniferous tree resin, to make it impermeable, and then used to process meat and possibly other foods. Alternatively, the pot may have been used to process both meat and resin over its lifetime. It is also possible that the tree resin may have been used as a flavoring agent or medicine. The uses of tree resin in native North American cultures were multifarious, and varied from place to place. They were often used to attach arrowheads to arrows, to fletch arrows, to render baskets or bark containers watertight, although most of these uses are mentioned primarily for conifer resins (Fox et al. 1995; Swanton 1946). Some non-coniferous resins and oils are still used as flavoring agents such as birch oil and persimmon bark was used by some native peoples as a mouthwash and possibly as a tonic and febrifuge (Lewis and Elvin-Lewis 1977). Various trees were symbolic across the southeast, and the sap or resin from trees may have had a special meaning. The lack of a species or genus specific identification, at present, precludes the identification of the tree used, and therefore of its symbolism, but this possibility should be kept in mind in future studies.

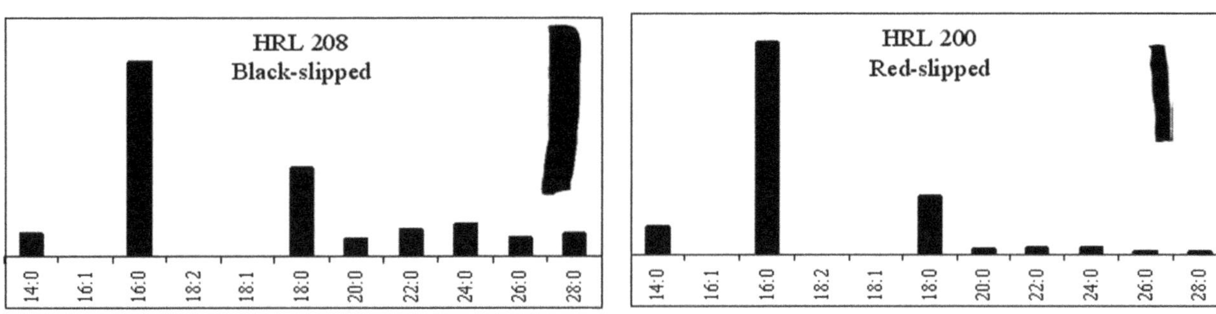


Figure 4: Relative abundance of fatty acids in red and black-slipped sherds from Osceola, the shapes of their rims are given to the right.

Such a complex pattern of special use in specially slipped pots does not appear at the Emerson site. This may be because only three sherds were contributed from the site or because Emerson was a small farming hamlet. Two of the three sherds from Emerson contained large amounts of residue interpreted as plant or fish remains. Maize was not identified in any of the three sherds from the site, despite its presence in the floral record from the site. Moreover, the fatty acids in the sherds were so varied that the sherds must have been used to process a variety of resources. Maize does not contain C20-C24 in measurable quantity, but these compounds were common in the Emerson sherds. This suggests that the original LMV diet was supplemented rather than supplanted by the adoption of maize horticulture into the region. This is in accordance with the archaeobotanical data on the subject. The burnished pot from Emerson did contain a ω-hydroxycosanoic acid, which suggests the processing of a tuber or a woody plant in the pot. With such a small sample size, it is impossible to determine whether this represents differential use between plain and burnished vessels, improved preservation in burnished sherds or a statistically insignificant coincidence.

The St. Gabriel Mound was a single mound village dated to the transition between the Coles Creek and Plaquemine periods. Residues from the site neither show the relatively high abundance of C16:1 present at Morgan Mounds, nor does there appear to be a difference between the residue extracted from the black-slipped sherd and those extracted from the other sherds (Figure 6). Further analysis, on a larger number of samples, is needed to properly determine whether slipping was as important in southern Louisiana as in the Tensas basin. Given the cultural variation between the two regions, it would not be surprising for slipping to mean different things in different places. The lack of variation in residues could be due to the poorer residue preservation at St. Gabriel, particularly when compared to the residue preservation in the sherd from the Tensas region, which was remarkably good.

Preservation of residues was also a problem for sherds from Bayou des Familles, the most recent site in this study. Given the damp nature of the site, poorer preservation is not surprising and resulted in one of the two residues remaining uninterpreted; the other only interpretable as plant or fish. Because there were fewer sherds, and no rimsherds, from the three southern

Louisiana sites, the data cannot be fully broken down. However, there were some interesting similarities and differences among the residues from these sites. Remarkable, when compared with the residues from the Tensas River basin, was an unusually large abundance of the unsaturated fatty acids C16:1 and C18:1, and the limited number of fatty acids with more than 20 carbon atoms. Both probably indicate greater utilization of marine or lacustrine resources in southern Louisiana than in the Tensas basin. This trend was particularly marked in the four residues from Morgan Mounds, as shown in Figure 6. Given the geographic variation between Morgan and the Tensas basin, it seems reasonable to interpret the Morgan sherds as having been used to process a larger amount of fish than the Tensas sherds.

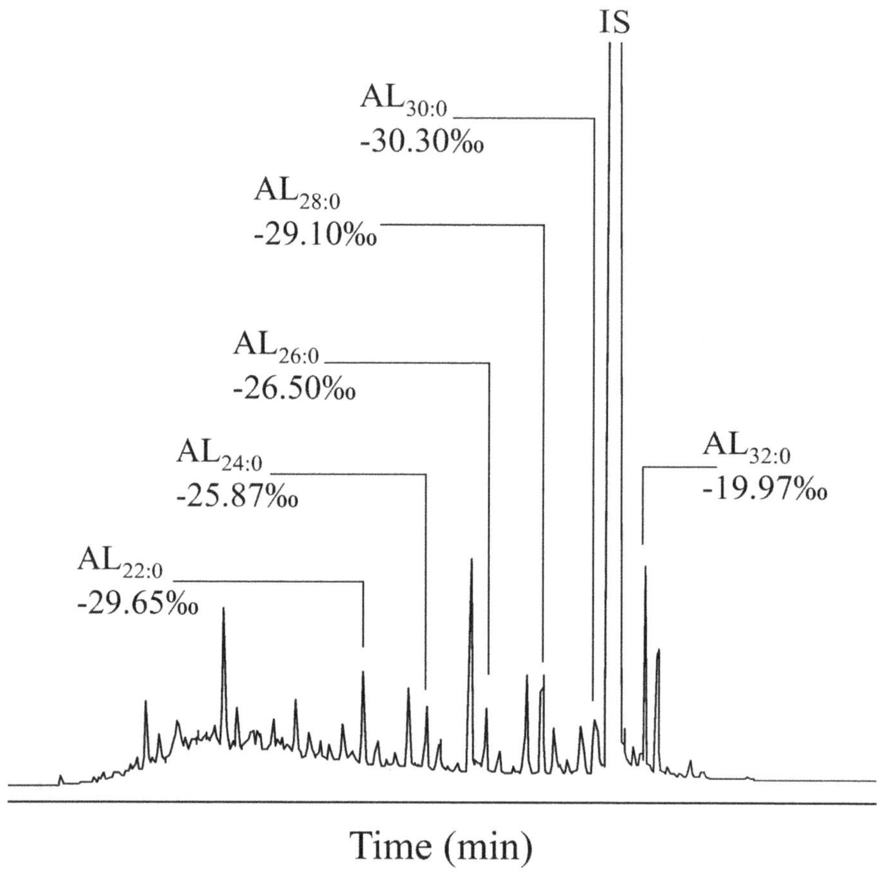


Figure 5: A chromatogram of the neutral lipids extracted from HRL 208, the only black-slipped sherd analyzed from the Osceola site. Identifications and stable carbon isotope data are appended to each compound. IS = internal standard. Note particularly the stable carbon isotope value for *n*-dotriacontanol ($AL_{32:0}$) which indicates the presence of maize components in the residue.

## Discussion

Interesting relationships between vessel form and function, as demonstrated by absorbed residue analysis, can be inferred from this analysis. Similarities in vessel shape, as reflected in rim shape and treatment, seem to echo similarities in vessel function. Rim shape across a single site, such as Reno Brake or Osceola, varies remarkably, but the residues within the vessels are fairly similar within the groups of plain and burnished jars. Differences in neck shape, a vessel trait often believed to have a functional purpose (Skibo and Schiffer 1995), did not seem to make a noticeable difference in vessel function as detected in this study. Variation in relative vessel thickness (thickness/diameter), often described as a trait to make a vessel suitable to cook foodstuffs for longer periods of time (Braun 1983; Skibo and Schiffer 1995), also did not seem to make a difference in vessel function in this study.

It is worth emphasizing, however, that interpretation of absorbed residue analysis is heavily dependent on the detection of biomarkers to show specific contents of pottery. For example, the fatty acids and neutral components present in the slipped sherds from Osceola were not strikingly different from those in the plain and burnished sherds. The contents of these vessels were identified as unique based on the presence of biomarkers. Just as there have been no baseline studies on residue preservation variations between plain and burnished vessels, there have also been no studies on such variations between slipped and unslipped vessels. Given the importance of color symbolism across the southeast, both before and after contact, it seems most likely that colored vessels were used in unique ways, compared with their unslipped cousins. Until studies on residue preservation are carried out, however, it is impossible to dismiss the alternative hypothesis that residues, including biomarkers, preserve better in slipped vessels.

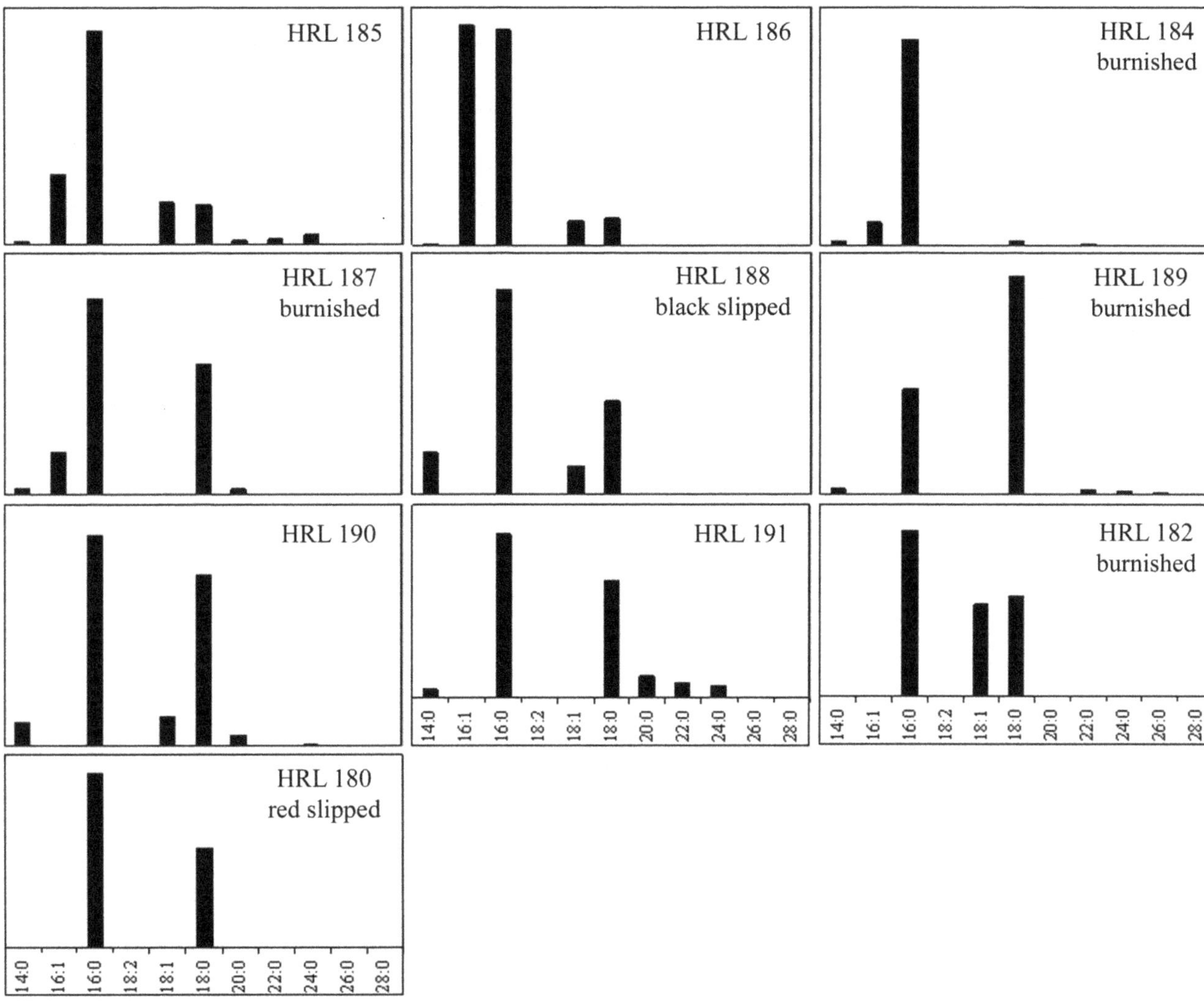


Figure 6: Relative abundance of fatty acids in sherds from sites in southern Louisiana. HRL 180 and 182 are from Bayou des Familles, HRL185-187 are from Morgan Mound and HRL 188-191 are from St. Gabriel Mound.

A clear relationship between form and function was shown by the identification of the only bowl sherd in the study as a possible lamp or fire carrier. Although the use of bowls as oil lamps is attested at from many sites in the Old World, bowls in North America are usually interpreted as serving vessels. A serving vessel, however, would not be subjected to any noticeable degree of heat, while the bowl from Osceola underwent temperatures in excess of 300-400°C, hotter even than the temperatures to which the cooking jars are exposed (Evershed et al. 1995). This evidence may be a hint that bowls in many parts of North American may have been used for a range of activities, including fire carrying or lamp bases as well as serving. To the best of my knowledge this is the first clear evidence of a lamp or fire carrier in the archaeological record of Midwestern or southeastern North America.

It is worth noting that differences in diet clearly discerned by archaeobotanical and faunal analyses did not show up in this study. These more common techniques clearly showed a greater dependence on meat at Reno Brake than at Osceola, which was more dependent on fish resources. This did not appear in the lipids absorbed in the pottery, which may be partly explained by the small number of sherds from Reno Brake.

It is not surprising that results from these different types of analyses do not completely tally. It has to be kept in mind that residue analysis tells us only about those portions of the resources at a site that were processed and preserved in pottery vessels. Resources that are not cooked in pots will never show up in pottery residue analysis, and non-food resources that are processed in pots could easily be confused with comestibles producing similar lipid profiles. Furthermore, a pot probably belonged only to a single family or family group, while the floral and faunal remains investigated from the site as a whole reflect the overall diet of the group much better than the residues from a relatively small number of potsherds. In conclusion, residue analysis does not tell us about diet (Barnard et al. 2007). It tells us about pottery at a site, which is equally important, but related to diet only in a tangential way.

The results of this study, however, emphasize the similarity in pot use in the LMV over time. Residues from all the sampled sites and from all time periods showed a mixture of contents including a wide variety of plants and fish. No one resource dominated the residues that built up in the sherds sampled for this study, despite the changes in diet and culture that took place. Part of the emphasis on the similarities of absorbed residues over time could be due to the importance of a variety of nut species across the LMV, as established by archaeobotanical analysis (Kidder and Fritz 1993). Very few residue studies have examined nut lipids, but the lipid profile of nuts can be extrapolated from the studies that have been made as well as the published lipid profiles of commercial nut oils (Gunstone et al. 1994; Malainey et al. 1999b). These extrapolations suggest that nut lipids strongly resemble other plant lipids, and would thus be almost undistinguishable from lipids from other $C_3$ plant sources in an absorbed pottery residue. The importance of nuts, paired with the high lipid content of nuts in general, could have masked changes in diet over time, particularly changes in resources with a noticeably lower fat content, such as other plant resources.

## Conclusions

An important conclusion from this study is the reminder that ancient people did not use their pots to process only one resource. All samples in this study showed strong evidence of resource mixing in pottery. This corresponds with ethnographic studies of southeastern pottery use, in which foodstuffs from the same 'aspect of the world' were cooked together, but foodstuffs from separate aspects of the world were cooked separately. For example, plants and meat from terrestrial animals were often processed together, while meat from birds, seen as coming from the sky rather than the earth, were often processed in separate vessels (Hudson 1976). The result of this would be common mixing of plant resources with meat from terrestrial animals. We should remember that ancient pottery was not used in the way that would be easiest for us to interpret, but in the way that made the most sense to the ancient user.

Another important conclusion is that more residue studies need to take place in order to make a statistically significant judgment on the relationship between vessel color and vessel function. A baseline study on the effect of surface treatment on residue preservation also needs need to be performed, despite the difficulty of performing a relatively accurate experimental study.[3] More archaeological studies of absorbed residues in rimsherds need to take place as well. The number of sherds in this study is clearly too small to conclude anything other than the way towards the idea that slip color was more important in vessel function than neck shape, rim treatment or relative thickness. Despite the problems inherent in destructive sampling of rimsherds, the potential payoff in terms of learning more about the relationship between the form and function of ceramic vessels is more than worthwhile. Furthermore, though many more samples will need to be analyzed to complete this study, particularly of rimsherds, this study shows that archaeological typologies can be given an extra dimension through absorbed residue analysis.

## References

**Aillaud**, S. (2001). *Field and Laboratory Studies of Diagenetic Reactions Affecting Lipid Residues Absorbed in Unglazed Archaeological Pottery Vessels. Organic Geochemistry Unit, Biogeochemistry Research Center.* Bristol: University of Bristol.

**Barnard**, H., S.H. Ambrose, D.E. Beehr, M.D. Forster, R.E. Lanehart, R.E. Parr, M.E. Malainey, M. Rider, C. Solazzo and R.M Yohe II (2007). Mixed Results of Seven Methods for Organic Residue Analysis Applied to One Vessel with the Residue of a Known Foodstuff. *Journal of Archaeological Science* **34**: 28-37.

**Braun**, D.P. (1983) Pots as Tools. In J.A. Moore and A.S. Keene (eds.). *Archaeological Hammers and Theories*. New York: Academic Press.

**Bull**, I.D., P.F.V. Bergen, C.J. Nott, P.R. Poulton and R.P. Evershed (2000). Organic Geochemical Studies of Soils from the Rothamsted Classical Experiments. The

[3] Described in heartrending detail by many researchers, including Aillaud (2001), Bull et al. (2000), Dudd et al. (1998), Evershed and Charters (1995), Malainey et al. (1999a) and Reber and Evershed (2004b).

Fate of Lipids in Different Long-term Experiments. *Organic Geochemistry* **31**: 389-408.

**Charters**, S., R.P. Evershed, L.J. Goad, C. Heron and P.W. Blinkhorn (1993). Identification of an Adhesive used to Repair a Roman Jar. *Archaeometry* **35**: 91-101.

**Cushman**, J.C. (2001). Cassulacean Acid Metabolism. A Plastic Photosynthetic Adaptation to Arid Environments. *Plant Physiology* **127**: 1439-1448.

**Dudd**, S.N., M. Regert and R.P. Evershed (1998). Assessing Microbial Lipid Contributions during Laboratory Degradations of Fats and Oils and Pure Triacylglycerols absorbed in Ceramic Potsherds. *Organic Geochemistry* **29**: 1345-1354.

**Eerkens**, J.W. (2002). The Preservation and Identification of Piñon Resins by GC-MS in Pottery from the Western Great Basin. *Archaeometry* **44**: 95-105.

**Evershed**, R.P. and S. Charters (1995). Simulating the Degradation of Animal Fats in Archaeological Ceramics. In J.O. Grimalt and C. Dorronsorov (eds.). *Organic Geochemistry. Developments and Applications to Energy, Climate, Environment and Human History.* Donostia-San Sebastian: A.I.G.O.A.

**Evershed**, R.P., A.W. Stott, A. Raven, S.N. Dudd, S. Charters and A. Leyden (1995). Formation of Long-chain Ketones in Ancient Pottery Vessels by Pyrolysis of Acyl Lipids. *Tetrahedron Letters* **36**: 8875-8878.

**Fox**, A., C. Heron and M.Q. Sutton (1995). Characterization of Natural Products on Native American Archaeological and Ethnographic Materials from the Great Basin Region, U.S.A. A Preliminary Study. *Archaeometry* **37**: 363-375.

**Fuller**, R.S. and D.S. Fuller (1987). *Archaeology at the Morgan Site Mound Complex in Southwest Louisiana, 1986.* Baton Rouge: State of Louisiana, Division of Archaeology, Office of Cultural Development, Department of Culture, Recreation and Tourism.

**Gunstone**, F.D., J.L. Harwood and F.B. Padley (eds.) (1994). *The Lipid Handbook*, London: Chapman and Hall.

**Hall**, R.L. (1997). *An Archaeology of the Soul. North American Indian Belief and Ritual.* Urbana: University of Illinois Press.

**Hayek**, E.W.H., P. Krenmayr and H. Lohninger ((1990). Identification of Archaeological and Recent Wood Tar Pitches using Gas Chromatography/Mass Spectrometry and Pattern Recognition. *Analytical Chemistry* **62**: 2038-2043.

**Hemsley**, S. (1997). *The Analysis of Lipid Residues from Iron Age Pottery. An Examination of the Diagnostic Potential of the Fatty Acids of the Surviving Triacylglycerides.* Bradford: University of Bradford.

**Heyne**, B. (1815). On the Deoxidation of the Leaves of *Coltyledon calycina. Transactions of the Linnean Society of London* **11**: 213-215.

**Hudson**, C.M. (1976). *The Southeastern Indians.* Knoxville: University of Tennessee Press.

**Kidder**, T.R. (1992). Timing and Consequences of the Introduction of Maize Agriculture in the Lower Mississippi Valley. *North American Archaeologist* **13**: 15-41.

**Kidder**, T.R. (1998). Mississippi Period Mound Groups and Communities in the Lower Mississippi Valley. In R.B. Lewis and C. Stout (eds.). *Mississippian Towns and Sacred Spaces.* Tuscaloosa: The University of Alabama Press.

**Kidder**, T.R. (1995). *Archaeological Data Recovery at 15 JE 218.* Jefferson Parish: Earth Search, Inc.

**Kidder**, T.R. and G.J. Fritz (1993). Subsistence and Social Change in the Lower Mississippi Valley. The Reno Brake and Osceola Sites, Louisiana. *Journal of Field Archaeology* **20**: 281-297.

**La Flesche**, F. and G.A. Bailey (1995). *The Osage and the Invisible World. From the Works of Francis La Flesche.* Norman: University of Oklahoma Press.

**Lewis**, W.H. and M.P.F. Elvin-Lewis (1977). *Medical Botany. Plants Affecting Man's Health.* New York: Wiley.

**Malainey**, M.E., K. Przybylski and B.L. Sherriff (1999a). The Effects of Thermal and Oxidative Degradation on the Fatty Acid Composition of Food Plants and Animals of Western Canada. Implications for the Identification of Archaeological Vessel Residues. *Journal of Archaeological Science* **26**: 95-103.

**Malainey**, M.E., K. Przybylski and B.L. Sherriff (1999b). The Fatty Acid Composition of Native Food Plants and Animals of Western Canada. *Journal of Archaeological Science* **26**: 83-94.

**Mallavadhani**, U.V., A.K. Panda and Y.R. Rao (1998). Review Article Number 134. Pharmacology and Chemotaxonomy of Diospyros. *Phytochemistry* **49**: 901-951.

**Mills**, J.S. and R. White (1977). Natural Resins of Art and Archaeology. Their Sources, Chemistry and Identification. *Studies in Conservation* **22**: 12-31.

**Reber**, E.A., S.N. Dudd, N.J. Van der Merwe and R.P. Evershed (2004). Direct Detection of Maize Processing in Archaeological Pottery through Compound-specific Stable Isotope Analysis of *n*-dotriacontanol in Absorbed Organic Residues. *Antiquity* **78(310)**: 682-691.

**Reber**, E.A. and R.P. Evershed (2004a). How did Mississippians prepare Maize? The Application of Compound-specific Carbon Isotope Analysis to Absorbed Pottery Residues from Several Mississippi Valley Sites. *Archaeometry* **46**: 19-33.

**Reber**, E.A. and R.P. Evershed (2004b). Identification of Maize in Absorbed Organic Residues. A Cautionary Tale. *Journal of Archaeological Science* **31**: 399-410.

**Regert**, M., S. Vacher, C. Moulherat and O. Decavallas (2003). Adhesive Production and Pottery Function during the Iron Age at the Site of Grand Aunay (Sarthe, France). *Archaeometry* **45**: 101-120.

**Skibo**, J.M. and M.B. Schiffer (1995). The Clay Cooking Pot. An Exploration of Women's Technology. In J.M. Skibo, W.H. Walker and A.E. Nielson (eds.).

*Expanding Archaeology*. Salt Lake City: University of Utah Press.
**Soxhlet**, F. von (1879). Die gewichtsanalytische Bestimmung des Milchfettes. *Dinglers Polytechnisches Journal* **232**: 461.
**Swanton**, J.R. (1946). *The Indians of the Southeastern United States*. Washington, D.C.: Smithsonian Institution Press.
**Walton**, T.J. (1990). Waxes, Cutin and Suberin. In J.L. Harwood and J.R. Bowyer (eds.). *Lipids, Membranes and Aspects of Photobiology*. London: Academic Press.

## CHAPTER THIRTEEN
## Analysis of Lipid Residues in Archaeological Artifacts: Marine Mammal Oil and Cooking Practices in the Arctic

C. Solazzo and D. Erhardt

Caroline Solazzo; Laboratoire de Chimie Organique et Macromoléculaire (LCOM); Université des Sciences et Technologies de Lille; UFR de Chimie, Bat C4, porte 210; 59655 Villeneuve d'Asq Cedex; France <solazzo.c@gmail.com> and David Erhardt; Smithsonian Center for Material Research and Education (SCMRE); Washington DC, which changed its name to Museum Conservation Institute (MCI) in March 2006. This work is supported by a Smithsonian Institution grant. We are grateful to the National Museum of Natural History for the access to its collections, to Dr. W. Fitzhugh from the Arctic Studies Center, to Dr. S. Loring and to the Department of Anthropology for the loan and discussions of the artifacts. Thanks are due to Dr. C. Potter, collection manager at the Vertebrate Zoology Department of the National Museum of Natural History and to Dr. O. Oftedal from the Department of Zoological Research at the Smithsonian National Zoo for provision of the seal and whale samples. Finally, we would like to thank everyone at the SCMRE and the LCOM for their help with the analysis.

Pottery sherds, stone lamps and artifacts from archaeological sites in Alaska and Canada were selected from the archives of the National Museum of Natural History in Washington DC. Ancient organic residues were collected on the sherds, often incrusted as a thick layer. Lipids were extracted from these residues followed by derivatization and analysis by high temperature gas chromatography (HTGC) and combined gas chromatography mass spectrometry (GC/MS). Modern samples of whales and seals were also analyzed. Some of these fresh specimens were thermally degraded to simulate cooking and aging. Alkaline treatment was sometimes necessary to hydrolyze the insoluble lipid fraction.

Thousands of years ago, Eskimos settled on the American coasts around the Arctic Circle and developed a fascinating culture in one of the most extreme climates of the world.[1] The ancient inhabitants of the arctic and sub-arctic coastal areas adapted to an economy relying primarily on the sea. In regions where land mammals are rare and plant resources sparse, marine mammals (seals, whales and walruses) provide an essential source of raw materials: skins and furs for clothing, boots, kayaks and tents; bones and ivory for tools, weapons and decorative objects; meat and fat for food and fuel. The muscles and blubber of large sea mammals were important sources for lipids and proteins and one individual could feed a family for days.

Blubber is a thick layer of fat and connective tissue located between the skin and the muscle of marine mammals. Blubber functions as insulation and as an energy store. It has a complex structure of fat cells and structural fibers made of collagen and elastin embedded in the fat matrix to give it both rigidity and elasticity. It is a good source of lipids but also provides proteins. Lipids composing the blubber are subject to variation and depend directly on the source of food. Studies show seasonal and geographical variations as well as compositional differences by sex, age and stage of pregnancy and lactation (Dahl et al. 2000; Grahl-Nielsen et al. 2003; Hamilton et al. 2004; Lockyer et al. 1984; Olson and Grahl-Nielsen 2003; Pabst et al. 1995). The oil rendered from the blubber was also used for fueling lamps (Blumer 1997; Lucier and Vanstone 1991; Nelson 1899; Spencer 1959).

### Cooking Practices in the Arctic Coastal Areas

The inhabitants of arctic and sub-arctic coastal areas relied heavily on sea mammals (walrus, seals and whales) for their subsistence (Dumond 1998). Carbohydrates were absent in the diet and had to be compensated for by increased consumption of lipids and proteins. For example, for a family of 10 people with 8 dogs at Point Barrow, 60 kg of meat were necessary every day (Spencer 1959). Fish, birds and other small games were a complementary source of food, as well as roots, berries and seaweeds. Caribou was hunted when available, but this could only be found inland and the migration patterns of caribou are more unpredictable than those of the sea mammals.

In Alaska, wooden dishes were used to store water, oil or fermented food (mixtures of meat, fat and berries); fish and meat could then be eaten raw, dipped in oil. Food

---

[1] The general term Eskimo refers to all groups inhabiting the Arctic and Sub-arctic regions from North Eastern Siberia to Greenland. With the Aleuts, occupying the Aleutian Islands, they form the Eskimo-Aleut family from which are derived the different language groups: Aleut in the Aleutian Islands, Yupik in Eastern Siberia and Southwestern Alaska, and Inupiaq-Inuktitut, a continuum of dialects spoken from the coasts of Northern Alaska and Canada to Labrador and Greenland. Inuit is the term in use for Arctic populations in Canada and Greenland.

was cooked by boiling or simmering, usually with oil rendered from blubber, in clay vessels to make stews and soups. Meat and fish were also roasted on an open fire when wood was available, or eaten frozen, dried or aged. 'Muktuk', a mix of skin and blubber that was consumed raw or cooked and that apparently tasted like coconut, was eaten with relish (Geist and Rainey 1936; Nelson 1899; Spencer 1959). Cooking by means of hot stones was practiced at Barrow (Reid 1990) and in southern Alaska (De Laguna 1956).

Open oil-burning lamps were an important utensil in daily life. Oil rendered from blubber provided fuel for the lamp. Made of clay or soapstone, the flat-bottom lamps were used for lighting, heating and cooking. On St. Lawrence Island a second pot fashioned of clay was hung above the lamp and used as a cooking pot. Another clay or wooden pot, in the shape of the lamp, could be placed under the wick to collect any dripping of oil (Collins 1937; Nelson 1899).

Oil-burning lamps were reproduced in the laboratory. Blubber was cut in 2 x 2 cm squares, placed in a modern clay saucer and lit with a straw wick. The blubber rendered oil and provided a constant flame for up to two hours (using about 10 ml of oil). A small cup was placed 5 cm above the lamp with 20 ml of water. The heat from the lamp was sufficient for the water to simmer but not to boil.

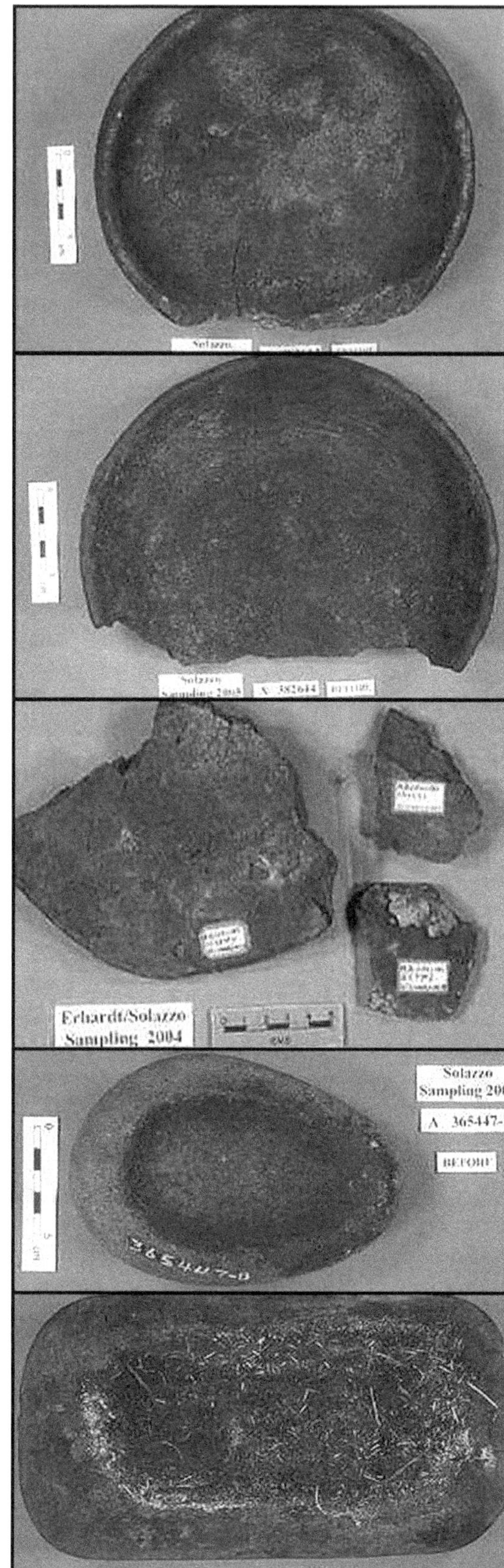

Figure 1: Examples of artifacts from which residues were taken. From top to bottom: saucer-shaped clay lamps numbers 382643 and 382644 from Nunivak Island, potsherds from Miyowagh on St. Lawrence Island, soapstone lamp number 365447 from Kodiak Island and wooden dish number 342583 from the vicinity of Wales.

## The Artifacts

Pottery was developed in Alaska and western Canada as early as 1500 BCE and persisted until European contact and the historic period. Arctic pottery is made of a coarse black paste tempered with mineral or organic materials, depending on their availability. It is described as thick, crude and brittle. As it crumbles easily few complete vessels have been found. They were generally flat bottomed containers with thick walls and wide openings, compared to the size of the pot. Usually, oil-lamps were shallow with thick walls, while cooking pots were deeper and had thinner walls. Decorations are very rare (Ackerman 1961; Collins 1937; De Laguna 1947; Geist and Rainey 1936; Reid 1990; Spencer 1959; Stimmel and Stromberg 1986). Beach gravel was used as a mineral temper and could make up as much as 50% of the paste. Marine mammal blood and blubber were sometimes added to the temper. Fiber temper (animal hair, grass, wood, feathers and baleen) was responsible for the high porosity. Ceramics could be coated with burned grease and fat on both sides, sealing the pores before each use. The limited availability of wood restricted the firing to low temperatures. These types of vessels are good insulators, making them suitable for simmering, stewing or the rendering of oil by stone boiling (Reid 1990).

No pottery was manufactured in the Aleutian Islands. Its absence is due to the lack of suitable clay and the abundance of easily carved volcanic rock (McCartney 1970). The lamps have thick walls and are ovoid, narrowing at one end where the wick was placed (DeLaguna 1956; Laughlin 1980). Figure 1 presents a few of the artifacts chosen for analysis.

Archaeological residues were collected from artifacts excavated in Alaska and Canada and kept by the National Museum of Natural History of the Smithsonian Institution. Because of their geographical context it is fairly certain that the ceramic pots and the soapstone lamps must have contained primarily marine products. Lipids analysis was conducted using chromatographic techniques. Lipids usually survive well in archaeological contexts because of their hydrophobic nature. The conservation of fatty acids appeared particularly enhanced by the specific conditions of the Arctic. The composition of the fatty acid profile resulting from hydrolyzed and oxidized acylglycerides, observed after processing and burial of fats and oils, was characterized. Biomarkers, such as products of the oxidation of cholesterol and isoprenoid acids, were also observed.

## Archaeological Samples

Archaeological soapstone lamps from Canada as well as clay, wooden and stone artifacts from archaeological sites in Alaska were examined for the presence of lipids. A series of 18 sherds, between 1000 and 1500 years old, were selected from the archaeological site of Miyowagh near Gambell in the northwestern part of the St. Lawrence Island.[2] Miyowagh is the second most ancient village at Gambell and dates back from the Old Bering Sea culture and the beginning of the Punuk. At this time, the hunting of walrus and seals was well developed; whales were obtained from stranded animals (Collins 1937). St. Lawrence Island is situated at 63°N in the Bering Sea, about 150 km (100 miles) from the coast of Alaska and 60 km (40 miles) from Siberia. The terrestrial fauna is restricted to arctic foxes, arctic squirrels and sometimes polar bears; the island is often frequented by birds. Because of its isolated location (St. Lawrence Island is encircled by the ice more than half the year) the inhabitants relied essentially on marine animals for subsistence: whales, seals and in particular walrus, abundant in the north of the island (Collins 1933; 1937; Dumond 1998; Geist and Rainey 1936).

Twelve clay sherds, eight stone lamps and tools, one wooden dish and one bone scraper from the coast of Alaska between Point Barrow and the Aleutian Islands were also sampled. At 71°N Point Barrow is the northernmost point of Alaska, protruding into the Beaufort Sea. St. Lawrence Island and Point Barrow are located along sea mammal migration routes and are still active whaling villages. Coastal Eskimos would hunt along the shore, chiefly in search of bowhead whales during their spring migration (Spencer 1959). The Aleutian Islands are situated west of Alaska. The archipelago consists of about 100 islands spread out over 1450 km (900 miles). The water around the islands is particularly rich in sea mammals attracted by the milder weather (Laughlin 1980). The other localities from which samples were investigated include the Seward Peninsula, Norton Sound, Nunivak Island and Kodiak Island (Figure 2, Table 1).

2 The name Miyowagh was introduced by Collins (1937), other designations include Mayughwaaq (Badten et al. 1987) and Miyowaghameet (Collins 1932).

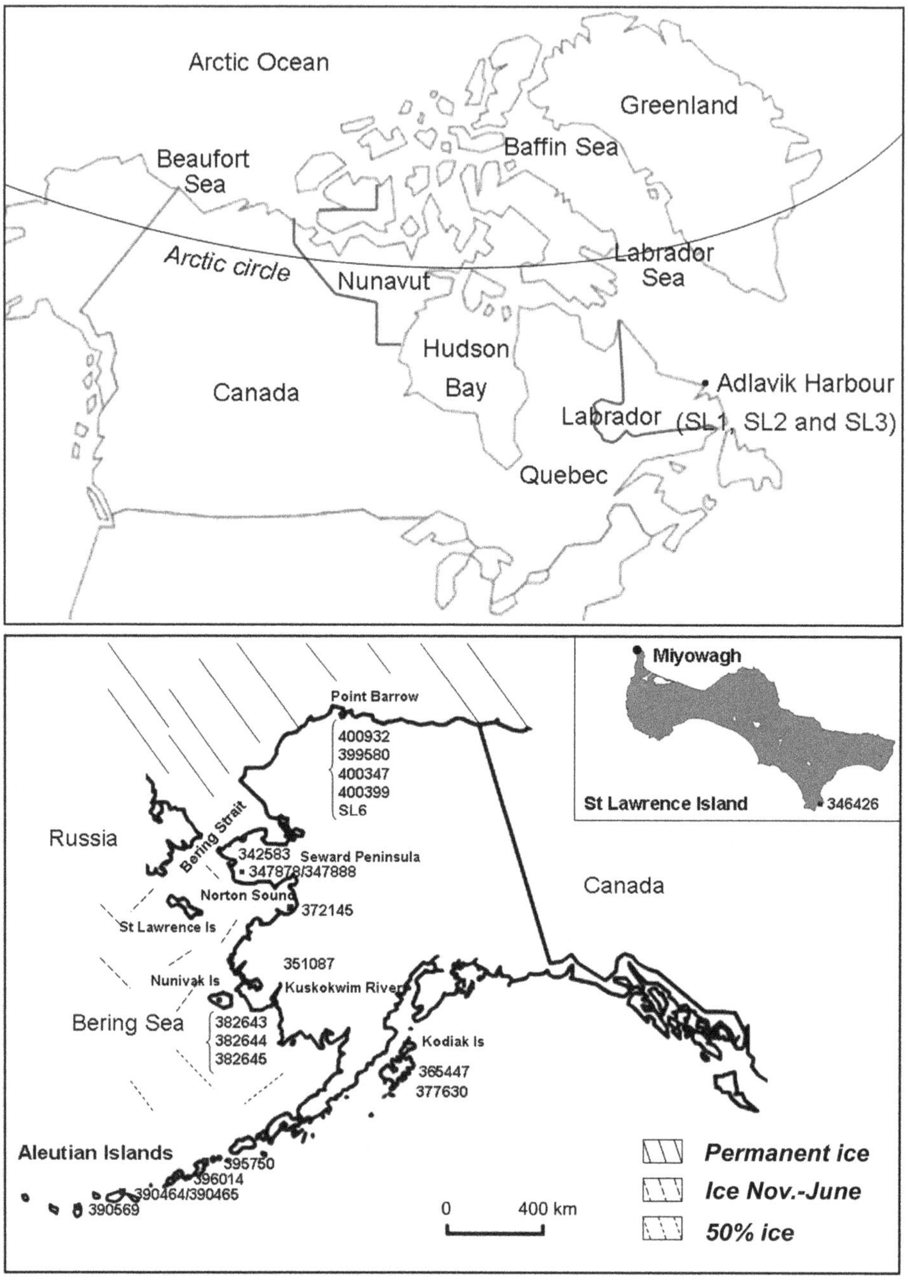


Figure 2: Maps showing the archaeological sites and provenance of the artifacts sampled in Alaska and Canada.

As described in the available archaeological literature the clay potsherds were made from a coarse and crude paste. The walls of the vessels varied from 0.5 to 2.7 cm and the surfaces were smoothed and undecorated. Most of the vessels had a high porosity and were certainly subjected to low temperature firing. They were tempered with mineral and vegetal matter like grass, feathers and maybe mammal hair (Collins 1937; Oswalt 1955). These sherds preserved amorphous organic residues on the inner or the outer surface, or both. In the latter case both sides were sampled when possible. The residues on the potsherds were thick, carbonized deposits. Two sherds (U1 and SL6) did not have a visible residue. In these cases the entire sherds were ground and any absorbed organic residue extracted. The stone lamps contained a thin layer of residue, strongly adhering to the surface.

Three other objects were also sampled: a stone 'bola' (used to kill seals), a wooden dish and a bone scraper. The first two were covered with a thick layer of decomposing fat, still fresh and mixed with grass in the dish. The bone scraper was covered with a black material that proved difficult to remove. Three additional samples were taken from soapstone vessels excavated at Adlavik Harbour, a coastal Inuit community (dated to about 1750 CE) in Labrador at 55°N. The samples were of a brown waxy material. The Thule people, who are the ancestors of the Inuit in Arctic Canada, began using soapstone lamps and pots about 1000 to 500 BCE.

The samples from St. Lawrence Island were excavated in 1930 and 1931 by Collins (1937); the other samples from Alaska were collected in the first half of the 20th century CE for the National Museum of Natural History and were stored at the Smithsonian Institution, most recently in an environment-controlled storage area. The samples from Labrador were unearthed between 1999 and 2004 and were stored in plastic bags and wooden drawers at the Smithsonian Arctic Studies Center.

## Modern Specimens

Some species that are traditionally hunted, or otherwise obtained, by the Eskimos were studied as reference. These included ringed seal, harbour seal, grey seal, beluga whale and narwhal (narwhal and grey seal are native to eastern Canada). In order to compare these fresh specimens with the archaeological samples they were subjected to high temperature storage to reproduce aging. By analyzing cooking residues from Arctic archaeological contexts the degradation of organic compounds resulting from the processing of foodstuffs can be understood and correlated with the anthropological data. Samples were taken from the blubber of a *Phoca hispida* (ringed seal), provided by the National Zoo of Washington DC; from the blubber, meat and skin of a *Phoca vitulina* (harbour seal) and a *Halichoerus grypus* (grey seal), hunted in Canada in 2005 and provided by the Vertebrate Zoology Department of the National Museum of Natural History; from the blubber and skin of the neck of a male *Delphinapterus leucas* (beluga whale) from the Churchill River (Manitoba, Canada) and conserved at the Smithsonian Institution; and from the blubber near the ear of a female *Monodon monoceros* (narwhal) stranded in Alaska in 2005. All samples were stored frozen.

| Sample | Location on the excavation site | Material | Remarks |
|---|---|---|---|
| 352986 | Miyowagh Cut 1M Section 6 | Clay potsherd | Flat, thick with a rim, sand and gravel tempered |
| 353010 | Miyowagh (no information) | Clay potsherd | Slightly curved, thick |
| 353036 | Miyowagh Cut 1M Section 8 | Clay potsherd | Slightly curved with a low curved wall |
| 353109 | Miyowagh Cut 2M Section 4 | Clay potsherd | Flat, thick, sand and gravel tempered |
| 353140-1 | Miyowagh Cut 2M Section 6 | Clay potsherd | Curved, thick |
| 353140-2 | Miyowagh Cut 2M Section 6 | Clay potsherd | Curved, thick, shaped as a saucer, sand and gravel tempered |
| 353197 | Miyowagh Cut 3M Section 4 | Clay potsherd | Flat, thick with a low, curved wall (rim), sand and gravel tempered |
| 353233-1 | Miyowagh Cut 3M Section 7 | Clay potsherd | Slightly curved, thick, sand and gravel tempered |
| 353233-2 | Miyowagh Cut 3M Section 7 | Clay potsherd | Slightly curved, thick, sand and gravel tempered |
| 353274 | Miyowagh Cut 3M Section 10 | Clay potsherd | Flat, thick, sand and gravel tempered |
| 353356 | Miyowagh Cut 4M Section 5 | Clay potsherd | Slightly curved, thick |
| 353917 | Miyowagh Cut 11M Section 5 | Clay potsherd | Flat, thick |
| 353961 | Miyowagh Floors 1, 2, Cuts 8,11,13 | Clay potsherd | Flat, thick |
| 354168 | Miyowagh above roof timbers of entrance to House B, Cut 15M, Section 5 | Clay potsherd | Flat, thick with a curved and low wall, sand and gravel tempered |
| 354380 | Miyowagh Cut 9(a), Cache site East of House B, below level of House B | Clay potsherd | Slightly curved, thin |

| | | | |
|---|---|---|---|
| **U1-U5** | Miyowagh | Clay potsherds | --- |
| **346426** | Cape Kialegak, St. Lawrence Island | Clay potsherd | Curved, thin rim, sand and gravel tempered |
| **347878** | Seward Peninsula Kowieruk, 3 miles east of Imaruk Basin | Clay potsherd | Curved, thin rim, sand, gravel and grass tempered |
| **347888** | Seward Peninsula Kowieruk, 3 miles east of Imaruk Basin | Clay potsherd | Curved, thin rim, sand, gravel and grass tempered |
| **372145** | Whale Island, St. Michael, Norton Sound | Clay potsherd | Curved, thin, sand, gravel and grass tempered |
| **351087** | Kuskokwim River | Clay potsherd | Curved, thin rim, sand, gravel and grass tempered |
| **382643** | Old Eskimo burials on Nunivak Island | Earthenware lamp / bowl | Diameter 22 cm, thick |
| **382644** | Old Eskimo burials on Nunivak Island | Earthenware lamp / bowl | Diameter 22 cm, thick |
| **382645** | Nunivak Island, Old Eskimo burials | Clay potsherd | Thick, big pot with high walls |
| **365447** | 'Our site', Jones point, Uyak Bay, Kodiak Island | Stone lamp | 14 x 10 cm |
| **377630** | 'Our point', village site on Uyak Bay, Kodiak Island | Stone lamp | 19 x 15 cm |
| **390463** | Atka Island, Andreanof group, Aleutian Islands | Stone lamp | 15 x 13 cm |
| **390464** | Atka Island, Andreanof group, Aleutian Islands | Stone lamp fragment | 20 x 12 cm |
| **390569** | 'Site C', east end of Sweeper Cove, North of road (shell and bone heap), Adak Island, Aleutians, Andreanof group | Hearth stone fragment | Flat slab<br>21 x 9 cm |
| **395750** | Amaknak Island, northeast end of Unalaska, Fox Island, Aleutians group | Stone lamp | 19 x 16 cm |
| **396014** | Umnak P-A Fox Island group | Stone vessel fragment | Curved thick wall with beginning of bottom |
| **342583** | Wainwright Mitliktavik site, 22 miles northeast of Wales | Wooden dish | 45 x 25 cm, crystals and grass |
| **400347** | From a locality in the southern part of Utkiavik, Point Barrow, purchased | Bone kayak bailer-scraper | --- |
| **400399** | From a place in south end of Utkiavik, Point Barrow, purchased | Stone bolas | Diameter 7 cm |
| **399580** | Piginik (Barrow Birnick site), Mound J, cut 1, southwest corner of house A | Clay potsherd | Curved, thick, sand and gravel tempered |
| **400932** | Nunagiak (Barrow Nunakiat?), above house X | Clay potsherd | Curved, thick, sand and gravel tempered |
| **SL6** | School Collection, from Point Barrow, Alaska | Clay potsherd | Rim, gravel tempered |
| **SL1** | Adlavik Harbour-1 House-1 Unit 29 | Soapstone (steatite) body vessel sherd, appears to be a lamp fragment | --- |
| **SL2** | Adlavik Harbour-1 House-1 Test-Pit#2 Cut#2 | Soapstone (steatite) body vessel sherd, vessel type indeterminable | --- |
| **SL3** | Adlavik Harbour-1 House-1 Unit 3a | Soapstone (steatite) body vessel sherd, side of a large cooking kettle | --- |

Table 1 (on this and the previous page): Description of the collection of artifacts sampled for analysis.

Because archaeological samples have suffered severe conditions through time, it is necessary to try and reproduce an artificial degradation of fresh samples. Cooking pots are exposed to heat during their use and modifications of organic molecules will also occur during the long-term burial of the artifacts. The usual method for reproducing the degradation of foodstuffs in ceramic is to store them at high temperatures as heat accelerates the oxidation process (Copley et al. 2005; Malainey et al. 1999a; b). Samples of seal and whale products in modern clay saucers were artificially aged in an oven at an average temperature of 100°C (Patrick et al. 1985). A few grams of blubber and meat were boiled with water in saucers for two hours at 100°C and left in the oven for several weeks. Residues were sampled each week. The saucer that contained meat of beluga whale was cut in pieces and one of them was ground and treated as the ancient samples.

## Analytical Methods

Fatty acid standards, *n*-tetratriacontane ($C_{34}H_{70}$) and the derivatization agent N,O-bis(trimethylsilyl)trifluoro-acetamide (BSTFA), containing 1% trimethyl-chlorosilane (TMCS), were purchased from Supelco; chloroform, methanol and diethyl ether from Sigma-Aldrich. Archaeological samples as well as modern specimens were extracted in a 2:1 mix of chloroform and methanol. When enough material was available, 10 mg of the fresh specimens and 150 mg of the carbonized residues were mixed with 0.5 ml of the extraction solution. After sonication for 40 min and centrifugation for 20 min, the supernatants were concentrated to a volume of 0.5 ml, the total lipid extract (TLE). Of this, 50-100 μl was derivatized by trimethylsilylation. 20 μl *n*-tetratriacontane was added to the extracts and, after evaporation of the solvent under nitrogen, the samples were heated for 30 min at 70°C with 50 μl of BSTFA. Another 50 μl of the TLE was saponified with 0.5 ml 2N KOH in methanol for one hour at 70°C (Colombini 2002). Acidified with HCl (pH = 3), the acid fraction was recovered with diethyl-ether, evaporated under nitrogen and derivatized as described above.

HTGC was performed in a 6890N Agilent instrument with a J&W DB-1HT fused-silica capillary column (15 m × 0.32 mm, 0.10 μm) with a 100% dimethyl-polysiloxane stationary phase. Set at an initial oven temperature of 50°C, the temperature was increased to 380°C at 10°C/min and held at 380°C for 5 min. CC/MS analysis was performed on a Finnigan Trace GC 2000 Thermo-electron with AS2000 Autosampler, coupled with a PolarisQ Ion Trap GC/MSn. The GC was equipped with an Rtx®-5MS fused-silica capillary column (30 m × 0.25 mm, 0.25 μm) with a 5% diphenyl/95% dimethyl-polysiloxane stationary phase. The oven temperature was programmed for a 2.5 min isothermal at 80°C, then increased to 260°C at 10°C/min, held at 260°C for 0.1 min, then increased to 300°C at 5°C/min and finally held at 300°C for 10 min.

## Fatty Acid Profiles in the Modern Specimens

Whale and seal blubber is mainly composed of triacylglycerides (TAGs). Because of post-mortem hydrolysis, free fatty acids and mono- and diacylglycerides (DAGs) were recovered in addition to TAGs. TAGs ranged from 44 to 62 total acid carbons, with a peak at 50 or 52 for all species except for grey seal for which TAGs ranged from 44 to 66 and reached a maximum at 56. DAGs ranged from 28 to 40, with two main peaks at 34 and 36. Unsaturated fatty acids were also present in small amounts, including C14:1, C16:1, C18:1, C20:1, C20:4, C20:5, C22:1, C22:5 and C22:6.
Alkaline treatment of the fresh samples allowed identification of the saturated fatty acids as their trimethylsilylester derivatives. The fatty acids and their relative proportions for the studied seal and whale species are presented in Table 2. Fish oil and blubber are rich in unsaturated fatty acids, characteristic of their marine origin (Morgan 1983). Eicosapentaenoic acid (EPA, C20:5ω3) and docosahexaenoic acid (DHA, C22:6ω3) are the major polyunsaturated fatty acids. Other fatty acids found in marine oils are gadoleic (C20:1ω11), gondoic (C20:1ω9), erucic (C22:1ω9), cetoleic (C22:1ω11), arachidonic (C20:4ω6), clupadonic (22:5ω3) and selacholic (C24:1) acids and to a lesser extend nisinic (C24:6), shibic (C26:5) and thynnic (C26:6) acids (Dahl et al. 2000; Grahl-Nielsen et al. 2003; Olsen and Grahl-Nielsen 2003).

After hours at an average temperature of 100°C, polyunsaturated fatty acids are the first components to oxidize. Table 2 also shows the changes in fatty acid composition of the species tested in the laboratory. Depending on the species, the results show that polyunsaturated fatty acids disappear in the first hours or days, leaving the monounsaturated as the main components. For grey seal (12 days) and ringed seal (7 days), for example, monounsaturated fatty acids have higher levels. This may be caused by the saturation of the polyunsaturated fatty acids (Malainey et al. 1999a; b; Morgan 1983). Only after one or two weeks the relative levels of monounsaturated fatty acids start to decrease while the percentage of saturated fatty acids increase. In the tests on beluga whale products, only monounsaturated and saturated fatty acids, ranging from C12:0 to C18:0, are left after 24 hours. Dicarboxylic acids with 4-9 carbon atoms and short-chain saturated fatty acids appeared after 7 days. At the same time the concentration of monounsaturated fatty acids reduced dramatically (Table 2, Figure 3a). Polyunsaturated fatty acids can also oxidize to form diacid (dicarboxylic acids) by oxidative cleavage of the double bonds of unsaturated fatty acids. Different mechanisms lead to the formation of different diacids from unsaturated fatty acids (Colombini et al. 2005; Copley et al. 2005; Passi et al. 1993; Regert et al. 1998). The most common transformation is the oxidation of oleic acid, or any other fatty acid with a double bond in position 9, generating azelaic acid (C9 diacid). Diacids with 9 or 7 carbon atoms can be formed from polyunsaturated fatty acids, such as linoleic acid (C18:2ω6) and linolenic acid (C18:3ω3), or oleic (C18:1ω9) and palmitoleic (C16:1ω7). Diacids with 5 carbons can be formed from EPA with 4 carbons from DHA (Passi et al. 1993). Long-chain monounsaturated fatty acids (such as gadoleic, cetoleic and erucic acids with double bonds in position 11 or 13) are the primary source for diacids with more than 9 carbons.

| Compound | Grey seal blubber | Grey seal blubber (12 days) | Harbor seal blubber | Harbor seal blubber (50 days) | Ringed seal blubber | Ringed seal blubber (7 days) |
|---|---|---|---|---|---|---|
| C8:0 | 0.124 | 0.067 | -- | 0.205 | -- | -- |
| C9:0 | 0.290 | 0.074 | -- | 0.245 | -- | -- |
| C10:0 | 0.184 | 0.085 | -- | 0.202 | -- | -- |
| C11:0 | -- | -- | -- | 0.541 | -- | -- |
| C12:0 | 0.508 | 0.118 | -- | 0.553 | -- | -- |
| C13:0 | -- | -- | -- | 1.392 | -- | 1.018 |
| C14:0 | 16.968 | 9.677 | 5.156 | 10.286 | 2.297 | 4.609 |
| C15:0 | -- | 0.971 | 0.327 | 1.201 | 0.400 | 1.302 |
| C16:0 | 31.462 | 11.556 | 14.458 | 27.683 | 3.759 | 11.218 |
| C17:0 | -- | -- | -- | 1.357 | 0.547 | 1.072 |
| C18:0 | 2.378 | 0.533 | 1.558 | 6.876 | 1.565 | 6.617 |
| C20:0 | -- | 3.949 | -- | -- | 0.665 | -- |
| C14:1 | 1.949 | 2.078 | 2.495 | 1.535 | 2.485 | 1.814 |
| C16:1 | 4.481 | 17.588 | 34.823 | 17.760 | 21.048 | 25.939 |
| C18:1 (ω9) | 10.794 | 31.302 | 36.667 | 16.762 | 27.51 | 24.060 |
| C18:1 (ω7) | -- | -- | -- | -- | 3.051 | 11.253 |
| C20:1 | 10.160 | 11.866 | 2.273 | 5.549 | 6.615 | 8.343 |
| C22:1 | 4.073 | 5.345 | -- | 1.462 | 0.715 | 1.376 |
| C24:1 | -- | -- | -- | 0.240 | -- | -- |
| C20:4 | 1.529 | 0.670 | -- | -- | -- | -- |
| C20:5 | 7.824 | 1.282 | 2.243 | -- | 10.796 | 0.288 |
| C22:5 | 2.496 | 0.247 | -- | -- | 5.243 | -- |
| C22:6 | 2.928 | 0.440 | -- | -- | 13.302 | -- |
| C4 diacid | -- | -- | -- | 0.932 | -- | -- |
| C5 diacid | 0.449 | 0.234 | -- | 0.269 | -- | -- |
| C6 diacid | -- | 0.177 | -- | 0.284 | -- | -- |
| C7 diacid | 0.652 | 0.210 | -- | 0.383 | -- | 0.724 |
| C8 diacid | -- | 0.216 | -- | 0.841 | -- | 0.366 |
| C9 diacid | 0.745 | 1.317 | -- | 0.756 | -- | -- |
| Total diacids | 1.846 | 2.154 | -- | 3.465 | -- | 1.090 |
| Total monounsaturated fatty acids | 31.457 | 68.179 | 76.258 | 41.773 | 52.813 | 72.785 |
| P/S ratio | 13.226 | 2.926 | 9.277 | 4.026 | 2.401 | 1.695 |

| Compound | Narwhal blubber | Narwhal blubber (50 days) | Beluga whale blubber | Beluga whale blubber (1 day) | Beluga whale blubber (7 days) | Beluga whale blubber (90 days) | Archaeological samples |
|---|---|---|---|---|---|---|---|
| **C8:0** | -- | 0.557 | -- | -- | 0.973 | 0.777 | **0<A<4** |
| **C9:0** | -- | 0.693 | -- | -- | -- | 0.968 | **0<A<4** |
| **C10:0** | 0.073 | 0.759 | -- | -- | 1.749 | 1.192 | **0<A<2** |
| **C11:0** | -- | 1.165 | -- | -- | -- | 2.099 | **0<A<2** |
| **C12:0** | 0.820 | 2.330 | 1.522 | 0.836 | 4.189 | 2.112 | **0<A<1** |
| **C13:0** | -- | 2.318 | -- | -- | 1.048 | -- | **0<A<1** |
| **C14:0** | 5.529 | 13.094 | 9.100 | 9.630 | 15.057 | 12.517 | **4<A<11** |
| **C15:0** | 0.484 | 1.573 | 0.815 | -- | 0.851 | 1.746 | **0<A<2** |
| **C16:0** | 9.530 | 22.515 | 10.248 | 11.933 | 15.958 | 15.168 | **10<A<35** |
| **C17:0** | -- | 3.393 | 0.459 | -- | 0.403 | 1.522 | **0<A<1** |
| **C18:0** | 2.135 | 11.369 | 0.880 | 1.810 | 4.461 | 4.175 | **5<A<10** |
| **C20:0** | 0.885 | -- | 0.260 | -- | 0.414 | 1.135 | **0<A<2** |
| **C14:1** | 1.350 | 0.650 | 3.225 | 2.178 | 1.193 | 1.507 | **trace** |
| **C16:1** | 22.788 | 10.505 | 23.056 | 40.022 | 17.404 | 18.021 | **<8** |
| **C18:1 (ω9)** | 28.449 | 8.939 | 27.389 | 22.766 | 13.306 | 12.470 | **7<A<14** |
| **C18:1 (ω7)** | -- | -- | 2.583 | -- | -- | -- | |
| **C20:1** | 15.229 | 4.846 | 12.974 | 8.896 | 6.603 | 7.035 | **5<A<10** |
| **C22:1** | 5.799 | 2.453 | 4.843 | 1.929 | 6.641 | 4.908 | **2<A<8** |
| **C24:1** | 0.119 | 0.911 | 0.483 | -- | -- | 1.690 | **trace** |
| **C20:4** | 0.348 | -- | -- | -- | -- | -- | -- |
| **C20:5** | 1.955 | -- | 0.860 | -- | -- | -- | -- |
| **C22:5** | 0.688 | -- | 0.373 | -- | -- | -- | -- |
| **C22:6** | 1.644 | -- | 0.929 | -- | -- | -- | -- |
| **C4 diacid** | -- | 2.167 | -- | -- | 0.848 | -- | -- |
| **C5 diacid** | -- | 0.530 | -- | -- | 1.090 | 0.243 | -- |
| **C6 diacid** | -- | -- | -- | -- | 1.036 | 0.880 | -- |
| **C7 diacid** | -- | 0.676 | -- | -- | 1.617 | 1.998 | -- |
| **C8 diacid** | 0.821 | 1.385 | -- | -- | 2.714 | 5.817 | -- |
| **C9 diacid** | 1.349 | 3.171 | -- | -- | 9.747 | 2.019 | -- |
| **Total diacids** | 2.17 | 7.929 | -- | -- | 8.656 | 10.957 | **3<A<40** |
| **Total monounsaturated fatty acids** | 73.734 | 28.304 | 74.07 | 75.791 | 45.147 | 45.631 | -- |
| **P/S ratio** | 4.464 | 1.980 | 11.652 | 6.593 | 3.577 | 3.633 | -- |

Table 2 (on this and the previous page): Relative percentage composition of fatty acids in marine mammal blubber after simulated decomposition; in the last column the range is given of the fatty acids found in the archaeological samples (A). Diacids are dicarboxylic fatty acids; P/S ratio = C16:0/C18:0.

In the aged samples of beluga whale, the formation of diacids is clearly the result of the degradation of polyunsaturated fatty acids. Although the polyunsaturated fatty acids quickly decompose, in 24 hours, diacids appear only after 7 days at 100°C. The profile observed at 24 hours is an intermediate between the beginning of decomposition and its final stage. Before 7 days, diacids must be present only as their, partially oxidized, precursors still attached to the glycerol molecule.

The P/S ratio (C16:0/C18:0) can be an indicator to discriminate between categories of residues. The ratio calculated from the modern sample of beluga whale is relatively high, more than 11. It progressively decreases with the time of exposure to heat to reach a value of 3.6 after 7 days at 100°C and seems to stabilize after that. In all species, this ratio decreases to values between 1.7 and 4.0. Although a general pattern was observed, it is not possible to distinguish between species, or even between seals and whales. Basically the same profile was obtained for all kinds of marine oils and fats. Other species, such as walrus, should be tested and added to this study.

### HTGC Results of the Archaeological Samples

In the archaeological samples polyunsaturated fatty acids and TAGs appeared to have degraded. DAGs were rarely detected, but monoacylglycerides (MAGs), such as monopalmitin and monostearin, were encountered frequently. High amounts of glycerol were found in many samples. Its presence indicates hydrolysis over time. Monounsaturated fatty acids with an even carbon chain from 14 to 22, and sometimes 24, were detected. Although the gas chromatographic method used does not separate gadoleic from gondoic acid, or erucic from cetoleic acid, the presence of long-chain monounsaturated fatty acids must be considered evidence for a marine origin.

The observed degradation of the lipids in fresh specimens demonstrated that polyunsaturated fatty acids can oxidize in a couple of hours at 100°C or higher, and in a couple of days at lower temperatures. This is followed by the formation of diacids. These were found in abundance in all archaeological samples. The profile of diacids was similar in most samples, ranging from 4-14 carbon atoms with azelaic acid (C9 diacid) the most abundant. Azelaic acid is the primary degradation product in mammal fat, due to the abundance of C18:1, C18:2 and C16:1. Longer chain diacids are from the degradation of long-chain monounsaturated fatty acids, such as C20:1 and C22:1, also abundant in marine mammal fat. Oxidizing conditions or high temperatures are responsible for the cleavage of the double bond in unsaturated fatty acids (Colombini et al. 2005; Copley et al. 2005; Regert 1998).

Cooking or burning of the fat has probably occurred repeatedly. Tests in our laboratory reproducing a blubber-burning lamp showed that the heat above the flame reached a temperature of about 200°C. Although some contamination may occur from the surroundings after immediate discarding, the high concentration of polyunsaturated fatty acids in marine fat are certainly responsible for most diacids. The conditions of preservation of the archaeological samples in permafrost can provide an explanation for the excellent recovery of diacids, preventing them from leaching into the soil. The P/S ratio in all archaeological samples is lower than in the fresh mammal samples, but similar to the degraded samples (Tables 2 and 3). The process of degrading mammal blubber by heating appears to lower this ratio to reach about the same value as the archaeological residues.

### GC/MS Results of the Archaeological Samples

When possible, samples were also analyzed by GC/MS. Diacids were observed ranging from 4 to 14 carbon atoms and saturated fatty acids from C6:0 to C18:0, and in some cases also C20:0 and C22:0. Even and odd middle-chain fatty acids were also present, mostly C6:0 to C13:0. C15:0 and C17:0, present in small quantities in animal and marine mammal fat, were also detected. The short- and middle-chain fatty acids may result from the oxidation of unsaturated long-chain fatty acids. For instance, C9:0 is a product of the oxidative scission of oleic acid. A few oxidation products were identified by GC/MS: 2-hydroxydecanoic acid, 3-hydroxydecanoic acid and 12-hydroxyoctadecanoic acid. 12-hydroxyoctadecanoic acid is formed from monounsaturated fatty acids with the double bond in position 11, such as gondoic acid or cetoleic acid, after oxidation and dehydration.

Phytanic acid (3,7,11,15-tetramethyl-hexadecanoic acid) was also encountered (Table 3). This compound is found in fish and marine mammals and it could be an important biomarker. Two other isoprenoid fatty acids are considered characteristic of oils from marine mammals: 4,8,12- trimethyltridecanoic acid (TMTD) and pristanic acid (2,6,10,14-tetramethylpentadecanoic acid). All three are proposed to come from the phytoplankton and krill ingested by marine mammals (Ackman 1968; 1989; Hansen 1970). These isoprenoid compounds are found in higher concentrations in marine than in terrestrial animals. Their detection in pots can be used as a good biomarker to prove the presence of a marine source of food, especially whale oil (Hansen 1970). Pristanic acid and TMTD have been detected in only a few samples, which may be due to the naturally higher abundance of phytanic acid (Copley et al. 2005). Cholesterol has been detected, as well as two of its oxidation products: cholesta-3,5-dien-7-one and cholest-5-en-7-one. Cholesta-3,5-dien-7-one is known to be thermally derived from 7-ketocholesterol, which is often the main product of auto-oxidation of cholesterol (Rose-Salin 1996). It has been shown that the auto-oxidation of polyunsaturated TAGs accelerates the oxidation of cholesterol (Kim and Nawar 1991; Osada et al. 1993). These unstable compounds are quickly oxidized and produce radicals that accelerate the oxidation of cholesterol (peroxidation of the TAGs by heating).

## The Samples from Miyowagh

Archaeological residues from Miyowagh, St. Lawrence Island, are probably the oldest samples analyzed in this study. They were found in kitchen middens from a 1500 year old site on a gravel beach. Residues are characterized by monounsaturated fatty acids, especially monounsaturated fatty acids with 20 carbon atoms, gadoleic acid (C20:1ω9) and gondoic acid (C20:1ω11); and with 22 carbon atoms, erucic acid (C22:1ω9) and cetoleic acid (C22:1ω11). A variable amount of C22:1 was present, with an average of 4.5% of the total composition in fatty acids. C20:1 was quite abundant, from 5% to 10% of all fatty acids. Only two samples had low levels of C20:1 (354380-2(E) and 353274), these did not contain C22:1. C18:1 varied from 7% to 14%. Palmitoleic acid, C16:1 was present at a maximum of 8% of all fatty acids. C16:1 is less stable than C18:1 and during degradation its quantity is more likely to decrease and palmitoleic acid will disappear first. C16:0, with an average of 22% of all fatty acids, varied between 35% in sample 354380-2(E) and 9%, 10% and 12% in samples 353961(I), U3(I) and U5(E) respectively. Stearic acid had an average relative abundance of 7% and appeared more consistent among the samples tested. In the last column of Table 2 the ranges of percentages of fatty acids found in all archaeological samples (A) is presented and compared to those found in sea mammals.

The P/S ratios varied little, from 1.2 to 4 with an average value of 2.9, a range typical of mammal fat. Only 353274 has a relatively high ratio of 7.2. MAGs were detected in all but three samples. When present, monostearin was the main MAG, usually accompanied by mono-olein. Monopalmitin was sometimes identified by HTGC, but can be confused with C22:1. GC/MS is more sensitive and allows the identification and separation of monopalmitin and monomyristin. When comparing all ceramics sampled on both interior and exterior, 353917, 353961, 354380-1, U2 and U3 have more degraded organic residues towards the inside (Table 4). These results corroborate the assumption that these sherds are from oil-burning lamps. The oil burning formed a layer of accumulated soot exposed to repeated heating. The residues found on the exterior were either spilled from inside the lamp or applied to create a waterproof coating. This fat was certainly exposed to a less intense heat.

| Sample | Mass (g) | Glycerol | Mono acyl-glycerides | Mono unsaturated fatty acids | Diacids | Iso-prenoid acids | Sterols | P/S ratio |
|---|---|---|---|---|---|---|---|---|
| *Miyowagh, Gambell, St. Lawrence Island* | | | | | | | | |
| 352986 I | 0.011 | 8.311 | 3.614 | 26.111 | 16.264 | | | 1.889 |
| 353010 E | 0.100 | 0.239 | 2.109 | 32.636 | 7.091 | * | * | 2.919 |
| 353036 I | 0.019 | 6.100 | 6.213 | 24.051 | 17.29 | | | 2.468 |
| 353109 E | 0.152 | 4.353 | 0.609 | 25.996 | 20.134 | | | 2.586 |
| 353140-1 E | 0.163 | 3.781 | 3.392 | 36.707 | 10.853 | * | * | 2.847 |
| 353140-2 E | 0.152 | 10.188 | 5.418 | 24.515 | 21.200 | * | - | 2.557 |
| 353197 E | 0.159 | 5.776 | 3.331 | 39.872 | 10.338 | | | 2.802 |
| 353233-1 E | 0.152 | 12.554 | 4.552 | 31.809 | 10.997 | * | * | 2.858 |
| 353233-2 E | 0.150 | 4.553 | 5.082 | 33.274 | 14.037 | * | * | 2.768 |
| 353274 E | 0.152 | 19.916 | 0 | 5.238 | 17.053 | | | 7.163 |
| 353356 E | 0.157 | 5.664 | 2.988 | 34.898 | 10.888 | | | 3.616 |
| 353917 E | 0.150 | 3.976 | 0.670 | 30.991 | 11.997 | | | 3.268 |
| 353917 I | 0.052 | 4.251 | 0.090 | 21.768 | 22.140 | | | 4.042 |
| 353961 E | 0.152 | 4.338 | 8.348 | 28.863 | 13.408 | * | * | 3.154 |
| 353961 I | 0.028 | 13.632 | 1.843 | 15.262 | 31.813 | | | 1.223 |
| 354168 E | 0.153 | 3.944 | 3.550 | 30.884 | 14.485 | * | - | 3.207 |

| | | | | | | | | |
|---|---|---|---|---|---|---|---|---|
| 354380-1 E | 0.150 | 4.008 | 2.556 | 28.365 | 14.908 | | | 3.280 |
| 354380-1 I | 0.050 | 5.280 | 4.309 | 26.59 | 24.167 | | | 2.885 |
| 354380-2 E | 0.081 | 2.316 | 0 | 12.936 | 12.195 | | | 3.277 |
| 354380-2 I | 0.108 | 2.314 | 3.340 | 23.692 | 13.144 | | | 3.122 |
| U1 Ceramic | 0.948 | 2.307 | 1.842 | 18.556 | 17.829 | | | 2.959 |
| U2 E | 0.165 | 4.441 | 1.219 | 29.739 | 17.051 | * | * | 3.748 |
| U2 I | 0.101 | 4.917 | 0 | 24.999 | 29.151 | | | 3.898 |
| U3 E | 0.154 | 4.120 | 4.786 | 34.268 | 14.537 | * | * | 2.671 |
| U3 I | 0.100 | 10.765 | 5.785 | 24.167 | 24.466 | | | 1.818 |
| U4 E | 0.150 | 6.096 | 1.817 | 23.177 | 10.548 | * | * | 3.672 |
| U5 E | 0.153 | 2.115 | 2.197 | 24.114 | 31.911 | * | * | 1.954 |
| *Kialegak, St. Lawrence Island* | | | | | | | | |
| 346426 CLAY | 0.040 | 2.038 | 0 | 44.443 | 11.304 | * | * | 2.409 |
| *Seward Peninsula* | | | | | | | | |
| 347878 CLAY | 0.035 | 12.146 | 2.881 | 29.073 | 16.116 | - | - | 2.222 |
| 347888 CLAY | 0.100 | 10.717 | 6.786 | 25.486 | 18.778 | - | - | 3.252 |
| 372145 CLAY | 0.042 | 7.865 | 11.478 | 35.659 | 7.982 | - | - | 2.116 |
| *Kuskokwim River* | | | | | | | | |
| 351087S1 CLAY | 0.100 | 6.989 | 4.307 | 36.971 | 9.810 | * | * | 2.176 |
| 351087S2 CLAY | 0.104 | 10.139 | 2.335 | 28.465 | 11.390 | | | 2.817 |
| *Nunivak Island* | | | | | | | | |
| 382643 CLAY | 0.093 | 11.447 | 3.236 | 14.997 | 31.535 | * | - | 3.782 |
| 382644 CLAY | 0.090 | 19.159 | 0.925 | 29.475 | 17.948 | * | * | 3.720 |
| 382645 CLAY | 0.145 | 11.983 | 8.152 | 30.546 | 12.651 | * | * | 2.636 |
| *Kodiak Island* | | | | | | | | |
| 365447 STONE | 0.034 | 7.871 | 36.217 | 16.507 | 6.042 | - | - | 2.769 |
| 377630 STONE | 0.033 | 7.710 | 13.2 | 20.689 | 10.425 | * | * | 3.272 |
| *Aleutian Islands* | | | | | | | | |
| 390463 STONE | 0.019 | 4.128 | 18.293 | 19.579 | 7.524 | * | - | 3.260 |
| 390464 STONE | 0.060 | 33.157 | 0 | 18.052 | 10.282 | | | 3.603 |
| 390569 STONE | 0.113 | 0.676 | 12.138 | 20.232 | 16.354 | - | - | 1.381 |
| 395750 STONE | 0.033 | 10.713 | 35.237 | 14.863 | 4.904 | | | 2.196 |

| | | | | | | | | |
|---|---|---|---|---|---|---|---|---|
| 396014 STONE | 0.093 | 0.697 | 3.038 | 18.738 | 4.846 | - | * | 4.46 |
| *Northern Alaska* | | | | | | | | |
| 342583 WOOD | 0.023 | 1.335 | 4.875 | 41.473 | 10.293 | - | - | 7.648 |
| *Point Barrow* | | | | | | | | |
| 400347 WOOD | 0.034 | 1.344 | 28.859 | 11.754 | 42.883 | * | - | 1.605 |
| 400399 STONE | 0.030 | 0.778 | 5.384 | 11.631 | 27.339 | * | - | 3.149 |
| 399580 CLAY | 0.100 | 5.104 | 21.640 | 31.925 | 8.734 | | | 2.415 |
| 400932 CLAY | 0.100 | 2.744 | 7.062 | 26.67 | 10.620 | - | - | 1.506 |
| SL6 CLAY | 0.156 | 1.106 | 2.750 | 25.062 | 4.979 | * | * | 2.552 |
| *Adlavik Harbour* | | | | | | | | |
| SL1 STONE | 0.150 | 11.029 | 8.018 | 24.752 | 19.715 | | | 2.282 |
| SL2 STONE | 0.150 | 12.983 | 13.603 | 42.984 | 3.265 | | | 2.572 |
| SL3E STONE | 0.150 | 11.113 | 30.404 | 13.927 | 17.080 | | | 1.196 |
| SL3I STONE | 0.150 | 11.835 | 13.719 | 23.212 | 14.354 | * | * | 1.721 |

Table 3 (above and on the previous pages): Fatty acid profiles of the archaeological samples. Glycerol, monoacylglycerides, monounsaturated fatty acids, diacids and the P/S-ratio (C16:0/C18:0) were calculated from HTGC data. The values indicate the relative percentages calculated from the sum of all detected fatty acids. Isoprenoid acids and sterols (cholesterol and products of oxidation) were positively detected (*) or not (-) by GC/MS (blanks indicate that GC/MS was not performed).

DAGs were found in only two samples from St. Lawrence Island. U5 contained diacids, mostly C8 and C9, with reduced amounts of monounsaturated fatty acids, C16:0 and C18:0. Because the hydrolysis of the DAGs was incomplete, monounsaturated fatty acids and C14:0, C16:0 and C18:0 are less abundant as free fatty acids. It seems that the decomposition of the residue is less advanced; this vessel may have been used to collect dripping oil instead of being used as a cooking pot or a lamp.

### Artifacts from Alaska and Labrador

The artifacts from coastal archaeological sites in Alaska were classified according to their geographic origin and their material (clay, stone, wood or bone). HTGC chromatograms of a clay potsherd (399580) and the bone scraper (400347) are shown in Figures 3b and 3c, GC/MS chromatograms of a soapstone lamp (SL3), the bola (400399) and the wooden dish (342583) are shown in Figures 3d, e and f. Samples from ceramics found at Kialegak, Seward Peninsula, Kuskokwim River, Nunivak Island and Point Barrow had a composition similar to the samples described above. On average, the P/S ratio is 2.6. One sample seems to be more degraded: 382643 from Nunivak only had 14.7% unsaturated fatty acids, but more than 30% diacids. Independent of their geographical provenance, all clay vessels have the same fatty acids profile, and the slight differences can be explained by a more or less advanced degradation or slight differences in the initial lipid composition. The latter can be due to the species or the feeding period, factors that can not be accounted for by this kind of analysis.

Samples from stone artifacts from the Aleutian Islands and Labrador are distinguished from the samples of clay artifacts as they have much less C18:1, C16:1 and diacids while MAGs are abundant. Possibly, residues have less adherence on stone and some components may leach out more easily, especially in the milder climate of the Aleutian Islands. Another possibility is that the oil used on the Aleutian Island comes from different marine mammals than further north. However, the P/S ratio is normal for a marine source with an average 3.26. Figure 4 shows the percentage of monounsaturated fatty acids plotted against the percentage of diacids in these artifacts.

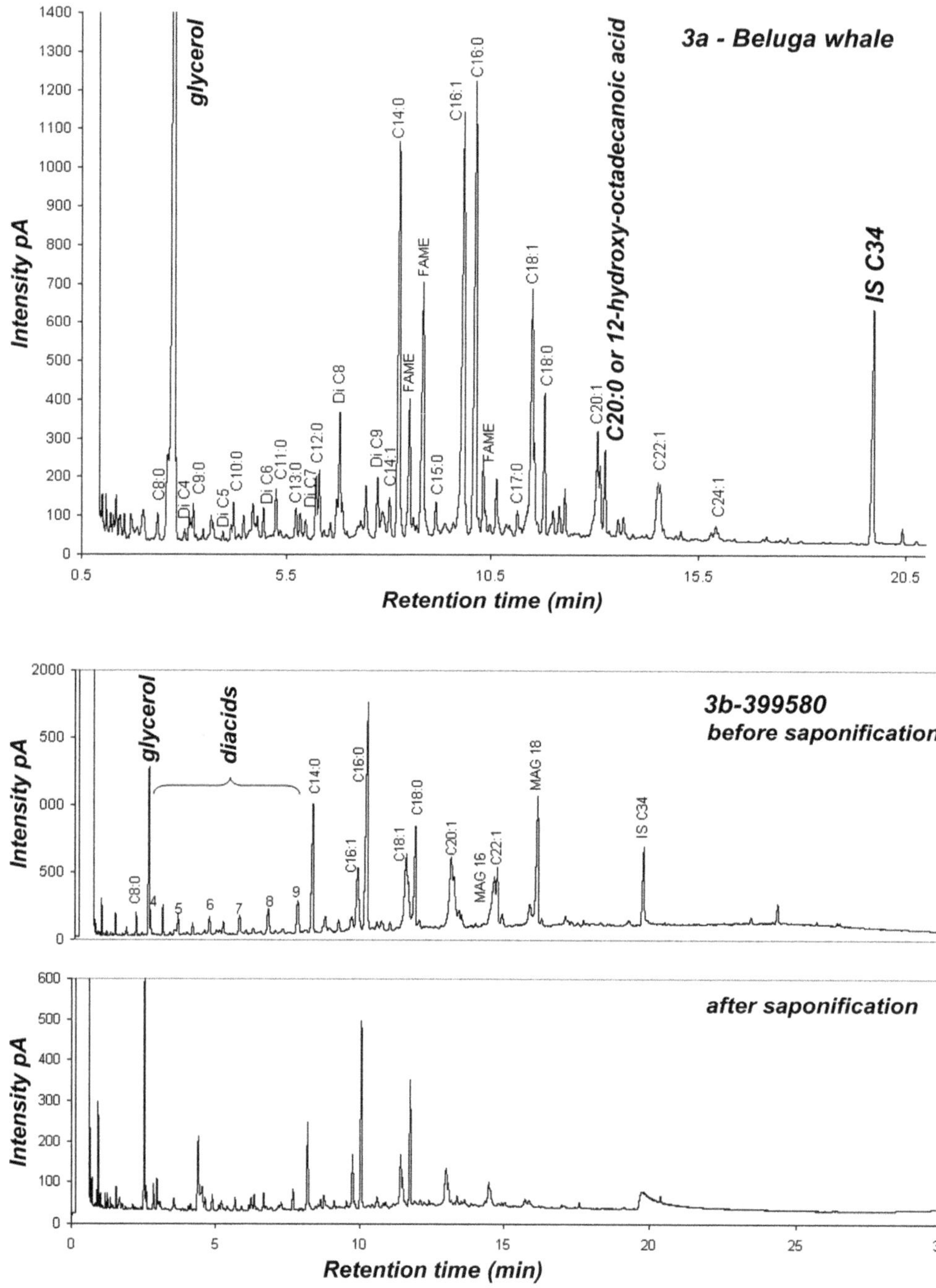


Figure 3a and 3b: High temperature gas chromatograms of Beluga whale blubber (a) and clay potsherd 399580 (b).

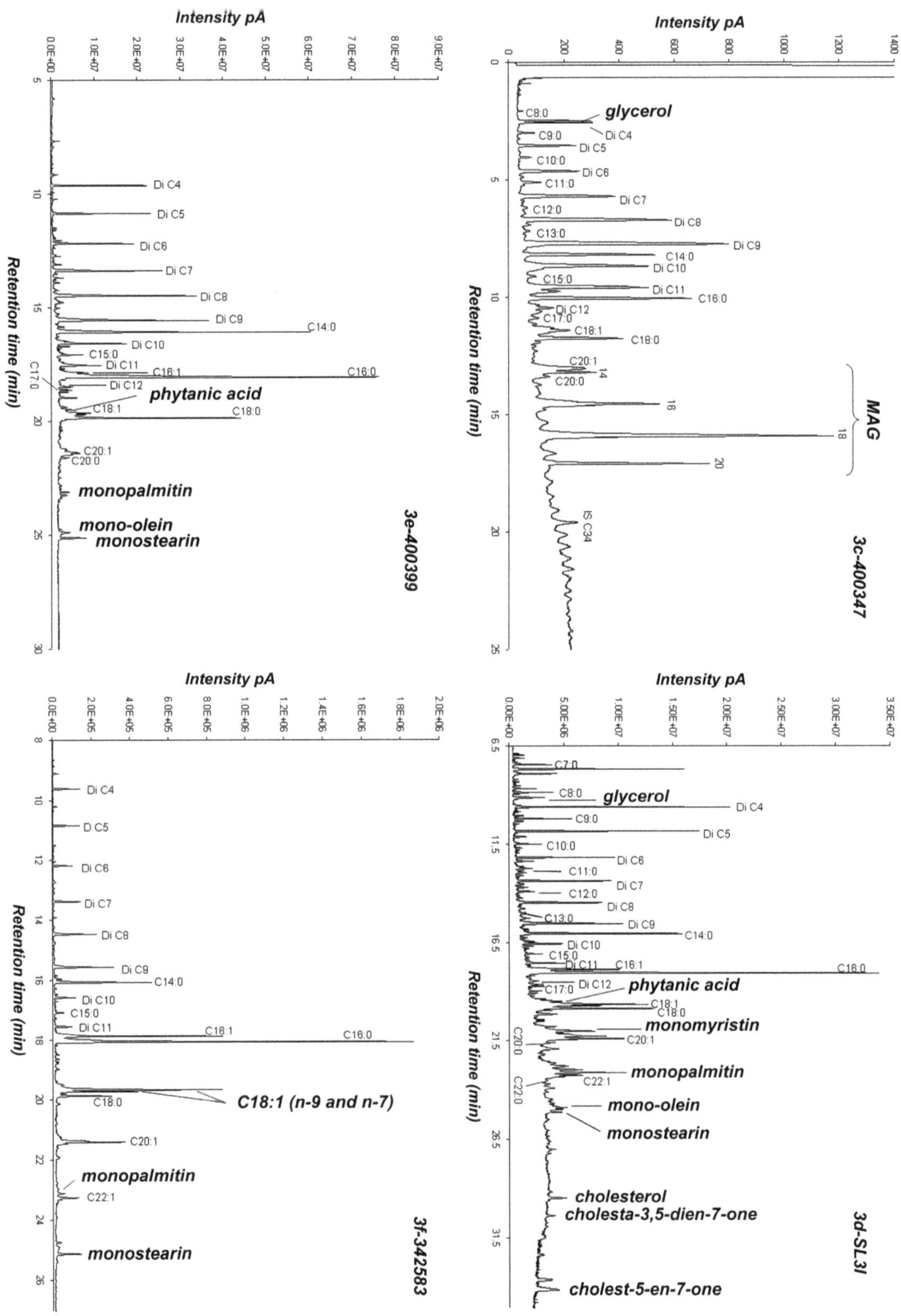


Figure 3c-f: High temperature gas chromatogram of bone scraper 400347 (c) and GC/MS chromatograms of soapstone lamp SL3 (d), bola 400399 (e) and wooden dish 342583 (f).

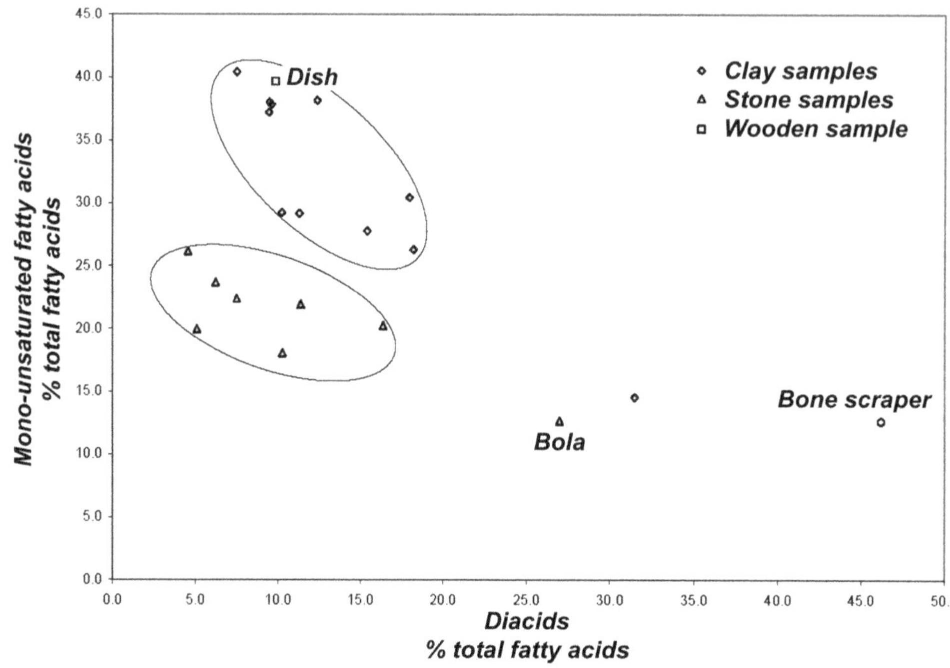


Figure 4: Monounsaturated fatty acids plotted versus dicarboxylic fatty acids for artifacts from coastal Alaska.

The stone bola 400399 from Point Barrow was covered with a thick layer of decomposed fat, low in palmitoleic and oleic acid, but very high in diacids. This is remarkable as the object was probably not exposed to heat, but used to kill seals. The presence of diacids could be due to biodegradation by micro-organisms of the fresh remains before the object was discarded. The sample from bone scraper 400347 from Point Barrow was similar to the bola, high in diacids (about 40%) and low in C16:1 and C18:1. It was also low in C16:0 and C18:0. On the chromatogram (Figure 3c), the high concentration of MAGs is striking. MAG-C18:0 is the most abundant, although C16:0 is the main free fatty acid. Scrapers were used to remove fat from intestines and skins from seals, caribous or small land mammals. The sample from wooden dish 342583 from northern Alaska had a very high amount of monounsaturated fatty acids and a quite high P/S ratio of 7.6. This sample was collected in 1929 when it was described as a 'flat wooden dish used beneath stone lamp to catch drippings, found in place in an abandoned house'. The blubber was still fresh, which may explain the high P/S ratio (typical of fresh blubber).

| | 353917 | | 353961 | | 354380-1 | | U2 | | U3 | |
|---|---|---|---|---|---|---|---|---|---|---|
| | E | I | E | I | E | I | E | I | E | I |
| MAGs | 0.670 | 0.090 | 8.348 | 1.843 | 2.556 | 4.309 | 1.219 | --- | 4.786 | 5.785 |
| Monounsaturated fatty acids | 30.964 | 21.768 | 28.863 | 15.262 | 28.365 | 26.59 | 29.739 | 24.999 | 34.268 | 24.167 |
| Diacids | 11.997 | 22.140 | 13.408 | 31.813 | 14.908 | 24.167 | 17.051 | 29.151 | 14.537 | 24.466 |
| P/S ratio | 3.268 | 4.042 | 3.154 | 1.223 | 3.280 | 2.885 | 3.748 | 3.898 | 2.671 | 1.828 |

Table 4: Comparison between samples taken from the exterior (E) and from the interior (I) of the sherd. MAGs are monoacylglycerides; diacids are dicarboxylic fatty acids; P/S ratio = C16:0/C18:0.

## Discussion

The archaeological residues recovered in the coastal arctic regions proved to have been remarkably well preserved, most likely due to a particularly favorable environment. To summarize, our samples were characterized by the presence of high concentrations monounsaturated fatty acids and diacids, a P/S ratio of about 3 and the presence of MAGs as well as phytanic acid. Archaeological residues showed important similarities with the fat of marine mammals. Most potsherds did not retain a recognizable shape that would enable to recognize them as fragments of a lamp or a cooking pot. Lamps are likely to have contained marine mammal fat exclusively; cooking pots may occasionally have been used to prepare land mammal meat, for example caribou. It was not possible to test samples from such species, but if land mammals were used, palmitic, stearic and oleic acids would be more abundant (Malainey et al. 1999 a; b).

In the Northern Slope of Alaska, caribou could be encountered during its migration. It was also bartered for sea mammal oil with the land Eskimos. Two samples from clay vessels from Point Barrow had about twice the average percentage of C18:0 (7%): SL6 with 13.57% and 400932 with 14.87%. In SL6 myristic and palmitic acids were also much more abundant than in any other archaeological sample. In these two samples, terrestrial mammal meat could have been mixed with sea mammal fat. Indeed, long-chain monounsaturated fatty acids were more abundant in these samples compared to the other artifacts tested.

Many factors influence the biochemical modifications of foodstuffs. Were they cooked, fried or stored? How and where were they discarded? Repeated exposure to heat, biodegradation by micro-organisms, contamination by the surroundings and leaching of the residue are some of the processes that may take place over the hundreds of years before the artifacts are eventually unearthed. Decomposition of lipids occurs through hydrolysis and oxidation. Low temperatures and an anaerobic environment can temporarily protect compounds from extensive degradation.

Even though the lipid profiles in most of our samples are typical of burned blubber, the species that provided the fat still have to be determined. It may not be possible to determine what kind of oil was used in the lamp as it is likely that both seal and whale oil were used, depending on their availability. Diacids are exceptionally abundant, meaning that the residues were not leached by water. Detailed information of the archaeological context is lacking for many samples, especially of those artifacts obtained as gifts or bought from Eskimos. Even though this means that the period of use and the nature of the soil in which they were buried may remain unclear, the extreme cold of Alaska and northern Canada evidently provides an ideal environment for the biochemical study of their former contents.

## References

**Ackerman**, R.E. (1961). *Archaeological Investigations into the Prehistory of St Lawrence Island, Alaska.* University of Pennsylvania (PhD dissertation).

**Ackman**, R.G. (1968). Examination of Isoprenoid Fatty Acids as Distinguishing Characteristics of Specific Marine Oils with Particular Reference to Whale Oils. *Comparative Biochemistry and Physiology* **24**: 549-565.

**Ackman**, R.G. (1989) *Marine Biogenic Lipids, Fats, and Oils. Volume II.* Boca Raton, CRC Press Inc: 210-261.

**Badten**, L.W., V. Kaneshiro, M. Oovi and S.A. Jacobson (ed.) (1987). L. Badten, *Dictionary of the St. Lawrence Island/Siberian Yupik Eskimo Language.* Fairbanks, Alaska Native Language Center.

**Collins**, H.B. Jr. (1932). Prehistoric Eskimo Culture on St. Lawrence Island. *Geographical Review* **22(1)**: 107-119.

**Collins**, H.B. Jr. (1937). *Archeology of St Lawrence Island, Alaska. Smithsonian Miscellaneous Collections Volume 96(1).* Washington DC, Smithsonian Institution.

**Colombini**, M.P., F. Modugno, R. Fuoco and A. Tognazzi (2002) A GC/MS Study on the Deterioration of Lipidic Paint Binders. *Microchemical Journal* **73, 1-2**: 175-185.

**Colombini**, M.P., G. Giachi, F. Modugno and E. Ribechini (2005). Characterisation of Organic Residues in Pottery Vessels of the Roman Age from Antinoe (Egypt). *Microchemical Journal* **79, 1-2**: 83-90.

**Copley** M. (2004). Organic Residue Evidence for the Processing of Marine Animal Products in Pottery Vessels from the Pre-colonial Archaeological Site of Kasteelberg D East, South Africa. *South African Journal of Science* **100**: 279-283.

**Copley**, M., H. Bland, P.J. Rose, M. Horton and R.P. Eversherd (2005). Gas Chromatographic, Mass Spectrometric and Stable Carbon Isotopic Investigations of Organic Residues of Plant Oils and Animal Fats Employed as Illuminants in Archaeological Lamps from Egypt. *The Analyst* **130**: 860-871.

**Dahl**, T.M., C. Lydersen, K. Kovacs, S. Falk-Petersen, J. Sargent, I. Gjertz and B. Gulliksen (2000). Fatty Acid Composition of the Blubber in White Whales (*Delphinapterus leucas*). *Polar Biology* **23**: 401-409.

**De Laguna**, F. (1947). The Prehistory of Northern North America as seen from the Yukon. *Society for American Archaeology Memoir* **3.**

**De Laguna**, F. (1956). *Chugach Prehistory. The Archaeology of Prince William Sound, Alaska. University of Washington Publications in Anthropology. Volume 13.* Seattle, University of Washington Press.

**Dumond**, D.E. (1998). The Hillside Site, St Lawrence Island, Alaska. An examination of collections from the 1930s, *University of Oregon Anthropological Paper* **55**.

**Geist** O.W. and F.G. Rainey (1936). Archaeological Excavations at Kukulik, St. Lawrence Island, Alaska. Preliminary Report Prepared under the Direction of the Commissioner of Indian Affairs in Cooperation with the University of Alaska. *Miscellaneous Publications of the University of Alaska* **2**: 391.

**Grahl-Nielsen**, O., M. Andersen, A. Derocher, C. Lydersen, O. Wiig and K. Kovacs (2003) Fatty Acid Composition of the Adipose Tissue of Polar Bears and of their Prey: Ringed Seals, Bearded Seals and Harp Seals. *Marine Ecology Process Series* **265**: 275-282.

**Hansen**, R.P. (1970). Isoprenoid Fatty Acids in Antarctic Krill (*Euphasia superba*). *Journal of the Science of Food and Agriculture* **21**: 203-206.

**Hamilton**, J.L., R.M. Dillaman, W.A. McLellan and D.A. Pabst (2004). Structural Fiber Reinforcement of Keel Blubber in Harbor Porpoise (*Phocoena phocoena*). *Journal of Morphology* **261**: 105-117.

**Kim**, S.K. and W.W. Nawar (1991). Oxidative Interactions of Cholesterol with Triacylglycerols. *Journal of the American Oil Chemists Society* **68, 12**: 931-934.

**Laughlin**, W.S. (1980). *Aleuts. Survivors of the Bering Land Bridge*. New York, Hold, Rinehart and Winston.

**Lockyer**, C.H., L.C. McConnell and T.D. Waters (1984). The Biochemical Composition of Fin Whale Blubber. *Canadian Journal of Zoology* **62**: 2553-2562.

**Lucier** C.V. and J.W. Vanstone (1991). The Traditional Oil Lamp among Kangigmiut and Neighboring Inupiat of Kotzebue Sound, Alaska. *Arctic Antropology* **28, 2**: 1-14.

**Mc Cartney**, A.P. (1970). 'Pottery' in the Aleutian Islands. *American Antiquity* **35, 1**: 105-108.

**Malainey**, M.E., R. Przybylski and B.L. Sherriff (1999a). The Effects of Thermal and Oxidative Degradation on the Fatty Acid Composition of Food Plant and Animals of Western Canada. Implications for the Identification of Archaeological Vessel Residues. *Journal of Archaeological Science* **26**: 95-103.

**Malainey**, M.E., R. Przybylski and B.L. Sherriff (1999b). Identifying the Former Contents of Late Precontact Period Pottery Vessels from Western Canada using Gas Chromatography. *Journal of Archaeological Science* **26**: 425-438.

**Morgan**, E.D. (1983). The Composition of Fatty Materials from a Thule Eskimo Site on Herschel Island. *Arctic* **36, 4**: 356-360.

**Nelson**, E.W. (1899). *The Eskimo about Bering Strait*. Washington DC, 18th Annual Report of the Bureau of American Ethnology.

**Olsen**, E. and O. Grahl-Nielsen (2003). Blubber Fatty Acids of Minke Whales. Stratification, Population Identification and Relation to Diet. *Marine Biology* **142**: 13-24.

**Osada**, K., T. Kodama, L. Cui, K. Yamada and M. Sugano (1993). Levels and Formation of Oxidized Cholesterols in Processed Marine Food. *Journal of Agriculture and Food Chemistry* **41**: 1893-1898.

**Oswalt**, W. (1955). Alaskan Pottery. A Classification and Historical Reconstruction. *American Antiquity* **21, 1**: 32-43.

**Pabst**, D.A., W.A. McLellan, W.A. Gosline and P.M. Piermarini (1995). Morphology and Mechanics of Dolphin Blubber. *American Zoologist* **35**: 44A.

**Patrick**, M., A.J. de Koning and A.B. Smith (1985). Gas Liquid Chromatography Analysis of Fatty Acids in Food Residues from Ceramics found in the Southwestern Cape, South Africa. *Archaeometry* **27, 2**: 231-236.

**Passi**, S., M. Picardo, C. De Luca, M. Nazzaro-Porro, L. Rossi and G. Rotilio (1993). Saturated Dicarboxylic Acids as Products of Unsaturated Fatty Acid Oxidation. *Biochimica et Biophysica Acta* **1168**: 190-198.

**Regert**, M., H.A. Bland, S.N. Dudd, P.F. Van Bergen and R.P. Eversherd (1998). Free and Bound Fatty Acid Oxidation Products in Archaeological Ceramic Vessels. *Proceedings of the Royal Society of London B* **265**: 2027-2032.

**Reid**, K.C. (1990). Simmering Down. A Second Look at Ralph Linton's 'North American Cooking Pots'. *Anthropological Papers, Nevada State Museum* **23**: 7-17.

**Rose-Salin**, C. (1996). Mécanismes d'oxydation du cholestérol. Oéagineux, Corps gras. *Lipides* **3, 3**: 227-235.

**Spencer**, R.F. (1959). *The North Alaskan Eskimo. A Study in Ecology and Society*. Washington DC, Bureau of American Ethnology Bulletin 171.

**Stimmel**, C. and R.L. Sromberg (1986). A Reassessment of Thule Eskimo Ceramic Technology. In W.D. Kingery (ed.). *Ceramics and Civilization. Technology and Style*. Westerville, American Ceramic Society: 237-250.

# CHAPTER FOURTEEN
# The Archaeology of Alkaloids
## S.M. Rafferty

Sean Rafferty; Department of Anthropology, University at Albany, State University of New York; 1400 Washington Avenue; Albany, NY 122222; USA; <rafferty@albany.edu>. I would like to thank Amy Burns and Jennifer Newman, for bringing several valuable references to my attention, and Laurie Miroff, for her comments on the original manuscript of this chapter. Any omissions or errors of fact, however, are my own.

Chemical analysis of ancient residues is one of the most promising techniques in archaeology today. The variety of case studies in this volume is testament to the potential of this developing field of inquiry. That said, I would like to draw attention to a type of evidence that has seen comparatively little attention within the field of residue analysis: alkaloids and other molecules that are not related to subsistence resources, but are of great cultural importance nonetheless. In my conclusions I will identify issues relating to ritual practices and ideology, on which the study of prehistoric alkaloid-use can provide information. Most of my examples are drawn from North American contexts, but this geographic limitation is solely for the sake of brevity. Old World examples of the use of alkaloids abound, and some examples are presented towards the end of this chapter.

Alkaloids can be defined based on the following characteristics: they are organic molecules, generally containing at least one nitrogen atom, which is contained in a heterocyclic ring structure. Additional groups, such as methyl radicals, give each alkaloid its unique chemical properties. Alkaloids are comparatively small for organic compounds, and atomic weights range typically between 100 and 300. Alkaloids are basic in pH, and the resulting bitter taste is likely a primary evolutionary factor as a defense mechanism in the variety of vascular plants and fungi that contain them. Figure 1 depicts the molecular structure of several alkaloid compounds discussed in this chapter.

These molecules are significant to archaeology because they occur widely in wild plants as natural toxins, so any prehistoric population likely had a variety of alkaloid-bearing plants in their local environment. Of the available alkaloids, many are known to have a variety of biological effects on the human body and mind when ingested, and can act as narcotics, stimulants, or hallucinogens. Therefore, alkaloid-bearing plants were widely exploited as medicinal or psychoactive agents by prehistoric cultures worldwide. Many of these compounds are still utilized by modern societies; many more were used in prehistory.

Like foodstuffs, alkaloid-bearing plants could leave organic residues behind containing trace amounts of the active compounds. Some studies have indicated that alkaloids, or characteristic metabolites and decay products, can survive in prehistoric samples in detectable concentrations despite the solubility of some compounds (Raffuf and Morris 1960). This is important as there have been allegations that the study of prehistoric alkaloids is a fruitless endeavor due to the high potential of loss or decay of alkaloid biomarkers in archaeological contexts. I would argue that this is a matter to be evaluated experimentally rather than assumed a priori, and that the limited available experimental results appear promising for archaeological research.

An advantage of the study of alkaloids is that they are relatively easy to deal with in terms of chemical analysis. These small molecules are relatively easy to extract with organic solvents, such as methanol, chloroform, or diethyl-ether, and can be retained in solution for some time pending analysis. In fact, the retention of dissolved alkaloids can be an analytical step as it can produce characteristic decay products in a reasonable amount of time, weeks to months, which would be impractical to derive through natural aging. Dissolved alkaloids can be identified with standard instrumental methods, such as gas chromatography (GC), combined gas chromatography mass spectrometry (GC/MS) or high performance liquid chromatography (HPLC). Because many of these compounds are used in modern contexts, often as legally controlled substances, protocols for their detection and comparative samples are often available from food and drug regulatory agencies. Databases of alkaloid mass spectra are readily available and do not need to be produced *de novo*.

Despite the relative simplicity of the analysis and the high data potential of the topic, there has been comparatively little archaeological research into the ancient uses of alkaloids. Most studies of prehistoric plant use, whether through archaeobotanical approaches or through chemical analysis of organic residues, have focused on subsistence resources. In terms of residue analysis this has most commonly proceeded through the identification of characteristic fatty acids (Evershed et al. 2001). There has been precious little effort expended in analyzing organic residues that may have derived from plants with medicinal or psychoactive uses, which could have played roles in prehistoric ritual practices. This should not be taken as a criticism of the excellent work on subsistence that has been conducted, but rather as the identification of untapped sources of data than can potentially be used to address archaeological and anthropological questions of great significance.

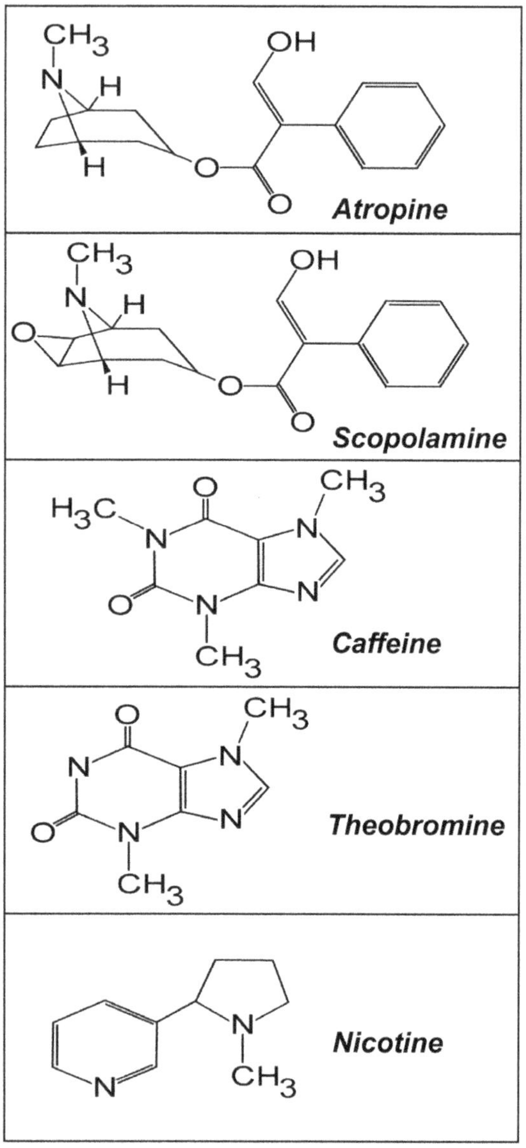


Figure 1: Structures of the alkaloid compounds discussed in this chapter.

In this chapter I will not primarily present analytical results, but offer suggestions for further lines of inquiry instead. It is my hope that colleagues with similar interests will follow up on the potential research questions that I will present, or develop new ones. I will illustrate my argument with a short discussion of one applied study in alkaloid residue analysis that I have conducted, the results of which have previously been published (Rafferty 2002; 2006).

## Examples of Alkaloid Archaeology

I will first describe in brief several alkaloids that were often used in prehistoric times, focusing on their context of use and their prehistoric significance. This is intended as a means of illustrating what can be learned from studying the archaeology of alkaloids. I will conclude with a discussion of the implications for archaeology in general and some suggestions for future investigations.

Probably the most ubiquitous of alkaloids used by both modern and ancient human cultures is caffeine (trimethylxanthine). Caffeine is found in several well-known tropical plants, such as *Caffea arabica* (coffee) and *Theobroma cacao* (cacao, the basic ingredient in chocolate). It is soluble in water, as well as ethanol, acetone, chloroform and diethyl ether. By binding to receptor sites for the neurotransmitter adenosine, caffeine acts as a neurostimulant, increasing neuronal activity and stimulating the secretion of epinephrine by the pituitary gland. Caffeine also has the effect of increasing dopamine production and thereby stimulating the pleasure center in the brain.

Of the tropical caffeine-bearing plants, cacao has had the greatest cultural importance in the prehistory of the Americas. Wild cacao is native to South America, and was a cultivated garden crop in prehistoric Mesoamerica. Recent analyses of the early use of cacao have pointed to possible use of the plant by the early Pre-Classic Olmec, perhaps as early as 1,500 BCE (Coe and Coe 2000). This early date is based on linguistic evidence, 'cacao' being a word derived from the Olmec language. The earliest physical evidence of cacao in Mesoamerica dates to the Middle Pre-Classic Period Maya, some thousand years later. HPLC analysis of ceramic vessels from the Mayan site of Colha, dating between 600 BCE and 250 CE, identified both caffeine and the related alkaloid theobromine (which is present in cacao in a much higher concentration then caffeine and thus makes a superior biomarker). It is likely that these vessels were used in the preparation of a cacao-based beverage (Hurst et al. 1989; 2002). The vessels were found primarily in elite burials raising the possibility that the practice of drinking cacao beverages was socially restricted within Mayan society.

While this work is groundbreaking, pushing the known history of cacao back by 1,000 years, there is much we don't know about cacao use in Mesoamerica, and further residue analysis can provide additional information. Was the use of cacao widespread within prehistoric society, or restricted to certain classes or social contexts? Ethnohistoric evidence indicates that cacao was limited to elite consumption during the time of the Spanish conquest (Emboden 1979, 137). We also know that the Mayan vessels so far identified as containing cacao were primarily elite wares, but commoner emulation of elite practices cannot as yet be ruled out. Is the linguistic evidence of a pre-Mayan origin supported by physical evidence? Analysis of Olmec vessels would address this question. Were other substances added to cacao and if so, which substances? Ethnohistoric accounts show a variety of substances that could have been added, including maize, chili or honey (Hurst et al. 2002, 289). Cacao was sometimes taken in conjunction with hallucinogenic fungi, according to the 16th century CE observations of the Franciscan Fray Bernadino de Sahagún (Rudgley 1993, 73). Cacao has also been described as an additive to other substances, such as the hallucinogenic snuff made from the bark of *Virola calophylla* in the Amazon region (Rudgley 1993, 64). Only additional analysis of the organic residues from larger samples, from additional sites, will begin to answer these questions.

Caffeine is also found in a variety of plants of the holly family, genus Ilex. There are numerous individual Ilex species, over two dozen of which were used by Native Americans in infusions (Hu 1979), three of which are known to contain caffeine. The plants have a variety of common names, which vary by region, but are most commonly referred to by anthropologists as 'yaupon'. Despite the diversity of species the one most often associated with Native American use is *Ilex vomitoria*, a shrub that is native to the American southeast from Texas to Virginia (Merrill 1979, 40). Ilex species plants were documented, in historic times, to have been used by Native American cultures in the Southeast, such as the Cherokee or the Creek, to brew a beverage generically referred to as the 'Black Drink' (Hudson 1976). This was drunk under a variety of circumstances depending on the culture in question. In most cases its use was confined to ritual occasions requiring spiritual purification. In some societies it was used on a daily basis by male warriors while conducting village business; in others it was restricted to specific individuals of high rank. Some accounts show the beverage being used to facilitate social interaction between neighboring and potentially hostile tribes, in much the same way as tobacco. In all cases its use was a highly formalized affair, even when used often by many people in a society, Black Drink was never a secular beverage as coffee and tea are today (Merrill 1979).

As with the Mayan use of cacao, there are characteristic material correlates for the use of yaupon tea by Native Americans than can aid in its identification. Historic observations of cultures using Black Drink repeatedly state that this was often done using characteristic drinking vessels made of conch shell. Similar cups to those identified ethnohistorically, often adorned with ritually significant decoration including motifs relating to the Southeastern Ceremonial Complex (also known as the 'Southern Cult'), were widely traded throughout the Mississippi Valley during late prehistoric times, and are found in the burials of high-ranking individuals from that time.

Drinking the infusion of *Ilex vomitoria*, or related species, is documented to have had a profound physiological effect. Aside from the potent dosage of caffeine, and all its associated stimulatory and diuretic effects, the beverage has been described as having a powerful emetic effect (causing vomiting), as well as a diaphoretic effect (causing profuse sweating), both of which were believed to have a purifying effect on the body and spirit. However, whether these extreme effects are solely due to the ilex plants or to the combination of additives is a matter open to question.

Despite the abundant historical sources on the use of ilex infusions in ritual settings by Southeastern native societies, there is still much unknown and residue analysis has much to offer in this regard. The effects of caffeine on the nervous system are well known, but do not explain all the documented effects of ilex infusions, such as vomiting. Anecdotal evidence indicates that other plants were added to the infusions, such as *Eryngium aquaticum* (button snakeroot), other accounts have indicated that some of the physiological effects, especially vomiting, were due to mixing seawater with the infusion, or simply to drinking a large volume of the liquid (Swanton 1946). The emetic role of *Ilex vomitoria*, despite the species designation, is further called into question by ethnohistoric accounts of the Seminole brewing an infusion that did cause vomiting, but did not include *Ilex vomitoria* as an ingredient (Sturtevant 1954).

Residue analysis of experimentally reproduced infusions could answer some of the questions regarding the active ingredients in the Black Drink, and what its constituent plants may have been. Archaeological identification of prehistoric instances of the historically-documented ritual would be more straightforward: the identification of caffeine in pottery or shell vessel residues would be a likely indicator of the Black Drink ritual. Analysis of shell cups and other ceramics for the presence of caffeine would provide powerful corroborative evidence that these artifacts did have the ritual function they have been assumed to have had. It is yet unproven if the more highly symbolically decorated cups of the Mississippians were functional vessels for Black Drink consumption, or if they were more symbolic representations, as a modern chalice is symbolic of beverage consumption but is usually not actually used for that purpose. Comparisons of shell cup residues to residues from more utilitarian vessels could also provide evidence for whether the Black Drink practice was truly a socially restricted practice in prehistory as the ethnohistoric evidence indicates.

Another plant with significant cultural implications for Native American cultures is jimsonweed or *Datura*. *Datura* is a member of the *Solonacea* family, which includes other psychoactive plants like nightshade, belladonna and tobacco (as well as more benign plants like tomatoes). *Datura* contains several exceptionally potent tropane alkaloids, including scopolamine and atropine. These compounds, when ingested, have a variety of physiological effects, primarily through blocking muscarinic acetylcholine receptors. In small doses this can have a sedative effect, and also functions as an antispasmodic and reduces nausea. In higher doses, however, these alkaloids induce powerful hallucinations, disorientation, paranoia and distortion of the sense of motion, resulting in a flying or falling sensation. Unlike many hallucinogenic compounds, where the subject retains some sense of reality (they hear, see or feel things that they know are not really there), the hallucinations caused by *Datura* intoxication are totally convincing to the subject, leading to an induced form of psychotic break in extreme cases. *Datura* is dangerous even beyond these extreme effects, as the lethal toxic dose is only slightly higher than the minimum effective dose, and inadvertent poisoning is a real concern in its use.

*Datura* species, such as *Datura stramonium* and *Datura meteloides*, grow wild throughout the North America, and other species are common throughout the Old World.[1] *Datura* use has been documented in a number of Native American cultures. It can be ingested in a variety of ways, but infusions or smoking appears the most common. *Datura* use was especially prevalent among Southwestern, Californian and Great Basin Native American tribes, often under the name 'toloache', a term of Aztec derivation, usually in the context of vision quests, or youth initiation ceremonies (Emboden 1979, 80; Winter 2000b). Use of *Datura* for these purposes extended into Mexico and was observed historically among the Huichol (Winter 2000c, 281), and also among the Aztec (Winter 2000c, 298). *Datura* had other uses as well. Southwestern tribes, such as the Hopi, Zuni or Navajo, used *Datura meteloides* as a ceremonial hallucinogen for healers to achieve a trance state in order to make diagnoses of illness, as a poison or as a narcotic (Stevenson 1915; Vestal 1952; Whiting 1939). Eastern tribes such as the Iroquois or Algonquians (who called the plant 'wyscoean') also made use of *Datura* (Emboden 1979, 97; Herrick 1995, 200).

A full list of all the historically documented uses of *Datura* would be beyond the scope of this paper. That said, the prehistoric roots of the use of *Datura* are poorly understood. Archaeobotanical evidence is scant, and most ethnobotanists would not be likely to identify macrobotanical remains as related to cultural use even when they are present. Inferential evidence has been proposed based on certain stylistic attributes of vessels with spiked exteriors that are believed to be evocative of the morphology of the fruit of the *Datura* plant (evoking the alternative common name for the plant, 'Thorn Apple' (Litzinger 1981). There has been some ethnobotanical identification of *Datura* amongst prehistoric pueblo farmer cultures in the American Southwest (Yarnell 1959). More recently, the remains of a ceremonial structure at the BBB Motor Site in Illinois contained pottery with charred *Datura* residues (Emerson 1989, 48-49; Wagner 2000, 196). These roughly contemporary, late prehistoric occurrences of *Datura* use, widely separated by geography, attest to what was likely a widespread practice, but one about which we have very little empirical data.

Residue analysis of pottery from late prehistoric cultures in North America can answer a variety of questions in regards to the use of *Datura*. The alkaloid biomarkers for *Datura*, scopolamine and atropine, are easily extracted from a solid phase residue through infusions of ethanol, methanol or chloroform and are easily identifiable in modern residues. Their archaeological survivability is as yet undemonstrated. Scopolamine is hygroscopic and readily water-soluble, raising the possibility of leaching out in archaeological contexts. Atropine is less soluble in water and might make a more viable biomarker for archaeological residues of *Datura*.

A variety of research questions are possible for *Datura* residue analysis. Is the ceramic stylistic evidence for *Datura* use in North America supported by empirical data? The analysis of possible residues on these spiked form vessels for *Datura* biomarkers would address this question. Comparisons of archaeological contexts of *Datura* residues could determine if the ceremonial context that has been documented historically had prehistoric roots. Are only high-status vessels associated with *Datura*, or was its use more widespread? How was *Datura* ingested in prehistory, by infusion or smoking, or by both? Early smoking pipes in North America predate the earliest chemical evidence of tobacco by centuries (Rafferty 2002; 2006). Could the possible pre-tobacco pipes have been used for smoking *Datura* instead of tobacco? There is evidence that *Datura* and tobacco were used in conjunction in some instances, possibly mixed together in an infusion (Winter 2000a, 3).

Tobacco is, of all the known plants used for non-subsistence practices, most associated with Native American use although wild species of tobacco were also used by aboriginal Australians prior to European contact (Rudgley 1993, 135). While the prehistory of tobacco is better known than that of other non-subsistence plants, there is still much to learn and residue analysis of

---

1 The genus name in fact derives from the name of the plant in Sanskrit (Emboden 1979, 65).

tobacco alkaloids is the most promising approach. Ethnohistoric evidence indicates that the use of tobacco was ubiquitous to Native American cultures at the time of contact, used from California to eastern Canada, even among cultures that did not otherwise cultivate plants (Springer 1981). Tobacco can be ingested in a variety of ways, including chewing leaves, snuffing ground leaves, in infusions and even by enema. However, smoking is the widely preferred mode for the ingestion of tobacco, with the use of smoking pipes dominating in North America. These provide a useful indirect form of evidence for the use of tobacco, though one not without its problems. Cigars were more popular than pipes among Native populations in South America making the prehistory of South American tobacco more difficult to address due to a lack of durable evidence (Wilbert 1987).

Tobacco is the generic name for a variety of species of the genus Nicotiana. Many of these species grow within North America, though the most widespread in the region, *Nicotiana rustica*, is of South American origin (Pearsall 1992). Nicotiana species are defined by their high concentration of the alkaloid nicotine, and *Nicotiana rustica* has the highest nicotine content of any of them. Nicotine functions by binding to nicotinic acetylcholine receptors, causing increased heart rate, vasoconstriction and increased alertness. In sufficient doses, nicotine can have hallucinogenic effects equivalent to indole alkaloids such as ibogaine and harmaline (Joniger and Dobkin de Rios 1973; 1976). Lethal overdose is possible in extreme cases, though the likelihood is far lower for nicotine than for the alkaloids in other Native American psychoactive plants such as *Datura*. It has been argued that one reason for the adoption of tobacco was that it was safer and more controllable than *Datura* (Goodman 1993, 24; Winter 2000b, 31), and did not require the intervention of ritual specialists (Von Gernet 1992, 178), though tobacco was viewed as more powerful than *Datura* by some Native cultures (Winter 2000b, 50).

There is a vast amount of indirect evidence for the prehistoric use of tobacco by Native Americans in the form of smoking pipes of ground stone or of ceramic. I would argue, however, that this evidence has been something of a distraction from the study of the use of tobacco. This is because, especially for the earliest examples, we have only the assumption of tobacco, but no physical evidence, and there is a wide variety of plants that could be smoked besides tobacco. Archaeobotanical evidence for tobacco is scanty. The oldest known dates for tobacco are from the North Coast of Peru, with dates ranging between 2500 and 1800 BCE (Pearsall 1992, 178). The earliest evidence of the use of Nicotiana species tobacco comes from the central Mississippi Valley between 100 and 200 CE (Asch 1991; 1994, 45; Haberman 1984, 271; Wagner 2000; Winter 2000a). Early dates for the rest of North America fall between 400 and 800 CE (Asch and Asch 1985, 196; Haberman 1984, 272–273; Wagner 1998, 840).

These data leave a sizeable temporal gap between the archaeobotanical evidence for tobacco and the archaeological evidence of smoking pipes. The earliest known tobacco pipes in North America date to the Late Archaic Period, over a thousand years prior to the earliest known archaeobotanical evidence. Residue analysis has the potential to close this gap and to determine what, other than tobacco, may have filled it.

I have been involved in one pilot study which demonstrated that nicotine can be extracted from organic residues and detected using GC/MS. The analytical protocol used was adapted from existing protocols for the extraction and detection of nicotine from human hair as a means of analyzing the effect of environmental tobacco smoke (Zahlsen and Nilsen 1994). Nicotine was extracted from smoking pipe residue with diethyl ether. The resulting extract was then concentrated through evaporation under nitrogen. The concentrated extract was analyzed using a HP 5890 gas chromatograph coupled to an HP-1 mass selective detector. Under the instrumental parameters used in this analysis nicotine has a GC retention time of 12.9 minutes (Figure 2). The integrated chromatographic peaks were analyzed by select ion monitoring (SIM) mass spectrometry focusing on ions with m/z 84, 133, 161 and 162. The resulting mass spectra were then matched against a library of spectra to yield a probability match for nicotine.[2] This protocol was successfully applied to samples of modern *Nicotiana rustica* ash, initially spiked with pure nicotine, then to unadulterated ash. A sample of 19$^{th}$ century CE pipe residue was also successfully analyzed, indicating that nicotine could survive in archaeological context at least over a relatively short period. With the analytical technique proven effective, it was then applied to a small sample of very old North American smoking pipes.

Two smoking pipes, one from the Adena Culture Cresap Mound site in West Virginia and the other from the Middlesex Culture Boucher site in Vermont, each dating to between 300 and 500 BCE, were tested in this study. Both of these pipes were tubular in form and were some of the oldest smoking pipes in eastern North America (though they do post-date the oldest pipes of the Late Archaic Period). Both were recovered from burial contexts, which is the most common context for the earliest pipes.

---

[2] Identified as pyridine, 3-(1-methyl-2-pyrrolidinyl) by the International Union of Pure and Applied Chemistry (IUPAC).

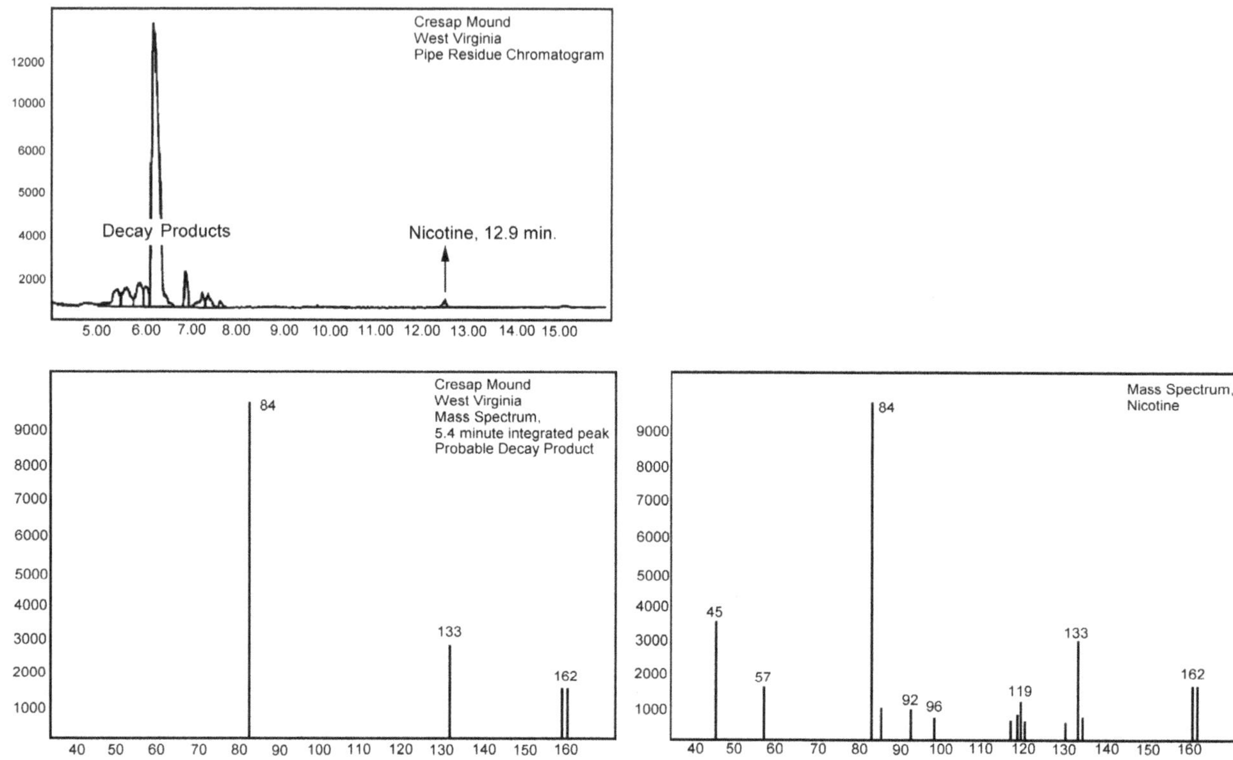


Figure 2: GC/MS results of an early tobacco residue from eastern North America.

Characteristic nicotine decay products, principally the alkaloid anabasine, were successfully detected in both pipes, indicating that they were used to smoke a species of tobacco, most likely *Nicotiana rustica* (Rafferty 2002; 2006). Figure 2 depicts the relevant chromatograms and mass spectra. These results extended back the known chronology of tobacco by several centuries though clearly more work is required on this topic. Continuation of this research is complicated by several factors, including the ready solubility of nicotine in water, leading to biomarker decay for older samples. While 12.9 minute chromatographic peaks were identified in both test samples, the abundance of nicotine was so low that the instrument could not integrate them, causing the mass spectrometry determination to analyze degradation products of nicotine instead. However, arguments that nicotine would not survive at all (Fletcher 1979; Haberman 1984) are not supported by this research, and even when degraded nicotine decays into characteristic products that can be detected by this protocol.

Another issue of concern is the comparative rarity of smoking pipes in comparison to other artifacts like pottery or lithics. While nicotine could survive in the fabric of a smoking pipe in the absence of a visible residue, this would require destructive analysis which is unlikely to be allowed for rarer artifact types, and the identification of unwashed specimens possibly containing intact visible residue has proven challenging. These issues aside, funding is currently being sought to investigate the early use of tobacco in North America using residue analysis of larger samples of early pipes.

While I hope that the results of this research are deemed intriguing by other researchers in this area, they are primarily a proof of concept demonstration and present as many questions as they answer. Many more samples from early pipes are needed, despite the difficulties in obtaining these samples as outlined previously. There is a range of research questions that could be addressed through residue analysis of organic residues potentially containing nicotine. For instance, my limited study indicates that tobacco was likely present in eastern North America as early as 500 BCE, but the earliest known smoking pipes were used centuries, if not millennia, prior to that date. Do these very early pipes that predate current evidence for tobacco use contain nicotine residue, indicating an earlier adoption of the practice of tobacco smoking than has since been demonstrated? If so, this would be a significant finding beyond moving the origins of tobacco smoking back into the Late Archaic Period. Tobacco is not native to the Eastern Woodlands of North America, and its presence indicates human intervention, though this may fall short of full-fledged agriculture. Still, if tobacco was encouraged, tended or even planted, this would be one of the earliest domesticates in the region, showing that non-subsistence plants played a significant role in Archaic Period horticulture in addition to edible plants such as chenopodium or sunflower. If, on the other hand, nicotine is not identified in pre-Woodland Period pipes, were other plants, such as *Datura*, used instead? If this were to prove the case, it would be evidence that tobacco was adopted as a more benign replacement for *Datura* as some have speculated.

Another point to consider in conducting residue analysis on smoking pipes is that, even for historically documented Native American societies that used tobacco, tobacco is only one among many plants that were smoked, though it is the plant most often accorded the greatest status by Native American cultures. Other plants include sumac and dogwood, both of which were included in a smoking mixture known to the Algonquian as 'kinnickkinnick' that was used instead of tobacco, and also as an additive to tobacco. One study identified over a dozen different plants, in addition to *Nicotiana rustica* tobacco, that were smoked in northeastern North America (Yarnell 1964). It would be useful to know if there is a prehistoric depth to such practices, though this would be complicated by the need to identify a wide range of species-specific biomarkers in addition to nicotine.

The four alkaloid-containing plants discussed above, cacao, tobacco, *Datura* species and Ilex species are only a select sample of possible candidates for residue analysis study. There are a variety of other psychoactive plants used by South American cultures, such as *Banisteriopsis caapi* (ayahusaca or yajé) or *Virola calophylla* (epena). The prehistoric use of these substances could be addressed through the detection of their primary alkaloids (harmaline and dimethyltryptamine respectively).

Little is known about the use of psychoactive substances in Prehistoric European or Asian cultures, though there are a range of possible alkaloid-bearing plants that could have been used, such as *Atropa belladonna* (belladonna or deadly nightshade) which, like *Datura*, contained both atropine and scopolamine and *Cannabis sativa* (though the psychoactive compound for this plant, tetrahydrocannabinol, is not an alkaloid). *Papaver somniferum* (opium poppy) was used in several Old World cultures and the identification of the alkaloid morphine by chromatographic, mass spectrometric and radioimmunoassay techniques show evidence of opium use in the Mediterranean region as early as 1500 BCE (Bisset et al. 1996; Koschel 1996). Archaeobotanical evidence indicates that opium use may have an even greater temporal depth than that, with poppy seeds having been identified from a 3700 BCE context from Italy (Merlin 2003). The possibility of the use of opium and cannabis has been raised for European Neolithic cultures (Sherratt 1991, 52) though this has not been corroborated by physical evidence.

There is clearly a wide range of psychoactive plants used by traditional cultures, past and present, Old World and New. Most of these depend on alkaloid compounds for their mind-altering effects. Alkaloids have been proven to be viable biomarkers for residue analysis, though the use of this type of evidence is only in its fledgling stages.

## Discussion

I would like to discuss the implications for the study of alkaloids for archaeology in a more general frame than the identification of the prehistoric use of specific plants. From a methodological and 'middle range' level, the analysis and interpretation of alkaloid residues also serve to draw scientific attention beyond subsistence. Ethnological evidence shows that small-scale, traditional cultures make use of a vast array of plants in their local environments, and only a small percentage of these plants are used as food. In focusing only on what people eat, we are ignoring a wealth of knowledge that falls outside of that range of behaviors. We know very little about the non-subsistence plant use of prehistoric cultures world-wide. It has not been a question given any priority. While this lack of attention has in part been due to a perception that non-subsistence, medicinal and psychoactive plants are more part of a society's ritual and symbolic life, a region that has long been *terra incognita* for archaeology, there has also been a lack of methodological approaches that could shed light on the issue. Scholars do not ask questions for which there are no tools to answer, or at least they do not publish the results. The development of residue analysis techniques able to identify the alkaloid compounds that are the active agents in many non-subsistence plants provides such a tool. It is up to us as researchers to avail ourselves of it.

There are higher-order, theoretical issues to consider as well, which relate to the types of questions that archaeological inquiry is seen as appropriate to address. Archaeology has long been at its core a materialist discipline, focusing primarily on issues of technology, economy, settlement patterns and subsistence. This is, of course, due to the nature of our data: material culture, the physical residues left behind by ancient human actions. There is a disjunction between what archaeologists find, material things, and what we want to know, what people did in the past and why they did it. Archaeology is very good at getting at issues of technology and economy, but less successful at addressing issues of ritual and symbols, or essentially of 'culture' in the anthropological sense rather than in the specific archaeological sense. This was most famously expressed by Hawke, who is best known for his 'ladder of inference', and who noted that, as the ideational component of ancient practices increased in comparison to the material component, the ability of archaeologists to reliably interpret the nature of those practices becomes increasingly compromised (Hawke 1954). As noted recently, Hawke's seminal paper has been widely misread as a statement that the ancient mind is inaccessible to us, letting archaeologists off the proverbial hook when it comes to interpreting ancient ritual or religion (Evans 1998). In fact, Hawke meant nothing of the sort, he only said that it would be difficult and not that we should not try.

This has been a point of tension within the discipline ever since. How, given the essentially arbitrary nature of symbols, can archaeologists interpret them when the cultural context in which they were operative is now extinct? A promising approach in recent years has been given the name 'cognitive archaeology' and has as its central premise a reworking of the nature of symbols. Symbols as used by humans have an arbitrary 'cultural' component, but they also have a predictable 'biological' component, in that the engine of human symbolism, the brain, is a constant factor between cultures. Human neurology is a constant, even as culture varies widely.

It is in this area that I feel that the archaeological study of alkaloids can have their greatest impact. It is a given that small-scale societies typically use induced altered states of consciousness in ritual contexts. Imagery generated within these altered states of consciousness become part of the symbolic milieu of a culture. While altered states can be induced through a variety of means, including fasting, physical exertion, sensory deprivation and sensory overload, the most common means is pharmacological through the use of psychoactive plants. Alkaloids in their vast biological diversity are a common means by which small-scale societies world-wide, past and present, have induced altered states. We now have in our hands a body of analytical techniques that can identify what alkaloids were used, and pharmacological data on what the neurological effects of those alkaloids are likely to have been. This provides a back door into that non-arbitrary aspect of human symbolism, their neurological component. This can be of great utility in our interpretation of past belief systems in that it gives us an insight into the meaning of prehistoric symbols since we have an understanding of the source of at least some of their generation.

An example of this potential comes from aboriginal use of nicotine by smoking tobacco, the focus of my own research. There is a continual association of smoking pipes with bird-related iconography in Native America, from as early as the Early Woodland Period at 1000 BCE, through to the 16th century CE contact between Native Americans and Europeans. Raptorial bird images are widespread in a range of ritual contexts even beyond their association with smoking pipes, and are found in tablet inscriptions, figurines, spear-thrower weights and even in earthen effigy mounds. What are we to make of this fixation on avian symbols? Chemical identification of nicotine in early pipes proves that they were used as nicotine delivery devices, the possibility of their use to ingest *Datura* has been raised though not proven. And, one of the documented effects of the ingestion of large doses of nicotine (as well as the more potent alkaloids found in *Datura*) is a sense of disembodiment and of exosomatic flight. It is likely then, that the birds so vital to Native American religious imagery are, at least to some degree, the symbolic embodiment of the physiological effect of nicotine (Dobkin de Rios 1976). Our understanding of one type of alkaloid residue has opened an interpretive window into a symbolic frontier.

In summary, I have outlined a small number of particularly important compounds for which there is ethnographic evidence of use by Native Americans. There are many other plants that ethnobotanists have identified as being used in non-subsistence contexts by small-scale societies in the Americas and beyond, the chemistry of which we know very little. As my own interests deal with aboriginal smoking practices, I can provide the example of the several dozen plants known to have been smoked by Native Americans, in addition to tobacco (Yarnell 1964), of which we know next to nothing in terms of their physiological effect. This relates to one small region of recent history in North America. When one considers the vast pharmacopoeia that native cultures have availed themselves of through time across the globe, and considers the potential information on human thought and behavior that can be obtained through their study, it is clear that there is much 'alkaloid archaeology' yet to be done.

## References

**Asch**, D.L. (1994). Aboriginal Specialty-Plant Cultivation in Eastern North America. Illinois Prehistory and a Post-Contact Perspective. In W. Green (ed.). *Agricultural Origins and Development in the Midcontinent*. Iowa City: University of Iowa Press.

**Asch**, D.L. and N.B. Asch (1985). Prehistoric Plant Cultivation in West-Central Illinois. In R.I. Ford (ed.). *Prehistoric Food Production in North America*. Ann Arbor: University of Michigan Press, pp. 149–203.

**Bisset**, N.G., J.G. Bruhn and M.H. Zenk (1996). The Presence of Opium in a 3,500 Year Old Cypriote Base-Ring Juglet. *Ägypten und Levante* **6**: 203-204.

**Coe**, S. and M. Coe (2000). *The True History of Chocolate*. London: Thames and Hudson.

**Dobkin de Rios**, M. (1976). Suggested Hallucinogenic Motifs in New World Massive Earthworks. In B. du Toit (ed.). *Drugs, Rituals and Altered States of Consciousness*. Rotterdam: Balkema Press.

**Emboden**, W. (1979). *Narcotic Plants*. New York: Macmillan Publishing Co., Inc.

**Emerson**, T.E. (1989). Water, Serpents and the Underworld. An Exploration into Cahokia Symbolism. In P. Galloway (ed.). *The Southeastern Ceremonial Complex: Artifacts and Analysis*. Lincoln: University of Nebraska Press.

**Evans**, C. (1998). Historicism, Chronology And Straw Men. Situating Hawkes' 'Ladder Of Influence'. *Antiquity* **72,276**: 398-404.

**Evershed**, R.P., S.N. Dudd, M.J. Lockheart and S. Jim (2001). Lipids in Archaeology. In D.R. Brothwell and A.M. Pollard (eds.). *Handbook of Archaeological Sciences*. Chichester, New York: John Wiley & Sons, Ltd.

**Fletcher**, J. III (1979). Early Woodland Tubular Pipes from Eastern Tennessee. *Central States Archaeological Journal* **26**: 6–11.

**Goodman**, J. (1993). Tobacco in History. The Cultures of Dependence. London: Routledge.

**Haberman**, T.W. (1984). Evidence of Aboriginal Tobaccos in Eastern North America. *American Antiquity* **49, 2**: 268-287.

**Hawke**, C.F.J. (1954). Archaeological Theory and Method. Some Suggestions from the Old World. *American Anthropologist* **56**, 155-88.

**Herrick**, J.W. (1995). Iroquois Medical Botany. Syracuse: Syracuse University Press.

**Hu**, Shiu Ying (1979). The Botany of Yaupon. In C. Hudson (ed.). *The Black Drink. A Native American Tea.* Athens: University of Georgia Press.

**Hudson**, C. (1976). *The Southern Indians*. Knoxville: University of Tennessee Press.

**Hurst**, W.J. S.M. Tarka Jr., T.G. Powis, F. Valdez Jr. and T. Hester (2002). Cacao Usage by the Earliest Maya Civilization. *Nature* **418**: 289.

**Hurst**, W.J., R.A. Martin, S.M. Tarka Jr. and G.D. Hall (1989). Authentication of Cocoa in Maya Vessels Using High-Performance Liquid Chromatographic Techniques. *Journal of Chromatography* **466**: 279-289.

**Joniger**, O. and M. Dobkin de Rios (1973). Suggestive Hallucinogenic Properties of Tobacco. *Medical Anthropology Newsletter* **4,4**: 6-11.

**Joniger**, O. and M. Dobkin de Rios (1976). Nicotiana an Hallucinogen? *Economic Botany* **30**: 149-151.

**Koschel**, K. (1996). Opium Alkaloids in a Cypriote Base Ring I Vessel (Bilbil) of the Middle Bronze Age from Europe. *Ägypten und Levante* **6**: 159-166.

**Litzinger**, W.J. (1981). Ceramic Evidence for Prehistoric *Datura* Use In North America. *Journal of Ethnopharmacology* **4,1**: 57-74.

**Merlin**, M.D. (2003). Archaeological Evidence for the Tradition of Psychoactive Plant Use in the Old World. *Economic Botany* **57,3**: 295-323.

**Merrill**, W.L. (1979). The Beloved Tree. *Ilex vomitoria* Among the Indians of the Southeast and Adjacent Regions. In C. Hudson (ed.). *The Black Drink. A Native American Tea*. Athens: University of Georgia Press.

**Pearsall**, D.M. (1992). The Origins of Plant Cultivation in South America. In C.W. Cowan and P.J. Watson (eds.). *The Origins of Agriculture. An International Perspective*. Washington D.C.: Smithsonian Institution.

**Rafferty**, S. (2002). Chemical Analysis of Early Woodland Period Smoking Pipe Residue. *Journal of Archaeological Science* **29,8**: 897-907.

**Rafferty**, S. (2006). Evidence of Early Tobacco in Northeastern North America? *Journal of Archaeological Science* **33,4**: 453-458.

**Raffuf**, R.F., and A.M Morris (1960). Persistence of Alkaloids in Plant Tissues. *Science* **131**: 1047.

**Rudgley**, R. (1993). *The Alchemy of Culture. Intoxicants in Society*. London: British Museum Press.

**Sherratt**, A. (1991). Sacred and Profane Substances. The Ritual Use of Narcotics in Later Neolithic Europe. In P. Garwood, D. Jennings, R. Skeates and J. Toms (eds.). *Sacred and Profane. Proceedings of a Conference on Archaeology, Ritual and Religion, Oxford, 1989.* Oxford: Oxford University Committee for Archaeology.

**Springer**, J.W. (1981). An Ethnohistoric Study of the Smoking Complex in Eastern North America. *Ethnohistory* **28,3**: 217-235.

**Stevenson**, M.C. (1915). *Ethnobotany of the Zuni Indians. Bureau of American Ethnology Annual Report Number 30.* Washington D.C.: Bureau of American Ethnology.

**Sturtevant**, W.C. (1954). The Medicine Bundles and Busks of the Florida Seminole. *Florida Archaeologist* **7**: 52-55.

**Swanton**, J.R. (1946). *The Indians of the Southeastern United States. Bureau of American Ethnology Bulletin Number 137*. Washington D.C.: Smithsonian Institution.

**Vestal**, P.A. (1945). *The Ethnobotany of the Ramah Navaho. Papers of the Peabody Museum of American Archaeology and Ethnology Number 40, 4.* Cambridge: Peabody Museum of American Archaeology and Ethnology.

**Von Gernet**, A. (1992). Hallucinogens and the Origins of the Iroquoian Pipe/Tobacco/Smoking Complex. In C.F. Hays III (ed.). *Proceedings of the 1988 Smoking Pipe Conference*. Rochester: Rochester Museum and Science Service.

**Wagner**, G. E. (2000). Tobacco in Prehistoric Eastern North America. In J.C. Winter (ed.). *Tobacco Use by Native Americans: Sacred Smoke and Silent Killer.* Norman: University of Oklahoma Press.

**Whiting**, A.F. (1939). *Ethnobotany of the Hopi. Bulletin Number 15.* Flagstaff: Museum of Northern Arizona.

**Wilbert**, J.C. (1987). *Tobacco and Shamanism in South America*. New Haven: Yale University Press.

**Winter**, J.C. (2000a). Introduction to the North American Tobacco Species. In J.C. Winter (ed.). *Tobacco Use by Native North Americans. Sacred Smoke and Silent Killer.* Norman: University of Oklahoma Press.

**Winter**, J.C. (2000b). Traditional Uses of Tobacco. In J.C. Winter (ed.). *Tobacco Use by Native North*

*Americans. Sacred Smoke and Silent Killer*. Norman: University of Oklahoma Press.
**Winter**, J.C. (2000c). From Earth Mother to Snake Woman. The Role of Tobacco in the Evolution of Native American Religious Organization. In J.C. Winter (ed.). *Tobacco Use by Native North Americans. Sacred Smoke and Silent Killer*. Norman: University of Oklahoma Press.
**Yarnell**, R.A. (1959). Prehistoric Pueblo Use of *Datura*. *Palacio* **66**: 176-178.
**Yarnell**, R.A. (1964). *Aboriginal Relationships Between Culture and Plant Life in the Upper Great Lakes Region*. Ann Arbor: University of Michigan.
**Zahlsen**, K. and O.G. Nilsen (1994). Nicotine in Hair of Smokers and Non-Smokers. Sampling Procedure and Gas Chromatographic/Mass Spectrometric Analysis. *Pharmacology and Toxicology* **75**: 143–149.

# CHAPTER FIFTEEN
# Reconstructing Mississippian Diet in the American Bottom with Stable Isotope Ratios of Pot Sherd Residues
## D.E. Beehr and S.H. Ambrose

Dana Beehr; Department of Anthropology, University of Illinois at Urbana-Champaign; 607 South Matthews Avenue; Urbana, IL 61801; USA; <beehr@uiuc.edu> and Stanley Ambrose; Department of Anthropology, University of Illinois at Urbana-Champaign; 607 South Matthews Avenue; Urbana, IL 61801; USA; <ambrose@uiuc.edu>. We gratefully acknowledge the assistance of the following people in conducting this research: Sallie Greenberg at the University of Illinois at Urbana-Champaign Geological Survey for technical assistance, Dr. Tim Pauketat at the University of Illinois at Urbana-Champaign Anthropology Department for providing sherds from Halliday, Grossmann and Pfeffer, and Dr. Laura Kozuch and the staff of the Illinois Transportation Archaeological Research Program for providing access to the sherds from the Old Man site and the sub-Mound 51 borrow pit. Support for mass spectrometry instrumentation at the University of Illinois at Urbana-Champaign Environmental Isotope Paleobiogeochemistry Laboratory was provided by National Science Foundation (USA) grant SBR 98-71480.

In this chapter we report on experiments to evaluate the effects of soil organic matter contamination on carbonized residues in buried prehistoric pottery, using a method conventionally applied to purification of prehistoric bone collagen and plants for isotopic analysis (Ambrose 1990; DeNiro and Hastorf 1985). Experiments show that stable carbon and nitrogen isotope ratios of carbonized residues on ceramic vessel interiors reflect the isotopic composition of the foods cooked in them. Carbonized residues from the interior surfaces of 88 ceramic vessels found at five contemporaneous early Mississippian American Bottom sites were analyzed to determine the percentage of maize (*Zea mays*) cooked in them. Four of the five assemblages studied contained carbonized residues with high carbon isotope ratios, reflecting varied and sometimes high percentages of maize in cooked meals. Nitrogen isotope ratios of eight sherds from Cahokia showed that meat was cooked in at least one vessel. Absorbed soil organic matter and carbonates may contaminate carbonized residues in buried sherds, and in this case could lead to underestimates of the amount of maize. Therefore, chemical pretreatment methods conventionally used to decontaminate buried charcoal and bone for radiocarbon and stable isotope analysis were applied to a subset of 17 samples and compared to their untreated residues. Results show that pretreatment did not, in most cases, significantly alter stable isotope ratios of carbonized pottery residues. We conclude that carbonized potsherd residue isotopic analysis provides a simple and powerful tool for reconstructing dietary practices in such settings, and that simple chemical pretreatments should be used, where possible, for greater accuracy and precision of isotopic analysis.

Stable carbon and nitrogen isotopic analysis can provide quantitative data on proportions of foods with different isotopic composition and can thus be used to assess proportions of marine versus terrestrial, $C_3$ versus $C_4$ based food webs, as well as trophic levels. The basic principle underlying this approach to dietary reconstruction is that 'you are what you eat', in other words, the isotopic composition of an organism's diet controls that of its tissues (Ambrose 1993). This principle applies to any reservoir of organic matter, including soil carbon isotopes (Ambrose and Sikes 1991), and can be extended to reconstructing human diet indirectly through the isotopic composition of the food residues preserved in ceramic vessels (Hastorf and DeNiro 1985; Reber and Evershed 2004a; 2004b; Reber et al. 2004). Analysis of such residues can provide information about the isotopic composition of meals and vessel function as well as human diet. One could thus use pottery residue isotopic analysis to ask of humans: 'were they what they cooked?' (Beehr and Ambrose, in press).

Four basic preconditions must be fulfilled for diet reconstruction with stable isotopic analysis of archaeological ceramic residues: there are systematic differences in the isotopic composition of foods prepared by cooking; their isotopic composition is not significantly altered by cooking; their isotopic composition is not significantly altered during burial and contamination by soil carbon and nitrogen can be removed. Many food webs exhibit systematic differences in isotopic composition between major food classes (Van der Merwe 1982; 2005; DeNiro 1987), and heating experiments have shown that cooking temperatures do not significantly affect isotope ratios (DeNiro et al. 1985; Hastorf and DeNiro 1985; Marino and DeNiro 1987; Whitney 1992). DeNiro and Hastorf (1985) documented the effects of soil contaminants and methods of removal of contaminants from carbonized archaeological plants. These methods have been used to treat carbonized ceramic residues for accelerator mass spectrometry (AMS) radiocarbon dating (Lovis 1990). However, the effects of decontamination on carbonized residues in archaeological ceramics have not previously been investigated.

Carbonized organic matter readily absorbs dissolved soil organic matter such as humic acid and fulvic acid. Removal of these soil organic contaminants is necessary because the stable isotopic composition of soil organics can vary substantially in the tropics, and warmer temperate regions, where both $^{13}C$-depleted $C_3$ and $^{13}C$-enriched $C_4$ plants grow, and because the radiocarbon age of soil organic matter integrates that of plant communities that post-date the buried artifacts (Ambrose and Sikes 1991). Soil and shell carbonates generally have substantially higher $^{13}C/^{12}C$ ratios than soil organic matter (Cerling 1984; Fritz and Poplawski 1974), and failure to remove them will usually inflate estimates of the amount of $C_4$ carbon in a residue. Soil carbonates generally do not form in recent soils in our study region, so the residues were not treated with acids to remove carbonates. However, many sherds analyzed in our study have shell temper, and carbonate bedrock underlies the entire region and could be a component of the clay matrix, so carbonate contamination remains a possible source of error.

Maize was one of the most important components of later prehistoric agricultural diets in eastern North America. Because maize is the only major dietary staple that uses the $C_4$ pathway for photosynthesis, and thus the only staple with a high $^{13}C/^{12}C$ ratio, its consumption can be easily measured with stable carbon isotope analysis of food residues or consumer tissues. During Late Woodland and Mississippian times, reliance on maize agriculture may have increased along with social complexity (Fritz 1992; Kidder 1992; Lopinot 1992, 1994, 1997). Quantitative evidence for maize consumption can thus provide important evidence for testing hypotheses about developments in prehistoric economy and society. Most isotopic research in the American Bottom region of the Mississippi Valley in Illinois and Missouri has used human bones. Research at Mound 72 in Cahokia, suggests a link between maize consumption, protein nutrition, health and social status (Ambrose et al. 2003). Skeletal isotopic analysis is ideal for estimating long-term average diets of individuals. However, skeletal analysis is destructive and labor-intensive, skeletons are often poorly preserved or absent and access to human skeletal material is increasingly restricted.

Carbonized residues on potsherd interiors can also be used to reconstruct human diet. They preserve the carbon and nitrogen isotopic composition of foods even after they have been cooked to temperatures high enough to become carbonized (Hastorf and DeNiro 1985; Whitney 1992). Depending on the frequencies of burning and cleaning, and the stickiness of the foods, vessel residues may provide a 'single-meal' snapshot, or they may integrate the isotopic composition of a series of cooking episodes. Pottery is ubiquitous in Mississippian and Late Woodland sites in the American Bottom. Compared to skeletal isotopic analysis, sherd residue analysis is simple, fast, non-destructive and inexpensive. Because potsherds are often found in residential contexts, isotopic analysis may provide useful evidence for household culinary practices and vessel functions.

In this chapter we will summarize the results of carbonized sherd residue isotopic analysis from five sites in the American Bottom region, including one probable ceremonial locality on the valley floor at Cahokia Mounds, and four sites of the Richland Complex in the surrounding uplands (Beehr and Ambrose, in press). We will focus on a subset of these samples treated to remove potential soil organic contaminants. These results will be compared with results from the untreated sherd residues to determine the effects of this pretreatment procedure on stable carbon and nitrogen isotope ratios of sherd residues. The generally small shifts in isotopic composition inspire confidence in the accuracy of the analyses of the untreated sherd residues.

## Cahokia and its Neighbors

Cahokia was a farming village along the Mississippi river that arose to prominence around 1050 CE, in a process so rapid it has been called the 'Big Bang in the Bottom' (Pauketat 1997, 31). The rise, dominance, decline and eventual abandonment of Cahokia have been divided into four stages: the Lohmann, Stirling, Moorehead and Sand Prairie phases, of which only the first two are relevant to this study. Cahokia's fluorescence during the Lohmann and Stirling phases was accompanied by widespread changes in pottery, architecture, spatial organization, complexity, social organization and ritual, and may indicate that Cahokia assumed dominance over this region at this time (Pauketat and Emerson 1997; Emerson 1997a; 1997b; Kelly 1990, 136).

At the beginning of the Lohmann phase (1050-1100 CE), construction on a monumental scale began at Cahokia, including the massive Monks Mound, which was the largest earthen structure in North America. Monks Mound was probably built in annual construction episodes that may have been accompanied by feasting events (Pauketat 2000). Some of the sherds in this study are from the sub-Mound 51 feasting pit, a contemporaneous feature filled with feasting debris that may have served as a borrow pit for the construction of Monks Mound (Pauketat et al. 2002).

Pronounced social stratification may have accompanied the rise of Cahokia. With the advent of the Stirling phase (1100-1200 CE) a divine chiefdom may have been instituted at Cahokia (Pauketat 1994, 175, 185). Innovations in symbolic and ritual practice appeared at this time, including ritual forms of pottery, figurines that may embody mythological themes, the construction of mound centers throughout the American Bottom, and the construction of new forms of ritual architecture such as post-circle monuments or 'woodhenges' (Emerson 1997a; Pauketat and Emerson 1991; 1997). Mound 72 was constructed at roughly the same time as the reorganization of Cahokia's center. It contains dramatic evidence of social inequality. High status individuals receiving special treatment include 16 burials on cedar litters, and one adult male buried on a beaded blanket shaped like an eagle effigy, associated with large quantities of high status goods (Fowler et al. 1999). Mass burial pits contain what have been interpreted as low status female sacrifices. Dental evidence from the female sacrifices suggests that they came from a genetically distinct population (Fowler et al. 1999, 82).

Contemporary with the rise of Cahokia, a series of settlements known as the Richland Complex emerged in the uplands to the east. Richland settlements did not incorporate all new Mississippian features of Cahokia, and innovations were often adopted in incomplete ways. For example, shell temper was added to inappropriate clays, or cord marked paddles were used to shape pots after which surface traces of cord markings were removed (Alt 2001; Pauketat 2003). Richland settlements may have been involved in provisioning Cahokia (Pauketat 2003). They were established at the beginning of the Lohmann phase, at which point the population of Cahokia reached its greatest size, estimated to be 10,000 to 15,000. Resources within Cahokia's catchment area may have been insufficient to fully support Cahokia's large population (Pauketat and Lopinot 1997). Provisioning by hinterland populations may have been necessary. Evidence for provisioning comes from analysis of the faunal assemblage at Cahokia and in the surrounding countryside. At the beginning of the Lohmann phase, the numbers of deer parts in Cahokian faunal assemblages increase dramatically, and this increase continues during the Stirling phase, while at other settlements, faunal assemblages show decreasing frequencies of deer remains (Kelly 1997). This suggests a decreased reliance on deer by populations living outside of Cahokia, possibly because they were provisioning Cahokia with deer, and consequently were relying more on maize and perhaps other resources for subsistence.

The fact that meat consumption increased at the high status center of Cahokia throughout this era seems to indicate that diet was one marker of status. If high meat consumption correlated with high status, then low meat consumption, and possibly high maize consumption, may have been correlated with low status. Maize production intensified during the Mississippian era. At the beginning of the Lohmann phase measures of maize presence, such as feature ubiquity indices and frequencies of macrobotanical remains, suggest that production of maize and other starchy seed crops increased at this time (Lopinot 1992; 1994; 1997). In addition, changes in ratios of kernel fragments to cob fragments suggest that there was an increased emphasis on storable forms of maize for consumption (Lopinot 1992). The trend of intensified crop production continued during the Stirling phase (Johannsen 1984; Lopinot 1994). Increased reliance on farming could have been a subsistence strategy for commoners.

Stable isotopic analysis of human bone provides support for the hypothesis that increased maize consumption is one marker of low status. Status related differences in levels of maize consumption were first revealed by analysis of human bones from Mound 72 and at Fingerhut, a non-elite Cahokian cemetery (Bender et al. 1981). The reliability of these results is questionable for two reasons. First, collagen is synthesized mainly from dietary protein, while bone apatite carbonate accurately reflects the isotopic composition of all dietary macronutrients, including fats, carbohydrates and proteins (Ambrose and Norr 1993). Because maize is a low-protein resource (about 10% protein), its consumption is likely to be underestimated by isotopic analysis of collagen. Second, Bender et al. (1981) do not describe their criteria for assessing bone preservation and procedures for decontamination (Ambrose 1990; 1993). It is possible that some bone was poorly preserved or contained soil organic contaminants, which would have also led to underestimation of maize consumption.

Buikstra and Milner (1991) suggested that individuals at Cahokia relied less on maize than individuals at peripheral sites. However, small sample size and preservation were again unresolved issues. Buikstra et al. (1994) looked more directly at maize consumption of the skeletal population of Mound 72. A total of 51 individuals from a variety of contexts was analyzed in this study, including the male on the bead blanket, skeletons buried in close association with him that are considered to be his 'retainers,' four decapitated individuals interpreted as sacrifices, women from the mass burial pits, and one of the individuals from the cedar litter burials. Despite problems with preservation, Buikstra et al. (1994) found that individuals from high status contexts generally consumed less maize than individuals from low status contexts and suggested a link between maize consumption and status.

Ambrose et al. (2003) re-evaluated the data from Buikstra and Milner's earlier study, using previously unreported collagen carbon and nitrogen as well as apatite carbon isotopes of the nine well-preserved individuals from this site. Collagen carbon isotope ratios indicated that high status individuals consumed only slightly less maize than low status individuals, but nitrogen isotope ratios suggested most of the high status individuals consumed significantly more animal protein.

Apatite carbon isotopes indicated high status individuals consumed diets with an average of 45% maize, while diets of low status individuals averaged 80% maize. This significant difference in maize consumption is consistent with the status linked substantial difference in nitrogen isotope ratios.

## Diet Reconstruction

We investigated maize consumption at Cahokia Mound 51 and four Richland phase sites by carbon and nitrogen isotope analysis of carbonized residues on potsherd interior surfaces. Maize consumption can be quantified with carbon isotopes because the two main photosynthetic pathways by which plants fix atmospheric carbon dioxide result in different $^{13}C/^{12}C$ ratios (Van der Merwe 1982; Morton and Schwarcz 1988; 2004). Plants with the $C_3$ pathway assimilate proportionately less $^{13}C$ than plants using the $C_4$ pathway. In eastern North America, maize is the only dietary staple plant with the $C_4$ pathway. Therefore, it is possible to estimate the amount of maize consumed by an individual or cooked in a food residue by determining its carbon isotope ratio. This ratio is expressed using the delta (δ) notation as parts per thousand difference from a standard, in this case the $^{13}C/^{12}C$ ratio of the Peedee Formation Belemnite (PDB) carbonate, and for nitrogen, relative to the $^{15}N/^{14}N$ ratio of the AIR (atmospheric $N_2$) standard. The industrial-era mean $\delta^{13}C$ value for plants with the $C_3$ photosynthetic pathway is -26.5‰, while that for $C_4$ plants is -12.5‰ (Van der Merwe 1982). Pre-industrial values may have been closer to -26‰ for $C_3$ plants, and -11‰ for $C_4$ plants (Tieszen and Fagre 1993a; b).

Nitrogen isotope ratios reflect the isotopic composition of the dietary protein. Because $^{15}N$ increases with trophic level, a high $\delta^{15}N$ value indicates more animal protein in the diet (Ambrose 2000; Schoeninger and DeNiro 1984). Because animal foods have much more protein than plants, a small amount of animal protein can have a disproportionate effect on consumer $\delta^{15}N$ values. Nitrogen-fixing plants such as legumes, including domestic beans (*Phaseolus vulgaris*), have lower $\delta^{15}N$ values than non-fixing plants, and have significantly more protein than maize. The diversity and complexity of food web nitrogen isotope ratios and protein contents preclude accurate estimation of proportions of meat to plant foods. Nonetheless, higher $\delta^{15}N$ generally indicates greater animal protein consumption.

Carbonized potsherd residues may form episodically during cooking accidents or accumulate slowly in vessel hot spots. Skibo (1992) has studied the relationship between types of food cooked and the formation of carbonized residues. Different types of foods usually carbonize through different processes and in different locations. Meat and vegetables, for example, tend to form carbon patches during boiling when small particles of food adhere to the sides of the pot at or above the water line. In contrast, thick pasty foods such as porridge tend to form carbon patches near the base of the pot (Skibo 1992). Fats are lighter than water and will also accumulate preferentially at the water line. Fat $\delta^{13}C$ values are systematically lower, more negative, by approximately 5‰ than those of proteins and carbohydrates in the same foods (DeNiro and Epstein 1977; Ambrose and Norr 1993; Tieszen and Fagre 1993a). Therefore the location and $\delta^{13}C$ values of carbonized residues should help determine the type of foods cooked. An experiment with a ceramic vessel used to cook unhomogenized whole milk shows that uncarbonized organic residues absorbed in the pot have the lowest $\delta^{13}C$ values near the rim, and residues in this position also have the high carbon and low nitrogen content expected for fats (Barnard et al. 2007). Extremely low $\delta^{13}C$ values (-32.0‰ to -29.6‰) of residues from the necks of five Woodland period vessels are also consistent with fatty residues of $C_3$-based foods (Lovis 1990).

Cooking does not significantly change carbon isotope ratios of pot residues from single food types (Morton and Schwarcz 1988; 2004; Whitney 1992; Hastorf and DeNiro 1985), nor does short term burial (Whitney 1992). Experiments by Hart et al. (2003) involving isotopic analysis of residues formed from mixtures of rice ($C_3$) and maize ($C_4$) in a wide range of proportions found lower $\delta^{13}C$ values than expected for the amounts of maize in weighed mixtures. Under-representation of maize in carbonized residues in this experiment may be due to greater stickiness of rice.

Because most plants in eastern North America use the $C_3$ photosynthetic pathway, any contamination with soil organic matter in the American Bottom sites is likely to have low $\delta^{13}C$ values. Therefore $\delta^{13}C$ values of residues found in this study in sherds that have not been treated to remove soil organic contaminants should be considered minimum estimates of amounts of $C_4$ plants cooked in these vessels. The majority of the sherds used in our study were body sherds, and the exact locations of residues could not be accurately assessed. After the analysis of 88 untreated carbonized sherd residues, 22 of the largest remaining samples were treated to remove base-soluble soil contaminants, using procedures generally applied to decontaminating bone collagen for stable isotopic analysis (Ambrose 1990).

## The Archaeological Sites

Sherds in this study were collected from the sub-Mound 51 feasting pit at Cahokia and from four contemporaneous sites from the Richland Complex.

Richland sites varied in size, form, and perhaps function. They may have occupied different levels of the site hierarchy proposed by Fowler and later revised (Emerson 1997a, 1997b; Fowler 1978). The Old Man site is a small rural settlement that formed the lowest level of the Cahokian site hierarchy. It was probably a farmstead that supported one or two families, with an occupation span of five to ten years (Betzenhauser 2003; Emerson 1997a). Halliday was a large farming village with a mix of Mississippian and pre-Mississippian characteristics, including hybrid styles of architecture, pre-Cahokian style courtyard groupings, and house sizes within the range of Lohmann-phase houses elsewhere (Alt 2001, 146). Pfeffer was a single mound center of the third tier of the Cahokian settlement hierarchy. Pfeffer was like Halliday because it showed hybrid architectural forms combining both traditional single-post and novel wall-trench styles of construction. Unlike Halliday, Pfeffer did not maintain courtyard groupings (Kruchten 2000). Grossmann is a farming village similar to Halliday, but it has higher frequencies of exotic materials, a higher diversity of structures than would be expected from a site of this type and more evidence of rebuilding. Its most famous feature is undoubtedly the cache of over 70 ground stone axes (Alt 2002; Pauketat and Alt 2004). The sub-Mound 51 borrow pit on Cahokia's main plaza contains feasting debris that was most likely associated with ceremonies relating to mound construction. The fill in this pit contains large amounts of well-preserved food debris including a wide variety of plant and animal remains. Maize fragments are very rare, which is unexpected considering the importance of maize in subsistence at this time (Pauketat et al. 2002).

## Methods and Results on Untreated Samples

We tried to sample at least ten sherds with carbonized residues on vessel interiors from each site in order estimate the range of isotope ratios. In the case of the Old Man site, this was not possible due to the small size of the ceramic assemblage. To avoid duplicate sampling, only one sherd was sampled per feature from each site, unless selected sherds differed in obvious ways such as temper or surface treatment but no more than two sherds were taken per feature. In the case of sub-Mound 51, no more than three sherds from obviously different vessels were taken per level. Seven samples were taken from Old Man, 15 from Pfeffer, 28 from Grossmann, 32 from Halliday and 21 from sub-Mound 51 adding up to a total of 103 samples.

Residues were identified visually with a low-power stereomicroscope and removed by gently scraping the sherd with a scalpel blade, leaving the underlying ceramic surface intact, thus avoiding obvious contamination by shell temper. No chemical pretreatments were applied and the samples were not homogenized by grinding. Residues were weighed, compressed in tinfoil capsules and converted by combustion into purified $N_2$ and $CO_2$ in a Carlo Erba NA2500 elemental analyzer interfaced with a Finnegan MAT 252 mass spectrometer in the University of Illinois Environmental Isotope Paleobiogeochemistry Laboratory. The average sample weight was 193 µg, with a range of 10 to 746 µg. Weights for the pretreated samples that were replicated were higher, with an average run weight of 555 µg. In addition, thirty-five standards of known $\delta^{13}C$ and/or $\delta^{15}N$ values and weight percent C and N were run to monitor accuracy and precision of isotopic and elemental analysis. Standards included 21 samples of thiourea, eight of sucrose and six of phenylalanine. The mean $\delta^{13}C$ values for thiourea, sucrose and phenylalanine were -23.68‰ ± 0.18; -11.15‰ ± 0.08 and -25.48‰ ± 0.07. The mean $\delta^{15}N$ values for thiourea and phenylalanine were -0.78 ± 0.21 and 21.14‰ ± 0.10, respectively.

Replicate analyses of samples with insufficient carbon or nitrogen to generate a major ion beam of at least 0.25 V generally have low reproducibility of carbon and nitrogen isotope ratios. Although a total of 103 sherds were sampled, reliable carbon isotope data were obtained from 88 sherds. The $\delta^{13}C$ values of untreated residues for the sites are as follows: -23.9 ± 1.0‰ for the Old Man Site, -18.7 ± 3.6‰ for Halliday, -22.4 ± 4.0‰ for Pfeffer, -22.4 ± 4.0‰ for Grossmann and -19.6 ± 4.9‰ for the Cahokia sub-Mound 51 feasting pit. The frequency distributions of $\delta^{13}C$ values at four of the five sites analyzed indicate meals cooked in many vessels contained significant amounts of maize (Figure 1). Estimates of dietary percent $C_4$ (D%) were calculated using the equation below. Pots from the Old Man site were found to contain meals averaging 12% maize, Halliday had 41% maize, Pfeffer had 25% maize, Grossmann had 37% maize and the sub-Mound 51 sherds averaged 42% maize.

$$D = \frac{C_3 - x}{C_3 - C_4} \times 100\%$$

where x = the residue $\delta^{13}C$ value, $C_3$ = the average $\delta^{13}C$ value of -26‰ for $C_3$ plants and $C_4$ = -11‰ for $C_4$ plants (Tieszen and Fagre 1993b).

Three sites contained sherds with $\delta^{13}C$ values ranging from -26‰ to -11‰ (Figure 1), which span the full range from 100% $C_3$ to 100% $C_4$ plants. If sherd residues had absorbed significant amounts of soil organic matter, then the range of $\delta^{13}C$ values would be smaller and would converge on $\delta^{13}C$ value of the contaminants. These results should, however, still be regarded as somewhat inaccurate estimates of maize consumption because soil organic matter is probably derived mainly from $C_3$ plants. Also, if the work by Hart et al. (2003) is not anomalous, $\delta^{13}C$ values of residues from mixed $C_3$ and $C_4$ sources may be biased against maize and low $\delta^{13}C$ values may not indicate the complete absence of $C_4$ plants.

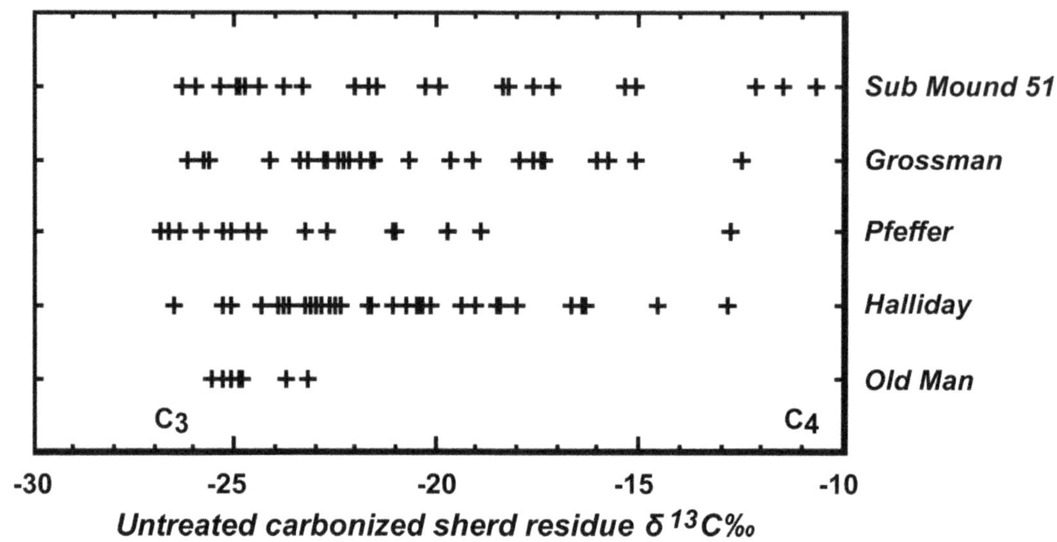


Figure 1: $\delta^{13}C$ values of untreated carbonized residues from Mississippian pottery sherds from five sites in the American Bottom region of southwest Illinois, based on data in Beehr and Ambrose (in press).

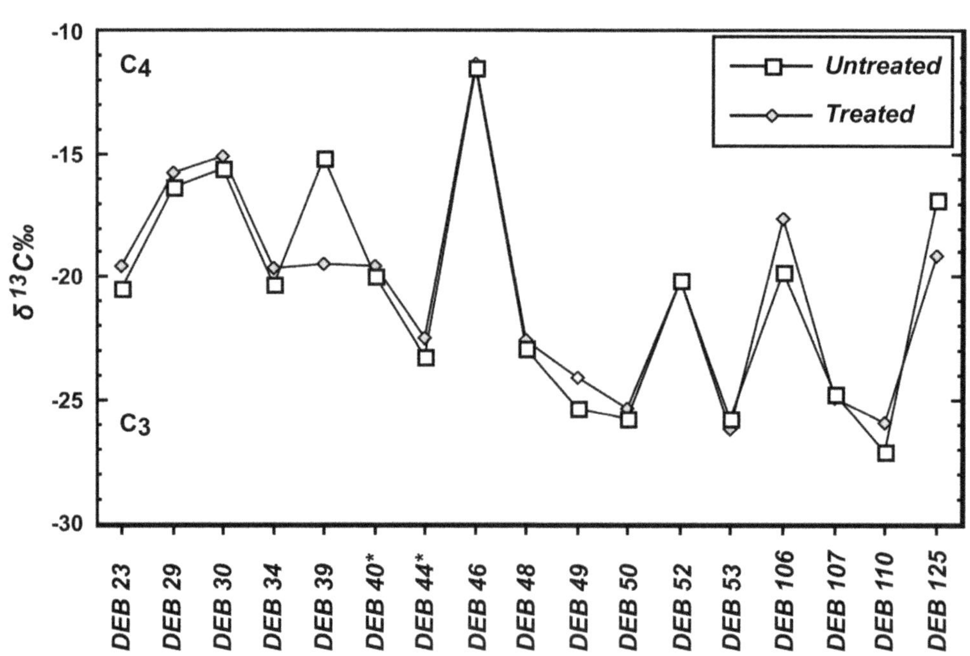


Figure 2: Treated versus untreated $\delta^{13}C$ values of carbonized residues from Mississippian pottery sherds.

Nitrogen isotope ratios were measurable on eight sub-Mound 51 sherds. Their mean $\delta^{15}N$ value was 6.0 ± 1.7‰, with a minimum of 4.7‰ and a maximum of 10.0‰. Meat was undoubtedly cooked in the pot with the highest $\delta^{15}N$ value. Low $\delta^{13}C$ values correlated with high $\delta^{15}N$ values, probably because animals in this area would have fed almost exclusively on $C_3$ plants.

## Methods and Results on Treated Samples

Treatment with a weak base solution to remove soil organic matter contamination was performed on the 21 largest sherd residue samples. The sherds selected for pretreatment are listed in Table 1. They comprise eight sherds from Halliday, seven from Pfeffer, two from Grossmann and four from Cahokia's sub-Mound 51 feasting pit. The sherd residues selected for base pretreatment were chosen on the basis of sample size. Samples greater than 500 µg were selected, when possible, in order to have sufficient material for nitrogen isotope analysis. The smallest sample was 460 µg and the largest 6600 µg (Table 1). The samples were treated with 1.5 ml 0.125 NaOH for 24 h, centrifuged and rinsed five times in distilled water and subsequently freeze-dried. The mean weight loss during treatment was 2.3% ± 22.4.

The mean $\delta^{13}C$ value for the treated sherds was -20.76‰ ± 4.12. That for the untreated samples was -20.97‰ ± 3.94. The difference between treated and untreated $\delta^{13}C$ values ranged from +4.9‰ to -2.2‰. The absolute mean difference was only 1.12‰ ± 1.33. When the extreme outlier DEB 91 is removed from consideration the two means become virtually identical, with a mean for the untreated sherds of -20.73‰ ± 3.88. On average the $\delta^{13}C$ values of the treated sherds are slightly less negative (Figure 2). This suggests that the treatment removed a small amount of $^{13}C$-depleted organic matter from the sherd residues. This is consistent with deposition in soils formed in a $C_3$-dominated floral habitat, such as eastern North America. Sample DEB 39 seems to have shifted to anomalously much less negative value, indicating that a substantial amount of $C_3$-based contaminants were removed with NaOH. Nevertheless, $\delta^{13}C$ values of the remaining treated sherd residue samples did not differ greatly from those of untreated ones, indicating soil organic contamination was minimal.

The mean $\delta^{15}N$ value for nine pretreated samples was 5.03‰ ± 2.29. Average C/N ratios for these sherds were 16.44 ± 6.21. Only four samples had enough N for reliable isotopic analysis, their mean was 6.46‰ ± 2.58. Figure 3 shows the difference between treated and untreated samples. Sample DEB 106 had low voltages for treated and untreated samples and may not be reliable. The average $\delta^{15}N$ value of the other treated sherds was 6.23‰, while that of the untreated samples was 4.95‰. This increase may indicate that the pretreatment removed contaminants with lower $\delta^{15}N$ values. However, only one of these three untreated samples had a sufficient ion beam voltage for accurate isotopic analysis so the mean difference estimate may be inaccurate. Soil organic matter tends to have lower $\delta^{15}N$ values (Ambrose 1991), so this difference, if real, was expected. Higher $\delta^{15}N$ values suggest the presence of greater amounts of animal matter in the sherd residues. This is consistent with C/N ratios that are higher than those of plants. Although the treated sample set is small, if this difference is real, we may have underestimated the amounts of meat cooked in vessels by using the generally lower $\delta^{15}N$ values of untreated sherd residues.

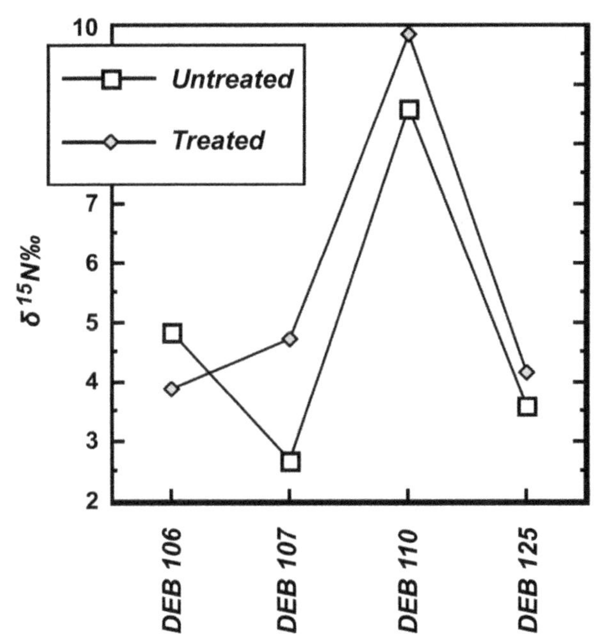


Figure 3: Treated versus untreated $\delta^{15}N$ values of carbonized residues from Mississippian pottery sherds. Sherd DEB 106 had low voltages for treated and untreated samples, so the difference between them may not have been accurately measured.

| Sample ID | Site | Wt. (μg) | $\delta^{13}C$‰ | Weight %C | Volts m/z=44 | $\delta^{15}N$‰ | Weight %N | Volts m/z=28 | Atomic C/N |
|---|---|---|---|---|---|---|---|---|---|
| **DEB 23** | 11S27 | 243 | -20.44 | 29.76 | 1.060 | | | | |
| **DEB 29** | 11S1131 | 335 | -16.40 | 28.39 | 1.394 | | | | |
| **DEB 30** | 11S27 | 530 | -15.60 | 7.26 | 0.552 | | | | |
| **DEB 31** | 11S27 | 155 | -22.57 | 4.54 | 0.128 | | | | |
| **DEB 32** | 11S27 | 68 | n/a | n/a | n/a | | | | |
| **DEB 34** | 11S27 | 207 | -20.33 | 7.76 | 0.237 | | | | |
| **DEB 39** | 11S27 | 469 | -15.19 | 20.49 | 1.394 | | | | |
| **DEB 40*** | 11S27 | 263 | -19.97 | 58.38 | 2.039 | 4.89 | 3.47 | 0.148 | 19.63 |
| **DEB 44*** | 11S27 | 620 | -23.25 | 8.55 | 0.745 | 5.85 | 1.17 | 0.122 | 8.53 |
| **DEB 46** | 11S204 | 279 | -11.47 | 36.35 | 1.475 | 3.07 | 2.27 | 0.110 | 18.68 |
| **DEB 48** | 11S204 | 370 | -22.91 | 4.68 | 0.244 | | | | |
| **DEB 49** | 11S204 | 1277 | -25.36 | 26.57 | 4.447 | 2.20 | 1.14 | 0.232 | 27.19 |
| **DEB 50** | 11S204 | 163 | -25.77 | 29.00 | 0.687 | | | | |
| **DEB 52** | 11S204 | 311 | -20.10 | 3.55 | 0.163 | | | | |
| **DEB 53** | 11S204 | 1939 | -25.76 | 18.71 | 5.180 | 7.12 | 1.83 | 0.667 | 11.93 |
| **DEB 54** | 11S204 | 175 | -20.75 | 18.68 | 0.499 | | | | |
| **DEB 91** | 11S1131 | 277 | -20.88 | 34.21 | 0.150 | 4.58 | 3.26 | 0.150 | 12.24 |
| **DEB 106** | 11S34/11MS2 | 660 | -19.83 | 26.14 | 2.441 | 3.87 | 1.62 | 0.189 | 18.83 |
| **DEB 107** | 11S34/11MS2 | 1208 | -24.77 | 34.61 | 5.837 | 4.72 | 2.31 | 0.511 | 17.48 |
| **DEB 110** | 11S34/11MS2 | 377 | -27.09 | 45.37 | 2.439 | 9.82 | 6.84 | 0.470 | 7.74 |
| **DEB 125** | 11S34/11MS2 | 1251 | -16.85 | 47.37 | 8.012 | 4.16 | 2.49 | 0.573 | 22.19 |

Table 1: Sample weights, $\delta^{13}C$ and $\delta^{15}N$ values, weight %C and weight %N, major ion beam voltages for mass/charge ratios of $CO_2$ (m/z = 44) and $N_2$ (m/z = 28) of treated carbonized residues from Mississippian pottery sherds. Samples DEB 40 and DEB 44, marked with *, were switched in position in the elemental analyzer carousel during the analyses of untreated samples.

## Discussion and Conclusions

These carbon isotope results can be compared to those of Reber and Evershed (2004a; Reber et al. 2004), who investigated the isotopic composition of uncarbonized molecular compounds of fatty residues absorbed in Mississippian ceramics. They used organic solvent extraction of lipids and fatty acids, followed by gas chromatography (to separate molecular compounds), coupled with isotope ratio mass spectrometry of $CO_2$ from combustion of individual molecular compounds (GC/C/IRMS). Their study involved 134 sherds from the American Bottom region, including one site (Halliday) also represented in our study. Perhaps because maize lipid residues degrade quickly after burial, they found isotopic evidence of maize in residues extracted from only a handful of sherds, although they found evidence of consumption of other plant substances as well as fish (Reber and Evershed 2004a). Nonetheless, the Halliday site had the highest frequencies of vessels with molecular and isotopic traces of maize. Compound specific isotopic analysis of absorbed residues can provide a qualitative indication of presence or absence of maize in ceramic vessels, but cannot provide accurate quantitative data on the proportions of maize cooked in vessels. The degree to which the absorbed residues analyzed were derived from soil organic contaminants cannot be easily estimated and methods of pretreatment to remove such contaminants remain to be developed.

Although a small average difference in $\delta^{13}C$ values between pretreated and untreated sherds was found in the present study, only a few samples were analyzed. A larger suite of treated and untreated samples should be analyzed. In this study, we expected the contaminants to have more negative $\delta^{13}C$ values, like those of $C_3$ plants, as well as lower $\delta^{15}N$ values. The general direction of the shift after pretreatment indicates that the contaminants had lower $\delta^{13}C$ and $\delta^{15}N$ values, which suggests a significant fraction of the soil organic matter was derived from $C_3$ plants.

Isotopic analysis of carbonized residues may be a deceptively simple procedure, but like with any other technique, methodological issues of sample preparation, purification and characterization must be explored before results can be considered accurate. This field is far from mature. Directions for future research include further exploration of pretreatment techniques, for example organic solvent extraction to remove lipids or treatment with HCl to remove carbonates and other contaminants. Isotopic analysis of the soils from the archaeological horizons of the ceramic samples would permit closer estimation of the expected composition of contaminants. Analysis of the isotopic composition of extracted contaminants would also be useful. Long-term burial experiments to monitor absorption of soil contaminants by carbonized residues should also be initiated, using sherds lacking organic carbon and carbonates from ceramic raw materials and firing.

The apparent bias against carbonization of maize compared to rice observed by Hart et al. (2003) suggests that residue-forming properties of cooked foods warrants further experimentation. If there is a bias against the preservation of carbonized maize residues, or if maize is masked by $C_3$ soil contaminants, then independent identification of maize by phytolith analysis (Hart et al. 2003), or compound-specific isotope analysis (Reber and Evershed 2004a; 2004b; Reber et al. 2004) will be helpful. Differences in the isotopic composition and the carbon and nitrogen content of fats, carbohydrates and proteins (Ambrose and Norr 1993; Tieszen and Fagre 1993a), as well as the vertical separation of fats in vessels during cooking (Barnard et al. 2007; Lovis 1990) should also be considered in sampling strategies and interpretation of results.

The results of this pilot study suggest that pretreatment to remove soil organic contaminants does not usually have a strong effect on carbon and nitrogen isotopic composition of carbonized residues inside American Bottom pottery vessels, so the conclusions drawn from results obtained from untreated sherds (Beehr and Ambrose, in press) are largely valid. The shift to higher $\delta^{15}N$ values we observed is based on an insufficient number of observations on treated samples, but if it is real, then we may have underestimated the amounts of animal matter cooked in these vessels. The isotopic ratios of sherd residues should be compared with those obtained from skeletal remains from the same sites and contexts, if possible. Sherd residues record the isotopic composition of meals, whereas skeletal values reflect the average diet over a decade or more. Pottery isotope analysis may thus provide an estimate of the variance in meal composition of the households with which the vessels were associated.

## References


**Alt**, S.M. (2001). Cahokian Change and the Authority of Tradition. In T.R. Pauketat (ed.). *The Archaeology of Traditions*. Gainesville: University Press of Florida, Gainesville, pp. 141-156.

**Alt**, S.M. (2002). Identities, Traditions and Diversity in Cahokia's Uplands. *Midcontinental Journal of Archaeology* **27**: 217-236.

**Ambrose**, S.H. (1990). Preparation and Characterization of Bone and Tooth Collagen for Isotopic Analysis. *Journal of Archaeological Science* **18**: 293-317.

**Ambrose**, S.H. (1991). Effects of Diet, Climate and Physiology on Nitrogen Isotope Abundances in Terrestrial Foodwebs. *Journal of Archaeological Science* **18**: 293-317.

**Ambrose**, S.H. (1993). Isotopic Analysis of Paleodiets. Methodological and Interpretive Considerations. In M.K. Sandford (ed.). *Investigations of Ancient Human Tissue. Chemical Analyses in Anthropology*. Langhorn: Gordon and Breach Science Publishers, pp. 59-130.

**Ambrose**, S.H. (2000). Controlled Diet and Climate Experiments on Nitrogen Isotope Ratios of Rats. In S.H. Ambrose and M.A. Katzenberg (eds.). *Biogeochemical Approaches to Paleodietary Analysis*. New York: Kluwer Academic, Plenum Publishers, pp. 243-259.

**Ambrose**, S.H., J.E. Buikstra and H.W. Krueger (2003). Gender and Status Differences in Diet at Mound 72, Cahokia, Revealed by Isotopic Analysis of Bone. *Journal of Anthropological Archaeology* **22**: 217-226.

**Ambrose**, S.H. and L.C. Norr (1993). Experimental Evidence for the Relationship of the Carbon Isotope Ratios of Whole Diet and Dietary Protein to Those of Bone Collagen and Carbonate. In J.B. Lambert and G. Grupe (eds.). *Prehistoric Human Bone. Archaeology at the Molecular Level*. Berlin: Springer Verlag, pp. 1-37.

**Ambrose**, S.H. and N.E. Sikes (1991). Soil Carbon Isotope Evidence for Holocene Habitat Change in the Kenya Rift Valley. *Science* **253**: 1402-1405.

**Barnard**, H., S.H. Ambrose, D.E. Beehr, M.D. Forster, R.E. Lanehart, R.E. Parr, M.E. Malainey, M. Rider, C. Solazzo and R.M Yohe II (2007). Mixed Results of Seven Methods for Organic Residue Analysis Applied to One Vessel with the Residue of a Known Foodstuff. *Journal of Archaeological Science* **34**: 28-37.

**Beehr**, D. and S.H. Ambrose (in press). Were They What They Cooked? Stable Isotopic Analysis of Mississippian Pottery Residues. In K. Twiss (ed.). *We Are What We Eat. Archaeology, Food, and Identity*. Carbondale: Center for Archaeological Investigations, Southern Illinois University.

**Bender**, M.M., D.A. Barreis and R.L. Stevenson (1981). Further Light on Carbon Isotopes and Hopewell Agriculture. *American Antiquity* **48**: 346-353.

**Betzenhauser**, A. (2003). Untitled Paper on Site 11MS542. University of Illinois at Urbana-Champaign.

**Buikstra**, J.E. and G.R. Milner (1991). Isotopic and Archaeological Interpretations of Diet in the Central Mississippian Valley. *Journal of Archaeological Science* **18**: 319-329.

**Buikstra**, J.E., J.C. Rose and G.R. Milner (1994). A Carbon Isotopic Perspective on Dietary Variation in Late Prehistoric Western Illinois. In W. Green (ed.). *Agricultural Origins and Development on the Midcontinent. Office of the State Archaeologist Report 19*. Iowa City: University of Iowa, pp. 155-170.

**Cerling**, T.E. (1984). The Stable Isotopic Composition of Modern Soil Carbonate and its Relationship to Climate. *Earth and Planetary Science Letters* **71**: 229-240.

**DeNiro**, M.J. (1987). Stable Isotopy and Archaeology. *American Scientist* **75**: 182-191.

**DeNiro**, M.J. and S.J. Epstein (1977). Mechanisms of Carbon Isotope Fractionation Associated with Lipid Synthesis. *Science* **197**: 261-263.

**DeNiro**, M.J. and C.A. Hastorf (1985). Alteration of $^{15}N/^{14}N$ and $^{13}C/^{12}C$ Ratios of Plant Matter During the Initial Stages of Diagenesis. Studies Utilizing Archaeological Specimens from Peru. *Geochimica et Cosmochimica Acta* **49**: 97-115.

**DeNiro**, M.J., M.J. Schoeninger and C.A. Hastorf (1985). Effect of Heating on the Stable Carbon and Nitrogen Isotope Ratios of Bone Collagen. *Journal of Archaeological Science* **12**: 1-7.

**Emerson**, T.E. (1997a). *Cahokia and the Archaeology of Power*. Tuscaloosa: University of Alabama Press.

**Emerson**, T.E. (1997b). Reflections from the Countryside on Cahokian Hegemony. In T.R. Pauketat and T.E. Emerson (eds.). *Cahokia. Domination and Ideology in the Mississippian World*. Lincoln: University of Nebraska Press, pp. 167-189.

**Fowler**, M.L. (1978). Cahokia and the American Bottom. Settlement Archaeology. In B.D. Smith (ed.). *Mississippian Settlement Patterns*. New York, San Francisco, London: Academic Press, pp. 455-478.

**Fowler**, M.L, J.C. Rose, B.V. Leest and S.R. Ahler (1999) *The Mound 72 Area. Dedicated and Sacred Space in Early Cahokia. Report of Investigations Number 54*. Springfield: Illinois State Museum Society.

**Fritz**, G.J. (1992). 'Newer,' 'Better' Maize and the Mississippian Emergence. A Critique of Prime Mover Explanations. In W.I. Woods (ed.). *Late Prehistoric Agriculture. Studies in Illinois Archaeology Number 8*. Springfield: Historic Preservation Agency, pp. 19-43.

**Fritz**, P. and S. Poplawski (1974). $^{18}O$ and $^{13}C$ in the shells of freshwater mollusks and their environments. *Earth and Planetary Science Letters* **24**: 91-98.

**Hart**, J.P., R.G. Thompson and H.J. Brumbach (2003). Phytolith Evidence for Early Maize (*Zea Mays*) in the Northern Finger Lakes of New York. *American Antiquity*, **68**: 619-640.

**Hastorf**, C.A. and M.J. DeNiro (1985). Reconstruction of Prehistoric Plant Production and Cooking Practices by a New Isotopic Method. *Nature* **315**: 489-491.

**Johannessen**, S. (1984). Paleoethnobotany. In C.J. Bareis and J.W. Porter (eds.). *American Bottom Archaeology*. Urbana, Chicago: University of Illinois Press, pp. 197-214.

**Kelly**, J.E. (1990). *The Emergence of Mississippian Culture in the American Bottom Region*. Washington: Smithsonian Institution Press.

**Kelly**, L.S. (1997). Patterns of Faunal Exploitation at Cahokia. In T.R. Pauketat and T.E. Emerson (eds.). *Cahokia. Domination and Ideology in the Mississippian World*. Lincoln: University of Nebraska Press, pp. 69-88.

**Kidder**, T.R. (1992). Timing and Consequences of the Introduction of Maize Agriculture in the Lower Mississippi Valley. *North American Archaeologist* **13**: 15-41.

**Kruchten**, J.D. (2000). *Early Cahokian Fluidity on the Fringe. Pfeffer Mounds and the Richland Complex*. Paper presented at the 57th Annual Southeastern Archaeology Conference, Macon, Georgia.

**Lopinot**, N.H. (1992). Spatial and Temporal Variability in Mississippian Subsistence. The Archaeobotanical Record. In W.I. Woods (ed.). *Late Prehistoric Agriculture. Studies in Illinois Archaeology Number 8*. Springfield: Illinois Historic Preservation Agency: pp. 44-94.

**Lopinot**, N.H. (1994). A New Crop of Data on the Cahokian Polity. In W.I. Green (ed.). *Agricultural Origins and Development in the Midcontinent. Office of the State Archaeologist Report 19*. Iowa City: University of Iowa, pp. 127-153.

**Lopinot**, N.H. (1997). Cahokian Food Production Reconsidered. In T.R. Pauketat and T.E. Emerson (eds.). *Cahokia. Domination and Ideology in the Mississippian World*. Lincoln: University of Nebraska Press, pp. 52-68.

**Lovis**, W.A. (1990). Curatorial Considerations for Systematic Research Collections. AMS Dating a Curated Ceramic Assemblage. *American Antiquity* **55**: 382-387.

**Marino**, B.D. and M.J. DeNiro (1987). Isotopic Analysis of Archaeobotanicals to Reconstruct Past Climates. Effects of Activities Associated with Food Preparation on Carbon, Hydrogen and Oxygen Isotope Ratios of Plant Cellulose. *Journal of Archaeological Science* **14**: 537-548.

**Morton**, J.D. and H.P. Schwarcz (1988). Stable Isotope Analysis of Food Residue from Ontario Ceramics. In R.M. Farquhar, R.G.V. Hancock and L.A. Pavlish (eds.). *Proceedings of 26th International Archaeometry Symposium*. Toronto: University of Toronto Press, pp. 89-93.

**Morton**, J.D. and H.P. Schwarcz (2004). Paleodietary Implications from Stable Isotopic Analysis of Prehistoric Ontario Ceramics. *Journal of Archaeological Science* **31**: 503-517.

**Pauketat**, T.R. (1994). *The Ascent of Chiefs. Cahokia and Mississippian Politics in Native North America.* Tuscaloosa, London: University of Alabama Press.

**Pauketat**, T.R. (1997). Cahokian Political Economy. InTimothy R. Pauketat and Thomas E. Emerson (eds.). *Cahokia. Domination and Ideology in the Mississippian World.* Lincoln: University of Nebraska Press, pp. 30-52.

**Pauketat**, T.R. (2000). The Tragedy of the Commoners. In M.-A. Dobres and J.E. Robb (eds.). *Agency in Archaeology.* London: Routledge, pp. 113-128.

**Pauketat**, T.R. (2003). Resettled Farmers and the Making of a Mississippian Polity. *American Antiquity* **68**: 39-66.

**Pauketat**, T.R. and S. Alt (2004). The Making and Meaning of a Mississippian Axe Head Cache. *Antiquity* **78**: 779-797.

**Pauketat**, T.R. and T.E. Emerson (1991). The Ideology of Authority and the Power of the Pot. *American Anthropologist* **93**: 919-941.

**Pauketat**, T.R. and T.E. Emerson (1997). Introduction. Domination and Ideology in the Mississippian World. In T.R. Pauketat and T.E. Emerson (eds.). *Cahokia. Domination and Ideology in the Mississippian World.* Lincoln: University of Nebraska Press, pp. 1-29.

**Pauketat**, T.R., L.S. Kelly, G.J. Fritz, N.H. Lopinot, S. Elias and E. Hargrave (2002). The Residues of Feasting and Public Ritual at Early Cahokia. *American Antiquity* **67**: 257-279.

**Pauketat**, T.R. and N.H. Lopinot (1997). Cahokian Population Dynamics. In T.R. Pauketat and T.E. Emerson (eds.). *Cahokia. Domination and Ideology in the Mississippian World.* Lincoln: University of Nebraska Press, pp. 103-123.

**Reber**, E.A., S.N. Dudd, N.J. van der Merwe and R.P. Evershed (2004). Direct Detection of Maize in Pottery Residues via Compound Specific Stable Carbon Isotope Analysis. *Antiquity* **78**: 682-691.

**Reber**, E.A. and R.P. Evershed (2004a). How did Mississippians Prepare Maize? The Application of Compound-Specific Carbon Isotope Analysis to Absorbed Pottery Residues from Several Mississippi Valley Sites. *Archaeometry* **46**: 19-33.

**Reber**, E.A. and R.P. Evershed (2004b). Identification of Maize in Absorbed Organic Residues. A Cautionary Tale. *Journal of Archaeological Science* **31**: 399-410.

**Schoeninger**, M.J. and M.J. DeNiro (1984). Nitrogen and Carbon Isotopic Composition of Bone Collagen from Marine and Terrestrial Animals. *Geochimica et Cosmochimica Acta* **48**: 625-639.

**Skibo**, J.M. (1992). *Pottery Function. A Use-Alteration Perspective.* New York: Plenum Press.

**Tieszen**, L.L. and T. Fagre (1993a). Effect of Diet Quality on the Isotopic Composition of Respiratory $CO_2$, Bone Collagen, Bioapatite and Soft Tissues. In J. Lambert and G. Grupe (eds.). *Prehistoric Human Bone. Archaeology at the Molecular Level.* Berlin: Springer Verlag, pp. 121-155.

**Tieszen**, L.L. and T. Fagre (1993b). Carbon Isotopic Variability in Modern and Archaeological Maize. *Journal of Archaeological Science* **20**: 25-40.

**Van der Merwe**, N.J. (1982). Carbon Isotopes, Photosynthesis, and Archaeology. *American Scientist* **70**: 596-606.

**Van der Merwe**, N.J. (2005). $CO_2$, Grasses and Human Evolution. In J.R. Ehleringer, T. Cerling and M.D. Dearing (eds.). *A History of Atmospheric $CO_2$ and its Effects on Plants, Animals, and Ecosystems.* New York: Springer, pp. 293-328.

**Whitney**, S.E. (1992). *A Stable Isotope Examination of Food Residues on Contemporary and Archaeological Pottery from Southern Somalia.* University of Georgia Ph.D. dissertation.

# CHAPTER SIXTEEN
## Results of Seven Methods for Organic Residue Analysis Applied to One Vessel with the Residue of a Known Foodstuff

**H. Barnard, S.H. Ambrose, D.E. Beehr, M.D. Forster, R.E. Lanehart, M.E. Malainey, R.E. Parr, M. Rider, C. Solazzo and R.M. Yohe II**

Hans Barnard, Research Associate; Cotsen Institute of Archaeology; University of California, Los Angeles; P.O.-box 951510; Los Angeles, CA 90025; USA; <wendrich@barnard.nl> to whom all correspondence should be directed and, in alphabetical order, Stanley Ambrose and Dana Beehr, Department of Anthropology, University of Illinois at Urbana-Champaign (USA); Marcus Forster, Department of Archaeological Sciences, University of Bradford (UK); Rheta Lanehart, Department of Anthropology, University of South Florida (USA); Mary Malainey, Department of Anthropology, Brandon University (Canada); Robert Parr and Robert Yohe, Laboratory of Archaeological Sciences, California State University, Bakersfield (USA); Micala Rider, Department of Anthropology, University of Arizona (USA) and Caroline Solazzo, Smithsonian Center for Materials Research and Education (now the Museum Conservation Institute, USA).
Support for stable isotope ratio mass spectrometry instrumentation at the Environmental Isotope Paleobiogeochemistry Laboratory was provided by the National Science Foundation (USA), grant SBR 98-71480. We would like to thank René Cappers, Alek Dooley, Jelmer Eerkens, Kym Faull, Mohsen Kamel, Wim van Neer, Willeke Wendrich and two anonymous reviewers for their help in preparing this chapter. A shorter version of this chapter was published in the *Journal of Archaeological Science 34*; 2007: pp. 28-37.

On Thursday 31 March 2005 we organized a symposium entitled 'Theory and Practice of Archaeological Residue Analysis'. With the assistance of Dr. Jelmer Eerkens (University of California, Davis) and Dr. Ran Boytner (University of California, Los Angeles), this symposium took place in Salt Lake City, during the 70th Annual Meeting of the Society for American Archaeology as sessions 21 and 46.[1] The participants were asked to prepare presentations on their own research, and to report on their analysis of a residue recently cooked in a new ceramic vessel. Eleven agreed to partake in the latter and were sent a segment of the wall of a vessel in which camel milk had been cooked. The accompanying letter and model report sheet did not provide information on what was prepared in the vessel, nor did it present a list of possibilities such as in Table 1. This made the analysis more challenging than is typical of most archaeological settings as the provenance, shape and date of a pot usually offer important clues to its former use and contents. Seven participants (almost 2/3) in this 'Round robin' managed to file a report before the meeting where the source of the residue was revealed. These individual reports will be briefly discussed here, preceded by some data on camel milk and followed by a discussion on the significance of our findings for the practice of archaeological residue analysis.

### Birth of the 'Round Robin'

Within the framework of a research project to investigate possible uses of Eastern Desert Ware, a corpus of 4th-6th century CE vessels found in Southeast Egypt and Northeast Sudan, one of the authors purchased two dozen new unglazed ceramic vessels in Luxor, Egypt, in November 2003. The chosen vessels were cheap porous bowls known as 'tagen' (طاجن), with a diameter of about 12 cm and a height of 6 cm, made of Nile 'silt' (غرين) by professional potters. Such vessels are frequently used, after first being 'seasoned' (made less porous) with oil or honey, to prepare stews with fish or meat in an oven.

In the following weeks, 22 foodstuffs that could have been encountered in Eastern Desert Ware were collected and prepared in the new vessels that were left unseasoned. Organic residues from these pots were to be extracted and compared with residues found in the ancient vessels. Two pots used in an experiment on meat preservation by Dr. Salima Ikram (American University in Cairo) were added to the series while one listed foodstuff (Table 1), camel milk, could not readily be found. In December 2003, about 200 g of each available foodstuff was introduced in its own marked vessel. This was then topped off with Hayat mineral water (from Siwa Oasis, Total Dissolved Solids: 190 mg/l), wrapped in aluminum foil and allowed to sit at room temperature for 24 hours (Figure 1). The next day the whole assemblage was put in a gas oven heated to approximately 200 °C and cooked for one hour. After cooling for four hours, the contents of the vessel were checked and water was added when necessary. Each vessel was then again wrapped and cooked for another hour, after which the assemblage was allowed to sit for 24 hours at room temperature. The vessels were then emptied, rinsed with cold water and air-dried for 10 days. Finally, the vessels were rinsed to remove the fungus that had developed in many, dried for 24 hours, broken and stored in sealed plastic bags. About a month later, mid-January 2004, small parts of the vessels were

[1] The programs, abstracts and a preliminary report of these can be found at the project's website http://www.archbase.org/residue/ (Barnard et al. 2007).

ground into a fine powder to facilitate extraction of the organic residues. Around the same time camel milk was finally acquired at the camel market in Daraw (southern Egypt). This was processed, in Aswan, in the same way as the other foodstuffs, with the exception that Baraka mineral water was used (بركة from Kafr el-Arbein, TDS: 430 mg/l). Due to time constraints, however, the whole vessel was transported to California for analysis, along with small amounts of powder of the other vessels. This quirk made it available for the exercise, first suggested by Dr. Ernestine Elster (University of California, Los Angeles) and christened 'Round robin' by Dr. Richard Evershed (University of Bristol, UK), which is the basis of this chapter. In May 2004 the vessel was machine-cut into 12 pieces, one being the base (Figure 2), after which the 11 wall fragments were distributed among the participants of our blind 'Round robin'.

| No. | Synonyms | Provenance |
|---|---|---|
| **1:** beans | *Vicia fava* - فول مجروش | Zam-zam No. 1 dried beans |
| **2:** barley | *Hordeum vulgare* - شعير | farmer in Luxor |
| **3:** wheat | *Triticum aestivum* - قمح | Isis organic wheat |
| **4:** sorghum | *Sorghum bicolor* - درة | farmer in Luxor |
| **5:** dates | *Phoenix dactylifera* - بلح عجوة | Isis organic dates |
| **6:** doam nuts | *Hyphaene thebaica* - دوم | Khan el-Khalili market |
| **7:** nabq fruits | *Ziziphus spina-christi* - نبق | street vendor in Cairo |
| **8:** sugar dates | *Balanites aegyptiaca* - بلح السكر | Harraz herbs and spices |
| **9:** coffee | *Coffea arabica* - بن / قهوة | roastery in Cairo |
| **10:** goat milk | حليب ماعزي | farmer in Luxor |
| **11:** camel milk | حليب جملي | Daraw camel market |
| **12:** cow milk | حليب بقري (pasteurized) | Labanita full cream |
| **13:** egg | بيض | Isis organic eggs |
| **14:** mutton | رقبة ضاني | Metro supermarket (Cairo) |
| **15:** camel meat | لحم جمل | Zagazig camel market |
| **16:** pork | لحم خنزير | Christian butcher in Cairo |
| **17:** veal | رقبة بتلو | Metro supermarket (Cairo) |
| **18:** fish | *Plectropomus maculatus* - ناجل | Metro supermarket (Cairo) |
| **19:** olive oil | *Olea europaea* - زيت زيتون | Borges extra vigin olive oil |
| **20:** water | مياه (حياة) -TDS: 190 mg/l | Hayat mineral water (Siwa) |
| **21:** broccoli | *Brassica oleracea* var. *italica* | Metro supermarket (Cairo) |
| **22:** beef and mutton | --- | Dr. Salima Ikram (AUC) |
| **23:** beef in sheep fat | --- | Dr. Salima Ikram (AUC) |
| **24:** lentils | *Lens culinaris* - عدس بني | Isis organic brown lentils |
| **25:** lupine | *Lupinus albus* - ترمس | el-Doha lupins |

Table 1: List of the foodstuffs prepared in new Egyptian 'tagen' in December 2003 and January 2004 (TDS = Total dissolved solids; AUC = American University in Cairo).

Figure 1: The preparation of organic residues of selected known foodstuffs in new and unseasoned Egyptian 'tagen' (Cairo, December 2003 and January 2004).

Figure 2: Egyptian 'tagen' number 11, with a residue of camel milk, just after it was cut (Los Angeles, May 2004). The wall fragments were subsequently distributed among the participants of our blind 'Round robin'.

## Camels and Camel Milk

Millions in Asia and Africa rely on camels (the one-humped *Camelus dromedarius* and the two-humped *Camelus bactrianus*) for transportation, milk, meat and leather, especially in areas that are too hot and dry for other large mammals such as cattle and horses (Ingram and Mount 1975; Schmidt-Nielsen 1964). Camels were introduced in Egypt, long after being domesticated in Central Asia, by invading armies during the 7th or 6th century BCE, towards the end of the Pharaonic period (Midant-Reynes and Braunstein-Silvestre 1977; Rowley-Conwy 1988). As they are eminently suitable for the Egyptian climate, camels were quickly adopted by the civilian population and had spread to Meroe in Sudan by the first century BCE (Bulliet 1975; Epstein 1971; Wilson 1984). Today they are still used as beasts of burden, riding being mostly limited to the military and tourists, and valued for their meat and leather. In modern Egypt, camel milk is only consumed for 'medical' or 'magical' reasons. It is thought to cure infectious diseases, possess aphrodisiac powers and make the skin impenetrable to scorpion stings and snake bites (Barnard 2000; Mares 1954; Rao et al. 1970). In the past, camel milk may have been more commonly consumed in Egypt, as it is in many other places, which is why it was entered into the study on Eastern Desert Ware.

A healthy lactating camel produces 5-15 kg milk daily for a period of 9-15 months (Alshaikh and Salah 1994; El-Bahay, 1962; Knoess 1977). The milk has a relatively high pH (6.5-6.7) and high concentrations of vitamin C and niacin (Elamin and Wilcox 1992; Farah et al. 1992; Sawaya et al. 1984). The absolute and relative composition of the milk depends on the fodder and the stage of lactation (Zhang et al. 2005), like in other mammals, but also on the state of hydration of the animal. In camels the latter can vary greatly as part of their adaptation to life in an arid environment (Yagil et al. 1974). Among other things, lack of drinking water will cause an increase in the mineral content of the milk and a decrease of fat, lactose and protein (Gorban and Izzeldin 1997, 2001; Yagil and Etzion 1980a, b).

The fatty acid composition of the fats in camel milk is also rather variable (Figure 3). In fresh milk the fats are suspended in globules (micelles), about 300 μm in diameter, in which they appear to be bound to proteins (Farah and Ruegg 1989; Ramet 2001; Yagil 1982). This may explain why the usual processing methods to preserve it (like fermenting, churning or curdling) are less effective (Yagil et al. 1984). Concentration of the milk by evaporation is obviously still possible and several methods have been developed to produce butter and cheese, at least in part from camel milk. Fermented camel milk, which can be carbonated or mildly alcoholic, is also commonly produced (Farah et al. 1989; Ramet 2001; Rao et al. 1970; Yagil 1982).

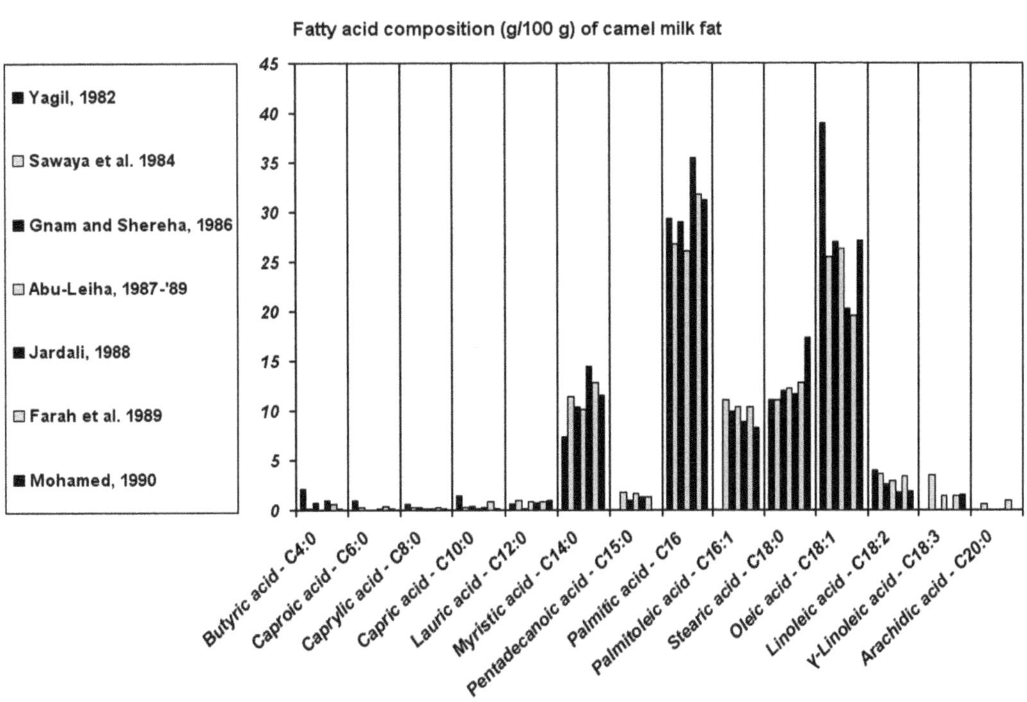


Figure 3: Comparison of seven fatty acid signatures of fresh camel milk (after Ramet 2001).

## The Analysis of Camel Milk Residues

The fatty acids found in the ceramic matrix of a vessel that was used to contain or process camel milk must have been those present in the milk or the products of oxidation, or other reactions, of these compounds. Their absolute and relative abundance will depend on many factors, including the affinity of the ceramic matrix for each of the various molecules, the stability of those molecules over time and the efficiency of the extraction and detection techniques to identify different molecules. Given what is known about the fatty acid signatures of fresh camel milk and organic residues typically recovered in archaeological contexts, the saturated fatty acids C16:0 (palmitic or hexadecanoic acid), C18:0 (stearic or octadecanoic acid) and C14:0 (myristic or tetradecanoic acid) were anticipated in the comparatively fresh residue of our blind 'Round robin'. Fair amounts of the mono-unsaturated C18:1 (oleic or octadecenoic acid) and C16:1 (palmitoleic or hexadecenoic acid), and their oxidation products (like dicarboxylic acids), were also expected (Figure 3). These fatty acids are common in fats and oils of vegetable and animal origin and their presence alone is unlikely to allow identification of an unknown residue such as that of camel milk. Making such an identification does require additional information and further manipulation of the data.

Another analytical technique, which could be used independently or in combination with the identification of specific compounds, entails the determination of the isotopic composition of organic residues, especially the ratios of the stable isotopes of carbon ($^{12}C$ and $^{13}C$) and nitrogen ($^{14}N$ and $^{15}N$, Appendix I). Families of plants have their own specific preference for one isotope over another and isotope ratios in living matter are therefore dissimilar to those in the environment. In the New World, for instance, maize (*Zea mays*) employs the $C_4$ pathway for photosynthesis, which causes higher $^{13}C/^{12}C$ ratios, while introduced European cereals employ the $C_3$ pathway, which produces lower $^{13}C/^{12}C$ ratios. This difference in $^{13}C/^{12}C$ ratios is carried up the food chain: herbivores feeding on $C_4$ plants, and carnivores feeding on such herbivores, will have higher $^{13}C/^{12}C$ ratios than those feeding on $C_3$ plants (Ambrose 1993; Morton and Schwarcz 2004). Egyptian camels typically feed on both C3 (most grasses and trees) and C4 plants (sorghum, sugar cane). Their milk contains 3-5 g protein (mostly casein) per 100 g (Abu-Leiha 1987; El-Bahay 1962; Knoess 1977; Ramet 2001; Sawaya et al. 1984) which should leave small amounts of nitrogen in its residue.

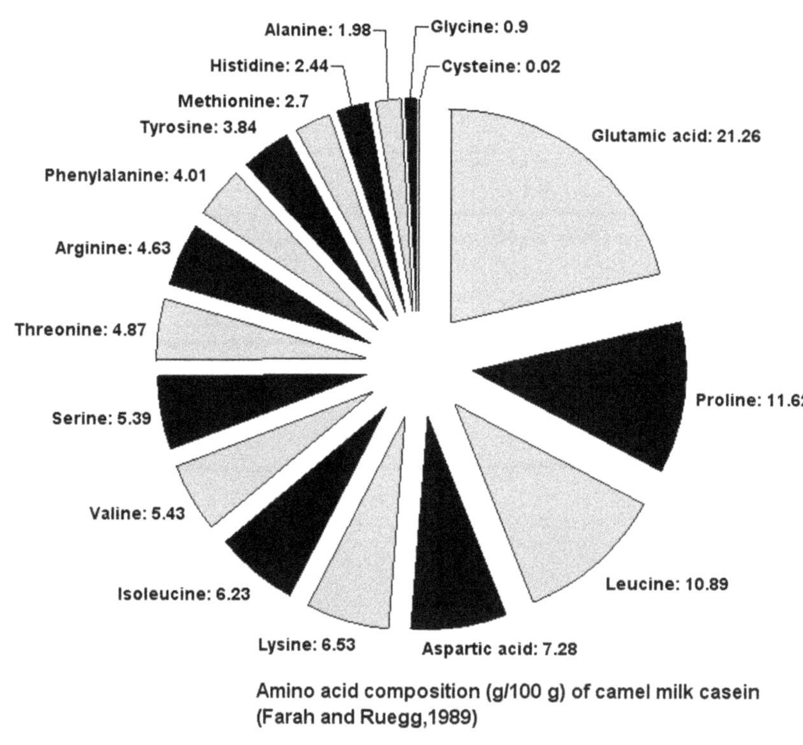


Figure 4: The amino acid signature of casein in fresh camel milk (after Farah and Ruegg, 1989).

The analysis of proteins such as casein, would potentially enable a secure identification of the source of most archaeological residues (Figure 4), as many proteins are species-specific (Craig and Collins 2002; Evershed and Tuross 1996, Appendix II). Ancient proteins have been isolated from archaeological samples with a variety of techniques (Allen et al. 1995; Craig and Collins 2000, 2002; Evershed and Tuross 1996; Kooyman et al. 1992, 2001). That this avenue has not yet been fully explored must be attributed to the facts that proteins are relatively difficult to handle in the laboratory and are expected to have denatured and fractured over time, especially after being heated. Furthermore, not all naturally occurring proteins have yet been fully analyzed, which would be necessary for comparison with unknown proteins (Fankhauser 1994, 1997). Specific antibodies are readily available for only a limited number of proteins and may not react with severely denatured or damaged proteins (Craig and Collins 2000). Counter or cross-over electrophoresis (CIEP), employing the principles of antigen-antibody reactions, has been applied as a method for identifying residual proteins on various archaeological materials in recent years (Allen et al. 1995; Henrikson et al. 1998; Kooyman et al. 1992; 2001; Shanks et al. 1999; Yohe et al. 1991), although some have expressed skepticism as to the reliability of this approach (Cattaneo et al. 1993; Downs and Lowenstein 1995; Eisele et al. 1995; Fiedel 1996). The rapidly evolving field of proteomics may soon develop new techniques that can also be used for archaeological residue analysis, especially with an increased cooperation between biochemists and archaeologists.

Several approaches to the analysis of ancient organic residues were represented in our blind 'Round robin' (Barnard et al. 2007; see Chapters 5 and 17 for more details on the analytical methods). Of the seven laboratories participating, one (Laboratory J) established stable isotope ratios of carbon and nitrogen; another (Laboratory A) identified proteins using antiserum; and the remaining five analyzed lipids, mostly fatty acids, using various methods to look for a recognizable signature. Laboratories that have in the past approached the subject through the analysis of starch residues (another relatively stable component of many foodstuffs), tartaric acid and calcium oxalate (present in beer and wine) were approached, but declined to participate.

### Results of the 'Round Robin': Stable Isotopes

Laboratory J Laboratory J removed the exposed surfaces of the sherd from the designated sampling area with a diamond abrasive rotary tool bit, to avoid possible contamination, after which a total of 12 samples were drilled with a ball-shaped diamond abrasive drill bit at evenly spaced intervals from the rim to the base of the vessel segment, six from the interior and six from the exterior. Aliquots of the ceramic powder, weighing an average of 4.9 mg, were manually compressed in tin foil capsules and combusted in a Carlo Erba elemental analyzer interfaced with a MAT 252 isotope ratio mass spectrometer (Table 2).

| Sample | Distance from rim | Sample weight | $\delta^{13}C$‰ | Weight %C | Volts mass 44 | $\delta^{15}N$‰ | Weight %N | Volts mass 28 | Atomic C/N |
|---|---|---|---|---|---|---|---|---|---|
| | | | | *Interior* | | | | | |
| **X11-131** | 3.79 | 4.770 | -25.11 | 3.15 | 2.648 | --- | --- | --- | --- |
| **X11-132** | 10.85 | 4.604 | -20.41 | 3.09 | 2.507 | --- | --- | --- | --- |
| **X11-133** | 16.17 | 4.898 | -22.20 | 2.87 | 2.477 | --- | --- | --- | --- |
| **X11-134** | 22.48 | 5.069 | -24.59 | 3.19 | 2.845 | 5.51 | 0.16 | 0.159 | 23.3 |
| **X11-135** | 30.13 | 4.784 | -22.80 | 2.34 | 1.965 | 4.21 | 0.19 | 0.183 | 14.4 |
| **X11-136** | 40.96 | 4.937 | -18.39 | 1.51 | 2.614 | 5.51 | 0.21 | 0.217 | 8.4 |
| | | | | *Exterior* | | | | | |
| **X11-137** | 4.14 | 4.744 | -21.57 | 1.61 | 1.340 | --- | --- | --- | -- |
| **X11-138** | 10.52 | 5.560 | -17.13 | 2.17 | 2.120 | --- | --- | --- | --- |
| **X11-139** | 16.19 | 4.611 | -20.17 | 1.56 | 1.255 | --- | --- | --- | --- |
| **X11-140** | 21.08 | 5.270 | -20.44 | 1.54 | 1.413 | --- | --- | --- | --- |
| **X11-141** | 29.20 | 5.127 | -17.89 | 1.49 | 1.300 | --- | --- | --- | --- |
| **X11-142** | 39.26 | 4.799 | -17.51 | 1,50 | 1.251 | --- | --- | --- | --- |

Table 2: Measured carbon and nitrogen isotopic and elemental concentrations. Missing values occur where nitrogen concentrations were too low for detection (the detection threshold is 0.1 volts on the major ion beam).

| Sample | Distance from rim | Sample weight | $\delta^{13}C$‰ | Weight %C | Volts mass 44 | $\delta^{15}N$‰ | Weight %N | Volts mass 28 | Atomic C/N |
|---|---|---|---|---|---|---|---|---|---|
| X11-131 | 3.79 | 10.501 | -24.75 | 3.16 | 5.990 | 6.24 | 0.06 | 0.130 | 61.4 |
| X11-132 | 10.85 | 10.531 | -19.97 | 3.20 | 6.074 | 3.80 | 0.05 | 0.101 | 74.7 |
| X11-133 | 16.17 | 10.374 | --- | 2.73 | --- | 5.34 | 0.06 | 0.127 | 53.0 |
| X11-134 | 22.48 | 10.528 | -24.41 | 3.45 | 6.518 | 6.08 | 0.15 | 0.349 | 26.8 |
| X11-135 | 30.13 | 10.263 | -22.58 | 2.43 | 4.514 | 5.72 | 0.18 | 0.399 | 15.8 |
| X11-136 | 40.96 | 10.463 | -18.29 | 1.59 | 3.005 | 6.20 | 0.20 | 0.469 | 9.3 |

Table 3: Measured carbon and nitrogen isotopic and elemental concentrations at six positions below the interior rim. The carbon concentration of sample X11-133 was not measured, due to an instrument error, so a weight was extrapolated from the smaller sample listed in Table 2.

Carbon concentrations averaged 2.69% on the vessel interior and 1.65% on the exterior. Vessel interior $\delta^{13}C$ values were systematically more negative than exterior values. The average $\delta^{13}C$ for the interior was -22.3‰, that for the exterior was -19.2‰. Interior and exterior $\delta^{13}C$ values varied greatly with the distance from the rim, first increasing by almost 5‰, then decreasing by 4‰ and then systematically increasing toward the base, with the exception of the lowest sample pair (Figure 5). Interior and exterior $\delta^{13}C$ values co-varied systematically with the distance below the rim (Figure 6).

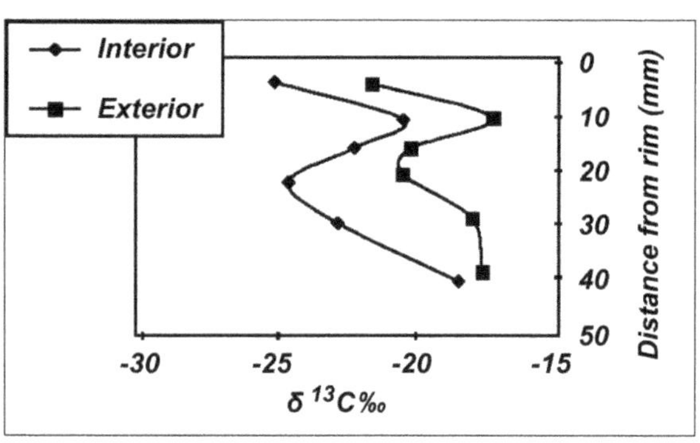


Figure 5: Variation in interior and exterior $\delta^{13}C$ values with distance below the rim.

Nitrogen concentrations were too low for detection in all vessel exterior samples and in the three interior samples closest to the rim. The amounts of nitrogen in the three lowest interior samples (>22 mm below the rim) increased systematically from 0.16 to 0.21% with the distance below the rim. Their average $\delta^{15}N$ value was 5.08‰, but these nitrogen samples were too small for accurate isotopic analysis. Therefore, the sample weights were doubled to an average of 10.4 mg to obtain better estimates of nitrogen concentrations and more reliable $\delta^{15}N$ values (Table 3). Nitrogen concentrations were 0.05 to 0.06% in the three samples closest to the rim and increased systematically from 0.05% to 0.20% between 16 and 41 mm below the rim. Atomic C/N ratios increased slightly and then decreased below the rim. The three lowest samples contained enough nitrogen for reliable isotopic analysis, their $\delta^{15}N$ values averaging 6.00‰.

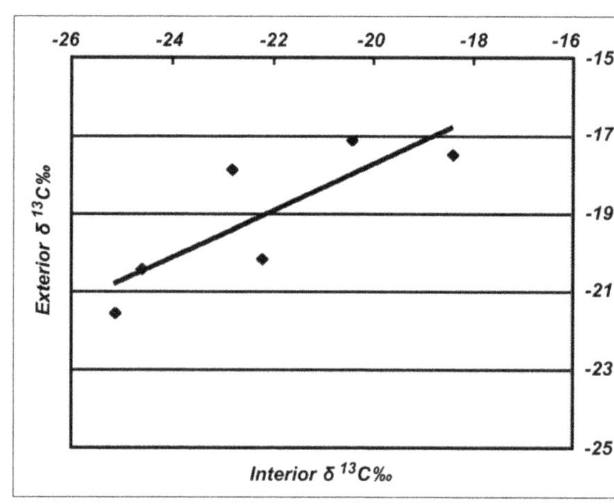


Figure 6: Covariance of interior and exterior $\delta^{13}C$ values ($y = 1.138 x - 0.497$ $R^2$ - 0.679).

The principles of archaeological residue analysis and diet reconstruction using stable carbon and nitrogen isotope ratios are discussed in detail elsewhere (Ambrose 1993; Morton and Schwarcz 2004; Reber and Evershed 2004). The co-variance of interior and exterior $\delta^{13}C$ values suggests that the vessel contents were absorbed throughout the thickness of the vessel, with the exception of the exterior sample closest to the base, which did not increase in step with its interior pair. The difference in $\delta^{13}C$ between the interior and exterior suggests that the vessel's contents had a lower $\delta^{13}C$ value than that of the fuel used to fire the vessel, or to heat it during cooking. Had the vessel been 'seasoned', then this pattern could also be due to application of a substance, to either the interior or the exterior of the vessel, with a $\delta^{13}C$ value that differed from that of the fuel. Carbon in the clay and temper may also have remained in the matrix after firing.

The distinct deflection to the most negative $\delta^{13}C$ values at 21-22 mm below the rim may indicate the 'boil line'. This hypothesis is based on the fact that the $\delta^{13}C$ value of fats are approximately 5‰ more negative than those of protein and carbohydrates in foods (DeNiro and Epstein 1977). If the food was cooked in a liquid, isotopically lighter fats would have floated to the surface and be absorbed at higher levels than heavier foodstuffs. The high C/N ratios near the rim are consistent with this hypothesis as fats have high carbon contents and lack nitrogen. The increase in nitrogen and decrease in atomic C/N ratios beginning 22 mm below the rim indicates that protein-rich food components were absorbed in lower parts of the vessel (figures 7 and 8).

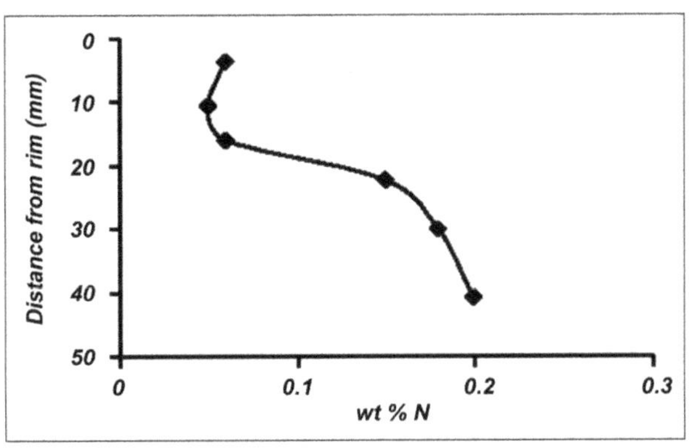


Figure 7: Change in nitrogen concentrations of the absorbed residue in powdered samples from six positions below the interior of the vessel rim.

The increase in $\delta^{13}C$ values between the rim and 10 mm below the rim is unlikely to be due to the isotopic composition of the food cooked in the vessel. Perhaps the vessel was fired upside down, with the rim in contact with fuel of $C_3$ origin. Alternatively, if the pot was seasoned with a $^{13}C$-enriched substance, the rim was not treated. Increasing $\delta^{13}C$ values beneath the boil line suggest that a mixture of $C_3$ and $C_4$ foodstuffs or proteins of a mixed-feeding animal were cooked in this pot. The $\delta^{15}N$ values are high enough to suggest the presence of meat, milk or blood of an herbivore, perhaps combined with plant foods. The C/N ratios of the three samples closest to the vessel base range from 9 to 27. These are higher than those of pure proteins, but are substantially lower than those of pure carbohydrates and lipids, thus indicating a mix of proteins and carbohydrates. As only one foodstuff was cooked in this vessel it was likely either a protein-rich seed, or a whole milk product. Moderately high $\delta^{13}C$ and $\delta^{15}N$ values, that lie between those of $C_3$ and $C_4$ plants, exclude derivation from a single plant food species.

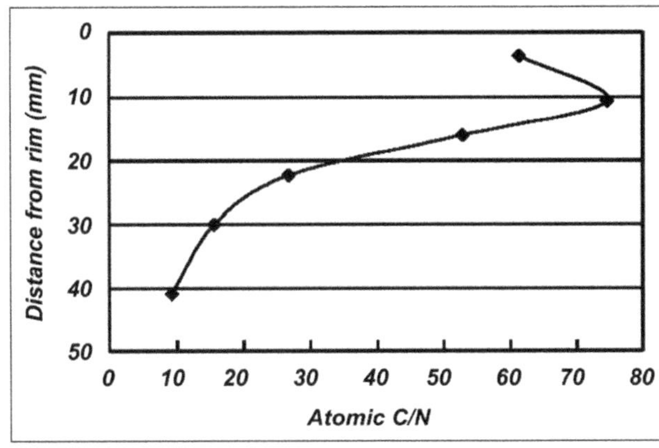


Figure 8: Change in atomic C/N ratios of the absorbed residue in powdered samples from six positions below the interior of the vessel rim.

The strong co-variance of the interior and exterior $\delta^{13}C$ values of the five samples closest to the rim suggests that the upper residues were readily absorbed in the ceramic matrix. It may, however, not have had the same capacity for absorption of higher density proteins or the carbohydrates that predominated near the vessel base. The systematic patterns of carbon and nitrogen concentrations and isotopic composition of the interior and exterior of this vessel were entirely unanticipated. However, these patterns are consistent with the preparation of a single foodstuff. Variation in carbon and nitrogen isotopic and elemental composition within a vessel used to cook a single substance can be substantial due to systematic differences in the composition of lipids, proteins and carbohydrates. Control over the sample position within the vessel is therefore essential for the adequate interpretation of carbon and nitrogen elemental and isotopic variation of absorbed organic matter in potsherds. The simple sampling strategy described here, combined with analysis of carbon and nitrogen elemental and isotopic composition, can provide a considerable amount of useful information about the foods prepared in porous ceramic vessels.

## Results of the 'Round Robin': Proteins

Laboratory A placed the sherd in a plastic dish with 0.5 ml ammonium hydroxide (5%) which was sonicated for 5 min. After this, both dish and contents were placed in a rotating mixer for 30 min. The resulting solution was transferred into a sterile plastic vial and stored at -20°C. About 3 µl of this ammonia solution was later transferred into a well punched in an agarose gel in a

Helena Laboratories Titan Gel electrophoresis chamber and paired with a second well with 3 μl antiserum. A control positive was prepared in another pair of wells after which an AC electric current of 105 V was passed through the gel for 45 min. This caused sample and antiserum to migrate, antigens towards the anode and antibodies (antiserum) towards the cathode, and come into contact. If there is protein in the sample which corresponds with the antiserum, an antigen-antibody reaction will occur resulting in the protein precipiting out in a specific pattern. After the run was completed the gel was pressed and dried. The dry gel was immersed in a Coomassie Blue R250 stain for 3 min. and then destained in a solution of ethanol, distilled water and acetic acid (5:5:1, v/v) until the background was clear and any positive responses visible. Sterile equipment was used throughout the analysis.

The antisera used in this study included agave, amaranth, bear, bovine, cactus, cat, cedar, chicken, *Capparidaceae*, *Chenopodiaceae*, *Compositae*, deer, dog, guinea pig, *Graminae*, legume, pine, rabbit, rat and sheep. All the animal antisera are from Cappel, purchased from MP Biomedicals, plant antisera were produced by the Department of Biological Sciences at the University of Calgary. All antisera are polyclonal: they recognize epitopes of closely related species. For example, anti-deer serum will react positively to other members of the family *Cervidae* such as elk, moose and caribou. Commercial antisera manufacturers provide product sheets listing cross reactions, the antisera from the University of Calgary are tested against a full suite of other species to ensure that there are no cross reactions.

This method of analysis is known as cross-over (or counter) electrophoresis (CIEP). Prior to the introduction of DNA fingerprinting this technique was commonly used in forensic laboratories to identify residues from crime scenes. Minor adaptations to the original method were made following procedures used by the Royal Canadian Mounted Police Serology Laboratory, Ottawa (1983) and the Centre for Forensic Sciences, Toronto. The solution used to remove residues, ammonium hydroxide, has proven to be the most effective extractant for old and denatured bloodstains without interfering with subsequent testing (Dorrill and Whitehead 1979; Kind and Cleevely 1969). There were no trace proteins detected in the sherd sent for analysis. This may be the result of the degradation of animal proteins that result from cooking. Experiments subsequent to this study have demonstrated that animal proteins, of known origin, heated to 100°C for 20 min yielded no positive readings using CIEP. This observation has important implications for the application of this particular form of residue analysis on cooking vessels or objects from within hearth features. Another issue to consider is that many of the taxonomic groups for which antisera currently exist, like those used in this study, would not be expected to react to the plant and animal residues found in this experimental sample (possible exceptions being bovine and sheep).

### Results of the 'Round Robin': Lipids

Laboratory B ground a small piece of the sherd in a mortar and pestle. The resulting powder was transferred into a sterile test tube with 8 ml acetonitrile (ACN). The mixture was sonicated and centrifuged, after which 5 ml was pipetted into a second test tube. The solvent was evaporated from this test tube by gentle heating under a stream of nitrogen. The dry residue was taken up in 100 μl ACN and treated with 30 μl trimethylsilyl-N-methyl-trifluoroacetamide (MSTFA). About 1 μl of the derivatized sample was sandwiched in air and injected, proceeded by 1 μl of ACN, into a Varian GC/MS instrument.

Two different pieces of the same sherd were analyzed, using the method of comparing the C18:0/C16:0 and C18:1/C16:0 ratios. One sample showed a C18:0/C16:0 ratio of 0.72 and a C18:1/C16:0 ratio of 1.15 (Figure 9). The second sample showed ratios of 1.03 and 0.32, respectively. Fatty acid amounts in known foodstuffs were obtained from the USDA Nutrient Data Laboratory website and used as a comparative data set. The peaks of all the fatty acids tested for (C12:0, C14:0, C16:0, C16:1, C18:0, C18:1, C18:2 and C20:0) were also compared with those in the known foodstuffs. The only peak that differed dramatically between the two spectra was the peak for C18:1 (oleic acid). This may be due to the oxidation of this unsaturated fatty acid, which may have also casued the difference in ratios between the two pieces tested. The chromatogram containing the higher amount of C18:1 was compared with the data of known foodstuffs. The results were tentatively interpreted as the residue of some animal product and most likely veal, eggs or goat milk.

Laboratory C received the sherd broken into two vertical strips. The upper portion of one vertical strip was selected as the 'time 0' sherd. The remainders of the pieces were stored in an oven at 75°C. After six days the lower portion of the first strip was removed from the oven and stored in a freezer at -20°C. The same was done, after 12 days, with the upper portion of the second strip. Residues were then extracted from the 'time 0' sherd and from those stored in the oven for six and 12 days. Contaminants were removed by grinding off the surfaces after which the sherd was crushed. The resulting powder was mixed with 30 ml of a chloroform and methanol mixture (2:1, v/v) and sonicated (2 x 10 min). Solids were removed by filtering into a separatory funnel. The solvent mixture was washed with 16 ml ultra-pure water and left until it separated into two phases.

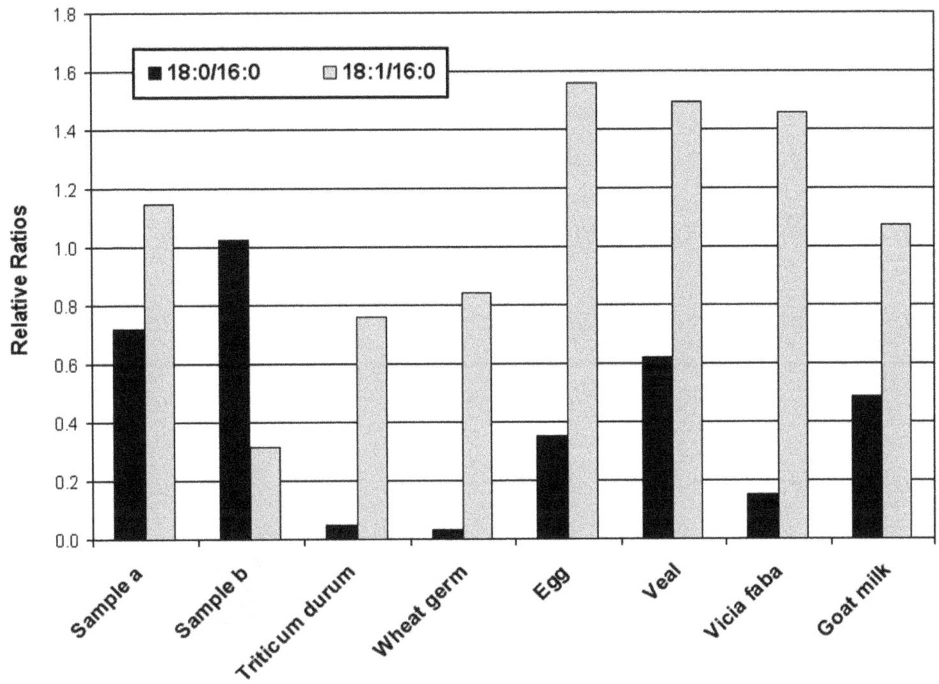


Figure 9: Graphic representation of the C18:0/C16:0 and C18:1/C16:0 ratios in two samples of the 'Round robin', established in Laboratory B, compared to those in six foodstuffs.

The lower chloroform-lipid phase was transferred into a flask from which the chloroform was removed by rotary evaporation. Any remaining water was removed by evaporation with 1.5 ml benzene. The dry residue was transferred into a vial with 1.5 ml of the chloroform and methanol mixture and stored. A 200 µl sample of the solution was dried under nitrogen and treated with 6 ml 0.5 N anhydrous hydrochloric acid in methanol. After cooling 4 ml ultra-pure water was added. The fatty acid methyl esters (FAMEs) were recovered with 3 ml petroleum ether and transferred into a vial. The solvent was removed by heat under a gentle stream of nitrogen. The dry residue was transferred into a GC vial with 1 ml iso-octane.

Analysis was performed on a Varian 3800 gas chromatograph fitted with a flame ionization detector. Chromatogram peaks were integrated using Varian Star Chromatography Workstation software and identified through comparisons with several external qualitative standards (NuChek Prep, Elysian, MN). To identify the residue the relative percentage composition was determined first with respect to all fatty acids present in the sample, and second with respect to the ten fatty acids utilized in the identification criteria: C12:0; C14:0; C15:0; C16:0; C16:1; C17:0; C18:0; C18:1ω9; C18:1ω11 and C18:2. Medium chain fatty acids represent the sum of C12:0, C14:0 and C15:0; while C18:1 is the sum of all isomers. It must be understood that the identifications given do not necessarily mean that those foods were actually prepared because different foods of similar fatty acid composition and lipid content can produce similar residues.

Significantly more fatty acids were recovered from the upper portions than of the lower portions of the sherd and, consequently, these provide the best information about the residue. The characterization is based on the relative fatty acid composition of the residue extracted from the upper portion of the second strip. Due to the low level of unsaturated fatty acids in the residues, it is possible to establish decomposition trends after only 12 days of oven storage. The sum of medium chain fatty acids exceeds 30%, while the C18:0 and C18:1 isomer levels are low. By applying published criteria it is possible to say that the partially decomposed residue is typical of decomposed plants (Malainey et al. 1999b, 2001). This category includes low fat-content plants such as roots or tubers, greens as well as certain berries and seeds. North American foods known to produce residues with high levels of C14:0 when boiled are biscuit root and *Chenopodium* seeds.

Laboratory G ground off about 0.1 g of the internal surface of the sherd was ground off, mixed the resulting powder with 1 ml of a chloroform and methanol mixture (2:1, v/v), sonicated and centrifuged. The supernatant was decanted and dried under a stream of nitrogen. The dry residue was treated with 30 µl N,O-bis(trimethylsilyl)trifluoro-acetamide (BSTFA) and about 1 µl of the derivatized sample was injected into a Hewlett Packard 5972 CG/MS. Next to a trace of C12:0, the sample appeared to contain a number of free fatty acids: C14:0, C16:0, C18:0 and C18:1 as well as the methyl-esters of C16:0 and C18:0 (Figure 10). Neither short chain fatty acids nor poly-unsaturated fatty acids were identified. No attempt was made to determine the ratios of the free fatty acids but C16:0 appeared predominant. There were traces of odd carbon number fatty acids, but these were in low abundance. Cholesterol was present in abundance along with cholesterol oxidation products and a trace of squalene. No phytosterols or alkanes were found. No terpenoids and no wax esters were identified, indicating that the source most likely did not contain a resin or a wax.

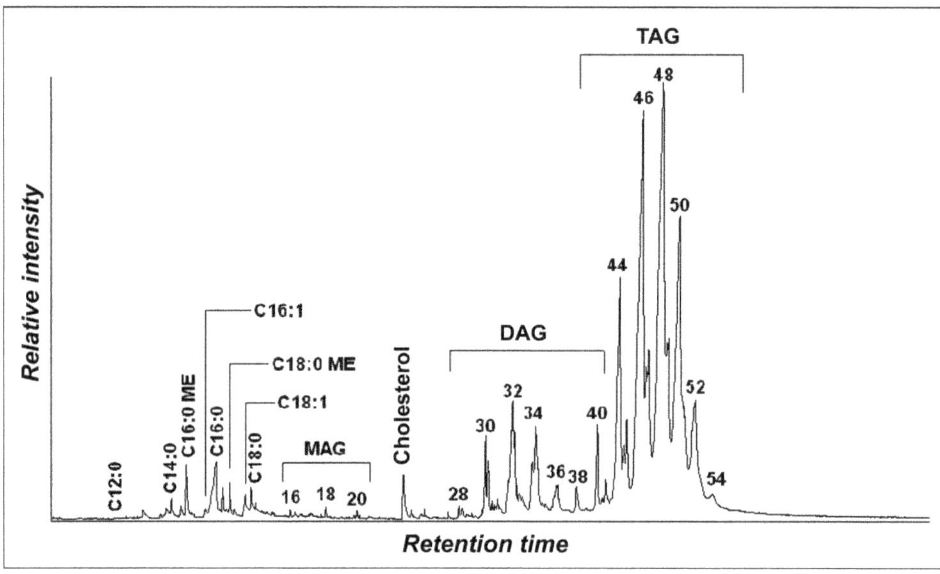


Figure 10: Chromatogram showing the lipid profile of the 'Round robin' as established by Laboratory G. ME: methyl-ester; MAG: monoacylglycerols; DAG: diacylglycerols; TAG: triacylglycerols (cf. Figure 11).

On the basis of these results it was suggested that the residue is of animal origin, the absence of short chain fatty acids pointing to a non-dairy source like veal or eggs. This identification should be confirmed by saponification of the sample, hydrolysis of the acylglycerols resulting in glycerol and a fatty acid salt, followed by an analysis of the result by gas chromatography combustion isotope ratio mass spectrometry (GC/C/IRMS) as discussed by Laboratory J above.

Laboratory H ground the sherd in a mortar and pestle. Lipids were extracted in a chloroform and methanol mixture by sonification. This was centrifuged and decanted after which the solvents were evaporated under nitrogen. The dry residue was treated with BSFTA, saponified with sodium methoxide in methanol and methylated with a $BF_3$-methanol complex. The sample was then analyzed in a GC-17A Shimadzu Chrom-Perfect gas chromatograph, with a 30 m Agilent DB-5 column raised 5°C/min from 40°C to 280°C, after a split injection. No activity was found in the received sherd.

Laboratory K cut the sherd into four pieces, two of which were crushed into a fine powder using a mortar and pestle. Lipids were extracted by sonification from 500 mg of the powder in 2 ml chloroform/methanol (2:1, v/v) after which the solvents were evaporated under a stream of nitrogen. 50 µl tetratriacontane ($CH_3$-$(CH_2)_{32}$-$CH_3$) was added as internal standard and the dry residue was then treated with BSTFA with 1% trimethyl-chlorosilane (TMCS). The lipids in another 500 mg were saponified with 0.5 ml 2N KOH, acidified with HCl and extracted with diethyl-ether. The solvent was then evaporated under a stream of nitrogen and treated with BSTFA with 1% TMCS after the internal standard was added. The derivitized residues were analyzed in a Hewlett Packard 5890 series II gas chromatograph (HTGC) and a Hewlett Packard 5890 gas chromatograph with a 5972 mass selective detector (GC/MS). The HTGC was used with a DB-1HT fused-silica capillary column (15 m x 0.32 mm, 0.10 µm). The HTGC oven temperature was initially set at 50°C and increased with 10°C/min to 380°C, at which temperature it was held for another 5 min. The GC/MS was equipped with a fused-silica capillary column (30 m x 0.25 mm,

0.25 µm). The GC oven temperature was programmed to start at 50°C, to increase with 10°C/min to 340°C and finally to be held at 340°C for 10 min.

The analysis of the total lipid extract, by HTGC, showed the presence of triacylglycerols with a total carbon number between 44 and 56, the main constituent having 52. After saponification and GC/MS, the main fatty acids appeared to be C12:0, C14:0, C16:1, C16:0, C18:1 and C18:0. The C16:0/C18:0 ratio (P/S ratio) for one part of the sherd was 3.175 by HTGC, and 5.741 by GC/MS. For the other part of the sherd the C16:0/C18:0 ratio was 3.09 by HTGC. The distribution of triacylglycerols indicates a residue of animal origin, a C16:0/C18:0 ratio of around 3 may be explained by the presence of egg. It was therefore concluded that an animal product was the source of the residue.

Analysis of samples taken eight months later, using another HTGC (an Agilent Technologies 6890N) with the same column, showed some interesting differences even though the sherd had been stored in a refrigerator (Figure 11). All fatty acids except C18:0 had decreased, while the most common triacylglycerol now had a total carbon number of 48 instead of 52. It seems that triacylglycerols containing mainly C18:0 had hydrolyzed to form free C18:0 and monoacylglycerol C18:0. The new C16:0/C18:0 ratio obtained after saponification appeared very low compared to the first ratio, which is concordant with an increase in C18:0. Other ratios, like C16:1/C18:1 and C16:0/C18:1, showed little variation after eight months. It therefore seems very likely that C18:0 was released, probably because of the temperature changes to which the sherd was exposed.

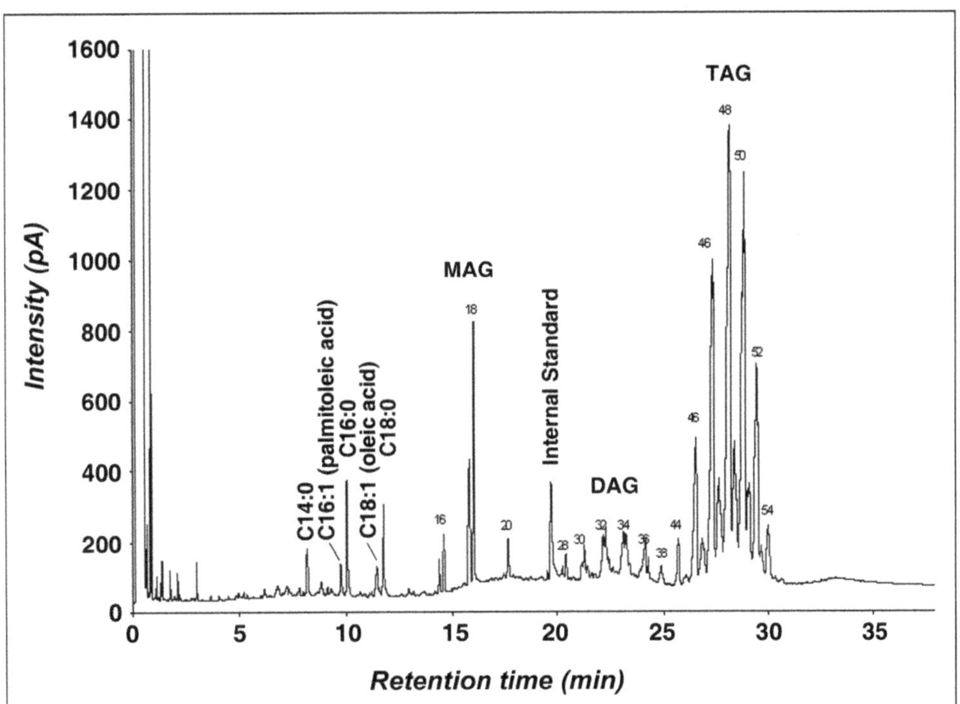


Figure 11: Chromatogram showing the lipid profile of the 'Round robin' after eight months of refrigerated storage as established by Laboratory K. The internal standard is tetratriacontane; MAG: monoacylglycerols; DAG: diacylglycerols; TAG: triacylglycerols (cf. Figure 10).

## Discussion

The first conclusion of our blind 'Round robin' must obviously be that none of us could pinpoint the source of the residue. This may seem disappointing but should be hardly surprising as our methods still only allow us to look at a small selection of molecules and not directly at foodstuffs themselves. In the reduction from camel milk to fatty acids too much information is lost and it is impossible to retrieve all of this. The driving motive behind our 'Round robin' was to get an overview of the methods currently available to recover some of this information in order to arrive at archaeologically relevant conclusions. We purposefully include the negative results of two of the laboratories as the trend to only publish successful projects, though fully understandable, has contributed to a biased representation of the state of the field and unrealistic expectations.

The second conclusion is that, with the current technology, it will only be possible in certain cases to assign archaeological residues a source more specific than 'meat of an herbivore' or 'low fat-content plant'. Rarely will it be feasible to infer such determinations directly from the chemical analysis alone. They usually require data from the analysis of the residues of, preferably artificially aged, known foodstuffs combined with archaeological and historical information. A residue of camel milk, for instance, need not be expected in vessels from North America, in pots from Pharaonic Egypt or even in the amphorae among which Eastern Desert Ware is usually found. That residue analysis can contribute to this assemblage of information is clear from the overview of our interpretations in Table 4. This illustrates the necessity of combining the chemical data with archaeological and historical information. Laboratory C, for instance, specialized in the study of North American hunter-gatherers who did not partake in dairying activities. In order to characterize their archaeological residues, using fatty acids, a collection was made of foodstuffs from that region and known residues were prepared and subjected to oven storage to simulate the effects of long-term decomposition. As Laboratory C was under renovation from Spring 2004 to Spring 2005 the preparation of new reference materials, relevant to the possible residue in the circulated sherd, was not possible.

| Lab. | Technique | Result | Interpretation |
|---|---|---|---|
| A | CIEP (20 antisera) | no proteins detected | proteins degraded after cooking? |
| B | GC/MS | C18:0/C16:0 = 0.72-1.03<br>C18:1/C16:0 = 0.32-1.15 | veal, egg or goat milk |
| C | GC | medium chain fatty acids > 30%, low C18:0 and C18:1 isomer levels | decomposed roots, tubers or berries |
| G | GC/MS | C14:0, C16:0, C18:0, 18:1 and cholesterol | veal or egg |
| H | GC | no activity | --- |
| J | IRMS | $\delta^{13}C$ = 19.2 (ext.) - 22.3‰ (int.)<br>$\delta^{15}N$ = 5.08-6.00‰ | milk of an herbivore |
| K | HTGC and GC/MS | triacylglycerols (44-56 C-atoms), C16:0/C18:0 = 3.175-5.741 | animal product |

Table 4: Overview of the interpretations after the analysis of the residue in our blind 'Round robin'.

In our blind 'Round robin' comparisons between unknown residues and those of known foodstuffs were made in two manners that are different but not mutually exclusive. Some residues preserve 'biomarkers', which are more or less specific for certain classes of foodstuffs. Such molecules can be alkaloids (such as caffeine), steroids (such as cholesterol), terpenoids (a large diverse group of mostly polycyclic compounds synthesized by plants), but also fatty acids like C22:1 (erucic or docosenoic acid) and phytanic (3,7,11,15-tetramethyl-hexadecanoic) acid. Respectively, these compounds naturally occur in coffee, animal products, certain families of plants (the terpenoids are well-studied and can be very specific, Connolly and Hill 2005; Hanson 2001), members of the *Brassicaceae (Cruciferae)* family (like mustard, nasturtium or rape seed) or fish and, finally, in the products of ruminants or fish (Hansel 2004). As is evident from the last two examples the same molecule is sometimes considered a biomarker for two apparently unrelated classes.

Some compounds can mark modern contamination of the sample. This is obvious in the case of man-made organic molecules like phthalates, in plastics to keep them flexible, but must also be considered for naturally occurring compounds like anthraquinones, commonly used as dye in textiles and paper, or 13-docosenamide (erucamide), coated on many plastics objects to prevent them sticking together. Proteins are inherently more specific than the relatively simple molecules mentioned above, and would theoretically be better biomarkers, but their isolation and identification from an archaeological context is still problematic in many cases. As some residues will not contain or preserve detectable concentrations of any biomarker, protein or otherwise, this method does not always yield helpful results.

The second way to match unknown with known residues is by comparing the ratios of the abundance of common fatty acids. Results can be sometimes be obtained with a simple C16:0/C18:0 ratio (P/S ratio) and sometimes after the two-dimensional plotting, on a double logarithmic scale, of other ratios such as C16:1/C18:1 versus (C15:0

+ C17:0)/C18:0 (Eerkens 2005). Such ratios tend to bring the residues from the same class of foodstuffs together rather well, but often fail to fully separate the different classes. The fact that these ratios may furthermore change over time, as different fatty acids oxidize at different rates, further limits the application of this method. One way to counter this problem is to monitor such changes in known residues in sherds that are stored in an oven to simulate the effects of long-term decomposition (Malainey et al. 1999a). The previously mentioned lipid biomarkers and fatty acids are usually found with the same analytical techniques, which enables the use of both methods on the same dataset. This approach would greatly benefit from a database, preferably accessible through the Internet, in which the raw data on residues from known sources is stored together with detailed information on the methods used to obtain these data.

The determination of food types by stable carbon and nitrogen isotope ratios is hampered by the similarities in isotopic composition of food types within a few broad classes such as $C_3$ or $C_4$ plant, or animals feeding on these classes of plants. As with most methods, additional archaeological and historical data on the studied material benefits the analysis. This technique requires specific instruments and combining it with those discussed above demands additional samples, time and funding. One way to cope with this would be cooperation and sharing of data, combined with the analysis of more known samples to evaluate cross-laboratory variation. Perhaps our most important conclusion is that such is not only necessary but also possible.

Other issues in need of attention by those working in the field include the formation processes related to organic residues. It is always assumed that the encountered residues are related to the food that was once present in the vessel. It is unclear, however, if the residue represents the first food to come into contact with the ceramics, after which the available binding sites are saturated, or the last, if older residues are continually replaced by new ones, or a combination of all food ever to have been inside the vessel, if the molecules that make up the residue compete for the available binding places. If the first is the case it must be taken into account that many unglazed vessels are 'seasoned', treated with oil or milk to make them less porous, before being used. This means that the ceramic matrix can indeed be saturated, but also that the residues will tell us more about the seasoning agent than about the actual use of the vessel. If we assume, on the other hand, that residues are easily replaced, we must account for the fact that vessels may end up on trash dumps or in graves where they can come into contact with organic materials that are not related to their original use. Finally, if there is competition for the available binding places it will be difficult to separate these different sources, especially if vessels have been in use for a long time for a variety of purposes.

Organic residues in potsherds may also originate from sources other than food but nevertheless related to the use of the vessel. Ceramics can have been employed for 'industrial' purposes (such as the preparation of organic dyes or glues) or as censers, smoking pipes or coffins (especially for infants), or simply to store a multitude of things. On the other end of the spectrum are foodstuffs that never come into contact with ceramic vessels but are eaten raw, roasted over a fire or prepared and consumed in other ways that do not call for ceramics. Archaeological residue analysis, as discussed here, can therefore never be more than complementary in the reconstruction of the diet of ancient peoples. For our laboratory techniques to reach their full potential as archaeological tools the study of more material with experimental or ethno-archaeological origins will prove as indispensable as a further cooperation and sharing of data by those working in this field.

## References

**Abu-Leiha**, I.H. (1987). Composition of Camel Milk. *Milchwissenschaft* **42**: 368-371.

**Abu-Leiha**, I.H. (1989). Physical and Chemical Characteristics of Camel Milk Fat and its Fractions. *Food Chemistry* **34**: 261-272.

**Allen**, J., M.E. Newman, M. Riford and G.H. Archer (1995). Blood and Plant Residues on Hawaiian Stone Tools from Two Archaeological Sites in Upland Kane'ohe, Ko' Olau Poko District, O'ahu Island. *Asian Perspectives* **34**: 283-302.

**Alshaikh**, M.A. and M.S. Salah (1994). Effect of Milking Interval on Secretion Rate and Composition of Camel Milk in Late Lactation. *Journal of Dairy Research* **61**: 451-456.

**Ambrose**, S.H. (1993). Isotopic Analysis of Paleodiets. Methodological and Interpretive Considerations. In M.K. Sandford (ed.). *Investigations of Ancient Human Tissue. Chemical Analyses in Anthropology.* Langhorne: Gordon and Breach Science Publishers, pp. 60-130.

**Barnard**, H. (2000). Geneeskunst geïnspireerd door armoede [Medicine inspired by poverty]. *Nederlands Tijdschrift voor Geneeskunde* **144**: 949-951 (in Dutch with English abstract).

**Barnard**, H., S.H. Ambrose, D.E. Beehr, M.D. Forster, R.E. Lanehart, M.E. Malainey, R.E. Parr, M. Rider, C. Solazzo and R.M Yohe II (2007). Mixed Results of Seven Methods for Organic Residue Analysis Applied to One Vessel with the Residue of a Known Foodstuff. *Journal of Archaeological Science* **34**: 28-37.

**Bulliet**, R.W. (1975). *The Camel and the Wheel.* Cambridge: Harvard University Press.

**Cattaneo**, C., K. Gelsthorpe, P. Phillips and R.J. Sokol (1997). Blood Residues on Stone Tools. Indoor and Outdoor Experiments. *World Archaeology* **25**: 29-43.

**Craig**, O.E. and M.J. Collins (2000). An Improved Method for the Immunological Detection of Mineral Bound Protein using Hydrofluoric Acid and Direct Capture. *Journal of Immunological Methods* **236**: 89-97.

**Craig**, O.E. and M.J. Collins (2002). The Removal of Protein from Mineral Surfaces. Implications for Residue Analysis of Archaeological Materials. *Journal of Archaeological Sciences* **29**: 1077-1082.

**Connolly**, J.D. and R.A. Hill (2005). Triterpenoids. *Natural Product Reports (Royal Society of Chemistry)* **22**: 230-248.

**DeNiro**, M.J. and S. Epstein (1977). Mechanisms of Carbon Isotope Fractionation associated with Lipid Synthesis. *Science* **197**: 261-263.

**Dorrill**, M. and P.H. Whitehead (1979). The Species Identification of Very Old Human Bloodstains. *Forensic Science International* **13**: 111-116.

**Downs**, E.F. and J.M. Lowenstein (1996). Identification of Archaeological Blood Proteins. A Cautionary Note. *Journal of Archaeological Science* **22**: 11-16.

**Eerkens**, J.W. (2005). GC/MS Analysis and Fatty Acid Ratios of Archaeological Potsherds from the Western Great Basin of North America. *Archaeometry* **47,1**: 83-102.

**Eisele**, J.A., D.D. Fowler, G. Haynes and R.A. Lewis (1995). Survival and Detection of Blood Residues on Stone Tools. *Antiquity* **69**: 36-46.

**Elamin**, F.M. and C.J. Wilcox (1992). Milk Composition of Majaheim Camels,' *Journal of Dairy Science* **75**: 3155-3157.

**El-Bahay**, G.M. (1962). Normal Contents of Egyptian Camel Milk. *Veterinary Medical Journal* **8**: 7-12.

**Epstein**, H. (1971). *The Origin of the Domestic Animals of Africa. Volume 2*. Leipzig: Africana Publication Corporation.

**Evershed**, R.P and N. Tuross (1996). Proteinaceous Material from Potsherd and Associated Soils. *Journal of Archaeological Sciences* **23**: 429-436.

**Farah**, Z., R. Rettenmaier and D. Atkins (1992). Vitamin Content of Camel Milk. *International Journal of Vitamin and Nutrition Research* **62**: 30-33.

**Farah**, Z. and M.W. Ruegg (1989). The Size Distribution of Casein Micelles in Camel Milk. *Food Microstructure* **8**: 211-212.

**Farah**, Z., T. Streiff, and M.R. Bachmann (1989). Manufacture and Characterization of Camel Milk Butter. *Milchwissenschaft* **44**, 412-414.

**Fiedel**, S.J. (1996). Blood from Stones? Some Methodological and Interpretive Problems in Blood Residue Analysis. *Journal of Archaeological Science* **23**: 139-147.

**Fankhauser**, B.L. (1994). Protein and Lipid Analysis of Food Residues. In J.G. Hather (ed.). *Tropical Archaeobotany. Applications and New Developments*. London: Routledge, pp. 227-250.

**Fankhauser**, B.L (1997). Amino Acid Analysis of Food Residues in Pottery. A Field and Laboratory Study. *Archaeology in Oceania* **32**: 131-140.

**Gnam**, S.O. and A.M. Shereha (1986). Composition of Libyan Camel's Milk. *Australian Journal of Dairy Technology* **41**: 33-36.

**Gorban**, A.M. and O.M. Izzeldin (1997). Mineral Content of Camel Milk and Colostrum. *Journal of Dairy Research* **64**: 471-474.

**Gorban**, A.M. and O.M. Izzeldin (2001). Fatty Acids and Lipids of Camel Milk and Colostrum. *International Journal of Food Science and Nutrition* **52**: 283-287.

**Hansel**, F.A., M.S. Copley, L.A.S. Madureira and R.P. Evershed (2004). Thermally Produced ω-(o-alkylphenyl)alkanoic Acids Provide Evidence for the Processing of Marine Produces in Archaeological Pottery Vessels. *Tetrahedron Letters* **45**: 2999-3002.

**Hanson**, J.R. (2001). Diterpenoids. *Natural Product Reports (Royal Society of Chemistry)* **18**: 88-94.

**Henrikson**, L.S., R.M. Yohe II, M.E. Newman and M. Druss (1999). Freshwater Crustaceans as an Aboriginal Food Resource in the Northern Great Basin. *Journal of California and Great Basin Anthropology* **20,1**: 72-87.

**Ingram**, D.L. and L.E. Mount (1975). *Man and Animals in Hot Environments*. Berlin-Heidelberg-New York: Springer-Verlag.

**Jardali**, Z. (1988). *Contribution à l'étude de la composition du lait de dromadaire*. Vandoeuvre-lès-Nancy: Institut National Polytechnique de Lorraine (unpublished thesis).

**Kind**, S.S. and R.M. Cleevely (1969). The Use of Ammoniacal Bloodstain Extracts in ABO Groupings. *Journal of Forensic Sciences* **15**: 131-134.

**Knoess**, K.H. (1977). The Camel as a Meat and Milk Animal. *World Animal Review* **22**: 39-44.

**Kooyman**, B., M.E. Newman and H. Ceri (1992). Verifying the Reliability of Blood Residue Analysis on Archaeological Tools. *Journal of Archaeological Science* **19**: 265-269.

**Kooyman**, B., M.E. Newman, C. Cluney, M. Lobb, S. Tolman, P. McNeil and L. Hill (2001). Identification of Horse Exploitation by Clovis Hunters Based on Protein Analysis. *American Antiquity* **66**: 686-691.

**Malainey**, M.E., R. Przybylski and B.L. Sherriff (1999a). The Effects of Thermal and Oxidative Degradation on the Fatty Acid Composition of Food Plants and Animals of Western Canada. Implications for the Identification of Archaeological Vessel Residues. *Journal of Archaeological Science* **26**: 95-103.

**Malainey**, M.E., R. Przybylski and B.L. Sherriff (1999b). Identifying the Former Contents of Late Precontact Period Pottery Vessels from Western Canada using Gas Chromatography. *Journal of Archaeological Science* **26**: 425-438.

**Malainey**, M.E., R. Przybylski and B.L. Sherriff (2001). One Person's Food. How and why Fish Avoidance may affect the Settlement and Subsistence Patterns of Hunter-Gatherers. *American Antiquity* **66**: 141-161.

**Mares**, R.G. (1954). Animal Husbandry, Animal Industry and Animal Disease in the Somaliland

Protectorate. *British Veterinary Journal* **110**: 411-423, 470-481.

**Midant-Reyes**, B. and F. Braunstein-Silvestre (1977). Le chameau en Egypte. *Or* **46**: 337-362.

**Mohamed**, M.A. (1990). *Camel Milk Composition.* Uppsala: Swedish University of Agricultural Science (unpublished thesis).

**Morton**, J. and H.P. Schwarcz (2004). Paleodietary Implications from Isotopic Analysis of Food Residues on Prehistoric Ontario Ceramics. *Journal of Archaeological Science* **31**: 503-517.

**Ramet**, J.-P. (2001). *The Technology of Making Cheese from Camel Milk (*Camelus dromedarius*).* Animal Production and Health Paper Number 113. Rome: Food and Agricultural Organization (UN).

**Rao**, M.B., R.C. Gupta and N.N. Dastur (1970). Camel's Milk and Milk Products. *Indian Journal of Dairy Science* **23**: 71-78.

**Reber**, E.A. and R.P. Evershed (2004). How did Mississippians Prepare Maize? The Application of Compound-specific Carbon Isotope Analysis to Absorbed Pottery Residues from Several Mississippi Valley Sites. *Archaeometry* **46**: 19-33.

**Rowley-Conwy**, P. (1988). The Camel in the Nile Valley. New Radiocarbon Accelerator (AMS) Dates from Qasr Ibrim. *Journal of Egyptian Archaeology* **74**: 245-248.

**Royal Canadian Mounted Police** (1983). Methods Manual. Serology Section. Ottawa (not paginated).

**Sawaya**, W.N., J.K. Kalil, A. Al-Shalhat and H. Al-Mohamed (1984). Chemical Composition and Nutritional Quality of Camel Milk. *Journal of Food Science* **49**: 744-747.

**Schmidt-Nielsen**, K. (1964). *Desert Animals. Adaptation and Environment.* Oxford: Oxford University Press.

**Shanks**, O.C., M. Kornfeld and D.D. Hawk (1998). Protein Analysis of Bugas-Holding Tools. New Trends in Immunological Studies. *Journal of Archaeological Science* **26**: 1183-1191.

**Wilson**, R.T. (1984). *The Camel.* London-New York: Longman.

**Yagil**, R. (1982). *Camels and Camel Milk.* Animal Production and Health Paper Number 26. Rome: Food and Agricultural Organization (UN).

**Yagil**, R. and Z. Etzion (1980a). Effect of Drought Conditions on the Quality of Camel Milk. *Journal of Dairy Research* **47**: 159-166.

**Yagil**, R. and Z. Etzion (1980b). Milk Yields of Camels (*Camelus dromedarius*) in Drought Areas. *Comparative Biochemistry and Physiology A* **67**: 207–209.

**Yagil**, R., A. Saran and Z. Etzion (1984). Camels' Milk: for drinking only? *Comparative Biochemistry and Physiology A* **78**: 263-266.

**Yagil**, R., U.A. Sod-Moriah and N. Meyerstein (1974). Dehydration and Camel Blood. *American Journal of Physiology* **226**: 298-308.

**Yohe**, R.M. II, M.E. Newman and J.S. Schneider (1991). Immunological Identification of Small-Mammal Proteins on Aboriginal Milling Equipment. *American Antiquity* **56**: 659-666.

**Zhang** H., J. Yao, D. Zhao, H. Liu, J. Li and M. Guo (2005). Changes in Chemical Composition of Alxa Bactrian Camel Milk during Lactation. *Journal of Dairy Science* **88**: 3402-3410.

# CHAPTER SEVENTEEN
# Introduction to the Analysis of Protein Residues in Archaeological Ceramics

**H. Barnard, L. Shoemaker, O.E. Craig, M. Rider, R.E. Parr, M.Q. Sutton and R.M. Yohe II**

Hans Barnard, Research Associate; Cotsen Institute of Archaeology; P.O.-box 951510; Los Angeles, CA 90095; USA; <wendrich@barnard.nl>; Lori Shoemaker; Pasarow Mass Spectrometry Laboratory, UCLA; Los Angeles, CA 90095; Micala Rider; Department of Anthropology, University of Arizona; Department of Anthropology; 1009 E. South Campus Drive; Tucson, AZ 85721-0030; Oliver Craig, Dipartimento di Biologica, Università di Roma 'Tor Vergata'; Via della Ricerca Scientifica; 100133 Rome; Italy; Robert Parr, Mark Sutton and Robert Yohe; Laboratory of Archaeological Sciences, California State University, Bakersfield; 9001 Stockdale Highway; Bakersfield, CA 93311; USA. We would like to thank Alek Dooley, Kym Faull, William Longacre and Willeke Wendrich for their help in preparing this chapter.

As shown from the increasing volume of publicatons on this subject (Chapter 1), including the chapters in this volume, organic residues in archaeological ceramics have the potential to yield important information that is not easily recovered by other, more traditional reseach techniques. The analysis of lipid residues and their association with food products is now relatively well established (Charters et al. 1997; Copley et al. 2005; Eerkens 2005; Evershed et al. 1997; Malainey et al. 1999a; b), although the interpretation of the results can still be problematic (Barnard et al. 2007). Several approaches have been suggested to expand the understanding of archaeological lipid residues, including the determination of fatty acid ratios (Eerkens 2005; Malainey et al. 1999a; b), stable isotope analysis of individual lipid compounds (Copley et al. 2005), or of the complete organic residue (Ambrose 1993; 2000; Morton and Schwarcz 1988; 2004). Organic compounds other than lipids that may survive in an archaeological context include polysaccharides (such as starches and cellulose), DNA and proteins.

Preserved ancient polysaccharides have been isolated from archaeological pottery (Oudemans et al. 2007), historical paints (Bersani et al. 2004; Stevanato et al. 1997; Vallance et al. 1998) and from the surfaces of stone tools (Del Pilar Babot and Apella 2003; Lombard and Wadley 2007). They could potentially serve as diagnostic markers for specific foodstuffs, but are generally soluble in water, causing them to leach out of context, and are rapidly digested by micro-organisms in the environment. Furthermore, the problems with the archaeological interpretation of the biochemical analysis will prove very similar to those concerning lipids. The amount of residual DNA, theoretically the most informative biomolecule, is very low in most food products, with the possible exception of meat products, and even lower in archaeological food residues, due to the loss of molecules by leaching and the activity of micro-organisms. Proteins, on the other hand, are generally much more abundant in natural substances than DNA. They could provide more specific information concerning their origin than lipids or polysaccharides because they are the direct products of the DNA of the plant or animal species in which they were synthesized.

In this chapter we present a short overview of the various methods available for the identification of protein residues from food in archaeological ceramics. This is preceded by some remarks on the fate of proteins in geological and archaeological contexts, and followed by five case studies, three of which represent our own research. This chapter is not a review of all the work done in the past nor a step-by-step manual, but rather an introduction for those planning to enter this promising field of study as well as those who seek a better understanding of the publications on this subject. For more details on protein biochemistry, we refer to Appendix II of this volume, which also has a small glossary, or to any textbook or website on this subject.

## The Preservation of Proteins

After deposition in an archaeological or geological context, proteins and peptides are subject to rapid attack by micro-organisms and relatively slow chemical decomposition (such as hydrolysis of the peptide bonds and racemization, decarboxylation and deamination of the amino acids, as discussed below). These will usually destroy the proteins, which will only survive under special circumstances. One of the best studied examples is collagen in buried bone (Collins et al. 2002). Collagen is a water-insoluble protein that appears to be protected from the extracellular enzymes produced by the micro-organisms in the soil (Child 1995; Hedges 2002). In favorable conditions, bone and collagen can be preserved for millennia due to physical and chemical changes resulting from a complex of interrelated processes collectively referred to as diagenesis (Collins et al. 2002; Hedges 2002). Other biomolecules, including amino acids (Jun 1974), heme (Schweitzer et al. 1997), and osteocalcin (Collins et al. 2000; Nielsen-March et al. 2002), have also been described in fossilized bones, as well as in ancient ostrich eggshell (Brooks et al. 1990), teeth enamel (Bada 1991), and geological contexts (Bada 1991; Belluomini et al. 1986; Cheng et al. 1975). Although this serves to show that the search for proteins in archaeological contexts is certainly warranted, the time scale and the nature of the above processes place them mostly in the fields of geology and paleontology.

The most important process concerning archaeological proteins from food sources, next to microbiological digestion, is denaturation as a result of food preparation. Denaturation alone does not alter the primary structure (the amino acid sequence) of a protein, but it will typically render proteins irreversibly non-functional and often insoluble. It may help to impede microbiological attacks, but at the same time promotes hydrolysis of the peptide bonds (Child 1995; Collins et al. 2002). In studies using immunological techniques, denaturated proteins may no longer be recognized by antibodies raised against their native form (Barnard et al. 2007; Downs and Lowenstein 1996; Fiedel 1996, but see Craig and Collins 2000; De Jong et al. 1974; Lowenstein 1981; Newman et al. 1997; Schweiter et al. 2002). Another process directly related to food preparation is the non-enzymatic, heat-induced condensation reaction between peptides and oligosaccharides (sugars). This process is known the Maillard reaction, commonly refered to as 'browning'. This reaction usually results in ketosamines, after Amadori rearrangement, and melanoidins, after polmerization. Browning obviously modifies the involved peptides, but also contributes to their preservation by inhibiting microbiological attacks and the hydrolysis of peptide bonds (Bada 1991).

Denaturation and browning are the basis of various food preparation techniques, but likewise occur as the result of many other events and certainly do not indicate an anthropogenetic origin of the organic residue. The conditions favorable for the preservation of proteins, primarily by reducing the extent of the digestion by micro-organisms, are generally the same as those during food preparation and conservation (Collins et al. 2002). Proteins from prepared food seem thus more likely to survive than those in other archaeological or geological contexts.

## Proteins as Archaeological Biomarkers

As proteins are present in substantial quantities in most organic materials (including food), are the direct products of DNA and appear to survive in the right environment for significant periods of time, they are potentially ideal biomarkers for the origin of an archaeological organic residue (Fankhauser 1994; 1997; Evershed and Tuross 1996). However, several factors frustrate the realization of this potential. First, as discussed above, archaeological proteins are likely to be denatured and fragmented by microbiological enzymes and hydrolysis. Also, some of the amino acids will have been chemically altered by decarboxylation, deamination and racemization; processes that advance over time and increase at higher temperatures (cooking). An anaerobe post-depositional environment will inhibit microbiological attacks and decarboxylation of the amino acids, while an arid environment will inhibit hydrolysis of the peptide bonds and racemization. Archaeological finds from cold and either very dry or water-logged (anaerobe) deposits can therefore be expected to yield better preserved peptides. Important to the study of archaeological proteins is that residual ancient proteins may not be intact proteins, but rather a mix of peptides with various modifications, as a result of changes over time as well as the complex depositional milieu.

Second, as a result of genetic conservation, molecules that serve similar functions in different species are often almost identical. Myoglobine, for instance, is thought to have evolved several hundred million years ago and parts of its amino acid sequence have never changed and are the same in virtually all animals. On the other hand, different proteins that have evolved to perform a similar task may have converged towards the same, most effective composition and shape. Convergent evolution explains why whales and dolphins (marine mammals) look like fish, and why echidnas (monotremes), porcupines (rodents) and hedgehogs (insectivores) are so similar in appearance and behavior. For the analysis of archaeological proteins, 'conservation' and 'convergent evolution' mean that diverse sources may leave residues of very similar proteins that may be impossible to distinguish, even after most of the amino acid sequence has been determined. Antibodies raised against a protein produced by one species may react with a similar protein from another species when both have epitopes, chemical groups that trigger an immune-response, in common. Such 'cross-reactivity' can make the interpretation of research using immunological techniques a hazardous affair (Fiedel 1996, but see Newman et al. 1997). Where the amino acid sequence is sufficiently specific, however, mass spectrometry techniques may allow differentiation for species assignment.

Third, oligopeptides and amino acids are a common constituent of the soil (humus). Here, peptides are preserved in anaerobic conditions or by browning, as melanoidins (Bada 1991; Child 1995; Evershed and Tuross 1996). Some of these molecules may leach into the buried ceramic vessel and contaminate the anthropogenetic residue that was present in the vessel before it was discarded. Organic molecules may also be introduced into the residue from other sources within the post-depositional environment, such as household refuse or a human burial, in case the vessel was part of funerary assemblage. Modern contamination can result from the handling of the sample during excavation or analysis, or from storage containers or laboratory equipment. It is impossible to eliminate all sources of contamination and not all contaminants will be easily distinguishable from the molecules of interest.

Furthermore, the techniques to isolate and analyze proteins from archaeological vessels usually require part of the object to be destroyed, which is obviously undesirable. Finally, the methods may be time-consuming and require expensive instruments or reagents (such as monoclonal antibodies). After all the above challenges have been overcome, the next issue that has to be addressed is the relationship between the encountered organic residues and the function of vessels within a society (Chapters 6, 12 and 15). Are organic residues the result of the first or the last use of a vessel, or rather an amalgam of residues from many use-phases during which specific molecules may preferentially replace others? What about vessels not used for food, or food not prepared in vessels? How do seasonal, social or ritual factors influence the formation and preservation of organic residues in ceramic vessels? Such questions will not be answered by biochemical research alone, but only by combining the results of biochemical analysis with archaeological, historical and ethnographical data (Chapters 1 and 16).

## Extraction of Archaeological Proteins

The analysis of proteinaceous residues in archaeological ceramics can be approached in two ways: immunological and proteomic. Each has their advantages and disadvantages and the choice between the two should be dictated by the research question (cf. Figure 7). All immunological techniques are based on the detection of specific previously selected compounds by labeled antibodies. They are generally very sensitive techniques and can be very specific, depending on the type of antibody used. As the use of antibodies depends on some preconceived notion of the proteins present in the sample, immunological approaches are not suitable for discovery-based identification of unknown proteins. Mass spectrometric techniques, on the other hand, aim to identify unknown peptides extracted from archaeological objects by their chemical structure in ways reminiscent of lipid analysis (Chapter 5): extraction followed by separation of the sample into its components and the identification of these components by comparing their mass spectra with those of known compounds in electronic reference libraries. This approach can be less sensitive than immunological techniques and its success depends on the amount of protein residue, the state of preservation of the sample, and the need for *de novo* sequencing.

Nearly all methods, both immunological and non-immunological, start with the extraction of the organic residue from the ceramic matrix. To facilitate this, a part of the archaeological object can be ground into powder, of which typically 5-10 g is used for each extraction. Sonication or prolonged periods of agitation, up to 12 h at temperatures around 40°C, are often applied to extract the proteins into solution. As most proteins of interest are water-soluble, ultra-pure water has been used as extraction solvent, but also aqueous solutions of ammonia (ammonium hydroxide), urea (in a relatively high concentration of 4-8 M), tris-(hydroxymethyl)-aminomethane hydrochloride (tris-HCl), detergents such as sodium dodecyl sulfate (SDS, also referred to as lauryl sulfate), alone or in various combinations. Another approach that has been successfully used is the destruction with HF (hydrofluoric acid) of part of the ceramic matrix to which the residue is attached (Craig and Collins 2000, see also Belluomini et al. 1986; Cheng et al. 1975). The extracted proteins will not be in their original native state, but are likely to be denatured, fractured (hydrolyzed) and chemically altered. Some proteins will go into solution more easily than others, while others will tend to attach to other molecules, including other peptides, but also possibly to the walls of extraction vessels. The last pitfall can be prevented by using specially coated test tubes or solubilization buffers, but it can also be used to facilitate analysis (see DACIA below).

## Immunological Detection

Techniques to detect proteins with immunological methods include radioimmunoassay (RIA), enzyme-linked immuno-sorbent assay (ELISA), counter immuno-electrophoresis (CIEP), and Western blot. In the first two, RIA and ELISA, specific compounds attach to a surface after which they are detected by labeled antibodies. The last two methods separate proteins in a sample by electrophoresis in a gel, prior to immunological detection. These techniques require some advance knowledge, or at least an educated guess, of the proteins under investigation, as the choice of antibodies depends on the expected proteins.

For RIA and ELISA the sample is placed into a special container. After an incubation period, usually several hours, some of the proteins in the sample will have attached themselves to the walls of the well (Figure 1, left). The sample is removed and replaced by a solution of primary antibodies against the protein of interest. After another incubation period, the well is washed, to remove the excess antibodies, and labeled secondary antibodies against the primary antibodies are introduced into the well. After a third incubation period, the well is washed again, this time to remove the excess antibody-antibodies. If the antigen of interest was not present in the sample, no primary/secondary antibody complex will have remained in the well (Figure 1, right). The complex is detected and quantified by a radio-active marker (RIA) or an enzyme that will cause a change in the color of a detection reagent (ELISA).

For CIEP, residues are usually extracted in ammonium hydroxide and placed in a well of an agarose gel that is paired with a second well containing a specific antibody (Dorrill and Whitehead 1979; Kind and Cleevely 1969). An electric current is applied to the gel, causing the (negative) antigens to move towards the anode and the (positive) antibodies towards the cathode, and come into contact. The antigen-antibody reaction of matching antigens and antibodies results in proteins precipitating out of the gel. After the run is completed the gel is pressed, dried and stained with a protein specific dye, such as $AgNO_3$, Coomassie Brilliant Blue or Sypro Ruby.

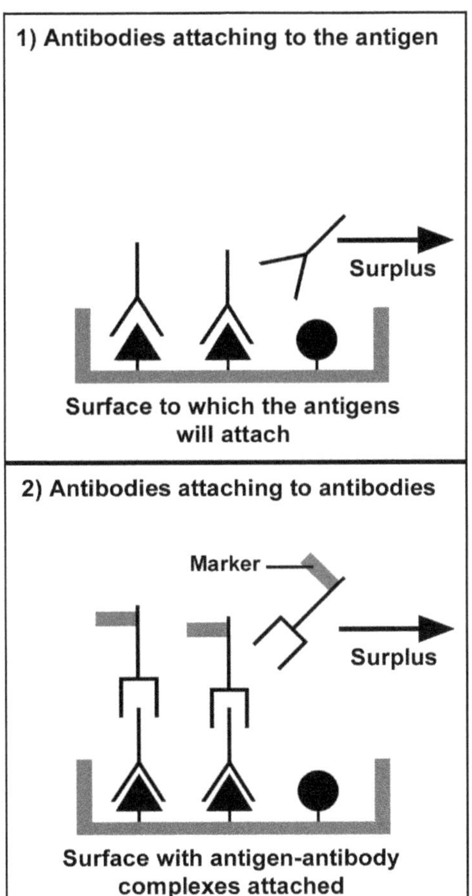


Figure 1: Schematic representation of a radio-immunoassay (RIA) or an enzyme-linked immuno-sorbent assay (ELISA). Antigens are allowed to attach themselves to the surface of a plate or a well. They react with specific antibodies (1), which in turn are detected by antibody-antibodies labeled with radio-active or chemical markers (2). The antigen on the right (•) is not recognized and not detected.

Before the development of DNA fingerprinting, CIEP was widely used to investigate bloodstains left at crime sites, and also to study archaeological protein residues (Allen et al. 1995; Cattaneo et al. 1993; Henrikson et al. 1998; Kooyman et al. 1992; 2001; Newman and Julig 1989; Shanks et al. 1999; 2001; Yohe et al. 1991). Most CIEP studies use polyclonal antibodies. These are produced by injecting the protein of interest into a laboratory animal, which stimulates an immune-response to the many epitopes on this protein. The resulting antibodies are then isolated from the animal's blood. Polyclonal antibodies will bind to many places on various proteins and are less specific, but also less expensive than monoclonal antibodies. To increase their specificity, polyclonal antibodies can be purified by exposing them to proteins known to cross-react before use. Monoclonal antibodies, produced by single lines of cloned B-cells, bind to a single epitope of one protein. A reliable commercial laboratory will test monoclonal antibodies for cross-reactivity, but the researcher must also test them in the environment of his or her own laboratory. Monoclonal antibodies are expensive, but obviously very specific for the selected epitope.

For Western blotting the proteins in the sample are first denatured by heating them in a buffer solution. A common extraction buffer is 'Laemmli's buffer' (named after U.K. Laemmli) that consists of tris-HCl, which keeps the solution as a fixed pH, a reducing agent, usually β-mercapto-ethanol or dithiothreitol (DTT), a detergent (usually SDS) and a dye. The SDS surrounds the denatured proteins with a negative charge and the DTT prevents the formation of random disulfide bonds and agglutination of the proteins. After denaturation, the sample is transferred into a well at the top of a thin layer of polymerized acrylamide between two glass plates submerged in a buffer solution. An electric potential is applied to the system that will pull the negatively charged proteins through the acrylamide gel towards the anode. Small proteins will move faster through the gel than larger proteins. The concentration of acrylamide in the gel determines its resolution; a higher percent acrylamide (14%) being optimal for separation of small to mid-sized proteins and lower percentages (7%) for the separation of large proteins. Gradient gels which provide a range of 4-20% acrylamide in the same gel offer the most flexibility. A mix of known proteins is usually placed in one of the wells to produce a 'ladder' of known molecular weights (Figure 2 shows such a ladder in a SDS-PAGE gel). After the run is completed, the gel is placed on a nitrocellulose membrane and an electric current is applied to transfer the peptides from the gel onto the membrane. The proteins on the surface of the membrane can then be visualized using a method similar to ELISA (Figure 1). To minimize non-specific binding, the membrane is first placed in a solution of known proteins, such as bovine serum albumin (BSA) or non-fat milk. The membrane is then incubated with specific antibodies, rinsed and incubated with labeled secondary antibodies (antibody-antibodies) that are finally detected by luminescence or fluorescence.

Many variations on these two techniques have been developed for specific applications (Figure 4). One of the better known examples is the home pregnancy test that uses a monoclonal antibody to detect human chorionic gonadotropin (HCG), a hormone produced by the placenta, in urine. Similarly, immunological detection can be employed to prove the presence of specific proteins from archaeological contexts, provided an epitope can be identified that is expected to survive and against which antibodies are available (Craig and Collins 2000; Malar et al. 2000; Newman and Julig 1989; Shanks et al. 1999; Reuther et al. 2006; Ubelaker 2004; Yohe et al. 1991; but see Dongoske et al. 2000; Downs and Lowenstein 1995; Fieldel 1996). Immunological detection techniques are very sensitive and can be very specific. They are widely used in medicine and scientific research and most components are readily commercially available. However, antibodies against proteins of interest to archaeologists are not always easily obtainable and may have to be custom designed, which is time consuming and expensive. Another significant limitation to these techniques is that the antigen to which the antibody is raised may or may not be left intact in ancient proteins, resulting in decreased detection of the protein. Finally, for this approach one must decide *a priori* which proteins to interrogate. The first three case studies later in this chapter provide examples where this was successfully done, although not always to identify proteins isolated from ceramic vessels.

## Gel Separation and Proteomics

Proteomics is an umbrella term that is used to describe the study of proteins; it is the most important alternative to the immunological approach (Aebersold and Mann 2003). Proteomic methods allow for discovery-based protein identification and do not rely on antibodies or intact antigens. A method with great promise for archaeological application entails separating a complex sample on a gel (by sodium dodecyl sulfate polyacrylamide gel electrophoresis, SDS-PAGE, or two-dimensional electrophoresis, 2-DE), followed by identification of the isolated proteins by mass spectrometry, a sensitive technique that allows accurate mass determinations of molecules (Gorg et al. 2004). The SDS-PAGE protocol is similar to that of Western blotting discussed above. After the electrophoresis is complete, however, the bands of proteins are not transferred onto a membrane and visualized by antigen-antibody reactions, but rather stained directly in the gel, with a protein specific dye such as $AgNO_3$, Coomassie Brilliant Blue or Sypro Ruby (Figure 2).

Comparing the position of the bands with those of a reference mix of known proteins gives an indication of the mass of the unknown proteins, while the intensity of the staining provides some indication of their relative abundance. The bands of proteins are cut out of the gel, digested by a site-specific cleavage reagent such as trypsin, and extracted for analysis by mass spectrometry. Trypsin is a commercially available enzyme that cleaves proteins (optimally at pH = 7-9) at the carboxyl-side (-$COO^-$) of lysine (K) and arginine (R), except where either is followed by proline (P). Trypsin digestion of the proteins in a sample results in short predictable peptide fragments.

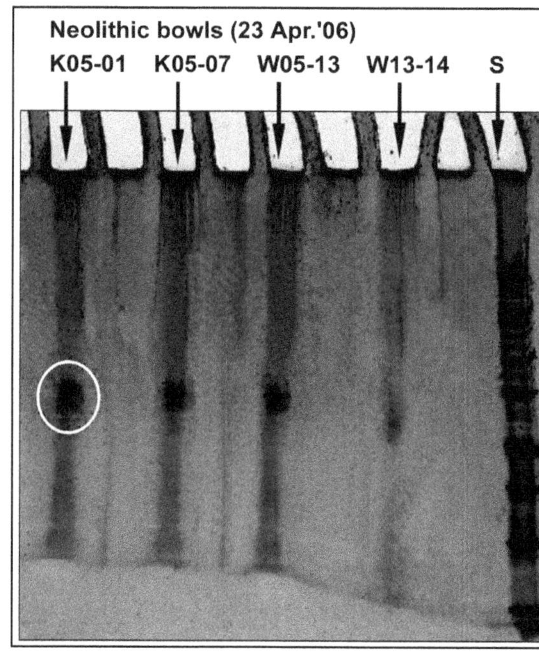


Figure 2: SDS-PAGE gel of four Neolithic residues, from ceramic bowls, stained with Sypro Ruby and viewed under ultraviolet light. Four lanes on the left are used for samples (K05-01, K05-07, W05-13 and W13-14), separated by empty lanes, the lane on the far right (S) shows a 'ladder', which results from a reference mix of known proteins. Horizontal bands, such as in the white circle, are peptides of similar relative mass. Preliminary LC-MS/MS analysis did not provide an identification of these peptides. Given the distinction of the bands, which indicates limited hydrolysis, they may well be modern contaminants.

A way to improve the resolution of gel separation is by separating the sample according to two different properties of the molecules, rather than just size, by two-dimensional gel electrophoresis (2-DE). This can be done by first separating proteins in a gel in one direction, usually by iso-electric point, then by turning the gel 90° and separate the same proteins along a second axis according to size (Görg et al. 2000; 2004). The latter is similar to the procedure for SDS-PAGE; the former is done by using a gel with a gradient pH. At all places in such a gel proteins will be charged, positive or negative, except at one specific pH. When an electric current is applied, the proteins will therefore move along the gel and accumulate where they have no charge, at their iso-electric point (Figure 3). After the separation is completed, the proteins are treated as in SDS-PAGE and separated by size. As it is unlikely that two different proteins have both the same iso-electric point and the same size, each spot on the resulting gel will contain a purified protein.

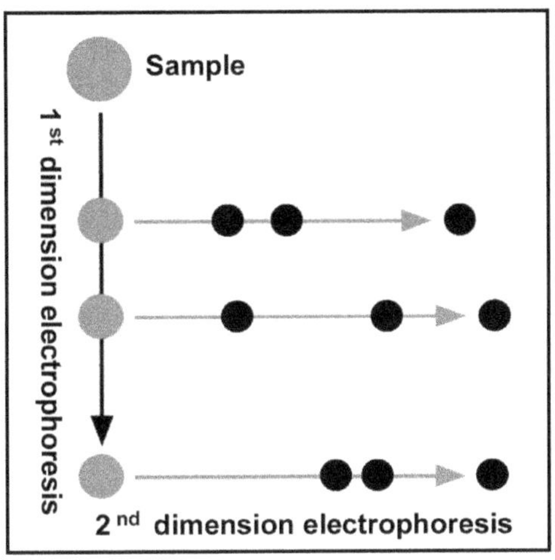


Figure 3: Schematic representation of two-dimensional gel electrophoresis (2-DE). The proteins in the sample (top-left) are first moved to their iso-electric point and subsequently separated by size. The nine small black spots on the resulting gel are expected to contain purified proteins from the sample.

Two approaches to identify proteins by mass spectrometry are relevant here: peptide mapping (also referred to as peptide-mass fingerprinting) and peptide sequencing (which includes *de novo* sequencing). The first is generally approached with matrix-assisted laser desorption and ionization time-of-flight (MALDI-TOF) mass spectrometry, the second by tandem mass spectrometry (MS/MS). For inherently unpredictable archaeological samples the combination of both mass spectrometric techniques should be considered to increase confidence in the protein identifications (Canas et al. 2006). The ancient protein extracts may contain both intact proteins as well as proteins that have been broken down into smaller peptides by microbial action, aging or the cooking process. Successful separation of these may require a special gel chemistry (tricine-based) that specifically retains small peptides. Alternatively, these extacts can circumvent the gel process entirely and be trypsin-digested in solution. The choice of the approach depends on consideration of the sample and the research question (Figure 7).

Peptide mapping is based on the accurate measurement of the mass of an unknown peptide, and comparing this with the masses of known sequenced peptides. For MALDI-TOF, the unknown peptides are taken up in a water/acetonitrile mixture and a matrix molecule (such as α-cyano-4-hydroxycinnamic acid or 2,5-dihydroxy-benzoic acid). As the accuracy of the mass measurement is crucial, an internal standard is also added to calibrate the instrument. The hydrophilic peptides will dissolve in the water and the hydrophobic peptides in the acetonitrile. A drop of the sample solution is placed in a well on a MALDI plate and the solvents evaporated, leaving a crystallized matrix in which the ancient peptides are trapped. The MALDI plate is then placed in the instrument and a laser is fired at the crystals. This produces a vapor cloud consisting of mostly singly protonated or deprotonated ions, $MH^+$ and $[M-H]^-$ respectively, which subsequently enter the TOF mass analyzer. The resulting masses can be compared to the measured or calculated masses of known peptides. Post-source decay of the peptides may provide some additional information on the sequence of the amino acids, but this data can be difficult to interpret.

The methods used for protein identification are evolving and improving rapidly. At present this is most frequently accomplished by deducing the amino acid sequence in pieces of the protein prepared by deliberate proteolytic digestion of the original material. The amino acid sequence of each piece is deduced by interpreting the results of the fragmentation of the pieces inside a mass spectrometer. This fragmentation process is referred to as collisionally induced dissociation (CID, also referred to as collisionally activated dissociation or CAD). The mass spectrometers that are used for these types of experiments are generally referred to as tandem or MS/MS instruments.

An extensive review of this process is beyond the scope of this chapter, but the fundamental process involves precisely measuring the molecular weight of the precursor peptide, inducing CID of the peptide by collision with an inert gas (a process that causes predictable fragmentation), precise measurement of the masses of the resulting fragments, and finally interpretation of the accumulated data (Aebersold and Mann 2003; Canas et al. 2006; Appendix II). In order to separate the peptides in a sample, MS/MS is often preceeded by liquid chromatography (LC-MS/MS) in which the sample, in a polar solvent (such as water or a mix of water and acetonitrile or methanol), is passed through a non-polar (silica-based) column. Competition between the stationary and the mobile phase separates the various components in time. The effluent of the column is fed into the ion-source of the MS/MS instrument and analyzed. All the collected data is pieced together and, with the help of reliable and readily available software programs and on-line databases, a list of probable proteins can be provided. As these experiments yield both the mass of the parent peptides as well as amino acid sequence information, relatively confident identifications of the peptides in the sample can usually be made. Several user-friendly algorithms have been developed for this purpose, including 'peptide sequence tagging' and 'probability based matching', all producing a 'hit list' of possible proteins in order of probability (Johnson et al. 2005; Yates 1998). One of the challenges of the field of proteomics is that proteins present in low amounts may not be observed (Aebersold and Mann 2003). This is important as the relevant peptides in archaeological samples may have low concentrations and can be easily obscured by contaminants. Furthermore, archaeological samples will contain many peptides and other impurities that can complicate the process.

The analytical methods described above are based on the accurate measurement of the mass of an unknown (ancient) protein and comparison of the results with the known masses of many proteins in large electronic databases. The success of these methods, therefore, depends on the presence of matching proteins in those databases. Many proteins that are relevant for archaeological research questions may not be in current protein databases, or may have been chemically altered over time (hydrolysis, decarboxylation, deamination, etc.) to the point where they no longer match their modern equivalent. This issue can be partially addressed by *de novo* sequencing, although this is generally not an automated process, requires relatively pure and abundant samples as well as an in-depth understanding of the biochemistry of proteins. As the recognition of a whole protein is often based on the identification of a much shorter peptide fragment (a tag), problems similar to cross-reactivity in immunological research may also occur. Because of genetic conservation and convergent evolution, large parts of proteins with a similar function are the same in different species. Distinguishing such species, for instance whether they were wild or domesticated, or from the Old World or the Americas, can be more important in archaeological studies than the identification of the protein. The field of proteomics advances fast and solutions to some of the problems and limitations sketched above may soon be found. A continuous discussion between biochemists and archaeologists will allow these improvements to reach their full scientific and scholarly potential.

## First Case Study: Cannibalism at Cowboy Wash

To illustrate the methods and techniques described above we present five well published case studies, three of which reflect our own field of research. All of these deal with the isolation, identification and interpretation of ancient proteins, but not always with proteins extracted from archaeological ceramics. Still, we feel that they demonstrate the potential of this rather recent, and not yet completely crystallized, addition to the archaeological toolbox. For more details on each of these case studies we refer to the original publications; for additional information on the scientific background see Appendix II at the end of this volume or any textbook or website on protein biochemistry.

A good example of how the analysis of ancient proteins can corroborate what was suspected on the basis of archaeological data is the study of pottery and human excrements from Cowboy Wash, an Anazasi site in the Four Corners region (USA) dating to around 1100 CE. Archaeological research of this site indicated that cannibalism may have been practiced; a suggestion that could be corroborated by the immunological detection of human myoglobin in 16 sherds of different ceramic vessels, as well as in a human coprolite (Billman et al. 2000; Malar et al. 2000).

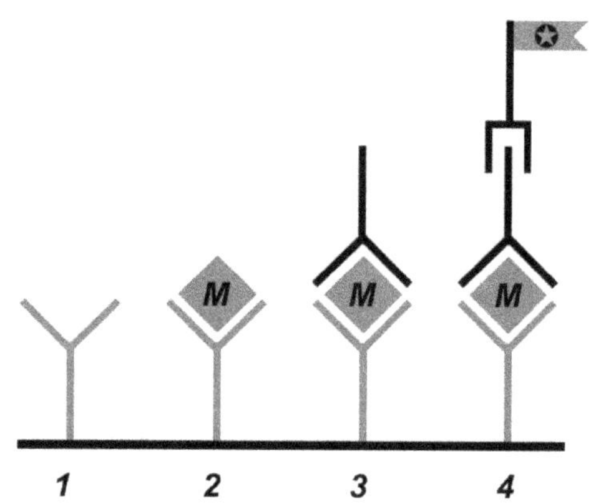


Figure 4: Schematic representation of the detection of human myoglobin using sandwich ELISA. 1: polyclonal rabbit anti-human myoglobin antibodies attach themselves to a plate; 2: myoglobin (human and possibly also of other origins) binds to the antibodies on the plate; 3: monoclonal mouse anti-human myoglobin binds exclusively to trapped human myoglobin (M); 4: the double antibody-antigen complexes are detected by labeled secondary antibodies (⋆).

Myoglobin is a protein that occurs in muscle tissue, including the heart. It will only occur in the stool if muscle or heart has been consumed. Human myoglobin in human faeces, or in a human coprolite, must therefore be considered firm evidence of cannibalism (although, to the best of our knowledge, the control experiment has not been performed). To confirm the archaeological hints of cannibalism at Cowboy Wash, indicated by butchering marks on human bones as well as the presence of human proteins on stone tools and in the sherds of cooking pots, a human coprolite was tested for human myoglobin using 'sandwich ELISA' and monoclonal mouse anti-human myoglobin antibodies (Figure 4). The stool from 25 healthy individuals were also tested as controls, as well as the faeces of 10 patients with blood in their stool and four individuals who had consumed considerable quantities of beef less than 24 hours earlier. No human myoglobin was found in any of the modern samples. The stool from the four individuals who had consumed beef tested positive for bovine myoglobin, proving that sufficient myoglobin survives passing through the human digestive tract to be detected by this method.

Human myoglobin was found in the ancient coprolite as well as in the sherds of cooking pots, strongly suggesting that human flesh had been prepared and consumed. Despite the scientific evidence, this conclusion was not accepted uncontested (Dongoske et al. 2000, but see Diamond 2000). The main objections included questions about the identification of the coprolite as human, the possibility of contamination of the sample or of non-specific binding of the antibodies to degraded other biomolecules, and the probability of other explanations than cannibalism for the archaeological observations. The discussion was further complicated by the cultural and emotional connotations attached to cannibalism (Sanday 1986; Sartore 1994).

## Second Case Study: Proteins on Stone Tools

The biochemical study of proteinaceous residues on ancient stone tools started in the late 1970s, around the same time as the study of lipid residues in ceramic vessels (Briuer 1976; Gurfunkel and Franklin 1988; Kooyman et al. 1992; Loy 1983; Newman and Julig 1989; Shanks et al. 2001). A compelling archaeological example of immunological techniques to identify proteins on stone tools was the study in the 1990s on artifacts from an aboriginal Tubatulabal village site in the southern Sierra Nevada of California, using custom antibodies produced in the Department of Biological Sciences of the University of Calgary. Numerous stone artifacts, primarily *manos* and *metates* known to have been used for plant processing from ethnographical comparison, as well as and one stone chopper, were convincingly shown to have positive responses to antibodies against pine, acorn and agave (Cuevas 2002), all important food resources to the Tubatulabal.

CIEP has been used to test for residues of the blood of small mammals on grinding stones (both *matates* and *manos*) from prehistoric southern California (Yohe et al. 1991), as well as to prove the use of freshwater crustaceans as a food source in the prehistoric northern Great Basin region, both in the USA (Henrikson et al. 1998). Unfortunately, this type of study has frequently been labeled 'blood residue analysis', even though proteins that elicit the antigen-antibody reaction are found in plants as well as animals, as those with plant allergies will be painfully aware. Furthermore, the proteins responsible for triggering an immune response are not limited to blood, but are found in all corporal fluids and tissues.

Concerns have been raised regarding the taxonomic specificity, or lack thereof, of CIEP results (Downs and Lowenstein 1995; Fiedel 1996). Given that the immunological differences between taxonomic groups are rather small, especially when polyclonal antibodies are applied, CIEP results such as 'chicken' or 'guinea pig' on artifacts from prehistoric North America, although bizarre from a historical point of view, are understandable and unavoidable. Polyclonal antibodies may provide the identification of proteins only as specific as the family level of taxonomic classification. Commercially produced antibodies raised against chickens, for example, will react positively with all gallinaceous fowl (grouse, quail, turkey, etc.) in addition to their south Asian brethren. A positive reaction to anti-bovine antibodies from an ancient North American locality would indicate bison rather than the modern domestic steer that was the source of the antibodies. If the use of immunologically purified polyclonal antibodies or, better still, monoclonal antibodies is not feasible, for practical or financial reasons, specific identifications can only be expected with those taxa that have only limited representatives of the family, such as the case with the American pronghorn antelope (*Antilocapra americana*), the only extant representative of the New World *Antilocapradae*.

CIEP is very sensitive; the detection limit is about $10^{-8}$ g protein. This results in a serious risk of contamination, both from modern and from ancient sources unrelated to the use of the objects. As with other immunological methods, CIEP does not identify unknown proteins, but is a method to prove the presence of *a priori* selected proteins, which partly defines the outcome of the research. Several series of blind tests with known fresh and artificially aged residues have been performed to test and improve the methodology (Cattaneo 1993; Eisele at al. 1995; Gurfinkel and Franklin 1988; Kooyman et al. 1992; Shanks et al. 2004). This has not yet resulted in a definite and generally excepted protocol. But, within its limitations and when employed with the appropriate care, this approach certainly has valuable applications (Shanks et al. 2001; Reuther 2006; Ubelaker et al. 2004). It could be that the ceramic matrix of unglazed pottery vessels is more apt to capture and preserve ancient proteins and other biomolecules than the surfaces of stone tools.

## Third Case Study: The Origins of Dairying

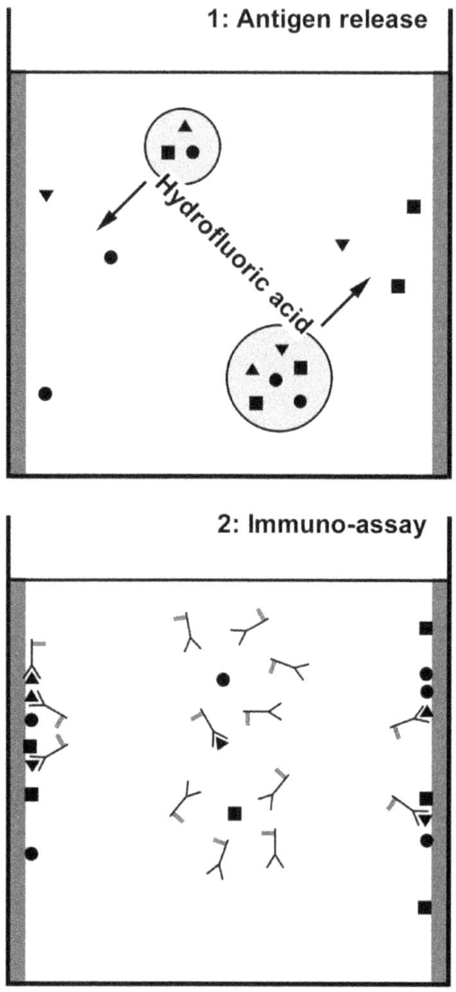


Figure 5: Digestion and capture immuno-assay (DACIA, after Craig and Collins 2000). Pottery powder is digested by hydrofluoric acid (HF) and releases the peptides previously attached to the ceramic matrix (■, ▼ and •). These peptides are captured by a polymer layer coating the walls of the test tube. After the contents of the test tube are removed, the peptide of interest (▼) is detected with marked monoclonal antibodies.

Dairying is often associated with specific slaughtering practices, more young animals are slaughtered by dairying societies to preserve more milk for use by humans, and with distinctive artifacts, such as strainers used for cheese productions (Chapter 4). Milk contains large amounts of proteins, such as casein, that could potentially remain attached to the ceramic matrix of such artifacts for long periods of time. Proteins that are sufficiently tightly attached to this matrix will survive longer, but are proportionally harder to disassociate (Keil et al 1994). Using a technique identified as digestion and capture immuno-assay (DACIA), evidence was found for dairying in the sherds of nine ceramic vessels, dated to around 500 BCE, from the Hebrides Archipelago, west of Scotland (Craig et al. 2000).

DACIA extracts the proteins from the pottery by dissolving parts of the ceramic matrix with the extremely corrosive hydrofluoric acid (HF), during the 'antigen release' phase (Figure 5). An amount of ground pottery is mixed with HF in a special container of which the walls are covered with a polymer that binds the released proteins. After incubation overnight the container is emptied and the remaining HF neutralized. The casein that remains attached to the walls of the container can now be detected using marked monoclonal antibodies raised against bovine casein, the 'immuno-assay' phase.

Positive responses strongly suggested that the vessels had been in contact with cow milk, thus indicating their use by a dairying society. The method was later successfully tested with experimental and ethnographic material (Craig and Collins 2002). It has been suggested that DACIA is the most reliable method to extract and analyze protein residues in pottery (Craig and Collins 2002). Limitations are the complications of working with HF and the dependence on the availability of antibodies and intact antigen epitopes.

## Fourth Case Study: Proteins in Paint Media

As briefly discussed above, the study of ancient proteins has mostly focused on the diagenesis of collagen (Child 1995; Collins et al. 2002; Hedges 2002), osteocalcin (Collins et al. 2000; Nielsen-March et al. 2002), and other biomolecules in buried bones (Bada 1991; Brooks et al. 1990; De Jong et al. 1974; Jun 1974; Lowenstein 1981; Schweiter et al. 2002; Schweitzer et al. 1997). Competing for second place, in terms of the number of publications and the progress made, are the study of organic residues on stone tools (discussed above as our second case study) and the research of proteinaceous binders in paint media (Birstein 1975; Chiavari et al. 1995; Colombini and Modugno 2004; Jones 1962; Jurado-López and Luque de Castro 2004; Karpowicz 1981; Ronca 1994). Artist's paints are a more or less complex mixture of a pigment (mineral or organic) suspended in a media (oil or water based) to which often compounds are added to change the behavior of the final product, such as its binding property to underlying layers, its curing time or its preservation over time. This last aspect explains the interest of conservators in the composition of historical paints as a basis for better preservation of works of art and crafts.

Only a limited suite of raw materials are traditionally used to produce proteinaceous paint binders, the most common being egg white, fish bones, milk casein and rabbit skin. Given this limited number of possible sources, immunological techniques seem the obvious approach to investigate these peptides (Heginbotham et al. 2004; Ramirez-Barat and De la Vina 2001), especially as they appear diverse enough to prevent cross-reactivity. A complicating factor is that the amount of material that can be investigated is often very limited and precious: it makes no sense to sacrifice a large part of an object for the sake of its preservation. The microscopic techniques that are used and developed to address this problem are of great interest to archaeologists who often have to work with equally limited amounts of material. Another challenge is the identification of proteins beyond the family level, as discussed above, for instance to determine whether the egg white is from chicken or goose eggs (or even a mix of the two).

Another technique that has been developed to interrogate the proteins in ancient paints is based on a statistical analysis of the relative abundance of the recovered amino acids (Aruga et al. 1999; Casoli et al. 1995; Colombini et al. 1998; Halpine 1992; Keck and Peters 1969). Because of the limited number of proteins that can reasonably be expected, it is possible to generate amino acid fingerprints of these proteins in a fresh specimen by GC/MS (Chapter 5). These reference profiles can then be used to identify the amino acid profile of the unknown ancient proteins in the sample. The advantage of this approach is that the results are not affected by denaturation, racemization or even hydrolysis of the proteins over time. It will obvioulsy not work if the proteins are similar, or for the mixture of proteins that might be expected in a ceramic vessel used for food. As with the immunological approach this method does not allow for discovery-based analysis, it can only show the presence of one of the *a priori* selected proteins. Proteins that are not on this list, or even the right mix of two or more proteins, may lead to false identification if the resulting profile coincides with one of the selected proteins.

## Fifth Case Study: the Round Robin

The 'Round robin' was comprised of a new clean vessel in which the milk of an Egyptian dromedary (*Camelus dromedarius*) was prepared, after which the vessel was cut into segments. These were circulated among archaeological laboratories that were interested in trying to identify the residue without prior knowledge of its source (Barnard et al. 2007; Chapter 16). Most participating laboratories isolated, identified and interpreted the lipid residue in the sherd; CIEP with 20 different antibodies did not result in a positive identification. After the source of the residue was revealed,[1] the protein residue in the sherd was analyzed by Laboratory B, which had previously done the same with the lipid residue by GC/MS. As the results of this study have not perviously been published they are discussed here in detail.

For the analysis the sherd was ground into a fine powder. To 7 g of this, 8.75 ml extraction buffer (1.25 ml/g sample) was added. The extraction buffer consisted of 1.5 g SDS and 3 ml tris-HCl (1M, pH=6.8), to which $H_2O$ was added to reach a total volume of 25 ml. The sample was mixed, on a Vortex mixer, for 5 minutes and sonicated for 20 minutes, after which it was rotated overnight at 35°C. The following day the sample was vortexed for 5 minutes and centrifuged at 3000 rcf (relative centrifugal force) for 10 minutes. Equal amounts of supernatant were pipetted into two smaller tubes and centrifuged at 13,200 rcf for 5 minutes. These supernatants were filtered using a 0.2 µm low protein binding syringe filter. The filtrate was refiltered using a 10,000 Dalton (molecular weight) centrifugal filter. The sample left in the upper chamber (greater than 10,000 Daltons) was precipitated using a TCA/acetone precipitation (TCA = trichloro-acetic acid, $Cl_3C$-COOH).

Sufficient amounts of TCA and acetone were added to reach a final proportion of 10% sample, 10% TCA and 80% acetone. The tube was inverted and vortexed gently, to mix the contents, and incubated over night at -20°C. The next day the sample was centrifuged at 20,000 rcf for 20 minutes to pellet the proteins. The liquid was carefully decanted leaving the pellet, which was dried in a vacuum chamber on dry ice. The proteins were resuspended in 50-100 µl loading buffer and vortexed in preparation for SDS-PAGE separation. The loading buffer consisted of 4 g SDS, 3 g DTT, 2 grains Bromophenol Blue, 12 ml tris-HCl (1M, pH=6.8) and 23 ml glycerol (87% v/v), to which $H_2O$ was added to reach a total volume of 100 ml.

A series of 30 µl resuspended protein was entered into the wells of a BioRad tris-HCl 4-20% gradient gel. This was run at 65 V for about 2.5 hours, until the dye front had completely run off the gel. The gel was then fixed and stained using the University of Arizona Proteomics Laboratory staining protocol (Breci et al. 2005). The protein was first fixed in the gel by washing with a 50% methanol/10% acetic acid solution for 30 minutes and then a diluted form of the solution (5% methanol and 1% acetic acid) for 15 minutes. During the washes, a

---

[1] This disclosure took place on Thursday 31 March 2005 during the symposium 'Theory and Practice of Archaeological Residue Analysis', which was the title of sessions 21 and 46 of the 70th Annual Meeting of the Society for American Archaeology in Salt Lake City, see http://www.archbase.org/residue/.

sufficient amount of solution was used to cover the gel. The container of the gel was covered and placed on a gentle rocker. The gel was then washed three times with Milli-Q $H_2O$, for 10 minutes each, and left over night in a fresh Milli-Q $H_2O$ bath.

The next day the proteins in the gel were sensitized to the silver stain by a sodium thiosulfate solution ($Na_2S_2O_3$•$5H_2O$, 2 mg/100 ml $H_2O$) for exactly 90 seconds. The gel was then rinsed with Milli-Q $H_2O$ three times, for 30 seconds each, to wash away the sensitizing solution. After pouring off the water from the last rinse, a silver nitrate solution ($AgNO_3$, 200 mg/100 ml $H_2O$), which was previously chilled at 4°C, was added and the gel was left on a rocker for about 45 minutes. The silver nitrate was properly disposed of and the gel was again rinsed three times with Milli-Q $H_2O$, 60 seconds each. Then a sodium carbonate developing solution was added consisting of $Na_2CO_3$ (6 g/100 ml $H_2O$), 37% formaldehyde (50 µl/100 ml $H_2O$) and $Na_2S_2O_3$•$5H_2O$ (2 mg/100 ml $H_2O$, as in the sensitizing solution). The gel was left in this solution until visible bands started to develop. Once bands began to form, the reaction was stopped with a 6% acetic acid solution. As this did not have an immediate effect, the resulting gel was slightly over-developed. The gel was left in the 6% acetic acid for about 10 minutes, washed a final time with Milli-Q $H_2O$ and stored in Milli-Q $H_2O$ until visual documentation (Figure 6).

The bands produced by the silver staining procedure were carefully cut out and submitted to the Proteomics Analysis Laboratory at the University of Arizona to be de-stained and digested with trypsin. A total of 15 fractions were produced. Each fraction was evaporated in a vacuum centrifuge to a volume of about 10 µl, transferred onto an autosampler plate and submitted for LC-MS/MS analysis at the University of Arizona Proteomics Analysis Laboratory.

A microbore high pressure liquid chromatography (HPLC) system was modified to operate at capillary flow rates using fused silica columns packed with 5 µm Xorbax C18 resin and 5 µm polyhydroxyethyl-A strong cation exchange resin. Samples were introduced onto the column using a Surveyor autosampler. The HPLC column eluted directly into the electrospray ionisation source of a ThermoFinnigan LCQ-DecaXP Plus ion trap mass spectrometer. Peptides were eluted in a 100% acetonitrile gradient at a flow rate of 200 nl/min (this varies with the used column). For each elution, masses were detected of all the peptides and the three most abundant were selected for further fragmentation and mass measurement of all their ionized fragments. As keratin contamination is quite frequent, the common peptide masses associated with keratin proteins were excluded from further fragmentation. Despite this precaution some keratin peptides still show up in the final dataset. The analysis took about 19 hours after which the data were ready to be processed.

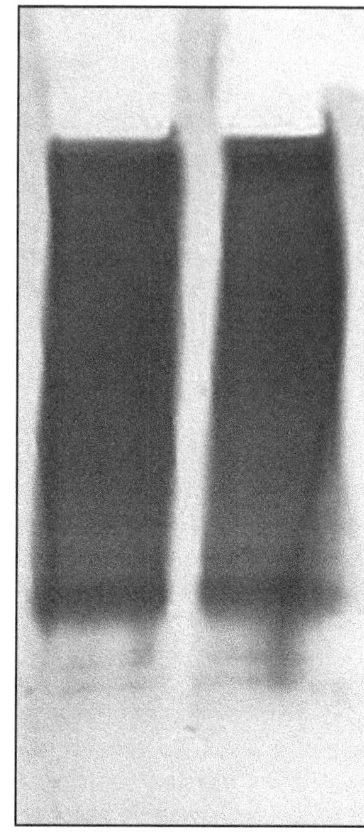

Figure 6: Silver-stained (slightly over-developed) SDS-PAGE gel of the protein residue in the 'Round robin'.

Both the SEQUEST and the X! Tandem MS/MS software programs were used to search for protein identifications. The results were used in conjunction to increase the confidence in single peptide identifications (Rohrbough et al. 2006). The X correlation cut-off score for the SEQUEST search results was 2.2. Three proteins were identified: lactophorin, lactalalbumin and casein. All these are present in camel milk. Five peptides were found, two from lactophorin:

R.NTATQSEETK.E
K.ELTPGAATTLEGK.L

two from lactalalbumin:

K.FLDDDLTDDK.M
R.EYGLFQINNK.I

and one from casein:

K.IYTFPQPQS.L

The X! Tandem peptide search yielded similar results, but only lactophorin and casein were identified. The peptides from these proteins were the same as those found in the SEQUEST search. The log(e) cut-off for these peptides was -3.0; the highest log(e) scores were -14.1 for casein and -29.5 for lactophorin.

The peptides from the three proteins found in both searches were compared against known peptide sequences using the NCBI BLAST search algorithm. The closest match for the lactophorin was *Bos tarus* (cow) with a 64% sequence match between the two peptides, meaning that 9 out of the 25 identified amino acids were different. There was more homology in the lactalalbumin. Both *Bos tarus* and *Sus scrofa* (pig) matched the identified peptides with 92% similarity, meaning that out of the 26 identified amino acids only 2 were different in these species. Other closely matching species had 3 or 4 additional differences. This region must therefore be considered highly conserved.

However, one of the mutations unique to *Camelus dromedarius* affects the way in which one of the lactalalbumin species was cleaved. One of the differences between dromedary lactalalbumin and that from other species is a change from threonine (T) to arginine (R). The detected peptide could only have come from the arginine mutation because trypsin, the digestive enzyme that is used to cleave the protein into its smaller peptide constituents, cleaves a protein between the amino acids lysine (K) and arginine (R), except where either is followed by proline (P). If the amino acid in the cleavage position belongs to a species without the arginine mutation, the peptide would not have been cleaved at that position. The resulting peptide would have been longer due to the fact that this particular cleavage site is missing in the dromedary peptide. Therefore, even though there are homologous proteins in other species, these peptides appear to be from a latalalbumin protein that originated in a dromedary. The results from the peptide search for the single casein protein yielded no closely related homologous matches.

The results from the SEQUEST and X! Tandem searches indicate that the origin of the residue in the 'Round robin' is the milk of a dromedary. It is interesting that the lactalalbumin protein only showed up in the SEQUEST results and not the X! Tandem results. As there were multiple peptides identified and the spectral data were very clear this could still be considered a positive identification. This discrepancy shows that there are important differences between the two search algorithms and highlights the dangers associated with using only one search engine to analyze the data collected. With a peptide identification from a single search algorithm it would seem more likely that the identification would be a false positive, especially with a relatively low score. Casein can be considered accurately identified as it not only showed up in both searches, but had a high score in each. Lactophorin was the most conclusively identified protein; it had multiple peptides identified in both searches with high scores. This protein seemed to show up in multiple bands, but this could also be due to protein diffusion between the time that the gel was developed and the time at which the bands were cut out for digestion. The fact the three milk proteins of *Camelus dromedarius* were identified in the Round robin indicates that protein analysis is potentially a powerful technique to identify the source of archaeological organic residues.

## Discussion

In this chapter we have shown not just the feasibility, but also the necessity developing techniques for the analysis of protein residues in archaeological ceramics. This analysis could follow the following framework (Figure 7). Based on a clear research question, first a decision should be made whether to follow the immunological or the proteomic pathway. If showing the presence or absence of one or two proteins can sufficiently address the research question, the immunological path is the obvious choice. Immunological techniques are sensitive and specific, certainly when using monoclonal antibodies, and relatively straightforward and economical. Care should be given to the selection and treatment of the samples, as with any other technological analysis. DACIA or CIEP are the techniques of choice and whenever possible monoclonal antibodies should be selected for the most specific epitope.

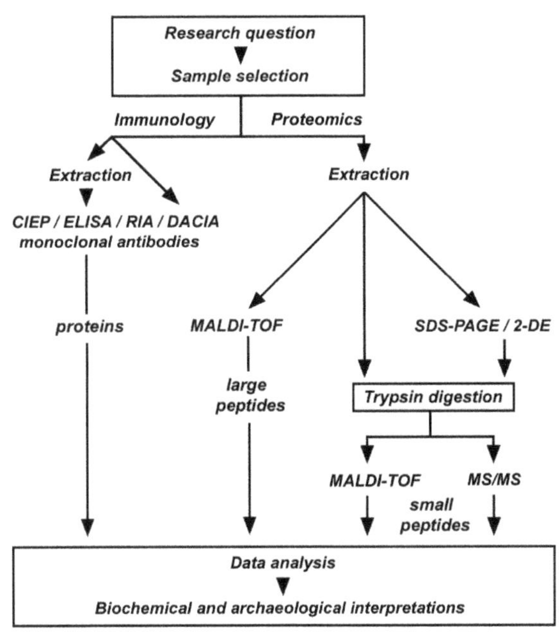


Figure 7: Suggested flow diagram for the analysis of protein residues in archaeological ceramics.

It has to be taken into consideration that antibodies against proteins of interest may not be readily available and that the production of specific monoclonal antibodies can be very costly. The modern antigens against which these antibodies are raised may not survive intact in archaeological contexts, resulting in a decreased detection. Proof of absence in a sample is no definitive proof of absence of the protein in ancient times. Cooperation between different research groups may permit the creation of a small pool of monoclonal antibodies of archaeological interest.

When the research question requires the identification of many unknown peptides, the proteomic pathway should be chosen, allowing for discovery-based investigation. Analytical techniques, however, can be time-consuming, require expensive and complex instruments and produce large amounts of data that are difficult to interpret. After extraction of the samples, the proteins (and large peptides) can be mass measured directly by MALDI-TOF or they can be separated into their components, typically on a gel, and analyzed by mass spectrometry. For this, the samples are trypsin-digested and analyzed by tandem MS/MS (as was the Round robin in our fifth case study) or by MALDI-TOF.

Similar to the immunological pathway, some proteins that are relevant for archaeological research questions will not be present for comparison in the available databases, while other proteins have been chemically modified and no longer match their modern equivalent. The latter problem can be addressed by optimizing potential post-translational modications in the MS/MS search programs. If the identification of the protein is based solely on a peptide mass tag (with no peptide sequence data), problems similar to cross-reactivity in immunological research may also occur. An on-line repository of proteins encountered during ethnographic and archaeological research may resolve some of these issues, but it must be expected that some ancient proteins will always remain unidentified. *De novo* sequencing offers an alternative to the lack of a database entry, but this requires substantial technical knowledge which is not always available.

The final steps of both the immunological and the proteomic pathways are data analysis and interpretation. If we assume that proteins have survived in archaeological ceramics for extended periods of time, and proteins from prepared food seem more likely to do so than those in other contexts, what then are the possible relationships between these proteins and the function of the original ceramic objects within a society? This question can only be addressed by combining biochemical analysis with archaeological, historical and ethnographical data, which requires a continuous dialogue between biochemists and archaeologists (Chapter 1).

## References

(see also Appendix II at the end of this volume)

**Aebersold**, R. and M. Mann (2003). Mass Spectrometry based Proteomics. *Nature* **422**: 198-207.

**Allen**, J., M.E. Newman, M. Riford and G.H. Archer (1995). Blood and Plant Residues on Hawaiian Stone Tools from Two Archaeological Sites in Upland Kane'ohe, Ko' Olau Poko District, O'ahu Island. *Asian Perspectives* **34**: 283-302.

**Ambrose**, S.H. (1993). Isotopic Analysis of Paleodiets. Methodological and Interpretive Considerations. In M.K. Sandford (ed.). *Investigations of Ancient Human Tissue. Chemical Analyses in Anthropology.* Langhorne: Gordon and Breach Science Publishers, pp. 60-130.

**Ambrose**, S.H. (2000). Controlled Diet and Climate Experiments on Nitrogen Isotope Ratios of Rats. In S.H. Ambrose and M.A. Katzenberg (eds.). *Biogeochemical Approaches to Paleodietary Analysis.* New York, Kluwer Academic, Plenum Publishers: 243-259.

**Aruga**, R., P. Mirti, A. Casoli and G. Palla (1999). Classification of Ancient Painting Media by the Joint Use of Pattern Recognition and Factor Analysis on GC/MS Data. *Fresenius Journal of Analytical Chemistry* **365**: 559-566.

**Bada**, J.L. (1991). Amino Acid Cosmogeochemistry. *Philosophical Transactions of the Royal Society, London* **B 333**: 349-358.

**Barnard**, H., S.H. Ambrose, D.E. Beehr, M.D. Forster, R.E. Lanehart, R.E. Parr, M.E. Malainey, M. Rider, C. Solazzo and R.M Yohe II (2007). Mixed Results of Seven Methods for Organic Residue Analysis Applied to One Vessel with the Residue of a Known Foodstuff. *Journal of Archaeological Science* **34**: 28-37.

**Belluomini**, G., M. Brance, G. Calderoni and M. Schnitzer (1986). Distribution and Geochemical Significance of Amino Acids and Amino Sugars in a Clay Suite of the Pliocene-Pleistocene Age from Central Italy. *Organic Geochemistry* **9,3**; 127-133.

**Bersani**, D., P.P. Lottici, G. Antonioli, E. Campani, A. Casoli and C. Violante (2004). Pigments and Binders in the Wall Painting of Santa Maria della Steccata in Parma (Italy). The Ultimate Technique of Parmigianino. *Journal of Raman Spectroscopy* **35**: 694-703.

**Billman**, B.R., P.M. Lambert and B.L. Leonard (2000). Cannibalism, Warfare, and Drought in the Mesa Verde Region during the Twelfth Century A.D. *American Antiquity* **65,1**: 145-178.

**Birstein**, V.J. (1975). On the Technology of Central Asian Wall Paintings. The Problem of Binding Media. *Studies in Conservation* **20**: 8-19.

**Breci**, L., E. Hattrup, M. Keeler, J. Letarte, R. Johnson and P.A. Haynes (2005). Comprehensive Proteomics in Yeast using Chromatographic Fractionation, Gas Phase Fractionation, Protein Gel Electrophoresis, and Isolectric Focusing. *Proteomics* **5, 8**: 2018-2028.

**Brooks**, A.S., P.E. Hare, J.E. Kokis, G.H. Miller, R.D. Ernst and F. Wendorf (1990). Dating Pleistocene

Archeological Sites by Protein Diagenesis in Ostrich Eggshell. *Science* **248, 4951**. 60-64.
**Briuer**, F.L. (1976). New Clues to Stone Tool Function. Plant and Animal Residues. *American Antiquity* **41**: 478-484.
**Canas**, B., D. Lopez-Ferrer, A. Ramos-Fernandez, E. Camafeita and E. Calvo (2006). Mass Spectrometry Technologies for Proteomics. *Briefings in Functional Genomics and Proteomics* **4,4**: 295-320.
**Casoli**, A.P., Mirti and G. Palla (1995). Characterization of Medieval Proteinaceous Painting Media using Gas Chromatography and Gas Chromatography - Mass Spetrometry. *Fresenius Journal of Analytical Chemistry* **352**: 372-379.
**Cattaneo**, C., K. Gelsthorpe, R. Phillips and R.J. Sokal (1993). Blood Residues on Stone Tools. Indoor and Outdoor Experiments. *World Archaeology* **25**: 29-43.
**Charters**, S., R.P. Evershed, A. Quye, P.W. Blinkhorn, and V. Reeves (1997). Simulation Experiments for Determining the Use of Ancient Pottery Vessels: the Behavior of Epicuticular Leaf Wax during Boiling of a Leafy Vegetable. *Journal of Archaeological Science* **24**:1-7.
**Cheng**, C.-N., R.C. Shufeldt and F.J. Stevenson (1975). Amino Acid Analysis of Soils and Sediments. Extraction and Desalting. *Soil Biology and Biochemistry* **7**: 143-151.
**Chiavari** G., D. Fabbri, G.C. Galletti and R. Mazzeo (1995). Use of Analytical Pyrolysis to characterize Egyptian Painting Layers. *Chromatographia* **40**: 594-600.
**Child**, A.M. (1995). Towards an Understanding of the Microbial Decomposition of Archaeological Bone in the Burial Environment. *Journal of Archaeological Science* **22**: 165-174.
**Collins**, M.J., A.M. Gernaey, C.M. Nielsen-Marsh, C. Vermeer and P. Westbroek (2000). Slow Rates of Degradation of Osteocalcin. Green Light for Fossil Bone Protein? *Geology* **28,12**: 1139-1142.
**Collins**, M.J., C.M. Nielsen-Marsh, J. Hiller, C.I. Smith, J.P. Roberts, R.V. Prigodich, T.J. Wess, J. Csapò, A.R. Millard and G. Turner-Walker (2002). The survival of Organic Matter in Bone. A Review. *Archaeometry* **44,3**: 383-394.
**Colombini**, M.P. and F. Modugno (2004). Characterization of Proteinaceous Binders in Artistic Paintings by Chromatographic Techniques. *Journal of Separation Science* **27**: 147-160.
**Colombini**, M.P., R. Fuoco, A. Gaicomelli and B. Muscatello (1998). Characterization of Proteinaceous Binders in Wall Painting Samples by Microwave-Assisted Acid Hydrolysis and GC-MS Determination of Amino Acids. *Studies in Conservation* **43,1**: 33-41.
**Copley**, M.S., R. Berstan, S.N. Dudd, V. Straker, S. Payne and R.P. Evershed (2005). Dairying in Antiquity I. Evidence from Absorbed Lipid Residues dating to the British Iron Age. *Journal of Archaeological Science* **32**: 485-503.
**Craig**, O.E. and M.J. Collins (2000). An Improved Method for the Immunological Detection of Mineral Bound Protein using Hydrofluoric Acid and Direct Capture. *Journal of Immunological Methods* **236**: 89-97.
**Craig**, O.E. and M.J. Collins (2002). The Removal of Protein from Mineral Surfaces. Implications for Residue Analysis of Archaeological Materials. *Journal of Archaeological Science* **29**: 1077-1082.
**Craig**, O.E., J. Mulville, M.P. Pearson, R. Sokol, K. Gelsthorpe, R. Stacey and M.J. Collins (2000). Detecting Milk in Ancient Pots. *Nature* **408**: 312.
**Crick**, F. (1970). Central Dogma of Molecular Biology. *Nature* **227**: 561-563.
**Cuevas**, K. (2002). *Archaeological Investigations at the Long Canyon Village Site, CA-KER-311, Kern County, California*. Department of Sociology and Anthropology, California State University, Bakersfield (unpublished Master thesis).
**De Jong**, E.W., P. Westbroek and J.W. Bruning (1974). Preservation of Antigenetic Properties of Macromolecules over 70 Myr. *Nature* **252**: 63-64.
**Del Pilar Babot**, M. and M.C. Apella (2003). Maize and Bone. Residues of Grinding in Northwestern Argentina. *Archaeometry* **45,1**: 121-132.
**Diamond**, J.M. (2000). Talk of Cannibalism. *Nature* **407**: 25-26.
**Dongoske**, K.E., D.L. Martin and T.J. Ferguson (2000). Critique of the Claim of Cannibalism at Cowboy Wash. *American Antiquity* **65,1**: 179-190.
**Dorrill**, M. and P.H. Whitehead (1979). The Species Identification of Very Old Human Bloodstains. *Forensic Science International* **13**: 111-116.
**Downs**, E.F. and J.M. Lowenstein (1996). Identification of Archaeological Blood Proteins. A Cautionary Note. *Journal of Archaeological Science* **22**: 11-16.
**Eisele**, J.A., D.D. Fowler, G. Haynes and R.A. Lewis (1995). Survival and Detection of Blood Residues on Stone Tools. *Antiquity* **69**: 36-46.
**Eerkens**, J.W. (2005). GC-MS Analysis and Fatty Acid Ratios of Archaeological Potsherds from the Western Great Basin of North America. *Archaeometry* **47,1**: 83-102.
**Evershed**, R.P., H.R. Mottram, S.N. Dudd, S. Charters, A.W. Stott and G.J. Lawrence (1997). New Criteria for the Identification of Animal Fats preserved in Archaeological Pottery. *Naturwissenschaften* **84**: 402-406.
**Evershed**, R.P. and N. Tuross (1996). Proteinaceous Material from Potsherds and Associated Soils. *Journal of Archaeological Science* **23**: 429-436.
**Fankhauser**, B.L. (1994). Protein and Lipid Analysis of Food Residues. In J.G. Hather (ed.). *Tropical Archaeobotany. Applications and New Developments*. London: Routledge, pp. 227-250.
**Fankhauser**, B.L. (1997). Amino Acid Analysis of Food Residues in Pottery. A Field and Laboratory Study. *Archaeology in Oceania* **32**: 131-140.
**Fiedel**, S.J. (1996). Blood from Stones? Some

Methodological and Interpretative Problems in Blood Residue Analysis. *Journal of Archaeological Science* **23**: 139-147.
**Görg**, A., C. Obermaier, G. Boguth, A. Harder, B. Scheibe, R. Wildgruber and W. Weiss (2000). The Current State of Two-dimensional Electrophoresis with Immobilized pH Gradients. *Electrophoresis* **21**: 1037-1053.
**Görg**, A., W. Weiss and M.J. Dunn (2004). Current Two-dimensional Electrophoresis Technology for Proteomics. *Proteomics* **4,12**: 3665-3685.
**Gurfinkel**, D.M. and U.M. Franklin (1988). A Study of the Feasibility of Detecting Blood Residues on Artifacts. *Journal of Archaeological Science* **15**: 83-97.
**Gurley**, L.R., J.G. Valdez, W.D. Spall, B.F. Smith and D.D. Gillette (1991). Proteins in the Fossil Bone of the Dinosaur, Seismosaurus. *Journal of Protein Chemistry* **10,1**: 75-90.
**Halpine**, S.M. (1992). Amino Acid Analysis of Proteinaceous Media from Cosimo Tura's The Annunciation with Saint Francis and Saint Louis of Toulouse. *Studies in Conservation* **37**: 22-38.
**Hedges**, R.E.M. (2002). Bone Diagenesis. An Overview of Processes. *Archaeometry* **44,3**: 319-328.
**Heginbotham**, A., V. Milay and M. Quick (2004). The Use of Immunofluorescence Microscopy (IFM) and Enzyme-Linked Immunosorbent Assay (ELISA) as Complementary Techniques for Protein Identification in Artists' Materials. 2004 Wooden Artifacts Group Postprints. Portland, Orgeon. *Wooden Artifacts Group (American Institute for Conservation) on-line journal* <http://aic.stanford.edu/sg/wag/> (accessed 31 December 2006).
**Henrikson**, L.S., R.M. Yohe II, M.E. Newman and M. Druss (1999). Freshwater Crustaceans as an Aboriginal Food Resource in the Northern Great Basin. *Journal of California and Great Basin Anthropology* **20,1**: 72-87.
**Johnson**, R.S., M.T. Davis, J.A. Taylor and S.D. Patterson (2005). Informatics for Protein Identification by Mass Spectrometry. *Methods* **35,3**: 223-236.
**Jones**, P.L. (1962). Some Observation of Methods for Identifying Proteins in Paint Media. *Studies in Conservation* **7**: 10-16.
**Jun**, K.K. (1974). Preserved Amino Acids from Silicified Protein in Fossil Radiolaria. *Nature* **252**: 690-692.
**Jurado-López**, A. and M.D. Luque de Castro (2004). Use of Near Infrared Spectroscopy in a Study of Binding Media used in Paintings. *Analytical and Bioanalytical Chemistry* 380: 706-711.
**Karpowicz** A. (1981). Ageing and Deterioration of Proteinaceous Media. *Studies in Conservation* **26**: 153-160.
**Kind**, S.S. and R.M. Cleevely (1969). The Use of Ammoniacal Bloodstain Extracts in ABO Groupings. *Journal of Forensic Sciences* **15**: 131-134.
**Keck**, S. and Th. Peters Jr. (1969). Identification of Protein-containing Paint Media by Quantitative Amino Acid Analysis. *Studies in Conservation* **14**: 75-82.
**Keil**, R.G., D.B. Montlucon, F.G. Prahl and J.I. Hedges (1994). Sorptive Preservation of Labile Organic Matter in Marine Sediments. *Nature* **370**: 549.
**Kooyman**, B., M.E. Newman and H. Ceri (1992). Verifying the Reliability of Blood Residue Analysis on Archaeological Tools. *Journal of Archaeological Science* **19**: 265-269.
**Kooyman**, B., M.E. Newman, C. Cluney, M. Lobb, S. Tolman, P. McNeil and L.V. Hills (2001). Identification of Horse Exploitation by Clovis Hunters based on Protein Analysis. *American Antiquity* **66,4**: 686-691.
**Kosten**, T. and S.M. Owens (2005). Immunotherapy for the Treatment of Drug Abuse. *Pharmacology and Therapeutics* **108,1**: 76-85.
**Lombard**, M., and L. Wadley (2007). The Morphological Identification of Micro-residues on Stone Tools using Light Microscopy. Progress and Difficulties based on Blind Tests. *Journal of Archaeological Science* **34**: 155-165.
**Lowenstein**, J.M. (1981). Immunological Reations from Fossil Material. *Philosophical Transactions of the Royal Society, London* **B 292**: 143-149.
**Loy**, T.H. (1983). Prehistoric Blood Residues. Detection of Tool Surfaces and Identification of Species Origin. *Science* **220**: 1269-1271.
**Malainey**, M.E., R. Przybylski and B.L. Sherriff (1999a). The Effects of Thermal and Oxidative Degradation on the Fatty Acid Composition of Food Plants and Animals of Western Canada: Implications for the Identification of Archaeological Vessel Residues. *Journal of Archaeological Science* **26**: 95-103.
**Malainey**, M.E., R. Przybylski and B.L. Sherriff (1999b). Identifying the Former Contents of Late Precontact Period Pottery Vessels from Western Canada using Gas Chromatography. *Journal of Archaeological Science* **26**: 425-438.
**Marlar**, R.A., B.L. Leonard, B.R. Billman, P.M. Lambert and J.E. Malar (2000). Biochemical Evidence of Cannibalism at a Prehistoric Site in Southwestern Colorado. *Nature* **407**: 74-78.
**Morton**, J.D. and H.P. Schwarcz (1988). Stable Isotope Analysis of Food Residue from Ontario Ceramics. In R.M. Farquhar, R.G.V. Hancock and L.A. Pavlish (eds.). *Proceedings of 26$^{th}$ International Archaeometry Symposium*. Toronto, University of Toronto Press: pp. 89-93.
**Morton**, J. and H.P. Schwarcz (2004). Paleodietary Implications from Isotopic Analysis of Food Residues on Prehistoric Ontario Ceramics. *Journal of Archaeological Science* **31**: 503-517.
**Newman**, M.E. and P. Julig (1989). The Identification of Protein Residues on Lithic Artifacts from a Stratified Boreal Forest Site. *Canadian Journal of Archaeology* **13**: 119-132.
**Newman**, M.E, R.M. Yohe, B. Kooyand and H. Ceri (1997). 'Blood' from Stones? Probably. A Response to Fiedel. *Journal of Archaeological Science* **24**: 1023-

1027.

**Nielsen-Marsh**, C.M., P.H. Ostrom, H. Gandhi, B. Shapiro, A. Cooper, P.V. Hauschka and M.J. Collins (2002). Sequence Preservation of Osteocalcin Protein and Mitochondrial DNA in Bison Bones older than 55kA. *Geology* **30,12**: 1099-1102.

**Oudemans**, T.F.M., G.B. Eijkel and J.J. Boon (2007). Identifying Biomolecular Origins of Solid Organic Residues Preserved in Iron Age Pottery using DTMS and MVA. *Journal of Archaeological Science* **34,2**: 173-193.

**Poinar**, H.N. and B.A. Stankiewicz (1999). Protein Preservation and DNA Retrieval from Ancient Tissues. *Proceedings of the National Academy of Sciences USA* **96,15**: 8426-8432.

**Ramirez-Barat**, B. and S. de la Vina (2001). Characterisation of Proteins in Paint Media by Immunofluorescence. A Note on Methodological Aspects. *Studies in Conservation* **46**: 282-288.

**Reuther**, J.D., J.M. Lowenstein, S.C. Gerlach, D. Hood, G. Scheuenstuhl and D.H. Ubelaker (2006). The Use of an Improved pRIA Technique in the Identification of Protein Residues. *Journal of Archaeological Science* **33**: 531-537.

**Rohrbough**, J., L. Breci, N. Merchant S. Miller and P. A. Haynes (2006). Verification of Single-peptide Protein Identifications by Application of Complementary Database Search Algorithms. *Journal of Biomolecular Techniques* **17**: 327-332.

**Ronca**, F. (1994). Protein Determination in Polychromed Stone Sculptures, Stuccoes and Gesso Grounds. *Studies in Conservation* **39**: 107-120.

**Sanday**, P.R. (1986). *Divine Hunger. Cannibalism as a Cultural System*. Cambridge, Cambridge University Press.

**Sartore**, R.L. (1994). *Humans eating Humans. The Dark Shadow of Cannibalism*. Notre Dame, Cross Cultural Publications.

**Schweiter**, M., C. Hill, J. Asara, W. Lane and S. Pincus (2002). Identification of Immunoreactive Material in Mammoth Fossils. *Journal of Molecular Evolution* **55(6)**: 696-705.

**Schweitzer**, M.H., M. Marshall, K. Carron, D.S. Bohle, S.C. Busse, E.V. Arnold, D. Barnard, J.R. Horner and J.R. Starkey (1997). Heme Compounds in Dinosaur Trabecular Bone. *Proceedings of the National Academy of Sciences USA* **94**: 6291-6296.

**Shanks,** O.C., R. Bonnichsen, A.T. Vella and E. Ream (2001). Recovery of Protein and DNA trapped in Stone Tool Microcracks. *Journal of Archaeological Science* **28**: 965-972.

**Shanks**, O.C., M. Kornfeld and D.D. Hawk (1999). Protein Analysis of Bugas-Holding Tools. New Trends in Immunological Studies. *Journal of Archaeological Science* **26**: 1183-1191.

**Shanks**, O.C., M. Kornfeld and W. Ream (2004). DNA and Protein Recovery from Washed Experimental Stone Tools. *Archaeometry* **46,4**: 663-672.

**Stevanato**, R., M. Rovea, M. Carbini, D. Favretto and P. Traldi. Curie-point Pyrolysis/Gas Chromatography/Mass Spectrometry in the Art Field. Part 3. The Characterization of some Non-proteinaceous Binders. *Rapid Communications in Mass Spectrometry* **11**: 286-294.

**Ubelaker**, D.H., J.M. Lowenstein and D.G. Hood (2004). Use of Solid-phase Double-antibody Radioimmunoassay to Identify Species from Small Skeletal Fragments. *Journal of Forensic Science* **49**: 924-929.

**Vallance**, S.L., B.W. Singer, S.M. Hitchen and J.H. Townsend (1998). The Development and Initial Application of a Gas Chromatographic Method for the Characterization of Gum Media. *Journal of the American Institute for Conservation* **37,3**: 294-311.

**Watson**, J.D. and F.H. Crick (1953). Molecular Structure of Nucleic Acids. A Structure for Deoxyribose Nucleic Acid. *Nature* **171**: 737-738.

**Yates** , J.R. 3rd (1998). Database Searching Using Mass Spectrometry Data. *Electrophoresis* **19,6**: 893-900.

**Yohe**, R.M., M.E. Newman and J.S. Schneider (1991). Immunological Identification of Small-mammal Proteins on Aboriginal Milling Equipment. *American Antiquity* **56,4**: 659-666.

# Appendix I: Common Isotopes of 99 Elements

| Number | Element | Isotope | Isotopic mas | Relative abundance | Average mass |
|---|---|---|---|---|---|
| 89 | **Ac** | 227 | 227.02775 | ? | ±227 |
| 47 | **Ag** | 107 | 106.90509 | 51.84 | 107.87 |
| | | 109 | 108.90476 | 48.16 | |
| 13 | **Al** | 27 | 26.98154 | 100 | 26.98 |
| 95 | **Am** | 241 | 241.05682 | S | ±243 |
| | | 243 | 243.06137 | | |
| 18 | **Ar** | 36 | 35.96755 | 0.34 | 39.95 |
| | | 38 | 37.96273 | 0.06 | |
| | | 40 | 39.96238 | 99.60 | |
| 33 | **As** | 75 | 74.92160 | 100 | 74.92 |
| 85 | **At** | 210 | 209.98713 | ? | ±210 |
| | | 211 | 210.98748 | | |
| 79 | **Au** | 197 | 196.96655 | 100 | 196.97 |
| 5 | **B** | 10 | 10.01294 | 19.97 | 10.81 |
| | | 11 | 11.00931 | 80.17 | |
| 56 | **Ba** | 130 | 129.90631 | 0.11 | 137.33 |
| | | 132 | 131.90506 | 0.10 | |
| | | 134 | 133.90450 | 2.42 | |
| | | 135 | 134.90568 | 6.59 | |
| | | 136 | 135.90457 | 7.85 | |
| | | 137 | 136.90582 | 11.23 | |
| | | 138 | 137.90524 | 71.70 | |
| 4 | **Be** | 9 | 9.01218 | 100 | 9.01 |
| 83 | **Bi** | 209 | 208.98038 | 100 | 208.98 |
| 97 | **Bk** | 247 | 247.07030 | S | ±247 |
| | | 249 | 249.07498 | | |
| 35 | **Br** | 79 | 78.91834 | 50.69 | 79.90 |
| | | 81 | 80.91629 | 49.31 | |
| 6 | **C** | 12 | **12 *)** | 98.93 | 12.01 |
| | | 13 | 13.00335 | 1.07 | |
| | **$^{14}C$** | 14 | 14.00324 | $10^{-10}$ | |
| 20 | **Ca** | 40 | 39.96259 | 96.94 | 40.08 |
| | | 42 | 41.95862 | 0.65 | |
| | | 43 | 42.95877 | 0.14 | |
| | | 44 | 43.95548 | 2.09 | |
| | | 46 | 45.95369 | 0.00 | |
| | | 48 | 47.95253 | 0.19 | |
| 48 | **Cd** | 106 | 105.90646 | 1.25 | 112.41 |
| | | 108 | 107.90418 | 0.89 | |
| | | 110 | 109.90301 | 12.49 | |
| | | 111 | 110.90418 | 12.80 | |
| | | 112 | 111.90276 | 24.13 | |
| | | 113 | 112.90440 | 12.22 | |
| | | 114 | 113.90336 | 28.73 | |
| | | 116 | 115.90476 | 7.49 | |
| 58 | **Ce** | 136 | 135.90714 | 0.19 | 140.12 |
| | | 138 | 137.90599 | 0.25 | |
| | | 140 | 139.90543 | 88.45 | |
| | | 142 | 141.90924 | 11.11 | |
| 98 | **Cf** | 249 | 249.07485 | S | ±251 |
| | | 250 | 250.07640 | | |
| | | 251 | 251.07958 | | |
| | | 252 | 252.08162 | | |
| 17 | **Cl** | 35 | 34.96885 | 75.78 | 35.45 |
| | | 37 | 36.96590 | 24.22 | 35.45 |
| 96 | **Cm** | 243 | 243.06138 | S | ±247 |
| | | 244 | 244.06275 | | |
| | | 245 | 245.06549 | | |
| | | 246 | 246.06722 | | |
| | | 247 | 247.07035 | | |
| | | 248 | 248.07234 | | |
| 27 | **Co** | 59 | 58.93320 | 100 | 58.93 |
| 24 | **Cr** | 50 | 49.94605 | 4.35 | 52.00 |
| | | 52 | 51.94051 | 83.79 | |
| | | 53 | 52.94065 | 9.50 | |
| | | 54 | 53.93888 | 2.37 | |
| 55 | **Cs** | 133 | 132.90545 | 100 | 132.91 |
| 29 | **Cu** | 63 | 62.92960 | 69.17 | 63.55 |
| | | 65 | 64.92779 | 30.83 | |
| 1**) | **H** | 1 | 1.00783 | 99.99 | 1.01 |
| | **D** | 2 | 2.01410 | 0.01 | |
| 66 | **Dy** | 156 | 155.92428 | 0.06 | 162.50 |
| | | 158 | 157.92441 | 0.10 | |
| | | 160 | 159.92519 | 2.34 | |
| | | 161 | 160.92693 | 18.91 | |
| | | 162 | 161.92680 | 25.51 | |
| | | 163 | 162.92873 | 24.90 | |
| | | 164 | 163.92917 | 28.18 | |
| 68 | **Er** | 162 | 161.92878 | 0.14 | 167.26 |
| | | 164 | 163.92920 | 1.61 | |
| | | 166 | 165.93029 | 33.61 | |
| | | 167 | 166.93205 | 22.93 | |
| | | 168 | 167.93237 | 26.78 | |
| | | 170 | 169.93546 | 14.93 | |
| 99 | **Es** | 252 | 252.08297 | S | ±252 |
| 63 | **Eu** | 151 | 150.91985 | 47.81 | 151.96 |
| | | 153 | 152.92123 | 52.19 | |
| 9 | **F** | 19 | 18.99840 | 100 | 19.00 |
| 26 | **Fe** | 54 | 53.93961 | 5.85 | 55.85 |
| | | 56 | 55.93494 | 91.75 | |
| | | 57 | 56.93540 | 2.12 | |
| | | 58 | 57.93328 | 0.28 | |
| 87 | **Fr** | 223 | 223.01973 | ? | ±223 |
| 31 | **Ga** | 69 | 68.92558 | 60.11 | 69.72 |
| | | 71 | 70.92471 | 39.89 | |

## Appendix I: Common Isotopes of 99 Elements

| | | | | | |
|---|---|---|---|---|---|
| 64 | **Gd** | 152 | 151.91979 | 0.20 | 157.25 |
| | | 154 | 153.92088 | 2.18 | |
| | | 155 | 154.92262 | 14.80 | |
| | | 156 | 155.92212 | 20.47 | |
| | | 157 | 156.92396 | 15.65 | |
| | | 158 | 157.92410 | 24.84 | |
| | | 160 | 159.92705 | 21.86 | |
| 32 | **Ge** | 70 | 69.92425 | 20.84 | 72.64 |
| | | 72 | 71.92208 | 27.54 | |
| | | 73 | 72.92346 | 7.73 | |
| | | 74 | 73.92118 | 36.28 | |
| | | 76 | 75.92140 | 7.61 | |
| 1**) | **H** | 1 | 1.00783 | 99.99 | 1.01 |
| | **D** | 2 | 2.01410 | 0.01 | |
| 2 | **He** | 3 | 3.01603 | 0.0001 | 4.00 |
| | | 4 | 4.00260 | 99.9999 | |
| 72 | **Hf** | 174 | 173.94004 | 0.16 | 178.49 |
| | | 176 | 175.94140 | 5.26 | |
| | | 177 | 176.94322 | 18.60 | |
| | | 178 | 177.94370 | 27.28 | |
| | | 179 | 178.94582 | 13.62 | |
| | | 180 | 179.94655 | 35.08 | |
| 80 | **Hg** | 196 | 195.96582 | 0.15 | 200.59 |
| | | 198 | 197.96675 | 9.97 | |
| | | 199 | 198.96826 | 16.87 | |
| | | 200 | 199.96831 | 23.10 | |
| | | 201 | 200.97029 | 13.18 | |
| | | 202 | 201.97063 | 29.86 | |
| | | 204 | 203.97348 | 6.87 | |
| 67 | **Ho** | 165 | 164.93032 | 100 | 164.93 |
| 53 | **I** | 127 | 126.90447 | 100 | 126.90 |
| 49 | **In** | 113 | 112.90406 | 4.29 | 114.82 |
| | | 115 | 114.90388 | 95.71 | |
| 77 | **Ir** | 191 | 190.96059 | 37.32 | 192.22 |
| | | 193 | 192.96292 | 62.72 | |
| 19 | **K** | 39 | 38.96371 | 93.26 | 39.10 |
| | | 40 | 39.96400 | 0.01 | |
| | | 41 | 40.96183 | 6.73 | |
| 36 | **Kr** | 78 | 77.92039 | 0.35 | 83.80 |
| | | 80 | 79.91638 | 2.28 | |
| | | 82 | 81.91348 | 11.58 | |
| | | 83 | 82.91414 | 11.49 | |
| | | 84 | 83.91151 | 57.00 | |
| | | 86 | 85.91061 | 17.30 | |
| 57 | **La** | 138 | 137.90711 | 0.09 | 138.91 |
| | | 139 | 138.90635 | 99.91 | |
| 3 | **Li** | 6 | 6.01512 | 7.59 | 6.94 |
| | | 7 | 7.01600 | 92.41 | |
| 71 | **Lu** | 175 | 174.94077 | 97.41 | 174.97 |
| | | 176 | 175.94268 | 2.59 | |
| 12 | **Mg** | 24 | 23.98504 | 78.99 | 24.31 |
| | | 25 | 24.98584 | 10.00 | |
| | | 26 | 25.98259 | 11.01 | |

| | | | | | |
|---|---|---|---|---|---|
| 25 | **Mn** | 55 | 54.93805 | 100 | 54.94 |
| 42 | **Mo** | 92 | 91.90681 | 14.84 | 95.94 |
| | | 94 | 93.90509 | 9.25 | |
| | | 95 | 94.90584 | 15.92 | |
| | | 96 | 95.90468 | 16.68 | |
| | | 97 | 96.90602 | 9.55 | |
| | | 98 | 97.90541 | 24.13 | |
| | | 100 | 99.90748 | 9.63 | |
| 7 | **N** | 14 | 14.00307 | 99.63 | 14.01 |
| | | 15 | 15.00011 | 0.37 | |
| 11 | **Na** | 23 | 22.98977 | 100 | 22.99 |
| 41 | **Nb** | 93 | 92.90638 | 100 | 92.91 |
| 60 | **Nd** | 142 | 141.90772 | 27.20 | 144.24 |
| | | 143 | 142.90981 | 12.20 | |
| | | 144 | 143.91008 | 23.80 | |
| | | 145 | 144.91257 | 8.30 | |
| | | 146 | 145.91311 | 17.23 | |
| | | 148 | 147.91689 | 5.71 | |
| | | 150 | 149.92089 | 5.62 | |
| 10 | **Ne** | 20 | 19.99244 | 90.48 | 20.18 |
| | | 21 | 20.99385 | 0.27 | |
| | | 22 | 21.99139 | 9.25 | |
| 28 | **Ni** | 58 | 57.93535 | 68.08 | 58.69 |
| | | 60 | 59.93079 | 26.22 | |
| | | 61 | 60.93106 | 1.14 | |
| | | 62 | 61.92835 | 3.63 | |
| | | 64 | 63.92797 | 0.93 | |
| 93 | **Np** | 237 | 237.04817 | S | ±237 |
| | | 239 | 239.05293 | | |
| 8 | **O** | 16 | 15.99491 | 99.76 | 16.00 |
| | | 17 | 16.99913 | 0.04 | |
| | | 18 | 17.99916 | 0.21 | |
| 76 | **Os** | 184 | 183.95249 | 0.02 | 190.23 |
| | | 186 | 185.95384 | 1.59 | |
| | | 187 | 186.95575 | 1.96 | |
| | | 188 | 187.95584 | 13.24 | |
| | | 189 | 188.95814 | 16.15 | |
| | | 190 | 189.95845 | 26.26 | |
| | | 192 | 191.96148 | 40.78 | |
| 15 | **P** | 31 | 30.97376 | 100 | 30.97 |
| 91 | **Pa** | 231 | 231.03588 | 100 | 231.04 |
| 82 | **Pb** | 204 | 203.97303 | 1.41 | 207.21 |
| | | 206 | 205.97445 | 24.11 | |
| | | 207 | 206.97588 | 22.11 | |
| | | 208 | 207.97664 | 52.41 | |
| 46 | **Pd** | 102 | 101.90561 | 1.02 | 106.42 |
| | | 104 | 103.90404 | 11.14 | |
| | | 105 | 104.90508 | 22.33 | |
| | | 106 | 105.90348 | 27.33 | |
| | | 108 | 107.90389 | 26.46 | |
| | | 110 | 109.90515 | 11.72 | |
| 61 | **Pm** | 145 | 144.91274 | ***) | ±145 |
| | | 147 | 146.91513 | | |

| | | | | | |
|---|---|---|---|---|---|
| 84 | **Po** | 209 | 208.98242 | ? | ±209 |
| | | 210 | 209.98286 | | |
| 59 | **Pr** | 141 | 140.90765 | 100 | 140.91 |
| 78 | **Pt** | 190 | 189.95993 | 0.01 | 195.08 |
| | | 192 | 191.96104 | 0.78 | |
| | | 194 | 193.96266 | 32.97 | |
| | | 195 | 194.96477 | 33.83 | |
| | | 196 | 195.96494 | 25.24 | |
| | | 198 | 197.96788 | 7.16 | |
| 94 | **Pu** | 238 | 238.04955 | S | ±244 |
| | | 239 | 239.05216 | | |
| | | 240 | 240.05381 | | |
| | | 241 | 241.05685 | | |
| | | 242 | 242.05874 | | |
| | | 244 | 244.06420 | | |
| 88 | **Ra** | 223 | 223.01850 | ? | ±226 |
| | | 224 | 224.02020 | | |
| | | 226 | 226.02540 | | |
| | | 228 | 228.03106 | | |
| 37 | **Rb** | 85 | 84.91179 | 72.17 | 85.47 |
| | | 87 | 86.90918 | 27.83 | |
| 75 | **Re** | 185 | 184.95296 | 37.40 | 186.21 |
| | | 187 | 186.95575 | 62.60 | |
| 45 | **Rh** | 103 | 102.90550 | 100 | 102.91 |
| 86 | **Rn** | 211 | 210.99059 | ? | ±222 |
| | | 220 | 220.01138 | | |
| | | 222 | 222.01757 | | |
| 44 | **Ru** | 96 | 95.90760 | 5.53 | 101.07 |
| | | 98 | 97.90529 | 1.87 | |
| | | 99 | 98.90594 | 12.76 | |
| | | 100 | 99.90422 | 12.60 | |
| | | 101 | 100.90558 | 17.06 | |
| | | 102 | 101.90435 | 31.55 | |
| | | 104 | 103.90543 | 18.62 | |
| 16 | **S** | 32 | 31.97207 | 94.93 | 32.07 |
| | | 33 | 32.97146 | 0.76 | |
| | | 34 | 33.96787 | 4.29 | |
| | | 36 | 35.96708 | 0.02 | |
| 51 | **Sb** | 121 | 120.90382 | 57.21 | 121.76 |
| | | 123 | 122.90422 | 42.79 | |
| 21 | **Sc** | 45 | 44.95591 | 100 | 44.96 |
| 34 | **Se** | 74 | 73.92248 | 0.89 | 78.96 |
| | | 76 | 75.91921 | 9.37 | |
| | | 77 | 76.91991 | 7.63 | |
| | | 78 | 77.91731 | 23.77 | |
| | | 80 | 79.91652 | 49.61 | |
| | | 82 | 81.91670 | 8.73 | |
| 14 | **Si** | 28 | 27.97693 | 92.23 | 28.09 |
| | | 29 | 28.97649 | 4.68 | |
| | | 30 | 29.97377 | 3.09 | |
| 62 | **Sm** | 144 | 143.91200 | 3.07 | 150.36 |
| | | 147 | 146.91489 | 14.99 | |
| | | 148 | 147.91482 | 11.24 | |

| | | | | | |
|---|---|---|---|---|---|
| | | 149 | 148.91718 | 13.82 | |
| | | 150 | 149.91727 | 7.38 | |
| | | 152 | 151.91973 | 26.75 | |
| | | 154 | 153.92221 | 22.75 | |
| 50 | **Sn** | 112 | 111.90482 | 0.97 | 118.71 |
| | | 114 | 113.90278 | 0.66 | |
| | | 115 | 114.90335 | 0.34 | |
| | | 116 | 115.90174 | 14.54 | |
| | | 117 | 116.90295 | 7.68 | |
| | | 118 | 117.90161 | 24.22 | |
| | | 119 | 118.90331 | 8.59 | |
| | | 120 | 119.90220 | 32.58 | |
| | | 122 | 121.90344 | 4.63 | |
| | | 124 | 123.90527 | 5.79 | |
| 38 | **Sr** | 84 | 83.91343 | 0.56 | 87.62 |
| | | 86 | 85.90926 | 9.86 | |
| | | 87 | 86.90888 | 7.00 | |
| | | 88 | 87.90561 | 82.58 | |
| 73 | **Ta** | 180 | 179.94747 | 0.01 | 180.95 |
| | | 181 | 180.94800 | 99.99 | |
| 65 | **Tb** | 159 | 158.92534 | 100 | 158.93 |
| 43 | **Tc** | 97 | 96.90637 | ***) | 98.00 |
| | | 98 | 97.90722 | | |
| | | 99 | 98.90625 | | |
| 52 | **Te** | 120 | 119.90402 | 0.09 | 127.60 |
| | | 122 | 121.90305 | 2.55 | |
| | | 123 | 122.90427 | 0.89 | |
| | | 124 | 123.90282 | 4.74 | |
| | | 125 | 124.90442 | 7.07 | |
| | | 126 | 125.90331 | 18.84 | |
| | | 128 | 127.90446 | 31.74 | |
| | | 130 | 129.90622 | 34.08 | |
| 90 | **Th** | 232 | 232.03805 | 100 | 232.04 |
| 22 | **Ti** | 46 | 45.95263 | 8.25 | 47.87 |
| | | 47 | 46.95176 | 7.44 | |
| | | 48 | 47.94795 | 73.72 | |
| | | 49 | 48.94787 | 5.41 | |
| | | 50 | 49.94479 | 5.18 | |
| 81 | **Tl** | 203 | 202.97233 | 29.52 | 204.38 |
| | | 205 | 204.97441 | 70.48 | |
| 69 | **Tm** | 169 | 168.93421 | 100 | 168.93 |
| 92 | **U** | 234 | 234.04095 | 0.01 | 238.03 |
| | | 235 | 235.04392 | 0.72 | |
| | | 238 | 238.05078 | 99.27 | |
| 23 | **V** | 50 | 49.94716 | 0.25 | 50.94 |
| | | 51 | 50.94396 | 99.75 | |
| 74 | **W** | 180 | 179.94671 | 0.12 | 183.84 |
| | | 182 | 181.94821 | 26.50 | |
| | | 183 | 182.95022 | 14.31 | |
| | | 184 | 183.95093 | 30.64 | |
| | | 186 | 185.95436 | 28.43 | |
| 54 | **Xe** | 124 | 123.90590 | 0.09 | 131.29 |
| | | 126 | 125.90427 | 0.09 | |

| | | | | | |
|---|---|---|---|---|---|
| | | 128 | 127.90353 | 1.92 | |
| | | 129 | 128.90478 | 26.44 | |
| | | 130 | 129.90351 | 4.08 | |
| | | 131 | 130.90508 | 21.18 | |
| | | 132 | 131.90415 | 26.89 | |
| | | 134 | 133.90539 | 10.44 | |
| | | 136 | 135.90722 | 8.87 | |
| 39 | **Y** | 89 | 88.90585 | 100 | 88.91 |
| 70 | **Yb** | 168 | 167.93389 | 0.13 | 173.04 |
| | | 170 | 169.93476 | 3.04 | |
| | | 171 | 170.93632 | 14.28 | |
| | | 172 | 171.93638 | 21.83 | |
| | | 173 | 172.93821 | 16.13 | |
| | | 174 | 173.93886 | 31.83 | |
| | | 176 | 175.94257 | 12.76 | |
| 30 | **Zn** | 64 | 63.92915 | 48.63 | 65.41 |
| | | 66 | 65.92604 | 27.90 | |
| | | 67 | 66.92713 | 4.10 | |
| | | 68 | 67.92485 | 18.75 | |
| | | 70 | 69.92533 | 0.62 | |
| 40 | **Zr** | 90 | 89.90470 | 51.45 | 91.22 |
| | | 91 | 90.90565 | 11.22 | |
| | | 92 | 91.90504 | 17.15 | |
| | | 94 | 93.90632 | 17.38 | |
| | | 96 | 95.90828 | 2.80 | |

Table 1: Data on the most common isotopes of 99 elements in alphabetical order of their abbreviated name (source: NIST). 'Number' is the atomic number (equals the number of protons); 'isotopic mass' is the exact mass in Daltons (relative to $^{12}C$ = 12 Da); 'isotope' is the number of the isotope (equals the integer mass in Da); 'relative abundance' is the percentage of each isotope in a sample of the natural element (S = synthetic); 'average mass' is the weighed mean mass in Da.

*) By definition, 12 Da equals the mass of $^{12}C$ ($19.9 \times 10^{-27}$ kg). The half-life of the rare (abundance about $10^{-10}$%) but scientifically important $^{14}C$ is 5730 years (the products of the decay are $^{14}N$ and β-particles).

**) Hydrogen (H), with one proton and one electron, and deuterium (D), with a proton, a neutron and an electron, are the same element (atomic number 1), but remain named differently for historical reasons. Deuterium was detected and named in 1931, before the discovery of the neutron, by Harold Urey for which he received the 1934 Nobel Prize in Chemistry. In 1933 Gilbert Lewis purified $D_2O$ (heavy water) and the first commercial plant to produce heavy water was built in 1934 in Vemork, Norway. During the Second World War Norway was under German control and it was feared by the Allies that Nazi Germany would use the Norwegian heavy water for a nuclear program resulting in an atomic bomb. Several attacks on the facilities, in February 1943, November 1943 and February 1944, effectively ended such German ambitions.

***) Of the elements with less than 82 protons, only three have no stable isotopes: technetium (Tc, atomic number 43), promethium (Pm, atomic number 61) and tungsten (W, atomic number 74). The search for 'element 43' has occupied many chemists and physicists for more than a century. Technetium was finally isolated by Carlo Perrier and Emilio Segrè in 1937, who found $^{97}Tc$ in an amount of molybdenum (Mo, atomic number 42) that was bombarded with deuterium (D, atomic number 1) for several months. As this was the first artificially produced element, they named it after the Greek 'technetos' (artificial). In 1952 Paul Merrill detected the spectral signature of technetium in 'red giant' stars. Technetium is now widely used in medical scintigraphic imaging. Promethium was discovered by Jacob Marinsky, Lawrence Glendenin and Charles Coryell, in 1945, who published their findings in 1947. Minute quantities of Tc and Pm are produced by the radio-active decay of uranium (U, atomic number 92), but these disappear relatively quickly (the half-lifes of the most stable $^{97}Tc = 2.6 \times 10^6$ years and $^{98}Tc = 4.2 \times 10^6$ years; the half-life of the most stable $^{145}Pm$ = 17.7 years). The half-lifes of the given isotopes of tungsten (also known as wolfram) are so long that they can be considered stable. Elements with more than 82 protons are either radio-active (unstable with highly variable half-lifes of the isotopes) or synthetic (S).

(H. Barnard)

# APPENDIX II
# A Short Overview of Protein Biochemistry

Proteins are long chains of amino acids (Figures 1 and 2, Table 1) connected by peptide bonds (Figure 1) and folded into a specific three-dimensional shape that defines their function.[1] Non-functional protein fragments, but also proteins shorter than 50 amino acids are usually referred to as polypeptides, or simply peptides.[2]

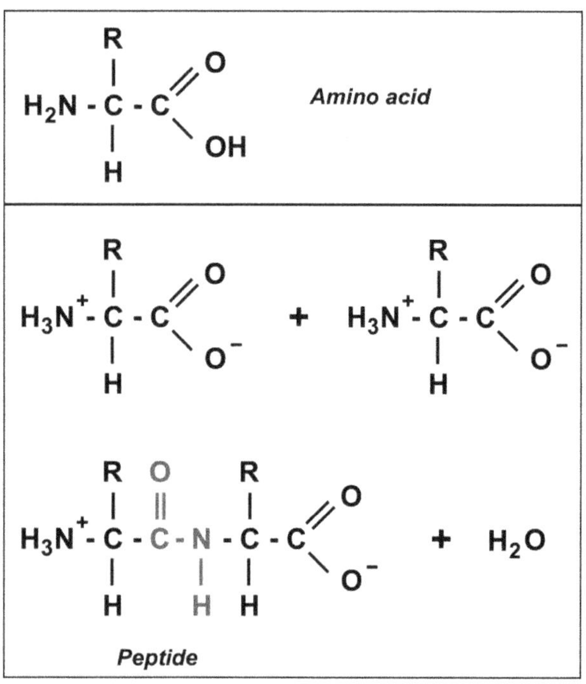


Figure 1: The general structure of the standard proteinogenic amino acids (top) and the formation of a peptide bond (bottom). R represents the side chain characteristic for individual amino acids and varies from a single H-atom (glycine) to the relatively complex side chain of tryptophan (Figure 2).

Proteins can be divided into three large classes: globular, fibrous and membrane proteins. Most globular proteins are soluble in water and many are enzymes (biocatalysts that facilitate or stimulate specific biochemical reactions). Most fibrous proteins are structural and provide rigidity (such as collagen, elastin and keratin in skin, nails and bones) or mobility (such as actin, myosin and kinesin in muscles, intestines and the heart) to organisms. Membrane proteins are incorporated into the lipid membranes of cells to serve as channels for the transport of molecules, in or out of the cell, or to act as receptors. Other proteins, such as insulin, serve as messenger molecules (hormones).

Ingested proteins are broken down into free amino acids that can be used for protein synthesis. They are also oxidized to provide energy, yielding amounts similar to carbohydrates (around 17 kJ/g, only about half the value provided by acylglycerides of around 37 kJ/g). There are 20 amino acids that occur in natural proteins, the 'standard proteinogenic amino acids' (Figure 2, Table 1). Because humans, like other mammals, cannot synthesize all of these amino acids, proteins are a necessary component of our food.

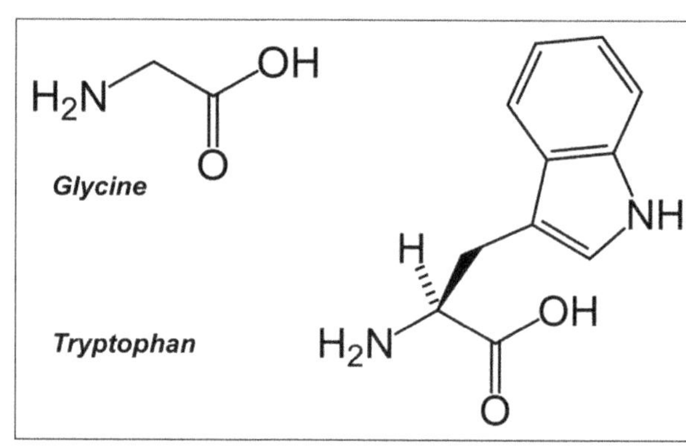


Figure 2: The smallest (glycine, with a single H-atom in the R position) and the largest (tryptophan) of the standard proteinogenic amino acids (Table 1).

---

1 The term protein is derived from the Greek 'proteios' (most important) and was coined by J.J. Berzelius in 1838 to indicate their significance.

2 The term peptide is derived from the Greek 'peptos' (digestible) and was coined by H.E. Fischer in 1901.

## Appendix II: A Short Overview of Protein Biochemistry

| Amino acid | Abbreviation | | Hydro-phobic | Polar | Mass (Da) | Charge | Codon | Abundance (%) |
|---|---|---|---|---|---|---|---|---|
| | 3 letters | 1 letter | | | | | | |
| **Alanine** | Ala | A | yes | no | 89.09 | 0 | GCA, GCC, GCG, GCU | 7.8 |
| **Arginine** | Arg | R | no | yes | 174.20 | pos | AGA, AGG, CGA, CGC, CGG, CGU | 5.1 |
| **Asparagine** | Asn | N | no | yes | 132.12 | 0 | AAC, AAU | 4.3 |
| **Aspartate** | Asp | D | no | yes | 133.10 | neg | GAC, GAU | 5.3 |
| **Cysteine** | Cys | C | yes | no | 121.16 | 0 | UGC, UGU | 1.9 |
| **Glutamate** | Glu | E | no | yes | 147.13 | neg | GAA, GAG | 6.3 |
| **Glutamine** | Gln | Q | no | yes | 146.15 | 0 | CAA, CAG | 4.2 |
| **Glycine** | Gly | G | yes | no | **75.07** | 0 | GGA, GGC, GGG, GGU | 7.2 |
| **Histidine** | His | H | no | yes | 155.16 | pos | CAC, CAU | 2.3 |
| **Isoleucine** | Ile | I | yes | no | 131.17 | 0 | AUA, AUC, AUU | 5.3 |
| **Leucine** | Leu | L | yes | no | 131.17 | 0 | CUA, CUC, CUG, CUU, UUA, UUG | **9.1** |
| **Lysine** | Lys | K | no | yes | 146.19 | pos | AAA, AAG | 5.9 |
| **Methionine** | Met | M | yes | no | 149.21 | 0 | **AUG** | 2.3 |
| **Phenylalanine** | Phe | F | yes | no | 165.19 | 0 | UUC,UUU | 3.9 |
| **Proline** | Pro | P | yes | no | 115.13 | 0 | CCA, CCC, CCG, CCU | 5.2 |
| **Serine** | Ser | S | no | yes | 105.09 | 0 | AGC, AGU, UCA, UCC, UCG, UCU | 6.8 |
| **Valine** | Val | V | yes | no | 117.15 | 0 | GUA, GUC, GUG, GUU | 6.6 |
| **Threonine** | Thr | T | yes | yes | 119.12 | 0 | ACA, ACC, ACG, ACU | 5.9 |
| **Tryptophan** | Trp | W | yes | no | **204.23** | 0 | **UGG** | **1.4** |
| **Tyrosine** | Tyr | Y | yes | yes | 181.19 | 0 | UAC, UAU | 3.2 |
| **Start of the peptide chain** | | | | | | | AUG (methionine) | |
| **Termination of the peptide chain** | | | | | | | UAA, UAG, UGA | |

Table 1: Overview of some properties of the standard proteinogenic amino acids, codons with a special function are given at the bottom (cf. Table 2). Mass is average mass in Daltons. Bold entries signify amino acids with the lowest (glycine) and the highest mass (tryptophan), those associated with a single codon (methionine, also the 'start'-codon, and tryptophan) as well as the most (leucine) and least abundant (tryptophan) amino acid in nature.

The genetic information necessary to create proteins is contained in DNA.[3] The actual synthesis of a protein starts with the transfer of this information to messenger RNA (mRNA) in a process called transcription (Figure 3).[4] This mRNA molecule is then used to generate the protein in a process called translation (Figure 4). As proteins thus express the genetic code, they are the main actors in nature.

## Transcription

Transcription is the first phase of protein synthesis, in biological systems, during which the genetic information contained within a part of a DNA molecule is transferred to a mRNA molecule. The functional part of DNA consists of a very long strand of the bases adenine (A), cytosine (C), guanine (G) and thymine (T), attached to a skeleton of sugars and phosphates. When not active, two strands of DNA form a double helix. These two strands are complementary: where one has adenine (A), the other has thymine (T) and where one has cytosine (C) the other has guanine (G). The strands are kept together by the hydrogen bonds between either cytocine and guanine, or adenine and thymine (Figure 3).[5] To read the genetic code for protein synthesis, the two strands are separated by RNA-polymerase and a copy of one of the strands is made on a RNA molecule (Figure 3). During the elongation of the RNA molecule, cytosine (C) is combined with guanine (G), and vice versa; thymine (T) is combined with adenine (A), but adenine (A) with uracil (U) that takes the position of thymine. After the mRNA molecule is completed, it is transported out of the nucleus of the cell into its cytoplasm for translation: the actual protein synthesis.

| Nucleotide triplet (codon) | | | | | |
|---|---|---|---|---|---|
| 1st (5' end) | 2nd | 3rd (3' end) | | | |
| | | A | C | G | U |
| A | A | Lys | Asn | Lys | Asn |
| | C | Thr | Thr | Thr | Thr |
| | G | Arg | Ser | Arg | Ser |
| | U | Ile | Ile | **Met** | Ile |
| | **AUG** = start of the chain | | | | |
| C | A | Gln | His | Gln | His |
| | C | Pro | Pro | Pro | Pro |
| | G | Arg | Arg | Arg | Arg |
| | U | Leu | Leu | Leu | Leu |
| G | A | Glu | Asp | Glu | Asp |
| | C | Ala | Ala | Ala | Ala |
| | G | Gly | Gly | Gly | Gly |
| | U | Val | Val | Val | Val |
| U | A | * | Tyr | * | Tyr |
| | C | Ser | Ser | Ser | Ser |
| | G | * | Cys | **Trp** | Cys |
| | U | Leu | Phe | Leu | Phe |
| | * = termination of the chain | | | | |

Table 2: The genetic code in which each triplet of nucleotides (codon) codes for one amino acid. A = adenine, C = cytosine, G = guanine, U = uracil (which in RNA take the place reserved for thymine in DNA). Bold entries signify amino acids with a single codon (methionine, also the 'start'-codon, and tryptophan).

[3] DNA stands for 'deoxyribonucleic acid', a long polymer of nucleotides (a heterocyclic base bound to a sugar and one or more phosphate groups) encoding for the sequence of amino acids in proteins. It was first isolated from the sperm of salmon, by Friedrich Miescher in 1869. Richard Altmann coined the name 'nucleic acid' in 1889. The components of DNA (the phosphate chain, sugars and four bases) were identified by Phoebus Levene in 1929. In 1937 Willian Astbury showed that DNA has a regular structure, although he could not determine what that structure was. Experiments by Oswald Avery, in 1943, and Alfred Hersey and Martha Chase, in 1953, did prove that DNA is the carrier of the genetic code. In 1951 Rosalind Franklin suggested, during a lecture in Cambridge, that DNA has a helical structure. A synthesis of all information (later resulting in a 'central dogma of molecular biology') was produced by James Watson and Francis Crick in 1953; work for which they received the 1962 Nobel Price for Physiology or Medicine. They had to share this honor with Maurice Wilkins, a colleague of Rosalind Franklin who had died in 1958. The central dogma was redefined by Francis Crick in 1970 (Crick 1970).

[4] RNA, or ribonucleic acid, is a DNA-like molecule but consisting of ribose instead of deoxyribose, and with uracil (U) in the position of thymine (T). The role of RNA was already suspected by Torbjörn Caspersson, Jean Brachet and Jack Schultz in 1939. The sequence of the nucleotides in a strand of RNA from yeast was determined by R.W. Holley in 1964, for which he was awarded the 1968 Nobel Prize for Physiology or Medicine.

[5] Hydrogen bonds are weak attracting forces between hydrogen atoms in a molecule (most importantly in a HO- or a HN-group) and other atoms (most importantly O or N) in the same or another molecule. Hydrogen bonds are weaker than covalent and ionic bonds but play a major role in maintaining the three-dimensional structure of DNA and proteins.

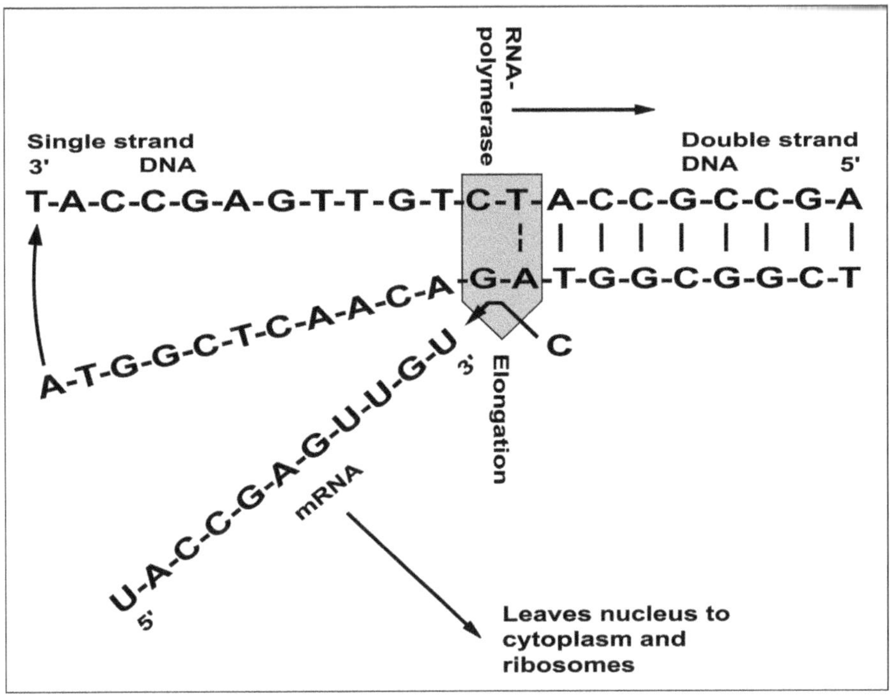


Figure 3: Schematic representation of the transcription of DNA into messenger RNA (mRNA) by RNA-polymerase. The DNA double helix is opened and a complementary copy of one of the strands is made on RNA, in which uracil (U) takes the place of thymine (T), opposite adenine (A).

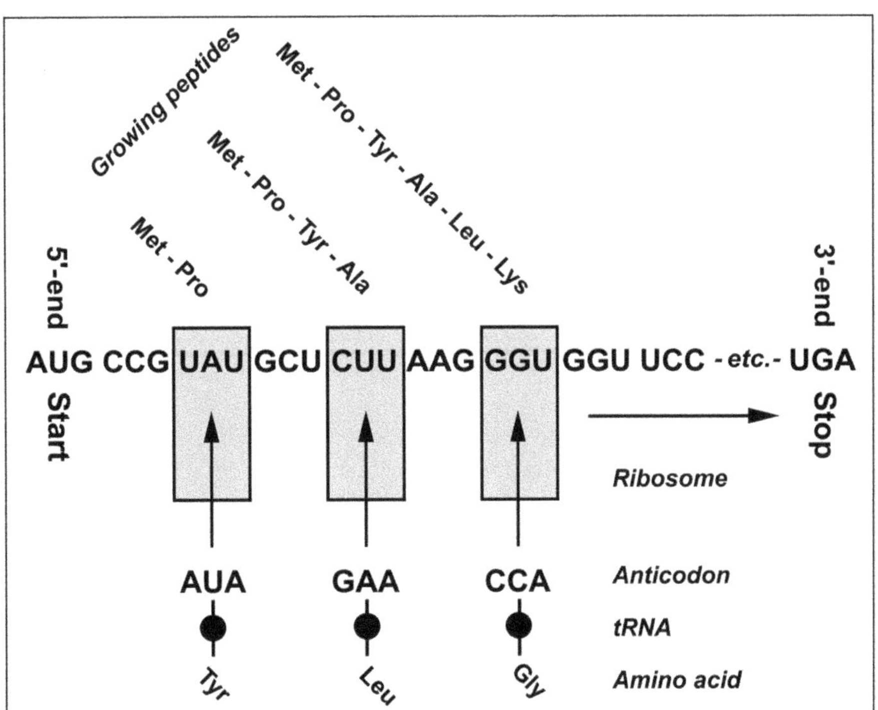


Figure 4: Schematic representation of the translation of messenger RNA (mRNA). Ribosomes read the genetic code and connect the amino acids, transported by transport RNA (tRNA), into a polypeptide. tRNA carries an anticodon to identify the correct place of insertion (C combines with G and A combines with U).

## Translation

Proteins are produced in ribosomes, small organelles in the cytoplasm. After mRNA is released into the cytoplasm, ribosomes slide over the molecule, from the 5'-end to the 3'-end,[6] to read the genetic code and create the associated protein (Figure 4). The sequence in which the amino acids are connected to form the polypeptide is directly determined by the sequence of the bases on the mRNA molecule. Three of these bases form a triplet, or codon, associated with a specific amino acid (Tables 1 and 2). As there are 64 possible codons and only 20 proteinogenic amino acids, most triplets are associated with more than one amino acid. There are also codons marking the start and the end of the sequence (Table 2).

Amino acids are transported to the right position by other strands of RNA called transport RNA (tRNA). The point of insertion is determined by an anti-codon (or 'nodoc') on the tRNA that recognizes the correct triplet on the strand of mRNA. After the peptide bond is made, and the amino acid is added to the growing polypeptide, the ribosome moves on to the next triplet, and so on until an 'end' codon is reached (Table 2, Figure 4).

## Post-translational Modifications

Following synthesis, proteins will fold into coils or sheets, which then folds again into a three dimensional structure (Figure 5), held together by hydrogen bonds (and other weak forces). The hydrophobic side-chains (Table 1) will generally face the interior of the protein. This usually brings specific functional groups into a specified position ensuring maximum activity of the molecule.[7]

Most proteins acquire alterations before they are fully biologically functional. Phosphorylation is a common mechanism for activating or inactivating a protein. Other post-translational modifications (PTMs) include adding functional groups (such as lipids or carbohydrates), changing the structure of specific amino acids, removing amino acids (from the amino end of the protein) or changing the structure of the peptide chain (by the formation of disulfide bonds or by cleavage of the chain into two or more parts).[8] PTMs are often effectuated by specialized proteins, such as kinases, phosphatases and chaperones. Table 3 gives an overview of the effect on the total average mass of a protein of some common PTMs.

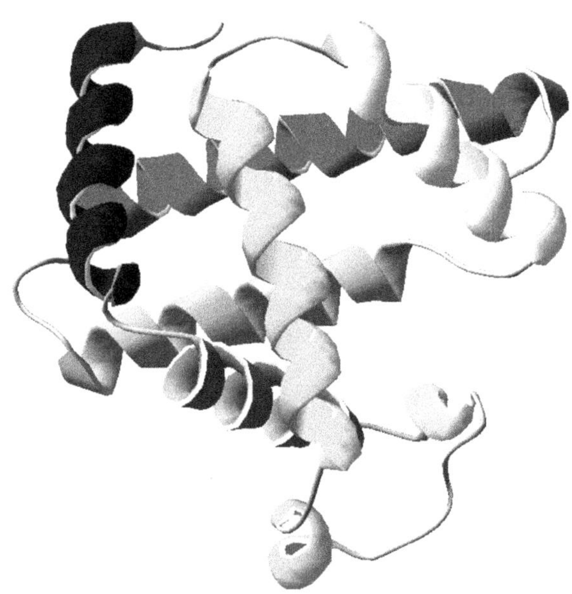

Figure 5: The three-dimensional structure of myoglobin, a single-chain protein of 153 amino acids (total molecular weight about 16,700 Da).[9]

[6] The terminal deoxyribose of a DNA strand, and the terminal ribose of a RNA strand, is referred to as the 3'-end (or three prime) and the terminal $-PO_3^-$ as the 5'-end (or five prime). Replication and transcription of DNA is performed in the 3' to 5' direction, the translation of RNA in the 5' to 3' direction.

[7] The primary structure of a protein is its amino acid sequence. The secondary structure refers to the local shape of a polypeptide, stabilized by hydrogen bonds. The tertiary structure is the overall shape of a protein usually stabilized by hydrophobic and hydrophilic forces, but also by hydrogen bonds, disulfide bonds or other PTMs to the primary structure. The quaternary structure refers to the interactions between several protein molecules meant to function more or less closely together.

[8] An example of how PTMs are needed to create a functional protein is the peptide hormone insulin. After synthesis of the polypeptide, a number of disulfide bonds are made and the peptide chain is cut twice. The central part of the chain is then removed, resulting in a functional protein that consists of two separate polypeptide chains connected by disulfide bonds.

[9] Myoglobin is the most important protein transporting oxygen in muscles. It was the first protein of which the structure was determined, in 1958, by Max F. Perutz and John C. Kendrew, who shared the 1962 Nobel Price in Chemistry for their research.

## Appendix II: A Short Overview of Protein Biochemistry

PTMs involving the addition of functional groups include:

- acetylation: the addition of an acetyl group, usually at the N-terminus of the protein;
- alkylation: the addition of an alkyl group;
- methylation: the addition of a methyl group, usually to lysine or arginine;
- glycosylation: the addition of a glycosyl group, usually to asparagine, serine or threonine, resulting in a glycoprotein;
- isoprenylation, the addition of an isoprenoid group such as farnesol or geranylgeraniol;
- lipoylation, the addition of lipoic acid;
- phosphorylation, the addition of a phosphate group, usually to serine, threonine or tyrosine;
- sulfation, the addition of a sulfate group, usually to serine, threonine or tyrosine.

PTMs involving changes in the structure of amino acids include deamination: the conversion of glutamine into glutamic acid or asparagine into aspartic acid. PTMs involving changes in the structure of the peptide chain include the formation of disulfide bonds (the covalent linking of two cysteine molecules) and proteolytic cleavage (the cleavage of a protein at a peptide bond).

| **PTM** | **Mass change** | **Common position** |
|---|---|---|
| Acetylation | + 42.04 | N-terminus or Lysine |
| Farnesylation | + 204.36 | |
| Formylation | + 28.01 | |
| Geranylgeraniolation | + 272.47 | |
| Lipoylation | + 188.31 | |
| Methylation | + 14.03 | |
| Myristoylation | + 210.36 | |
| Stearoylation | + 266.47 | |
| Deamination | + 0.98 | Asparagine or Glutamine |
| Deoxyhexose | + 146.14 | Asparagine, Serine or Threonine |
| Hexosamine | + 161.16 | |
| Hexose | + 162.14 | |
| N-acetylhexosamine | + 203.20 | |
| Pentose | + 132.12 | |
| Carboxylation | + 44.01 | Aspartate or Glutamate |
| Disulfide bond | – 2.02 | Cysteine |
| Cysteinylation | + 119.14 | |
| Glutathionylation | + 305.31 | |
| Oxidation | + 16.00 | |
| Hydroxylation | + 17.01 | Proline |
| Phosphorylation | + 79.98 | Serine, Threonine or Tyrosine |
| Sulfation | + 80.06 | |
| Amidation | – 0.98 | C-terminus |

Table 3: Overview of some common post-translational modifications (PTMs) with their usual place of occurrence and their effect on the total average mass of a polypeptide.

| | Gly (75.07) **57** | Ala (89.09) **71** | Ser (105.09) **87** | Pro (115.13) **97** | Val (117.15) **99** | Thr (119.12) **101** | Cys (121.16) **103** | Ile (131.17) **113** | Leu (131.17) **113** | Asn (132.12) **114** | Asp (133.10) **115** | Gln (146.15) **128** | Lys (146.19) **128** | Glu (147.13) **129** | Met (149.21) **131** | His (155.16) **137** | Phe (165.19) **147** | Arg (174.20) **156** | Tyr (181.19) **163** | Trp (204.23) **186** |
|---|---|---|---|---|---|---|---|---|---|---|---|---|---|---|---|---|---|---|---|---|
| **Gly 57** | --- | 14 | 30 | 40 | 42 | 44 | 46 | 56 | 56 | 57 | 58 | 71 | 71 | 72 | 74 | 80 | 90 | 99 | 106 | 129 |
| **Ala 71** | | --- | 16 | 26 | 28 | 30 | 32 | 42 | 42 | 43 | 44 | 57 | 57 | 58 | 60 | 66 | 76 | 85 | 92 | 115 |
| **Ser 87** | | | --- | 10 | 12 | 14 | 16 | 26 | 26 | 27 | 28 | 41 | 41 | 42 | 44 | 50 | 60 | 69 | 76 | 99 |
| **Pro 97** | | | | --- | 2 | 4 | 6 | 16 | 16 | 17 | 18 | 31 | 31 | 32 | 34 | 40 | 50 | 59 | 66 | 89 |
| **Val 99** | | | | | --- | 2 | 4 | 14 | 14 | 15 | 16 | 29 | 29 | 30 | 32 | 38 | 48 | 57 | 64 | 87 |
| **Thr 101** | | | | | | --- | 2 | 12 | 12 | 13 | 14 | 27 | 27 | 28 | 30 | 36 | 46 | 55 | 62 | 85 |
| **Cys 103** | | | | | | | --- | 10 | 10 | 11 | 12 | 25 | 25 | 26 | 28 | 34 | 44 | 53 | 60 | 83 |
| **Ile 113** | | | | | | | | --- | 0 | 1 | 2 | 15 | 15 | 16 | 18 | 24 | 34 | 43 | 50 | 73 |
| **Leu 113** | | | | | | | | | --- | 1 | 2 | 15 | 15 | 16 | 18 | 24 | 34 | 43 | 50 | 73 |
| **Asn 114** | | | | | | | | | | --- | 1 | 14 | 14 | 15 | 17 | 23 | 33 | 42 | 49 | 72 |
| **Asp 115** | | | | | | | | | | | --- | 13 | 13 | 14 | 16 | 22 | 32 | 41 | 48 | 71 |
| **Gln 128** | | | | | | | | | | | | --- | 0 | 1 | 3 | 9 | 19 | 28 | 35 | 58 |
| **Lys 128** | | | | | | | | | | | | | --- | 1 | 3 | 9 | 19 | 28 | 35 | 58 |
| **Glu 129** | | | | | | | | | | | | | | --- | 2 | 8 | 18 | 27 | 34 | 57 |
| **Met 131** | | | | | | | | | | | | | | | --- | 6 | 16 | 25 | 32 | 55 |
| **His 137** | | | | | | | | | | | | | | | | --- | 10 | 19 | 26 | 49 |
| **Phe 147** | | | | | | | | | | | | | | | | | --- | 9 | 16 | 39 |
| **Arg 156** | | | | | | | | | | | | | | | | | | --- | 7 | 30 |
| **Tyr 163** | | | | | | | | | | | | | | | | | | | --- | 23 |
| **Trp 186** | | | | | | | | | | | | | | | | | | | | --- |

Table 4: Overview of the effect of a single mutation, in which one amino acid is replaced by another, on the total integer mass of a polypeptide. The average mass of each amino acid is given at the top (Table 1), the bold figures indicate the integer masses of each amino acid when incorporated in a peptide (without $H_2O$ = 18 Da, Figure 1).

## Mutations

A mutation is a change in the base sequence of the functional part of DNA, and thus in the associated mRNA and protein. The addition or deletion of a base will completely change the part of the protein that is translated after the place where the change occurred, sometimes rendering the protein non-functional. For example, deletion of the first C in the triplet sequence
CCG-UAU-GCU-CUU-AAG-GGU (Figure 4),
will change the sequence in:
CGU-AUG-CUC-UUA-AGG-GU(G), resulting in:
(Met)-Arg-Met-Leu-Leu-Arg-Val, instead of:
(Met)-Pro-Tyr-Ala-Leu-Lys-Gly (Figure 4).
The substitution of one base by another will change only one amino acid in the final protein. For example, replacing the first C in CCG with an A will lead to the incorporation of Thr instead of Pro (Table 2), while the rest of the protein remains unchanged. Such a mutation will often have no effect on the functionality of the protein, but can also lead to an impaired function or, in very rare cases, to an improved function of the protein.[10]

---

[10] A good example is the many known mutations of the gene on chromosome 21 that codes for Cu/Zn superoxide dismutase (SOD1), an important anti-oxidant that neutralizes toxic free radicals that result from normal metabolism. Some of these mutations have no obvious negative effects, while others are associated with amyotrophic lateral sclerosis (ALS), a serious neurological disorder with famous patients like baseball player Lou Gehrig and astrophysicist Stephen Hawking.

As such, mutations are at the basis of both evolution and the large group of disorders known as 'inborn errors of metabolism'.[11] However, as many mutations have no clear positive or negative effect on the functionality of a protein, similar proteins synthesized by different species may have a slightly different amino acid sequence. The effects of single mutations on the total integer mass of a peptide are given in Table 4.

## Antibodies

One of the functions of proteins in biological systems is to act as antibodies. Antibodies are specialized proteins, immunoglobins, that bind to foreign bodies as part of the immune system. They are produced in plasma cells that differentiate from B-cells (a type of lymphocyte) upon contact with the foreign body.[12] Most foreign bodies comprise many different molecules that can activate the immune system; these are referred to as antigens. Macromolecular antigens, such as proteins and polysaccharides, usually carry more than one group that triggers the immune response. These are identified as epitopes and each will ultimately result in the production of a different antibody. The produced antibodies subsequently bind to their associated epitopes (Figure 6). In a biological system the major effects of the antigen-antibody reaction are agglutination: antigen-antibody complexes stick together into insoluble clusters; neutralization: the connection to the antibody renders the antigen biologically non-functional; activation of complement: the antigen-antibody complex triggers the complement system leading to its chemical and cellular destruction; activation of effector cells: the antigen-antibody complex activates specialized cells that ingest or destroy them.

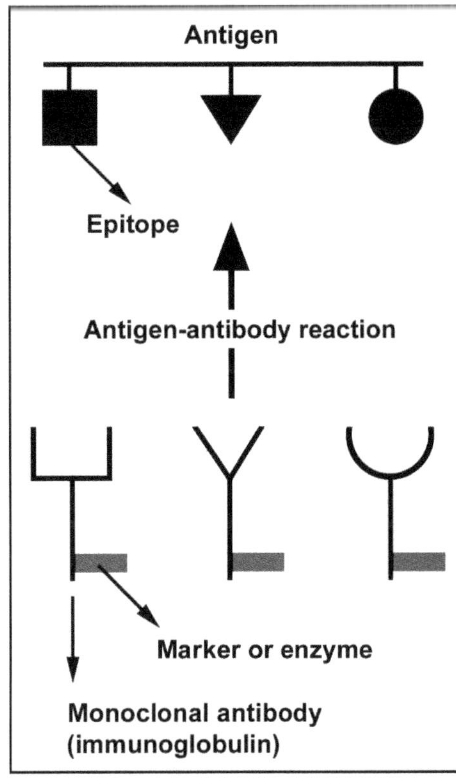


Figure 6: Schematic representation of the antigen-antibody reaction. Epitopes on an antigen (represented by ■, ▼ and •) provoke an immune response resulting in the production of specific antibodies (immunoglobulins) that bind to these epitopes, ultimately leading to the neutralization of the antigen. Monoclonal antibodies for research can be harvested from single B-cell lines and labeled with chemical or radio-active markers to enable detection and quantification.

Polyclonal antibodies are a mixture of antibodies against the various epitopes on one antigen. They can be harvested from laboratory animals that have been injected with the antigen of interest. Polyclonal antibodies are commercially available, at reasonable cost, against many compounds and are widely used for medical and scientific investigations.[13] To facilitate their detection and quantification they can be labeled with chemical or radio-active markers (Figure 6), or can be

[11] Most inborn errors of metabolism are caused by defects of single genes that code for enzymes. The term was coined by the British physician A.E. Garrod in his discussion of albinism, alcaptonuria, cystinuria, diaminuria, porphyrinuria, pentosuria and congenital steatorrhoea. Hundreds of similar disorders have since been described. Symptoms are due to either the accumulation of substances that are not efficiently metabolized, or the reduced ability to synthesize essential compounds. Early symptoms are failure to thrive, lethargy, apnoe or tachypnoe and recurrent vomiting; later toxic accumulations of metabolites cause jaundice, encephalopathy and death.

[12] In medicine, the presence of specific antibodies in the blood is considered proof of contact with a specific invader, such as the tuberculosis bacterium or the human immunodeficiency virus (HIV).

[13] For instance, as part of the 'Round robin' (Chapter 16) Laboratory A tested one of the sherds using polyclonal antibodies against agave, amaranth, bear, cactus, cat, cedar, chicken, *Capparidaceae*, *Chenopodiaceae*, *Compositae*, cow, deer, dog, guinea pig, *Graminae*, legume, pine, rabbit, rat and sheep. All the animal antisera were purchased from MP Biomedicals, while the plant antisera were produced by the Department of Biological Sciences at the University of Calgary.

detected with labeled secondary antibodies. Different antigens may have one or more epitopes in common. Polyclonal antibodies raised against one antigen can therefore partly react with other antigens, a phenomenon known as cross-reactivity. In a laboratory situation, cross-reactivity may result in false positive identifications. To increase the specificity of polyclonal antibodies, antibodies can be purified by passing them through a column containing proteins that are known to cross-react, a process known as affinity purification. Those antibodies that do will bind to these proteins, while those that are specific will pass through the column and are used for research. Monoclonal antibodies are antibodies that will react with only one specific epitope. They are harvested from a purified culture of cloned B-cells taken from a laboratory animal injected with the antigen of interest.[14] Tests using monoclonal antibodies against an epitope that is unique for the antigen of interest are very sensitive and highly specific, but can also be very expensive if the monoclonal antibodies have to be prepared especially.

## Chirality

The four residues attached to a single carbon atom are not in the same plane, but rather point towards the corners of a tetrahedron (a pyramid with a triangular base). When all four residues are different they can be arranged in two ways that differ much like a left-handed glove differs from a right-handed one (Figure 7). This property is called chirality and the two differently arranged molecules are referred to as enantiomers or optical isomers. Enantiomers behave identically during ordinary chemical reactions and most simple chemical reactions will produce an equal amount of both enantiomers, referred to as a racemic mixture. However, complex biochemical reactions, controlled by enzymes in biological systems, often produce only one enantiomer. A solution of such similar enantiomers rotates the plane of polarized light, either clockwise or anti-clockwise, which is why enantiomers are also referred to as optical isomers.[15]

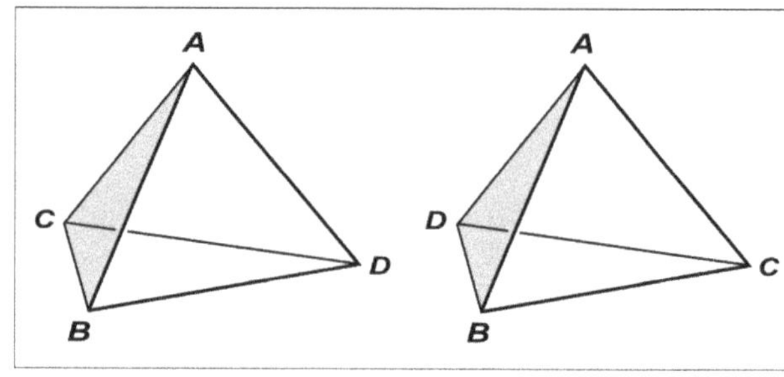


Figure 7: A schematic representation of chirality; the tetrahedron on the left cannot be turned so that its four corners (marked A, B, C and D) are arranged as in the tetrahedron on the right.

Most natural amino acids have the same spatial configuration as L-glyceraldehyde, which has arbitrarily been chosen as reference molecule, and are therefore referred to as L-amino acids. Most natural sugars are D-sugars.[16] This is biologically relevant as most enzymes are also enantiomers and have great preference for one enantiomer over another; much like a left-handed glove only fits a left hand comfortably.[17] The evolutionary advantage of the observed preference for L-amino acids and D-sugars is unclear.[18] The process of converting an optically active solution of enantiomers into a racemic mixture, with equal amounts of both optical isomers, is called racemization or epimerization. This can occur during post-translational modification, but is also part of the natural aging process of a protein.

---

14 The method to produce monoclonal antibodies was developed in the 1970's by Georges Köhler and César Milstein who shared the 1984 Nobel Prize in Physiology or Medicine for their work.

15 The effect of enantiomers on polarized light was first observed by Jean-Baptiste Biot in 1815. In 1848, Louis Pasteur described that salt of tartaric acid from wine rotated the plane of polarized light, but that salts from other sources could not ('racemic' is derived from the Latin word for grape). Jacobus van 't Hoff and Joseph Le Bel explained the optical activity in terms of the tetrahedral arrangement of the carbon atom in 1874.

16 This does not mean that all L-amino acids or all D-sugars have the same effect on polarized light as there is no direct relation between the configuration of an enantiomer and the direction of this effect.

17 Spearmint and caraway, for instance, contain different enantiomers of carvone and smell different because the olfactory receptors respond differently to different enantiomers. Taste difference between lemons and limes also result from different enantiomers of the same molecules.

18 The most infamous example of the significance of enantiomers is the thalidomide tragedy. From 1957 onward, thalidomide was prescribed to pregnant women to prevent morning sickness. Soon it was discovered that the drug also caused serious birth defects. It was later shown that one enantiomer of the drug was safe, while the other was responsible for the disastrous effects on early embryonic development. Even if only the safe enantiomer is ingested, the other is produced by metabolic reactions (racemization).

## Denaturation, Decomposition and Diagenesis

Temperature changes or changes in the chemical environment can result in irreversible changes in the structure of a protein molecule, a process called denaturation.[19] Denaturation starts with the breaking of the relatively weak hydrogen bonds and the unfolding of the peptide chain (the loss of secondary and tertiary structure). This results in both the hydrophobic side-chains as well as the hydrophilic peptide bonds becoming exposed to the surrounding environment. Water molecules will associate with the peptide bonds while the hydrophobic parts of the protein will associate with random hydrophobic parts on the same or another peptide chain. The result is a semi-solid, insoluble gel that consists of a random lattice of peptide chains in which water molecules are trapped. The primary structure of the peptide chains, which is the sequence of amino acids, is not affected by the process of denaturation.

Processes directly related to the aging of proteins include racemization (amino acid epimerization), hydrolysis and decarboxylation. Racemization is the transformation of L-amino acids into D-amino acids until a racemic mixture is reached. This has no effect on the primary structure of the polypeptide chain, but will render the protein less and less functional. Hydrolysis is the breaking, by taking up a $H_2O$-molecule (cf. Figure 1), of the peptide bonds holding the polypeptide chain together. The resistance of a peptide bond to hydrolysis partly depends on the adjoining amino acids. Peptide bonds involving serine or threonine will hydrolyze relatively easily, whereas those involving leucine or valine are relatively resistant to hydrolysis. As the protein is split into shorter polypeptide chains two other reactions become relevant: hydrolysis of the amino acid from the C-terminal of the polypeptide chain, yielding a free amino acid, and aminolysis at the N-terminal of the polypeptide chain, yielding a cyclic dipeptide (diketopiperazine). Racemization and hydrolysis need an aqueous environment and are mostly dependent on time and temperature. In a sterile aqueous environment at room temperature, racemization and hydrolysis of a protein will take 100,000 years or more,[20] at higher temperatures (>100°C) these reactions complete in minutes or seconds.

The rate of decarboxylation of amino acids, resulting in an amine and $CO_2$, depends on the presence of oxygen and on temperature. Alanine and glycine have the slowest decarboxylation rates, leucine and valine the fastest. In general, decarboxylation is a slower process than racemization or hydrolysis. Other decomposition processes depend on the presence of specific compounds, such as metal ions or ammonia, or are specific for certain amino acids. For instance, asparagine can be transformed into aspartic acid and glutamine into glutamic acid (deamination), arganine into ornithine and urea, threonine into α-aminobutyric acid, and serine into alanine.

Another process important to the ultimate fate of proteins is the non-enzymatic binding of the amino-group in a peptide with a sugar, a condensation reaction known as browning (Figure 8).[21] The rate of browning depends on the concentration and relative abundance of proteins and sugars (pentose sugars will react more easily than hexose sugars), and greatly increases with temperature.[22] After the initial binding of sugar and peptide, the new molecule rearranges itself into a ketosamine (the Amadori rearrangement, Figure 8). This can be followed by fission, dehydration or polymerization of the molecules. The final products of browning include a large array of molecules, many of which have a distinctive color (brown), smell and taste.[23]

---

19 Denaturation of proteins explains why a hard-boiled egg will not return to a liquid state after cooling; unlike butter, for instance, which will simply melt and re-solidify. A household example of chemical denaturation is the 'cooking' of fish by marinating in lime juice (*ceviche*).

20 The rate of racemization of ancient proteins can be used to date geological and archaeological features. This has been successful in fossils, teeth enamel, archaeological ostrich eggshells and bone. Complicating factors for the application of this dating technique are that racemization is not solely dependent on time, but also on temperature, humidity, the rate of hydrolysis of the protein and the presence of other compounds. Furthermore, in mammals racemization also occurs as a PTM, at a rate much faster than that which occurs after death.

21 Browning is also referred to as the Maillard reaction, after the French physician and chemist L.C. Maillard (1878-1936) who was the first to study the process scientifically.

22 Like racemization, enzymatically controlled browning occurs in mammals as a PTM.

23 Among many other things, browning is responsible for the changes that occur during the roasting of meat, the toasting of bread and the preparation of caramel (dehydration of the ketosamines) and butterscotch (fission of the ketosamines) from milk and sugar.

Humus, also referred to as humic acid or soil organic matter (SOM), is one of the largest reservoirs of amino acids in the world. It is formed out of the complex mixture of organic molecules, including peptides, originating from larger organic molecules in dead cells (mostly from plants). In the right environment, these molecules self-organize over time into a supramolecular structure characterized by hydrophobic (long-chain alkanes) and acidic (carboxyl) components. Other important components of SOM are melanoidins, which are the result of the browning and subsequent polymerization of proteins. Amino acids in such molecules appear more resistant to hydrolysis and they will survive longer than peptides and proteins.

Diagenesis is originally a geological term referring to the changes (physical, chemical or biological) affecting a sediment after its deposition, but excluding surface phenomena (weathering, erosion) and metamorphism (that occurs at high temperature and pressure). The term is also used in anthropology to describe the physical, chemical and biological changes of skeletal remains in an archaeological context. The complex interactions between the buried materials and their environment determine the ultimate destruction or preservation of the biological objects.

***Glucose + peptide***

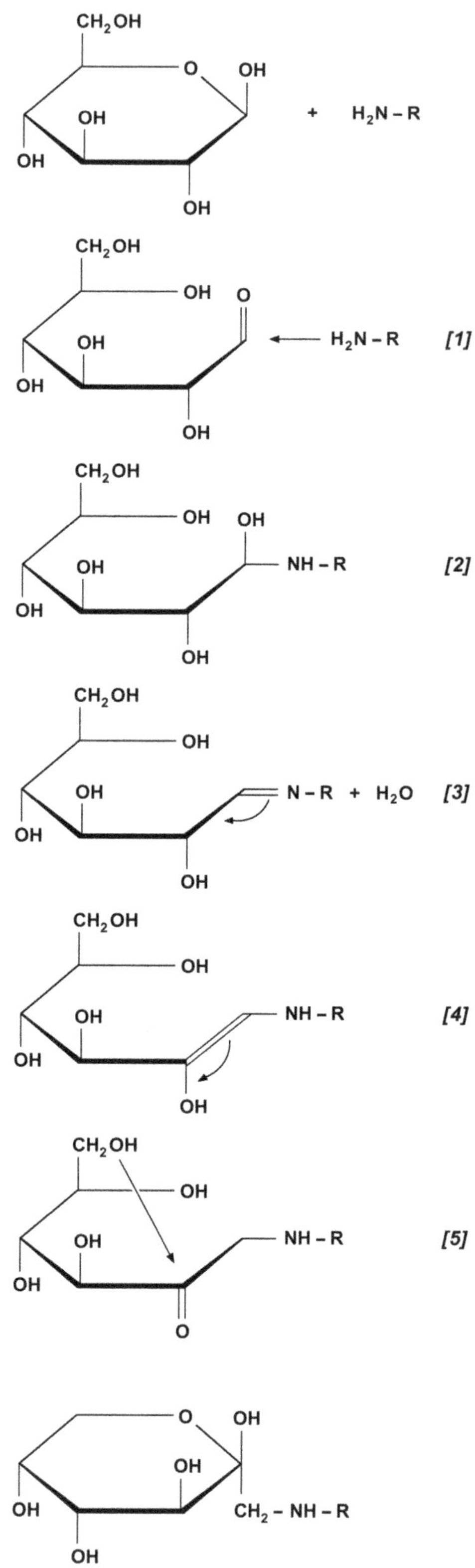


***Ketosamine (deoxy-amino-fructopyranose)***

Figure 8: Schematic representation of browning (the Maillard reaction), in which an amino compound (such as a peptide) binds to a sugar (such as glucose) with the loss of a $H_2O$ molecule [1-3], and the Amadori rearrangement [3-5] in which the unfavorable double bond between C and N moves along the opened glucose chain and a new cyclic molecule is formed.

## Appendix II: A Short Overview of Protein Biochemistry

Bone consists of a mineral matrix, mostly calcium phosphate (bioapatite), and organic material, mostly the proteins collagen and osteocalcin. After burial, changes will occur at all levels, from the loss or substitution of molecules to changes in the crystalline structure and the porosity of a bone. Diagenetic changes are the result of three interdependent processes: the chemical changes in the organic phase (hydrolysis of collagen); the chemical changes in the anorganic phase (dissolution and re-precipitation of minerals); and the activity of micro-organisms (affecting both the organic and the anorganic phases). These three pathways each have their own microscopic and macroscopic effects, which in turn influence the other pathways. The final result is usually complete destruction, rarely the mineralization (fossilization) of a bone.[24] As diagenesis will not be uniform throughout a bone, proteins and other biomolecules may survive more or less intact in isolated pockets and remain available for analysis.[25]

### Collisionally Induced Dissociation

Peptide sequencing is the deduction of the linear sequence of the amino acids in the molecule. This is accomplished using an instrument capable of performing tandem mass spectrometry (MS/MS) and collisionally induced dissociation (CID, also referred to as collisionally activated dissociation or CAD). There are many instrument designs available, with varying resolution and accuracy, all of which measure the mass of the precursor peptide, induce collision with an inert gas which causes predictable fragmentation, and measure the masses of the resulting fragments. Peptide identification is preceded by trypsin-digestion. Trypsin is a digestive enzyme that cleaves proteins (at pH = 7-9), at the carboxyl-side ($-COO^-$) of lysine (K) and arginine (R), except where either is followed by proline (P). Trypsin-digestion results in short (typically up to 4 kDa) peptide fragments that predicatably end in K or R.

A tandem MS/MS instrument is typically capable of making two mass measurements on either side of a collision event, separated either in space or in time. The first mass measurement records the mass of the precursor peptide. A small amount of inert gas (Ar, He or $N_2$) is added, and the peptides collide with these molucules. The second mass measurement records the masses of all the resulting fragments, or product ions.

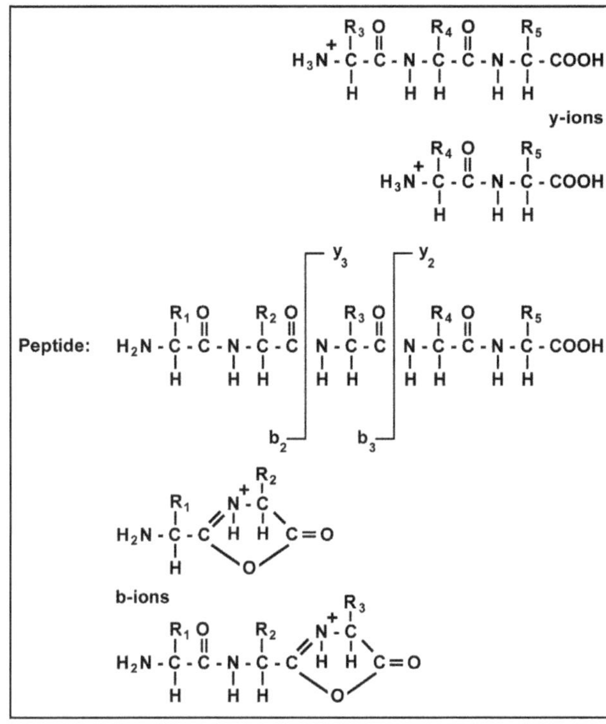


Figure 9: Schematic representation of the collisionally induced dissociation (CID) of a polypeptide chain ($R_1$-$R_5$ represent different side-chains of the amino acids). Cleavage of the OC-NH bonds occurs most frequently. Only one of the fragments carries a charge and can be monitored by mass spectrometry. The difference between subsequent y-ions (above) or b-ions (below) corresponds with the mass of one amino acid (R3).

The backbone of a polypeptide consists of three different bonds: HC-CO, OC-NH and HN-CH (Figure 9). The OC-NH bond is most frequently broken by the collision event. After the breaking of this bond all fragments carrying a (positive) charge will be detected and recorded by the second mass measurement. Fragments containing the terminal $NH_2$-group of the original peptide are identified as b-ions and those with the COOH-group as y-ions. The difference in mass between subsequent b-ions or y-ions represents the mass and the position of one of the amino acids in the polypeptide chain under investigation. Other ions that may be observed are those resulting from cleavage of the HC-OC or the HN-CH bond (identified as a-, c-, x- and z-ions) and immodium ions: ions with a low mass that can be diagnostic for the presence of specific amino acid residues in a peptide sequence.

---

24 An intermediate outcome of protein diagenesis, mostly due to the prevention of biodegradation, is the accidental (in for instance bog bodies) or deliberate mummification of biological remains. The conditions promoting this are similar to household preservation techniques such as drying, freezing, pickling, salting, smoking, tanning, etc.

25 Amino acids have been isolated from the enamel of a 3.5 million year old humanoid tooth found in Olduvai Gorge (Tanzania). The fact that these were almost completely racemized indicates that they were most likely ancient and not affected by contamination.

MS/MS offers both precursor mass as well as sequence information of the peptides, allowing a relatively confident identification of the peptides in the sample with the help of user-friendly software and on-line data bases. To separate the peptides in a sample, MS/MS is often preceeded by liquid chromatography (LC-MS/MS) in which the sample, in a polar solvent such as acetonitrile, methanol or water, is passed through a non-polar (silica based) column. Competition between the stationary and the mobile phase will separate the organic molecules in time. The effluent of the column is fed into the ion-source of a MS/MS instrument and analyzed.

## Glossary

The explanations in this glossary repeat the descriptions used in this Appendix and in Chapter 17, as they apply to archaeological protein analysis (see also the glossary at the end of Chapter 5). For more comprehensive definitions we refer to biochemical textbooks, such as Berg et al. 2002 or Voet and Voet 2004.

**2-DE**: two-dimensional electrophoresis, technique to separate a mixture of proteins in a gel by electrophoresis, according to isoelectric point in one direction and size in another direction (see SDS-PAGE)
**3'-end**: (three prime) the terminal deoxyribose of a DNA strand or the terminal ribose of a RNA strand; replication and transcription of DNA is performed in the 3' to 5' direction, the translation of RNA in the 5' to 3' direction
**5'-end**: (five prime) the terminal $-PO_3^-$ on a DNA or RNA strand (see 3'-end)
**A**: adenine, a base in DNA and RNA that pairs with thymine (T) in DNA and with uracil (U) in RNA
**A**: alanine, a proteinogenic amino acid
**acrylamide**: $CH_2$-CH-CO-$NH_2$, a toxic and carcinogenic substance that polymerizes into a gel used for Western blotting and SDS-PAGE
**Ala**: alanine (A), a proteinogenic amino acid
**Amadori rearrangement**: the rearrangement of a peptide-sugar molecule into a ketosamine (see browning)
**amino acid**: an organic compound containing both a $NH_2$-group and a COOH-group, amino acids can connect through peptide bonds to form polypeptides and proteins
**amino acid epimerization**: see racemization
**ammonium hydroxide**: a solution of ammonia in water causing a small fraction of the water to be deprotonated: $NH_3 + H_2O \leftrightarrow NH_4^+ + OH^-$
**amphoteric**: see iso-electric point
**anode**: the positive end of a direct current system (see cathode)
**antibody**: specialized protein, immunoglobin, that binds to a foreign body as part of the immunesystem; they are produced in plasma cells that differentiate from B-cells (lymphocytes)
**antibody-antibody**: (secondary antibody) antibodies raised against the constant, non-antigen recognizing, part of antibodies; labeled antibody-antibodies are used to detect antigen-antibody complexes
**anticodon**: (nodoc) the codon (triplet) on tRNA that recognizes the correct place of incorporation of the attached amino acid into a growing polypeptide
**antigen**: any molecule that can activate the immune system into producing antibodies (see epitope)
**Arg**: arginine (R), a proteinogenic amino acid
**Asn**: asparagine (N), a proteinogenic amino acid
**Asp**: aspartate (D), a proteinogenic amino acid
**AUG**: base triplet coding for methionine and the start of the peptide chain (see UAA, UAG, UGA)
**average mass**: see mass
**B-cell**: see antibody
**bioapatite**: the calcium phosphate matrix of bone
**b-ion**: positively charged peptide fragment, resulting from collisionally induced dissociation of the OC-NH bond, containing the $NH_2$-group of the original peptide (see y-ion)
**browning**: (Maillard reaction) the non-enzymatic binding of a peptide with a sugar molecule initially resulting in a ketosamine, after Amadori rearrangement, and often followed by fission, dehydration or polymerization (see melanoidins)
**BSA**: bovine serum albumin
**C**: cysteine, a proteinogenic amino acid
**C**: cytosine, a base in DNA and RNA that pairs with guanine (G)
**CAD**: collisionally activated dissociation, see CID
**calcium gluconate**: see HF
**cathode**: the negative end of a direct current system (see anode)
**chaperone**: protein that helps to fold another protein into an alternative form
**chirality**: property of a molecule of which two distinct three-dimensional configurations exist that each rotates the plane of polarized light in a different direction (see enantiomer)
**CID**: (CAD) collisionally induced dissociation of a peptide fragment into mostly b-ions and y-ions, allowing deduction of the original amino acid sequence after accurate mass measurement of both the parent molecule and the fragments
**CIEP**: counter immuno-electrophoresis, technique to immunologically detect selected target proteins after these have been separated by electrophoresis in a gel (see Western blot)
**codon**: (triplet) three consecutive bases (A, C, G, T or U) in DNA or RNA, each of the 64 possible codons codes for one of 20 amino acids (or for the end of the sequence)
**collagen**: a structural water-insoluble protein, the most common protein in bone (see osteocalcin)
**conservation**: the phenomenon that proteins with a similar function are almost identical from generation to generation and from species to species (see myoglobin)
**convergent evolution**: the phenomenon that proteins that have evolved from different sources to perform comparable tasks often have a very similar composition and shape (see conservation)

**Coomassie Brilliant Blue**: a synthetic heterocyclic organic dye that binds nonspecifically to virtually all proteins; used in CIEP, Western blotting and 2-DE gels
**cross-reactivity**: due to the presence of the same epitopes on different antigens, polyclonal antibodies raised against one antigen will also partly bind to others; in the laboratory such will result in false positive identifications
**C-terminus**: the terminal $COO^-$-group of a polypeptide chain (see N-terminus)
**cyclic dipeptide**: (diketopiperazine) two amino acids connected by two peptide bonds, the result of aminolysis at the N-terminal of a polypeptide chain
**Cys**: cysteine (C), a proteinogenic amino acid
**D**: aspartate, a proteinogenic amino acid
**Da**: Dalton, the unit of atomic mass, $^{12}C$ (19.9 x $10^{-27}$ kg) being 12.00 Da by definition
**DACIA**: digestion and capture immuno-assay, technique to immunologically detect selected target proteins after these have been released from a silicate-matrix by HF
**D-amino acid**: see L-amino acid
***de novo* sequencing**: refers to the analysis of MS/MS data when no database reference for that protein is available (see peptide sequencing)
**deamination of an amino acid**: loss of the $NH_2$-group, resulting in an organic acid and ammonia (see decarboxylation)
**decarboxylation of an amino acid**: loss of the COOH-group, resulting in an amine and carbon-dioxide (see deamination)
**denaturation**: permanent loss of the secondary and tertiary structure of a protein after a change in the temperature or the chemical environment dissolved the hydrogen bonds between different parts of the molecule
**diagenesis of bone**: all physical, chemical and biological changes of skeletal remains in an archaeological context; the ultimate outcome of the diagenetical processes is determined by the complex interactions between the buried materials and their environment
**diketopiperazine**: see cyclic dipeptide
**disulfide bond**: (S-bridge) the covalent link between the sulfur atoms in two cysteine molecules; disulfide bonds play an important role in maintaining the three dimensional structure of many proteins (see DTT)
**DNA**: deoxyribonucleic acid, a long polymer of nucleotides (a heterocyclic base bound to a sugar and one or more phosphate groups) containing the permanent genetic information necessary for protein synthesis; the sequence of the bases adenine (A), cytosine (C), guanine (G) and thymine (T) codes for the sequence of amino acids in proteins (see codon)
**D-sugar**: sugar with the same three-dimensional configuration as D-glyceraldehyde; most natural sugars are D-sugars (see chirality)
**DTT**: dithiolthreitol, HS-$CH_2$-$CH_2$-OH, a reagent that reduces of disulfide bonds in proteins
**E**: glutamate, a proteinogenic amino acid
**electrophoresis**: a separation technique based on the differences in rates of migration of proteins in the presence of an electric field (the mobile phase) within a stationary phase; this stationary phase can be a gel (gel electrophoresis) or a buffer solution in a capillary (capillary electrophoresis)
**ELISA**: enzyme-linked immuno-sorbent assay, technique to immunologically detect selected target proteins by chemically labeled antibodies (see RIA)
**enantiomer**: (optical isomer) molecule of which two distinct three-dimensional configurations exist that each rotates the plane of polarized light in a different direction (see chirality)
**enterokinase**: see trypsin
**enzyme**: biocatalyst, a protein that specifically accelerates a chemical reaction, in a biological system, without being used up
**epimerization**: see racemization
**epitope**: the part of a macromolecule that is recognized by the immune system (see antigen)
**ESI**: electrospray ionisation
**F**: phenylalanine, a proteinogenic amino acid
**five prime**: see 5'-end
**fossilization**: see mineralization
**G**: glycine, a proteinogenic amino acid
**G**: guanine, a base in DNA and RNA that pairs with cytosine (C)
**genetic code**: the information necessary to synthesize proteins as contained in the triplet sequence of DNA or mRNA molecules
**Gln**: glutamine (Q), a proteinogenic amino acid
**Glu**: glutamate (E), a proteinogenic amino acid
**Gly**: glycine (G), a proteinogenic amino acid
**glyceraldehyde**: optical active molecule of which the D- and L-enantiomers are used as reference for other molecules (see chirality)
**H**: histidine, a proteinogenic amino acid
**HCG**: human chorionic gonadotropin, a hormone produced by the human placenta
**HF**: see hydrofluoric acid
**His**: histidine (H), a proteinogenic amino acid
**hormone**: a circulating messenger molecule that binds to a specific receptor, triggering a particular response
**HPLC**: high pressure liquid chromatography, separation of a complex mixture by passing the sample through a non-polar column in a polar solvent; competition between the stationary and the mobile phase separates the molecules in time
**humic acid**: see humus
**humus**: soil organic matter, supramolecular compound in the soil, caracterized by long-chain alkanes, melanoidins and acidic components, formed out of the complex mixture of decaying organic molecules from dead cells
**hydrofluoric acid**: HF, a synthetic weak acid with highly corrosive properties; it will dissolve almost all inorganic metal and semi-metal oxides, including silicate compounds like glass and ceramics. HF can pass the skin

and harm internal tissues, especially in organs depending on calcium and magnesium such as the heart and the nerves. Exposure to small quantities of HF can be fatal, even with appropriate medical care (calcium gluconate is used as antidote)
**hydrogen bond**: the weak attracting force between a hydrogen atom in a molecule and atoms such as O or N in the same or another molecule; hydrogen bonds play an important role in maintaining the three dimensional structure of DNA and proteins (see denaturation)
**hydrolysis of a polypeptide**: the breaking, by addition of a water molecule, of the peptide bonds that bind the amino acids together
**hydrophilic**: water attracting
**hydrophobic**: water repelling
**I**: isoleucine, a proteinogenic amino acid
**Ile**: isoleucine (I), a proteinogenic amino acid
**immunoglobulin**: see antibody
**in vitro**: 'in glass', taking place in laboratory conditions (see *in vivo*)
**in vivo**: 'in life', taking place in a biological system (see *in vitro*)
**inborn errors of metabolism**: a large group of disorders each caused by a change in the DNA coding for an enzyme; symptoms are due to the toxic accumulation of substances or to the lack of essential compounds
**insulin**: protein hormone consisting of two polypeptide chains connected by disulfide bonds; after initial synthesis, the disulfide bonds are made and the peptide chain is cut twice after which the central part of the chain is removed leaving the functional protein (proteolytic cleavage)
**integer mass**: mass without decimals (see mass)
**iso-electric point**: (pI), the pH at which amphoteric molecules, which have both acidic and basic functional groups (such as amino acids), carry no net charge (see 2-DE)
**K**: lysine, a proteinogenic amino acid
**keratin**: protein in human skin and hair (and dust), frequent contaminant in protein identification analyses by mass spectrometry
**L**: leucine, a proteinogenic amino acid
**ladder**: see peptide ladder
**Laemmli**: several details of the methods for Western blot and SDS-PAGE are named after U.K. Laemmli, who developed the much of the protocol
**L-amino acid**: amino acid with the same three-dimensional configuration as L-glyceraldehyde; most natural amino acids are L-amino acids (see chirality)
**lauryl sulfate**: see SDS
**LC-MS/MS**: liquid chromatography tandem mass specrometry, see HPLC and MS/MS
**Leu**: leucine (L), a proteinogenic amino acid
**L-sugar**: see D-sugar
**Lys**: lysine (K), a proteinogenic amino acid
**M**: methionine, a proteinogenic amino acid
**Maillard reaction**: see browning
**MALDI-TOF**: matrix assisted laser desorption and ionization time-of-flight mass spectrometer, an instrument characterized by ability to measure high molecular masses, in which the the target proteins are ionized by a laser while trapped in a solid matrix
**mass**: the mass of an element can be the mass of its lightest isotope, the mono-isotopic mass, or the weighted average mass of all its isotopes, the average mass (see Da)
**melanoidins**: constituents of humus (SOM) resulting from the browning and polymerization of proteins
**Met**: methionine (M), a proteinogenic amino acid
**mineralization**: (fossilization) replacement of the collagen in bone by minerals (see diagenesis)
**monoclonal antibodies**: antibodies produced by a single cloned line of B-cells that will only bind to a single epitope (see polyclonal antibodies)
**mRNA**: messenger RNA that transfers the genetic code from DNA (in the nucleus) to the ribosomes (in the cytoplasm) where protein synthesis takes place
**MS/MS**: tandem mass spectrometry, an approach that allows precursor molecules to be mass measured, fragmented by collisions with gas molecules, and the resulting products to be mass measured
**mutation**: any change in the base sequence of the functional part of DNA, and thus in the associated mRNA and proteins
**MW**: molecular weight
**myoglobin**: the most important protein transporting oxygen in muscles; myoglobin has evolved several hundred million years ago and parts of the amino acid sequence and three-dimensional structure are the same in all species (see conservation)
**N**: asparagine, a proteinogenic amino acid
**nodoc**: see anticodon
**N-terminus**: the terminal $NH_3^+$-group of a polypeptide (see C-terminus)
**nucleotide**: a heterocyclic base bound to a sugar and one or more phosphate groups (see DNA)
**oligosaccharide**: a polymer of 3-10 sugars (see polysaccharide)
**optical isomer**: see enantiomer
**osteocalcin**: (bone Gla protein) second most abundant protein in bone (see collagen)
**P**: proline, a proteinogenic amino acid
**peptide bond**: the connection between two amino acids with the general formula $R_1$-C-CO-NH-C-$R_2$
**peptide ladder**: a reference mix of known proteins that produce reference bands on a gel
**peptide mapping**: peptide-mass fingerprinting, the identification of peptides by comparing their measured mass with the mass of known peptides in database (see peptide sequencing)
**peptide sequencing**: the identification of proteins by determining part of their amino acid sequence by tandem MS/MS and comparing this with the sequence of known proteins in a database (see peptide mapping)
**peptide**: see polypeptide

**peptide-mass fingerprinting**: see peptide mapping
**Phe**: phenylalanine (F), a proteinogenic amino acid
**phosphorylation**: a common PTM that affects the activity of a protein
**pI**: see iso-electric point
**plasma cell**: cell of the immune system that produces antibodies
**polyclonal antibodies**: a mixture of antibodies against the different epitopes on one specific antigen (see monoclonal antibodies)
**polypeptide**: a non-functional protein fragment, or a protein shorter than 50 amino acids
**polysaccharide**: a polymer of many monosaccharides (sugars) joined by glycosidic links, such as cellulose, chitin, glycogen and starch; polysaccharides are usually water-insoluble and have no sweet taste
**primary structure**: the amino acid sequence of a protein
**Pro**: proline (P), a proteinogenic amino acid
**protein**: a long chain of amino acids connected by peptide bonds and folded into a specific three-dimensional shape that defines its function; proteins are the main molecular actors in nature
**proteolytic cleavage**: the enzymatic cutting of a protein at a peptide bond
**proteomics**: an umbrella term to describe the study of proteins
**PTM**: post-translational modification, any change in a protein after synthesis (see insulin)
**Q**: glutamine, a proteinogenic amino acid
**quaternary structure**: the interactions between several protein molecules that function more or less closely together (for example dimerization)
**R**: arginine, a proteinogenic amino acid
**racemic mixture**: equal mix of the two different enantiomers of the same molecule (see chirality)
**racemization**: (epimerization) process during which single enantiomers of optically active molecules, such as L-amino acids, transform into an equal mix of L- and D-amino acids (see chirality)
**rcf**: relative centrifugal force, the force experienced by particles in a centrifuge expressed in units of gravity: rcf = $RPM^2$ x diameter (in cm) x (1.118 x $10^{-5}$)
**receptor**: a protein in the lipid cell membrane that initiates a specific reaction within the cell when it comes into contact with a specific compound outside the cell (see hormone)
**RIA**: radioimmunoassay, technique to immunologically detect selected target proteins by antibodies labeled with a radioactive compound (see ELISA)
**ribosome**: an organelle in the cytoplasm in which protein synthesis takes place according to the instructions on a strand of mRNA (see translation)
**RNA**: ribonucleic acid, a DNA-like polymer of nucleotides containing ribose instead of deoxyribose and uracil (U) instead of thymine (T); RNA functions as an intermediate in protein synthesis and contains parts of the genetic code (see DNA)
**RNA-polymerase**: large molecule in the nucleus that can make a complementary copy on RNA of part of a DNA molecule
**RPM**: revolutions per minute, see rcf
**S**: serine, a proteinogenic amino acid
**sandwich ELISA**: variant of the ELISA technique in which the antigens are immobilized by antigens
**S-bridge**: see disulfide bond
**SDS**: sodium dodecyl sulfate, $CH_3$-$(CH_2)_{11}$-$OSO_3Na$, also referred to as lauryl sulfate, a detergent used to help proteins into solution by surrounding them with negative charge, SDS is a common ingredient of shampoo
**SDS-PAGE**: sodium dodecyl sulfate polyacrylamide gel electrophoresis, technique to separate a mix of proteins by electrophoresis in a gel
**secondary antibody**: see antibody-antibody
**secondary structure**: the local shape of a polypeptide, stabilized by hydrogen bonds (and other weak forces)
**Ser**: serine (S), a proteinogenic amino acid
**silver nitrate**: $AgNO_3$, used in CIEP, Western blotting and 2-DE gels to stain virtually all proteins
**sodium dodecyl sulfate**: see SDS
**SOM**: soil organic matter, see humus
**sonication**: the application of ultrasound to stimulate residues into solution
**Southern blot**: technique, developed and named after E.M Southern, to immunologically detect selected target proteins after these have been separated by electrophoresis in a gel (see Western blot)
**Sypro Ruby**: a fluorescent dye, containing ruthenium (Ru, atomic number 44), which interacts noncovalently with most proteins; the stain is visualized with a UV or blue-light transilluminator, or a laser-based scanning instrument (Ru may be toxic)
**T**: threonine, a proteinogenic amino acid
**T**: thymine, a base in DNA that pairs with adenine (A)
**tandem mass spectrometry**: see MS/MS
**TCA**: trichloro-acetic acid, $Cl_3C$-COOH, used to precipitate proteins out of solution
**tertiary structure**: the overall shape of a protein usually stabilized by hydrophobic and hydrophilic forces, but also by hydrogen bonds, disulfide bonds or other PTMs
**tetrahedron**: a pyramid with a triangular base
**Thr**: threonine (T), a proteinogenic amino acid
**three prime**: see 3'-end
**transcription**: process in the nucleus during which a copy of part of a DNA molecule is made on a mRNA molecule
**translation**: synthesis of protein strands by a ribosome in the cytoplasm according to the instructions on a strand of mRNA
**trichloro-acetic acid**: see TCA
**triplet**: see codon
**tris-HCl**: $C_4H_{11}NO_3$ • HCl, used in solution as a buffer in SDS-PAGE and 2-DE gels
**tRNA**: transport RNA that transports individual amino acids to a ribosome where they are incorporated into a protein according to the instructions on mRNA

**Trp**: tryptophan (W), a proteinogenic amino acid
**trypsin**: a digestive enzyme, produced in the pancreas as trypsinogen and activated by enterokinase; trypsin cleaves proteins at lysine (K) and arginine (R), except where either is followed by proline (P). Trypsin digestion results in predictable peptide fragments
**Tyr**: tyrosine (Y), a proteinogenic amino acid
**U**: uracil, a base in RNA that pairs with adenine (A)
**UAA, UAG, UGA**: base triplets coding for the termination of the peptide chain (see AUG)
**V**: valine, a proteinogenic amino acid
**Val**: valine (V), a proteinogenic amino acid
**W**: tryptophan, a proteinogenic amino acid
**Western blot**: a technique to separate proteins on a ge after which antibodies are applied to determine the presence of known proteins (the technique was named by W.N. Burnette as a play on Southern blot)
**Y**: tyrosine, a proteinogenic amino acid
**y-ion**: positively charged peptide fragment, resulting from collisionally induced dissocation of the OC-NH bond, containing the COOH-group of the original peptide (see b-ion)

## References

(see also Chapter 17)
**Berg**, J.M., L. Stryer and J.L. Tymoczko (2002). *Biochemistry*. New York, Freeman and Company.
**Crick**, F. (1970). Central Dogma of Molecular Biology. *Nature* **227**: 561-563.
**Voet**, D. and J.G. Voet (2004). *Biochemistry. Volume 2*. New York, Wiley and Sons.

(H. Barnard, L. Shoemaker and O.E. Craig)

# List of Figures and Tables

**9**: *Acacia karroo* resin with woody residue on the surface (100x); insert (200x): woody residue;
**10**: *Acacia karroo* resin with woody residue and fibres on the surface (200x).

Figures 11 to 14 illustrate the combinations of residues occurring on tools that were used to replicate tasks.
**11**: Residues as a result of cutting a *Combretum zeyheri* branch (200x); a: resinous deposit, b: plant fibre, c: woody residue; insert (500x): d: starch grain in the woody residue, e: vessel in woody residue;
**12**: residues as a result of scraping the fibrous leaf of *Sanseviera pearsonii* (200x); a: plant fibre, b: leaf cell structure, c: plant tissue;
**13**: residues as a result of scraping the outside of an *Aepyceros melampus* hide (200x); a: hair, b: collagen, c: animal tissue;
**14**: residues as a result of scraping the inside of an *Aepyceros melampus* hide (200x); a: collagen fibre, b: animal tissue, c: collagen.

**CHAPTER 4**

***N.I. Shishlina, A.V. Borisov, A.A. Bobrov and M.M. Pakhomov***

Figure 1: Map showing the location of the sites mentioned in the text. 1: Sharackhalsun; 2: Mandjikiny-1; 3: Baga-Burul; 4: Ostrovnoy; 5: Mu-Sharet 1 and 4; 6: Zunda-Tolga-1; 7: Zunda-Tolga-3; 8: Chilgir-1.
Figure 2: Graves and vessels from the Caspian steppe. 1: Kurgan 5, Grave 6 at Baga-Burul; 2: vessel dating to the Early Catacomb Culture Period from Baga-Burul; 3: vessel from Kurgan 6, Grave 6 at Ostrovnoy; 4: Kurgan 3, Grave 8 at Ostrovnoy; 5: vessel dating to the Catacomb Culture Period from Ostrovnoy.
Table 1: Food identification criteria.
Figure 3: Two of the steppe plants used for food. 1: wormwood (*Artemisia*); 2: barley (*Hordeum*).
Table 2: Comparison of the data on residue in Bronze Age vessels from the northeastern Caspian steppe. The measurements to calculate the phosphorus residues were taken on samples from the bottom and from the upper part of the vessel (k: kurgan, g: grave).
Table 3: Organic residues in the stomach area of some of the skeletons from Baga-Burul and in vessels from the same graves.
Figure 4: Phytoliths from vessel residues. A, B and C: white wormwood (*Artemisia lerchiana*); D, E and F: cannabis (*Cannabis sativa*).
Figure 5: Phytoliths of steppe gramineous plants from Bronze Age Caspian vessels.
Figure 6: A:1: pollen of *Liliaceae*; A2 and 4: pollen of *Brassicaceae*; A3: pollen of *Salix*; B1: pollen of *Poaceae*; B2: pollen of *Alnus*; C, D and E: siliceous remains of the scales of gramineous plants; F and G: chitinous remains of the microplates of fish scales.

**CHAPTER 5**

***H. Barnard, A.N. Dooley and K.F. Faull***

Table 1: Details of a series of saturated fatty acids. Mass is the integer molecular mass in Daltons (Table 5).
Table 2: Details of a selection of mono-unsaturated fatty acids. Mass is the integer molecular mass in Daltons (Table 5). The single syllable that distinguishes the systematic name of an unsaturated fatty acid from that of its saturated counterpart is underlined.
Figure 1: Schematic representation of a fatty acid (on the left) and the way in which such molecules aggregate to form a membrane or a micelle when in an aqueous environment.
Table 3: Details of a series of dicarboxylic fatty acids. Mass is the integer molecular mass in Daltons (Table 5). Fatty acids with 6-10 C-atoms are collectively referred to as medium-chain fatty acids.
Figure 2: Saponification is the boiling of a lipid in a strong base, such as the reaction of triacylglycerol (fat) with potassium hydroxide (KOH) or sodium hydroxide (NaOH) resulting in the release of glycerol and fatty acid salts (soaps).
Table 4: Solubility of four common saturated fatty acids in six often-used solvents (in grams fatty acid per liter solvent at 20°C).
Figure 3: Schematic drawing of the GC-part of a GC/MS. The sample is injected into the top-left and carried by a gas through the column towards the detector that may be a mass spectrometer (MS).
Figure 4: Typical profile of the temperature in the oven of a GC/MS instrument during the analysis (temperature in °C, time in minutes). On the ramp the temperature is raised 12°C/min for 25 min, from 50°C to 350°C.
Figure 5: Part (10-33 min) of the chromatogram of the residue left in the matrix of a new, unglazed and unseasoned ceramic vessel after cooking a Red Sea *Plectropomus maculatus* (spotted coral grouper or red coral trout). Each peak, marked with its retention time (abscissa), represents a component of the residue (Figure 13). The tallest peak, at 23.28 min in this example, has a relative abundance of 100% (ordinate). $EI^+$: electron impact positive ionization; TOF: time-of-flight mass analyzer; TIC: total ion current (in arbitrary units).
Figure 6: Schematic representation of a mass spectrometer. One possible set-up is a gas chromatograph (GC), an electron impact ion source (EI), a time-of-flight mass analyzer (TOF) and a microchannel plate detector (MCP) respectively.
Figure 7: Schematic representation of the interface between a GC-column and an EI ion source. The molecules represented by squares are still in the stationary phase, while those represented by triangles are just moving into the mobile phase. Molecules represented by M have already been carried into the ion source where they are bombarded by high energy electrons (usually 70 eV). Molecules that lose an

**CHAPTER 6**

***M. Regert***



**CHAPTER 7**

***M.E. Malainey***

Figure 2: Plot of changes in amounts of C18:1 isomers and C18:2 in the cooking residue of deer and spanish dagger after 68 days oven storage. The logarithmic decay curve for relative amounts of C18:1 isomers in residues from the upper portion of the clay cylinder is C18:1 Upper = 38.95 - 1.46*log10(x) + e. Levels of C18:2 in residues from the same area is C18:2 Upper = 34.57 - 12.15*log10(x) + e.

Figure 3: Plot of changes in amounts of C18:1 isomers and C18:2 in the cooking residue of mescal seeds after 86 days oven storage. The logarithmic decay curve for relative amounts of C18:1 isomers in the residues from the upper portion of the clay cylinder is C18:1 Upper = 48.22 + 4.62*log10(x) + e. Levels of C18:2 in residues from the same area is C18:2 Upper = 33.12 - 14.54*log10(x) + e.

Figure 4: Relative fatty acid compositions of bison cooking residues from the upper portion of the clay cylinder stored in parkland soil compared to controls after one and two months of oven storage at 75°C. VLCS = very long chain (C20, C22 and C24) saturated fatty acids.

Figure 5: Relative fatty acid compositions of bison cooking residues from the middle portion of the clay cylinder stored in prairie soil compared to controls after one and two months of oven storage at 75°C. VLCS = very long chain (C20, C22 and C24) saturated fatty acids.

Figure 6: Relative fatty acid compositions of pike cooking residues from the upper portion of the clay cylinder stored in parkland soil compared to controls after ten months and after two months of oven storage at 75°C. VLCS = very long chain (C20, C22 and C24) saturated fatty acids.

Figure 7: Relative fatty acid compositions of pike cooking residues from the lower portion of the clay cylinder stored in forest soil compared to controls after ten months and after two months of oven storage at 75°C. VLCS = very long chain (C20, C22 and C24) saturated fatty acids.

Figure 8: Relative fatty acid compositions of Texas ebony cooking residues from the lower portion of the clay cylinder stored in forest soil compared to controls after one and two months of oven storage at 75°C. VLCS = very long chain (C20, C22 and C24) saturated fatty acids.

## CHAPTER 8

*J.W. Eerkens*

Figure 1: Simulated decomposition for four fatty acid ratios over 100 decomposition steps (based on deMan 1992).

Figure 2: Induced decomposition for three foods and two fatty acid ratios (based on Malainey et al. 1999b).

Figure 3: Biplot of two conservative fatty acid ratios for modern food products (ellipses represent subjective estimations of the range of values for each food group).

Figure 4: Biplot of two additional conservative fatty acid ratios for modern foods. Ellipses represent subjective estimations of the range of values for each food group (after Eerkens 2005).

## CHAPTER 9

*T.F.M. Oudemans and J.J. Boon*

Table 1: Overview of the archaeological samples discussed in this chapter.

Table 2: Compounds detected by GC/MS.

Figure 1: High temperature gas chromatograms of lipids from charred surface residues from sample 14-6-4.2b R from the Roman site Uitgeest-Groot Dorregeest (A) and sample 33.781 from the Neolithic site Hazendonk (B). IS = internal standard; M = monoacylglycerol; D = diacylglycerol; T = triacylglycerol, Ch = cholesterol and Ph = phthalate ester (numbers represent the total number of carbon atoms in the acyl moieties of lipids).

Table 3: Total lipid yield and preservation indices. *) Numbers refer to equations in the text.

Figure 2: Comparison between lipids from surface residues (R) and lipids absorbed in sherds (S) in cream colored crust 34-7-28 (A), brown residue 8-1 (B) and charred residue 14-6-4.3c (C). IS = internal standard; M = monoacylglycerols; DAG = diacylglycerols; TAG = triacylglycerols, Ch = cholesterol and Ph = phthalate ester (numbers represent the total number of carbon atoms in the acyl moieties of lipids).

Figure 3: Relative composition of acyl fragments in mass spectra of intact even carbon number TAGs of charred surface residue 34-0-30 from Uitgeest-Groot Dorregeest. The percentages are relative numbers based on the intensities of $R\text{-}COO^+$ fragments in the EI+ mass spectra of each TAG. Assuming that fragmentation is equivalent for all acyl chains, these figures can be seen as representing the acyl composition of intact TAGs.

Table 4a: Lipid composition extracts from surface (R) and absorbed (S) residues 34-0-30, 35-7-27 and 34-0-12 expressed in relative weight percentages $WP_i$ (in %). ECN is the calculated effective carbon number per compound. *) For sample 34-0-30 R the corrected $WP_i$ and the uncorrected $X_i$ are given per detected compound. The difference is expressed by Δ in percent relative to $WP_i$.

Table 4b: Lipid composition extracts from surface (R) and absorbed (S) residues 34-0-30, 35-7-27 and 34-0-12 expressed in corrected relative weight percentages $WP_i$ (in %). The numbers indicating the different DAGs and TAGs represent the total number of carbon atoms in the acyl fragments. ECN is the calculated effective carbon number per compound. *) For sample 34-0-30 R the corrected $WP_i$ and the uncorrected $X_i$ are given per detected compound. The difference is expressed by Δ in percent relative to $WP_i$.

### CHAPTER 10

***J. Cassidy***



### CHAPTER 11

***H.A. Hoekman-Sites***



### CHAPTER 12

***E.A. Reber***

Figure 5: A chromatogram of the neutral lipids extracted from HRL 208, the only black-slipped sherd analyzed from the Osceola site. Identifications and stable carbon isotope data are appended to each compound. IS = internal standard. Note particularly the stable carbon isotope value for *n*-dotriacontanol ($AL_{32:0}$) which indicates the presence of maize components in the residue.

Figure 6: Relative abundance of fatty acids in sherds from sites in southern Louisiana. HRL 180 and 182 are from Bayou des Familles, HRL185-187 are from Morgan Mound and HRL 188-191 are from St. Gabriel Mound.

### CHAPTER 13

***C. Solazzo and D. Erhardt***

Figure 1: Examples of artifacts from which residues were taken. From top to bottom: saucer-shaped clay lamps numbers 382643 and 382644 from Nunivak Island, potsherds from Miyowagh on St. Lawrence Island, soapstone lamp number 365447 from Kodiak Island and wooden dish number 342583 from the vicinity of Wales.

Figure 2: Maps showing the archaeological sites and provenance of the artifacts sampled in Alaska and Canada.

Table 1: Description of the collection of artifacts sampled for analysis.

Table 2: Relative percentage composition of fatty acids in marine mammal blubber after simulated decomposition; in the last column the range is given of the fatty acids found in the archaeological samples (A). Diacids are dicarboxylic fatty acids; P/S ratio = C16:0/C18:0.

Table 3: Fatty acid profiles of the archaeological samples. Glycerol, monoacylglycerides, mono-unsaturated fatty acids, diacids and the P/S-ratio (C16:0/C18:0) were calculated from HTGC data. The values indicate the relative percentages calculated from the sum of all detected fatty acids. Isoprenoid acids and sterols (cholesterol and products of oxidation) were positively detected (*) or not (-) by GC/MS (blanks indicate that GC/MS was not performed).

Figure 3: High temperature gas chromatograms of Beluga whale blubber (a); and clay potsherd 399580 (b); high temperature gas chromatogram of bone scraper 400347 (c); GC/MS chromatograms of soapstone lamp SL3 (d); bola 400399 (e); and wooden dish 342583 (f).

Figure 4: Monounsaturated fatty acids plotted versus dicarboxylic fatty acids for artifacts from coastal Alaska.

Table 4: comparison between samples taken from the outside (E) and from the inside (I) part of the sherd. MAGs are monoacylglycerides; diacids are dicarboxylic fatty acids; P/S ratio = C16:0/C18:0.

### CHAPTER 14

***S.M. Rafferty***

Figure 1: Structures of the alkaloid compounds discussed in this chapter.

Figure 2: GC/MS results of an early tobacco residue from eastern North America.

### CHAPTER 15

***D.E. Beehr and S.H. Ambrose***

Figure 1: $\delta^{13}C$ values of untreated carbonized residues from Mississippian pottery sherds from five sites in the American Bottom region of southwest Illinois, based on data in Beehr and Ambrose (in press).

Figure 2: Treated versus untreated $\delta^{13}C$ values of carbonized residues from Mississippian pottery sherds.

Figure 3: Treated versus untreated $\delta^{15}N$ values of carbonized residues from Mississippian pottery sherds. Sherd DEB 106 had low voltages for treated and untreated samples, so the difference between them may not have been accurately measured.

Table 1: Sample weights, $\delta^{13}C$ and $\delta^{15}N$ values, weight %C and weight %N, major ion beam voltages for mass/charge ratios of $CO_2$ (m/z = 44) and $N_2$ (m/z = 28) of treated carbonized residues from Mississippian pottery sherds. Samples DEB 40 and DEB 44, marked with *, were switched in position in the elemental analyzer carousel during the analyses of untreated samples.

### CHAPTER 16[1]

***H. Barnard, S.H. Ambrose, D.E. Beehr, M.D. Forster, R.E. Lanehart, M.E. Malainey, R.E. Parr, M. Rider, C. Solazzo and R.M. Yohe II***

Table 1: List of the foodstuffs prepared in new Egyptian 'tagen' in December 2003 and January 2004 (TDS = Total dissolved solids; AUC = American University in Cairo).

Figure 1: The preparation of organic residues of selected known foodstuffs in new and unseasoned Egyptian 'tagen' (Cairo, December 2003 and January 2004).

Figure 2: Egyptian 'tagen' number 11, with a residue of camel milk, just after it was cut (Los Angeles, May 2004). The wall fragments were subsequently distributed among the participants of our blind 'Round robin'.

Figure 3: Comparison of seven fatty acid signatures of fresh camel milk (after Ramet 2001).

Figure 4: The amino acid signature of casein in fresh camel milk (after Farah and Ruegg, 1989).

Table 2: Measured carbon and nitrogen isotopic and elemental concentrations. Missing values occur where nitrogen concentrations were too low for detection (the

[1] A shorter version of this chapter, including most illustrations, was published in the *Journal of Archaeological Science 34*; 2007: pp. 28-37.

well as the most (leucine) and least abundant (tryptophan) amino acid in nature.

Table 2: The genetic code in which each triplet of nucleotides (codon) codes for one amino acid. A = adenine, C = cytosine, G = guanine, U = uracil (which in RNA take the place reserved for thymine in DNA). Bold entries signify amino acids with a single codon (methionine, also the 'start'-codon, and tryptophan).

Figure 3: Schematic representation of the transcription of DNA into messenger RNA (mRNA) by RNA-polymerase. The DNA double helix is opened and a complementary copy of one of the strands is made on RNA, in which uracil (U) takes the place of thymine (T), opposite adenine (A).

Figure 4: Schematic representation of the translation of messenger RNA (mRNA). Ribosomes read the genetic code and connect the amino acids, transported by transport RNA (tRNA), into a polypeptide. tRNA carries an anticodon to identify the correct place of insertion (C combines with G and A combines with U).

Figure 5: The three-dimensional structure of myoglobin, a single-chain protein of 153 amino acids (total molecular weight about 16,700 Da).

Table 3: Overview of some common post-translational modifications (PTMs) with their usual place of occurrence and their effect on the total average mass of a polypeptide.

Table 4: Overview of the effect of a single mutation, in which one amino acid is replaced by another, on the total integer mass of a polypeptide. The average mass of each amino acid is given at the top (Table 1), the bold figures indicate the integer masses of each amino acid when incorporated in a peptide (without $H_2O$ = 18 Da, Figure 1).

Figure 6: Schematic representation of the antigen-antibody reaction. Epitopes on an antigen (represented by ■, ▼ and ●) provoke an immune response resulting in the production of specific antibodies (immunoglobulins) that bind to these epitopes, ultimately leading to the neutralization of the antigen. Monoclonal antibodies for research can be harvested from single B-cell lines and labeled with chemical or radio-active markers to enable detection and quantification.

Figure 7: A schematic representation of chirality; the tetrahedron on the left cannot be turned so that its four corners (marked A, B, C and D) are arranged as in the tetrahedron on the right.

Figure 8: Schematic representation of browning (the Maillard reaction), in which an amino compound (such as a peptide) binds to a sugar (such as glucose) with the loss of a $H_2O$ molecule [1-3], and the Amadori rearrangement [3-5] in which the unfavorable double bond between C and N moves along the opened glucose chain and a new cyclic molecule is formed.

Figure 9: Schematic representation of the collision induced dissociation (CID) of a polypeptide chain ($R_1$-$R_5$ represent different side-chains of the amino acids). Cleavage of the OC-NH bonds occurs most frequently. Only one of the fragments carries a charge and can be monitored by mass spectrometry. The difference between subsequent y-ions (above) or b-ions (below) corresponds with the mass of one amino acid (R3).

## The Authors

**S.H. Ambrose** *(Chapters 15 and 16)*
Department of Anthropology, University of Illinois at Urbana-Champaign;
607 South Mathews Avenue; Urbana, IL 61801; USA.
**H. Barnard** *(Chapters 1, 5, 16 and 17)*
Cotsen Institute of Archaeology; University of California, Los Angeles;
P.O.-Box 951510; Los Angeles, CA 90025; USA.
**D.E. Beehr** *(Chapters 15 and 16)*
Department of Anthropology, University of Illinois at Urbana-Champaign;
607 South Mathews Avenue; Urbana, IL 61801; USA.
**A.A. Bobrov** *(Chapter 4)*
Faculty of Soil Science, Moscow State University;
Leninskiye Gory; Moscow 119899; Russia.
**J.J. Boon** *(Chapter 9)*
Foundation for Fundamental Research on Matter (FOM), Institute for Atomic and Molecular Physics (AMOLF);
P.O.-Box 41883; 1009 DB Amsterdam; the Netherlands.
**A.V. Borisov** *(Chapter 4)*
Institute of Physical, Chemical and Biological Problems of Soil Science, Russian Academy of Sciences;
Institutskaya 2; Pushchino, Moscow 142290; Russia
**J.D. Cassidy** *(Chapter 10)*
Natural Resources and Environmental Affairs, Marine Corps Air Ground Combat Center;
P.O-Box 788110; Twentynine Palms, CA 92278; USA.
**O.E. Craig** *(Chapter 17)*
Dipartimento di Biologica; Università di Roma 'Tor Vergata';
Via della Ricerca Scientifica; 100133 Roma; Italy.
**A.N. Dooley** *(Chapter 5)*
European Managed Territories; AppliedBiosystems GmbH;
Frankfurter Straße 129$^{B}$; 64293 Darmstadt; Germany.
**J.W. Eerkens** *(Chapters 1 and 8)*
Department of Anthropology, University of California, Davis;
One Shields Avenue; Davis, CA 95616; USA.
**D. Erhardt** *(Chapter 13)*
Smithsonian Center for Material Research and Education (SCMRE, now the Museum Conservation Institute);
4210 Silver Hill Road; Suitland, MD 20746; USA.
**K.F. Faull** *(Chapter 5)*
Pasarow Mass Spectrometry Laboratory, The NPI-Semel Institute for Neuroscience and Human Behavior
and The Department of Psychiatry and Biobehavioral Sciences;
University of California, Los Angeles; 1217 Young Hall; Los Angeles, CA 90095; USA.
**M.D. Forster** *(Chapter 16)*
Department of Archaeological Sciences, University of Bradford;
Bradford, BD7 1DP; UK.
**H.A. Hoekman-Sites** *(Chapter 11)*
Department of Anthropology, Florida State University;
1847 West Tennessee Street; Tallahassee, FL 32306; USA.
**R.E. Lanehart** *(Chapter 16)*
Department, of Anthropology, University of South Florida;
4202 East Fowler Avenue; Tampa, FL 33620; USA.
**M. Lombard** *(Chapter 3)*
Natal Museum;
Private Bag 9070; Pietermaritzburg; 3200 South Africa.
**R.K. Lusteck** *(Chapter 2)*
Department of Anthropology, University of Minnesota;
395 Hubert H. Humphrey Hall; Minneapolis, MN 55455; USA.

**M.E. Malainey** *(Chapters 7 and 16)*
Department of Anthropology, Brandon University;
270-18th Street; Brandon, Manitoba, R7A 6A9; Canada.
**T.F.M. Oudemans** *(Chapter 9)*
Foundation for Fundamental Research on Matter (FOM), Institute for Atomic and Molecular Physics (AMOLF);
P.O.-Box 41883; 1009 DB Amsterdam; the Netherlands.
**M.M. Pakhomov** *(Chapter 4)*
Vyatka State University;
Krasnoarmeyskaya 26; Kirov, Nizniy Novgorod 610002; Russia.
**R.E. Parr** *(Chapters 16 and 17)*
Laboratory of Archaeological Sciences; California State University, Bakersfield;
9001 Stockdale Highway; Bakerfield, CA 93311; USA.
**S.M. Rafferty** *(Chapter 14)*
Department of Anthropology University at Albany, State University of New York;
1400 Washington Avenue; Albany, NY 12222; USA.
**E.A. Reber** *(Chapter 12)*
Anthropology Department, UNC Wilmington;
601 South College Road; Wilmington, NC 28403; USA.
**M. Regert** *(Chapter 6)*
Unité Mixte de Recherche 171, Centre National de la Recherche Scientifique (CNRS), Ministère de la Culture et de la Communication; Centre de Recherche et de Restauration des Musées de France (C2RMF);
14, Quai François Mitterrand; 75001 Paris Cedex; France.
**M. Rider** *(Chapters 16 and 17)*
Department of Anthropology, University of Arizona;
P.O.-Box 210030; Tuscon, AZ 85721; USA.
**N.I. Shishlina** *(Chapter 4)*
State Historical Museum;
Krasnaya Ploshchad 1/2; Moscow 103012; Russia.
**L. Shoemaker** *(Chapter 17)*
Pasarow Mass Spectrometry Laboratory, UCLA;
Los Angeles, CA 90095; USA.
**C. Solazzo** *(Chapters 13 and 16)*
Laboratoire de Chimie Organique et Macromoléculaire (LCOM); Université des Sciences et Technologies de Lille;
UFR de Chimie, Bat C4, porte 210; 59655 Villeneuve d'Asq Cedex; France.
**M. Sutton** *(Chapter 17)*
Laboratory of Archaeological Sciences; California State University, Bakersfield;
9001 Stockdale Highway; Bakerfield, CA 93311; USA.
**R.G. Thompson** *(Chapter 2)*
Interdisciplinary Archaeological Studies, University of Minnesota;
395 Hubert H. Humphrey Hall; Minneapolis, MN 55455; USA.
**L. Wadley** *(Chapter 3)*
Archaeology Department; School of Geography, Archaeology and Environmental Studies;
University of the Witwatersrand; Private Bag 3; WITS 2050 Johannesburg; South Africa.
**R.M. Yohe II** *(Chapters 16 and 17)*
Laboratory of Archaeological Sciences; California State University, Bakersfield;
9001 Stockdale Highway; Bakerfield, CA 93311; USA.

# INDEX

**--- C ---**

## Index



--- D ---

**--- N ---**

--- T ---

--- U ---



--- V ---



--- W ---

www.ingramcontent.com/pod-product-compliance
Ingram Content Group UK Ltd.
Pitfield, Milton Keynes, MK11 3LW, UK
UKHW061846041026
15591UKWH00015B/933

9 781407 300849